# Toyota Corolla & Geo/Chevrolet Prizm Automotive Repair Manual

## by Jay Storer and John H Haynes

Member of the Guild of Motoring Writers

---

**Models covered:**

All Toyota Corolla and Geo/Chevrolet Prizm models
1993 through 2001

---

(10D1 - 92036)

ABCDE
FGHIJ
KLMNO
PQR

AUTOMOTIVE
PARTS &
ACCESSORIES
ASSOCIATION   MEMBER

**Haynes Publishing Group**
Sparkford Nr Yeovil
Somerset BA22 7JJ England

**Haynes North America, Inc**
861 Lawrence Drive
Newbury Park
California 91320 USA

# About this manual

## Its purpose

The purpose of this manual is to help you get the best value from your vehicle. It can do so in several ways. It can help you decide what work must be done, even if you choose to have it done by a dealer service department or a repair shop; it provides information and procedures for routine maintenance and servicing; and it offers diagnostic and repair procedures to follow when trouble occurs.

We hope you use the manual to tackle the work yourself. For many simpler jobs, doing it yourself may be quicker than arranging an appointment to get the vehicle into a shop and making the trips to leave it and pick it up. More importantly, a lot of money can be saved by avoiding the expense the shop must pass on to you to cover its labor and overhead costs. An added benefit is the sense of satisfaction and accomplishment that you feel after doing the job yourself.

## Using the manual

The manual is divided into Chapters. Each Chapter is divided into numbered Sections, which are headed in bold type between horizontal lines. Each Section consists of consecutively numbered paragraphs.

At the beginning of each numbered Section you will be referred to any illustrations which apply to the procedures in that Section. The reference numbers used in illustration captions pinpoint the pertinent Section and the Step within that Section. That is, illustration 3.2 means the illustration refers to Section 3 and Step (or paragraph) 2 within that Section.

Procedures, once described in the text, are not normally repeated. When it's necessary to refer to another Chapter, the reference will be given as Chapter and Section number. Cross references given without use of the word "Chapter" apply to Sections and/or paragraphs in the same Chapter. For example, "see Section 8" means in the same Chapter.

References to the left or right side of the vehicle assume you are sitting in the driver's seat, facing forward.

Even though we have prepared this manual with extreme care, neither the publisher nor the author can accept responsibility for any errors in, or omissions from, the information given.

### NOTE

A **Note** provides information necessary to properly complete a procedure or information which will make the procedure easier to understand.

### CAUTION

A **Caution** provides a special procedure or special steps which must be taken while completing the procedure where the Caution is found. Not heeding a Caution can result in damage to the assembly being worked on.

### WARNING

A **Warning** provides a special procedure or special steps which must be taken while completing the procedure where the Warning is found. Not heeding a Warning can result in personal injury.

### Acknowledgements

Technical writers who contributed to this project include Bob Henderson, Eric Godfrey, Jeff Kibler, Rob Maddox, Mike Stubblefield and Larry Warren.

© Haynes North America, Inc. 2000

With permission from J.H. Haynes & Co. Ltd.

**A book in the Haynes Automotive Repair Manual Series**

**Printed in the U.S.A.**

**ISBN 1 56392 395 5**

**Library of Congress Catalog Card Number 00-104989**

# Contents

Haynes author, photographer, and mechanic with 1996 Toyota Corolla

# Introduction to the
# Toyota Corolla and Geo/Chevrolet Prizm

Toyota Corolla models are available in four-door sedan and station wagon body styles. Geo/Chevrolet Prizm models are available as four-door sedans only.

The transversely mounted inline four-cylinder engines used in these vehicles are equipped with electronic fuel injection.

The engine drives the front wheels through either a five-speed manual or a three or four-speed automatic transaxle via independent driveaxles.

Independent suspension, featuring coil spring/strut damper units, is used on all four wheels. The rack-and-pinion steering unit is mounted behind the engine with power-assist available as an option.

The brakes are disc at the front and drums at the rear, with power assist standard. An Anti-lock Brake System (ABS) became available on some models.

# Vehicle identification numbers

Modifications are a continuing and unpublicized process in vehicle manufacturing. Since spare parts manuals and lists are compiled on a numerical basis, the individual vehicle numbers are essential to correctly identify the component required.

## Vehicle Identification Number (VIN)

This very important identification number is stamped on the firewall in the engine compartment, on a plate attached to the dashboard inside the windshield on the driver's side of the vehicle and, on some models, the end of the driver's door (see illustration). The VIN also appears on the Vehicle Certificate of Title and Registration. It contains information such as where and when the vehicle was manufactured, the model year and the body style.

## Manufacturer's Certification Regulation label

The manufacturer's Certification Regulation label is attached to the driver's side door end or post. The plate contains the name of the manufacturer, the month and year of production, the Gross Vehicle Weight Rating (GVWR), the Gross Axle Weight Rating (GAWR) and the certification statement.

## VIN engine and model year codes

Two particularly important pieces of information found in the VIN are the engine code (Geo models only) and the model year code. Counting from the left, the engine code letter designation is the 8th character and the model year code letter designation is the 10th character.

On the Geo/Chevrolet models covered by this manual the engine codes are:

| | |
|---|---|
| 6 | 1.6L |
| 8 | 1.8L |

On all models covered by this manual the model year codes are:

| | |
|---|---|
| P | 1993 |
| R | 1994 |
| S | 1995 |
| T | 1996 |
| V | 1997 |
| W | 1998 |
| X | 1999 |
| Y | 2000 |
| Z | 2001 |

## Service parts identification label (Geo models only)

This label is located on the spare tire cover (see illustration). It lists the VIN, paint number, options and other information specific to the vehicle. Always refer to this label when you order parts.

## Engine number

The engine code number can be found on a pad on the front (radiator) side of the cylinder block (1997 and earlier models) or, 1998 and later models, the rear side of the block (see illustrations).

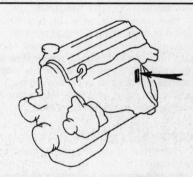

Location of the engine identification number (1997 and earlier models)

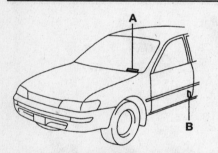

Locations of the Vehicle Identification Number (VIN) (A) and the Certification Regulation label (B)

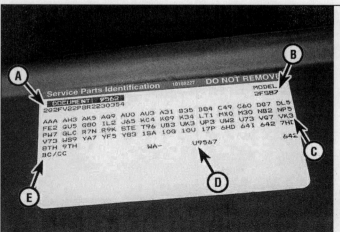

On Geo models, the service parts identification label is located on the spare tire cover (typical) or on the underside of the trunk lid

A  Vehicle identification number
B  Body type and style
C  Options
D  Paint codes
E  Paint type

On 1998 and later models, the engine identification number is located on the firewall side of the engine, below the exhaust manifold

# Buying parts

Replacement parts are available from many sources, which generally fall into one of two categories - authorized dealer parts departments and independent retail auto parts stores. Our advice concerning these parts is as follows:

**Retail auto parts stores:** Good auto parts stores will stock frequently needed components which wear out relatively fast, such as clutch components, exhaust systems, brake parts, tune-up parts, etc. These stores often supply new or reconditioned parts on an exchange basis, which can save a considerable amount of money. Discount auto parts stores are often very good places to buy materials and parts needed for general vehicle maintenance such as oil, grease, filters, spark plugs, belts, touch-up paint, bulbs, etc. They also usually sell tools and general accessories, have convenient hours, charge lower prices and can often be found not far from home.

**Authorized dealer parts department:** This is the best source for parts which are unique to the vehicle and not generally available elsewhere (such as major engine parts, transmission parts, trim pieces, etc.).

**Warranty information:** If the vehicle is still covered under warranty, be sure that any replacement parts purchased - regardless of the source - do not invalidate the warranty!

To be sure of obtaining the correct parts, have engine and chassis numbers available and, if possible, take the old parts along for positive identification.

# Maintenance techniques, tools and working facilities

## Maintenance techniques

There are a number of techniques involved in maintenance and repair that will be referred to throughout this manual. Application of these techniques will enable the home mechanic to be more efficient, better organized and capable of performing the various tasks properly, which will ensure that the repair job is thorough and complete.

## Fasteners

Fasteners are nuts, bolts, studs and screws used to hold two or more parts together. There are a few things to keep in mind when working with fasteners. Almost all of them use a locking device of some type, either a lockwasher, locknut, locking tab or thread adhesive. All threaded fasteners should be clean and straight, with undamaged threads and undamaged corners on the hex head where the wrench fits. Develop the habit of replacing all damaged nuts and bolts with new ones. Special locknuts with nylon or fiber inserts can only be used once. If they are removed, they lose their locking ability and must be replaced with new ones.

Rusted nuts and bolts should be treated with a penetrating fluid to ease removal and prevent breakage. Some mechanics use turpentine in a spout-type oil can, which works quite well. After applying the rust penetrant, let it work for a few minutes before trying to loosen the nut or bolt. Badly rusted fasteners may have to be chiseled or sawed off or removed with a special nut breaker, available at tool stores.

If a bolt or stud breaks off in an assembly, it can be drilled and removed with a special tool commonly available for this purpose.

Most automotive machine shops can perform this task, as well as other repair procedures, such as the repair of threaded holes that have been stripped out.

Flat washers and lockwashers, when removed from an assembly, should always be replaced exactly as removed. Replace any damaged washers with new ones. Never use a lockwasher on any soft metal surface (such as aluminum), thin sheet metal or plastic.

Grade 1 or 2          Grade 5          Grade 8

Bolt strength marking (standard/SAE/USS; bottom - metric)

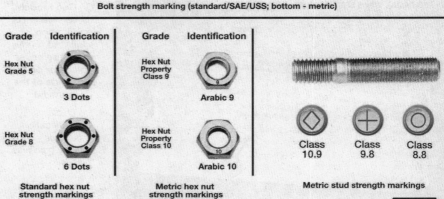

| Grade | Identification | Grade | Identification |
|---|---|---|---|
| Hex Nut Grade 5 | 3 Dots | Hex Nut Property Class 9 | Arabic 9 |
| Hex Nut Grade 8 | 6 Dots | Hex Nut Property Class 10 | Arabic 10 |

Standard hex nut strength markings

Metric hex nut strength markings

Class 10.9          Class 9.8          Class 8.8

Metric stud strength markings

00-1 HAYNES

## Fastener sizes

For a number of reasons, automobile manufacturers are making wider and wider use of metric fasteners. Therefore, it is important to be able to tell the difference between standard (sometimes called U.S. or SAE) and metric hardware, since they cannot be interchanged.

All bolts, whether standard or metric, are sized according to diameter, thread pitch and length. For example, a standard 1/2 - 13 x 1 bolt is 1/2 inch in diameter, has 13 threads per inch and is 1 inch long. An M12 - 1.75 x 25 metric bolt is 12 mm in diameter, has a thread pitch of 1.75 mm (the distance between threads) and is 25 mm long. The two bolts are nearly identical, and easily confused, but they are not interchangeable.

In addition to the differences in diameter, thread pitch and length, metric and standard bolts can also be distinguished by examining the bolt heads. To begin with, the distance across the flats on a standard bolt head is measured in inches, while the same dimension on a metric bolt is sized in millimeters (the same is true for nuts). As a result, a standard wrench should not be used on a metric bolt and a metric wrench should not be used on a standard bolt. Also, most standard bolts have slashes radiating out from the center of the head to denote the grade or strength of the bolt, which is an indication of the amount of torque that can be applied to it. The greater the number of slashes, the greater the strength of the bolt. Grades 0 through 5 are commonly used on automobiles. Metric bolts have a property class (grade) number, rather than a slash, molded into their heads to indicate bolt strength. In this case, the higher the number, the stronger the bolt. Property class numbers 8.8, 9.8 and 10.9 are commonly used on automobiles.

Strength markings can also be used to distinguish standard hex nuts from metric hex nuts. Many standard nuts have dots stamped into one side, while metric nuts are marked with a number. The greater the number of dots, or the higher the number, the greater the strength of the nut.

Metric studs are also marked on their ends according to property class (grade). Larger studs are numbered (the same as metric bolts), while smaller studs carry a geometric code to denote grade.

It should be noted that many fasteners, especially Grades 0 through 2, have no distinguishing marks on them. When such is the case, the only way to determine whether it is standard or metric is to measure the thread pitch or compare it to a known fastener of the same size.

Standard fasteners are often referred to as SAE, as opposed to metric. However, it should be noted that SAE technically refers to a non-metric fine thread fastener only. Coarse thread non-metric fasteners are referred to as USS sizes.

Since fasteners of the same size (both standard and metric) may have different

| Metric thread sizes | Ft-lbs | Nm |
| --- | --- | --- |
| M-6 | 6 to 9 | 9 to 12 |
| M-8 | 14 to 21 | 19 to 28 |
| M-10 | 28 to 40 | 38 to 54 |
| M-12 | 50 to 71 | 68 to 96 |
| M-14 | 80 to 140 | 109 to 154 |

| Pipe thread sizes | | |
| --- | --- | --- |
| 1/8 | 5 to 8 | 7 to 10 |
| 1/4 | 12 to 18 | 17 to 24 |
| 3/8 | 22 to 33 | 30 to 44 |
| 1/2 | 25 to 35 | 34 to 47 |

| U.S. thread sizes | | |
| --- | --- | --- |
| 1/4 - 20 | 6 to 9 | 9 to 12 |
| 5/16 - 18 | 12 to 18 | 17 to 24 |
| 5/16 - 24 | 14 to 20 | 19 to 27 |
| 3/8 - 16 | 22 to 32 | 30 to 43 |
| 3/8 - 24 | 27 to 38 | 37 to 51 |
| 7/16 - 14 | 40 to 55 | 55 to 74 |
| 7/16 - 20 | 40 to 60 | 55 to 81 |
| 1/2 - 13 | 55 to 80 | 75 to 108 |

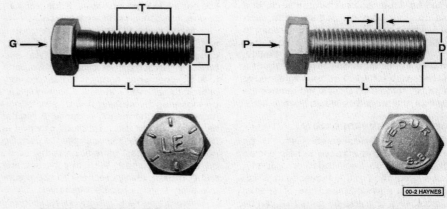

**Standard (SAE and USS) bolt dimensions/grade marks**

G   Grade marks (bolt strength)
L   Length (in inches)
T   Thread pitch (number of threads per inch)
D   Nominal diameter (in inches)

**Metric bolt dimensions/grade marks**

P   Property class (bolt strength)
L   Length (in millimeters)
T   Thread pitch (distance between threads in millimeters)
D   Diameter

strength ratings, be sure to reinstall any bolts, studs or nuts removed from your vehicle in their original locations. Also, when replacing a fastener with a new one, make sure that the new one has a strength rating equal to or greater than the original.

## Tightening sequences and procedures

Most threaded fasteners should be tightened to a specific torque value (torque is the twisting force applied to a threaded component such as a nut or bolt). Overtightening the fastener can weaken it and cause it to break, while undertightening can cause it to eventually come loose. Bolts, screws and studs, depending on the material they are

made of and their thread diameters, have specific torque values, many of which are noted in the Specifications at the beginning of each Chapter. Be sure to follow the torque recommendations closely. For fasteners not assigned a specific torque, a general torque value chart is presented here as a guide. These torque values are for dry (unlubricated) fasteners threaded into steel or cast iron (not aluminum). As was previously mentioned, the size and grade of a fastener determine the amount of torque that can safely be applied to it. The figures listed here are approximate for Grade 2 and Grade 3 fasteners. Higher grades can tolerate higher torque values.

Fasteners laid out in a pattern, such as cylinder head bolts, oil pan bolts, differential cover bolts, etc., must be loosened or tight-

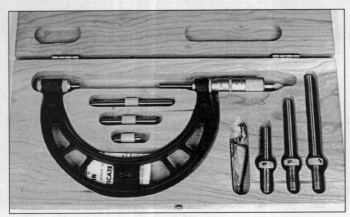

**Micrometer set**

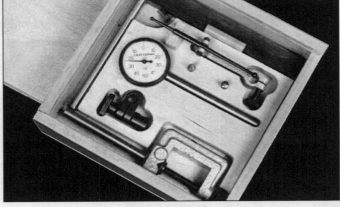

**Dial indicator set**

ened in sequence to avoid warping the component. This sequence will normally be shown in the appropriate Chapter. If a specific pattern is not given, the following procedures can be used to prevent warping.

Initially, the bolts or nuts should be assembled finger-tight only. Next, they should be tightened one full turn each, in a criss-cross or diagonal pattern. After each one has been tightened one full turn, return to the first one and tighten them all one-half turn, following the same pattern. Finally, tighten each of them one-quarter turn at a time until each fastener has been tightened to the proper torque. To loosen and remove the fasteners, the procedure would be reversed.

### Component disassembly

Component disassembly should be done with care and purpose to help ensure that the parts go back together properly. Always keep track of the sequence in which parts are removed. Make note of special characteristics or marks on parts that can be installed more than one way, such as a grooved thrust washer on a shaft. It is a good idea to lay the disassembled parts out on a clean surface in the order that they were removed. It may also be helpful to make sketches or take instant photos of components before removal.

When removing fasteners from a component, keep track of their locations. Sometimes threading a bolt back in a part, or putting the washers and nut back on a stud, can prevent mix-ups later. If nuts and bolts cannot be returned to their original locations, they should be kept in a compartmented box or a series of small boxes. A cupcake or muffin tin is ideal for this purpose, since each cavity can hold the bolts and nuts from a particular area (i.e. oil pan bolts, valve cover bolts, engine mount bolts, etc.). A pan of this type is especially helpful when working on assemblies with very small parts, such as the carburetor, alternator, valve train or interior dash and trim pieces. The cavities can be marked with paint or tape to identify the contents.

Whenever wiring looms, harnesses or connectors are separated, it is a good idea to identify the two halves with numbered pieces of masking tape so they can be easily reconnected.

### Gasket sealing surfaces

Throughout any vehicle, gaskets are used to seal the mating surfaces between two parts and keep lubricants, fluids, vacuum or pressure contained in an assembly.

Many times these gaskets are coated with a liquid or paste-type gasket sealing compound before assembly. Age, heat and pressure can sometimes cause the two parts to stick together so tightly that they are very difficult to separate. Often, the assembly can be loosened by striking it with a soft-face hammer near the mating surfaces. A regular hammer can be used if a block of wood is placed between the hammer and the part. Do not hammer on cast parts or parts that could be easily damaged. With any particularly stubborn part, always recheck to make sure that every fastener has been removed.

Avoid using a screwdriver or bar to pry apart an assembly, as they can easily mar the gasket sealing surfaces of the parts, which must remain smooth. If prying is absolutely necessary, use an old broom handle, but keep in mind that extra clean up will be necessary if the wood splinters.

After the parts are separated, the old gasket must be carefully scraped off and the gasket surfaces cleaned. Stubborn gasket material can be soaked with rust penetrant or treated with a special chemical to soften it so it can be easily scraped off. A scraper can be fashioned from a piece of copper tubing by flattening and sharpening one end. Copper is recommended because it is usually softer than the surfaces to be scraped, which reduces the chance of gouging the part. Some gaskets can be removed with a wire brush, but regardless of the method used, the mating surfaces must be left clean and smooth. If for some reason the gasket surface is gouged, then a gasket sealer thick enough to fill scratches will have to be used during reassembly of the components. For most applications, a non-drying (or semi-drying) gasket sealer should be used.

### Hose removal tips

**Warning:** *If the vehicle is equipped with air conditioning, do not disconnect any of the A/C hoses without first having the system depressurized by a dealer service department or a service station.*

Hose removal precautions closely parallel gasket removal precautions. Avoid scratching or gouging the surface that the hose mates against or the connection may leak. This is especially true for radiator hoses. Because of various chemical reactions, the rubber in hoses can bond itself to the metal spigot that the hose fits over. To remove a hose, first loosen the hose clamps that secure it to the spigot. Then, with slip-joint pliers, grab the hose at the clamp and rotate it around the spigot. Work it back and forth until it is completely free, then pull it off. Silicone or other lubricants will ease removal if they can be applied between the hose and the outside of the spigot. Apply the same lubricant to the inside of the hose and the outside of the spigot to simplify installation.

As a last resort (and if the hose is to be replaced with a new one anyway), the rubber can be slit with a knife and the hose peeled from the spigot. If this must be done, be careful that the metal connection is not damaged.

If a hose clamp is broken or damaged, do not reuse it. Wire-type clamps usually weaken with age, so it is a good idea to replace them with screw-type clamps whenever a hose is removed.

### *Tools*

A selection of good tools is a basic requirement for anyone who plans to maintain and repair his or her own vehicle. For the owner who has few tools, the initial investment might seem high, but when compared to the spiraling costs of professional auto maintenance and repair, it is a wise one.

To help the owner decide which tools are needed to perform the tasks detailed in this manual, the following tool lists are offered: *Maintenance and minor repair, Repair/overhaul* and *Special.*

The newcomer to practical mechanics

Dial caliper

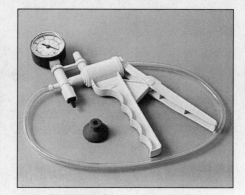

Hand-operated vacuum pump

Timing light

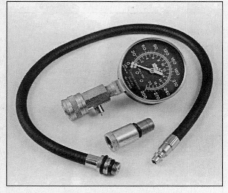

Compression gauge with spark plug
hole adapter

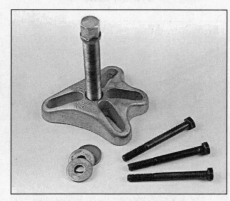

Damper/steering wheel puller

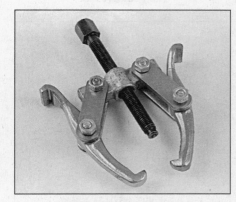

General purpose puller

Hydraulic lifter removal tool

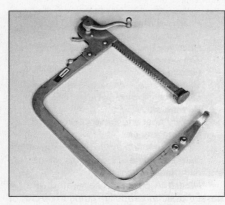

Valve spring compressor

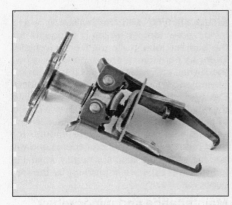

Valve spring compressor

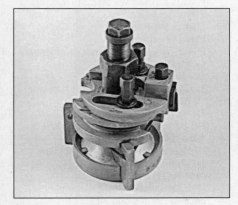

Ridge reamer

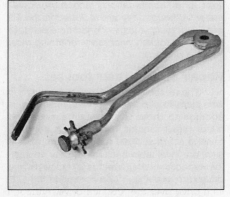

Piston ring groove cleaning tool

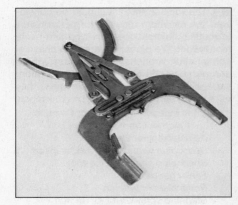

Ring removal/installation tool

Ring compressor

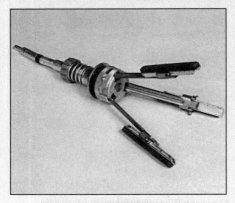

Cylinder hone

Brake hold-down spring tool

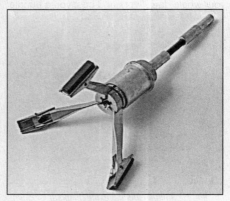

Brake cylinder hone

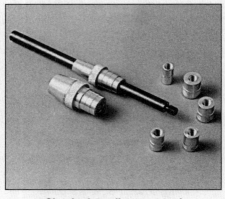

Clutch plate alignment tool

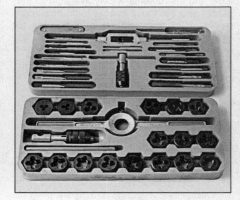

Tap and die set

should start off with the *maintenance and minor repair* tool kit, which is adequate for the simpler jobs performed on a vehicle. Then, as confidence and experience grow, the owner can tackle more difficult tasks, buying additional tools as they are needed. Eventually the basic kit will be expanded into the *repair and overhaul* tool set. Over a period of time, the experienced do-it-yourselfer will assemble a tool set complete enough for most repair and overhaul procedures and will add tools from the special category when it is felt that the expense is justified by the frequency of use.

## Maintenance and minor repair tool kit

The tools in this list should be considered the minimum required for performance of routine maintenance, servicing and minor repair work. We recommend the purchase of combination wrenches (box-end and open-end combined in one wrench). While more expensive than open end wrenches, they offer the advantages of both types of wrench.

*Combination wrench set (1/4-inch to
   1 inch or 6 mm to 19 mm)*
*Adjustable wrench, 8 inch*
*Spark plug wrench with rubber insert*
*Spark plug gap adjusting tool*
*Feeler gauge set*
*Brake bleeder wrench*
*Standard screwdriver (5/16-inch x
   6 inch)*

*Phillips screwdriver (No. 2 x 6 inch)*
*Combination pliers - 6 inch*
*Hacksaw and assortment of blades*
*Tire pressure gauge*
*Grease gun*
*Oil can*
*Fine emery cloth*
*Wire brush*
*Battery post and cable cleaning tool*
*Oil filter wrench*
*Funnel (medium size)*
*Safety goggles*
*Jackstands (2)*
*Drain pan*

**Note:** *If basic tune-ups are going to be part of routine maintenance, it will be necessary to purchase a good quality stroboscopic timing light and combination tachometer/dwell meter. Although they are included in the list of special tools, it is mentioned here because they are absolutely necessary for tuning most vehicles properly.*

## Repair and overhaul tool set

These tools are essential for anyone who plans to perform major repairs and are in addition to those in the maintenance and minor repair tool kit. Included is a comprehensive set of sockets which, though expensive, are invaluable because of their versatility, especially when various extensions and drives are available. We recommend the 1/2-inch drive over the 3/8-inch drive. Although the larger drive is bulky and more expensive,

it has the capacity of accepting a very wide range of large sockets. Ideally, however, the mechanic should have a 3/8-inch drive set and a 1/2-inch drive set.

*Socket set(s)*
*Reversible ratchet*
*Extension - 10 inch*
*Universal joint*
*Torque wrench (same size drive as
   sockets)*
*Ball peen hammer - 8 ounce*
*Soft-face hammer (plastic/rubber)*
*Standard screwdriver (1/4-inch x 6 inch)*
*Standard screwdriver (stubby -
   5/16-inch)*
*Phillips screwdriver (No. 3 x 8 inch)*
*Phillips screwdriver (stubby - No. 2)*
*Pliers - vise grip*
*Pliers - lineman's*
*Pliers - needle nose*
*Pliers - snap-ring (internal and external)*
*Cold chisel - 1/2-inch*
*Scribe*
*Scraper (made from flattened copper
   tubing)*
*Centerpunch*
*Pin punches (1/16, 1/8, 3/16-inch)*
*Steel rule/straightedge - 12 inch*
*Allen wrench set (1/8 to 3/8-inch or
   4 mm to 10 mm)*
*A selection of files*
*Wire brush (large)*
*Jackstands (second set)*
*Jack (scissor or hydraulic type)*

**Note:** *Another tool which is often useful is an electric drill with a chuck capacity of 3/8-inch and a set of good quality drill bits.*

## Special tools

The tools in this list include those which are not used regularly, are expensive to buy, or which need to be used in accordance with their manufacturer's instructions. Unless these tools will be used frequently, it is not very economical to purchase many of them. A consideration would be to split the cost and use between yourself and a friend or friends. In addition, most of these tools can be obtained from a tool rental shop on a temporary basis.

This list primarily contains only those tools and instruments widely available to the public, and not those special tools produced by the vehicle manufacturer for distribution to dealer service departments. Occasionally, references to the manufacturer's special tools are included in the text of this manual. Generally, an alternative method of doing the job without the special tool is offered. However, sometimes there is no alternative to their use. Where this is the case, and the tool cannot be purchased or borrowed, the work should be turned over to the dealer service department or an automotive repair shop.

*Valve spring compressor*
*Piston ring groove cleaning tool*
*Piston ring compressor*
*Piston ring installation tool*
*Cylinder compression gauge*
*Cylinder ridge reamer*
*Cylinder surfacing hone*
*Cylinder bore gauge*
*Micrometers and/or dial calipers*
*Hydraulic lifter removal tool*
*Balljoint separator*
*Universal-type puller*
*Impact screwdriver*
*Dial indicator set*
*Stroboscopic timing light (inductive pick-up)*
*Hand operated vacuum/pressure pump*
*Tachometer/dwell meter*
*Universal electrical multimeter*
*Cable hoist*
*Brake spring removal and installation tools*
*Floor jack*

## Buying tools

For the do-it-yourselfer who is just starting to get involved in vehicle maintenance and repair, there are a number of options available when purchasing tools. If maintenance and minor repair is the extent of the work to be done, the purchase of individual tools is satisfactory. If, on the other hand, extensive work is planned, it would be a good idea to purchase a modest tool set from one of the large retail chain stores. A set can usually be bought at a substantial savings over the individual tool prices, and they often come with a tool box. As additional tools are needed, add-on sets, individual tools and a larger tool box can be purchased to expand the tool selection. Building a tool set gradually allows the cost of the tools to be spread over a longer period of time and gives the mechanic the freedom to choose only those tools that will actually be used.

Tool stores will often be the only source of some of the special tools that are needed, but regardless of where tools are bought, try to avoid cheap ones, especially when buying screwdrivers and sockets, because they won't last very long. The expense involved in replacing cheap tools will eventually be greater than the initial cost of quality tools.

## Care and maintenance of tools

Good tools are expensive, so it makes sense to treat them with respect. Keep them clean and in usable condition and store them properly when not in use. Always wipe off any dirt, grease or metal chips before putting them away. Never leave tools lying around in the work area. Upon completion of a job, always check closely under the hood for tools that may have been left there so they won't get lost during a test drive.

Some tools, such as screwdrivers, pliers, wrenches and sockets, can be hung on a panel mounted on the garage or workshop wall, while others should be kept in a tool box or tray. Measuring instruments, gauges, meters, etc. must be carefully stored where they cannot be damaged by weather or impact from other tools.

When tools are used with care and stored properly, they will last a very long time. Even with the best of care, though, tools will wear out if used frequently. When a tool is damaged or worn out, replace it. Subsequent jobs will be safer and more enjoyable if you do.

## How to repair damaged threads

Sometimes, the internal threads of a nut or bolt hole can become stripped, usually from overtightening. Stripping threads is an all-too-common occurrence, especially when working with aluminum parts, because aluminum is so soft that it easily strips out.

Usually, external or internal threads are only partially stripped. After they've been cleaned up with a tap or die, they'll still work. Sometimes, however, threads are badly damaged. When this happens, you've got three choices:

1) *Drill and tap the hole to the next suitable oversize and install a larger diameter bolt, screw or stud.*
2) *Drill and tap the hole to accept a threaded plug, then drill and tap the plug to the original screw size. You can also buy a plug already threaded to the original size. Then you simply drill a hole to the specified size, then run the threaded plug into the hole with a bolt and jam*
nut. *Once the plug is fully seated, remove the jam nut and bolt.*
3) *The third method uses a patented thread repair kit like Heli-Coil or Slimsert. These easy-to-use kits are designed to repair damaged threads in straight-through holes and blind holes. Both are available as kits which can handle a variety of sizes and thread patterns. Drill the hole, then tap it with the special included tap. Install the Heli-Coil and the hole is back to its original diameter and thread pitch.*

Regardless of which method you use, be sure to proceed calmly and carefully. A little impatience or carelessness during one of these relatively simple procedures can ruin your whole day's work and cost you a bundle if you wreck an expensive part.

## Working facilities

Not to be overlooked when discussing tools is the workshop. If anything more than routine maintenance is to be carried out, some sort of suitable work area is essential.

It is understood, and appreciated, that many home mechanics do not have a good workshop or garage available, and end up removing an engine or doing major repairs outside. It is recommended, however, that the overhaul or repair be completed under the cover of a roof.

A clean, flat workbench or table of comfortable working height is an absolute necessity. The workbench should be equipped with a vise that has a jaw opening of at least four inches.

As mentioned previously, some clean, dry storage space is also required for tools, as well as the lubricants, fluids, cleaning solvents, etc. which soon become necessary.

Sometimes waste oil and fluids, drained from the engine or cooling system during normal maintenance or repairs, present a disposal problem. To avoid pouring them on the ground or into a sewage system, pour the used fluids into large containers, seal them with caps and take them to an authorized disposal site or recycling center. Plastic jugs, such as old antifreeze containers, are ideal for this purpose.

Always keep a supply of old newspapers and clean rags available. Old towels are excellent for mopping up spills. Many mechanics use rolls of paper towels for most work because they are readily available and disposable. To help keep the area under the vehicle clean, a large cardboard box can be cut open and flattened to protect the garage or shop floor.

Whenever working over a painted surface, such as when leaning over a fender to service something under the hood, always cover it with an old blanket or bedspread to protect the finish. Vinyl covered pads, made especially for this purpose, are available at auto parts stores.

# Jacking and towing

## Jacking

**Warning:** *The jack supplied with the vehicle should only be used for changing a tire or placing jackstands under the frame. Never work under the vehicle or start the engine while this jack is being used as the only means of support.*

The vehicle should be on level ground. Place the shift lever in Park, if you have an automatic, or Reverse if you have a manual transaxle. Block the wheel diagonally opposite the wheel being changed. Set the parking brake.

Remove the spare tire and jack from stowage. Remove the wheel cover and trim ring (if so equipped) with the tapered end of the lug nut wrench by inserting and twisting the handle and then prying against the back of the wheel cover. Loosen, but do not remove, the lug nuts (one-half turn is sufficient).

Place the scissors-type jack under the side of the vehicle and adjust the jack height until it fits between the notches in the vertical rocker panel flange nearest the wheel to be changed. There is a front and rear jacking point on each side of the vehicle **(see illustration)**.

Turn the jack handle clockwise until the tire clears the ground. Remove the lug nuts

and pull the wheel off. Replace it with the spare.

Install the lug nuts with the beveled edges facing in. Tighten them snugly. Don't attempt to tighten them completely until the vehicle is lowered or it could slip off the jack. Turn the jack handle counterclockwise to lower the vehicle. Remove the jack and tighten the lug nuts in a diagonal pattern.

Install the cover (and trim ring, if used) and be sure it's snapped into place all the way around.

Stow the tire, jack and wrench. Unblock the wheels.

## Towing

The manufacturer does not recommend towing except with a towing dolly under the front wheels. In an emergency the vehicle can be towed a short distance with a cable or chain attached to one of the towing eyelets located under the front or rear bumpers following the precautions above. The driver must remain in the vehicle to operate the steering and brakes (remember that power steering and power brakes will not work with the engine off).

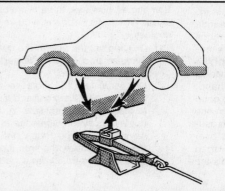

The jack fits over the rocker panel flange, between the two notches (there are two jacking points on each side of the vehicle)

# Anti-theft audio system

## General information

1    Some models are equipped with the anti-theft audio system which includes an anti-theft feature that will render the stereo inoperative if stolen. If the power source to the stereo is cut with the anti-theft feature activated, the stereo will be inoperative. Even if the power source is immediately re-connected, the stereo will not function. If your vehicle is equipped with this anti-theft system, do not disconnect the battery, remove the stereo or disconnect related components unless you have either turned off the feature or have the individual ID (code) number for the stereo.

2    Refer to your vehicle's owner's manual for more complete information on this audio system and its anti-theft feature.

## Disabling the anti-theft feature

3    Press the stereo's 1 and 4 buttons at the same time for five seconds with the ignition on and the radio power off. The display will show SEC, indicating the unit is in the secure mode (anti-theft feature enabled).

4    Press the SEEK left arrow until the first digit of your code appears.

5    Press the SEEK right arrow until the second digit of your code appears.

6    Press the TUNE left arrow until the third digit of your code appears.

7    Press the TUNE right arrow until the fourth digit of your code appears.

8    Press the AM/FM knob. "0000" will be displayed if you entered the correct code.

9    If Err is displayed, the code you entered was incorrect and the anti-theft feature is still enabled.

## Unlocking the stereo after a power loss

10    When power is restored to the stereo, the stereo won't turn on and LOC will appear on the display. Enter your ID code as follows; pause no more than 15 seconds between Steps.

11    Turn the ignition switch to ON, but leave the stereo off.

12    Press the SEEK left arrow until the first digit of your code appears.

13    Press the SEEK right arrow until the second digit of your code appears.

14    Press the TUNE left arrow until the third digit of your code appears.

15    Press the TUNE right arrow until the fourth digit of your code appears.

16    Press the AM/FM knob. If the time of day appears, the numbers you entered were correct and the stereo will work. If Err appears, the numbers you entered were not correct and the stereo is still inoperative.

# Booster battery (jump) starting

Observe these precautions when using a booster battery to start a vehicle:

a)  Before connecting the booster battery, make sure the ignition switch is in the Off position
b)  Turn off the lights, heater and other electrical loads.
c)  Your eyes should be shielded. Safety goggles are a good idea.
d)  Make sure the booster battery is the same voltage as the dead one in the vehicle.
e)  The two vehicles MUST NOT TOUCH each other!
f)  Make sure the transaxle is in Neutral (manual) or Park (automatic).
g)  If the booster battery is not a maintenance-free type, remove the vent caps and lay a cloth over the vent holes.

Connect the red jumper cable to the positive (+) terminals of each battery (see illustration).

Connect one end of the black jumper cable to the negative (-) terminal of the booster battery. The other end of this cable should be connected to a good ground on the vehicle to be started, such as a bolt or bracket on the body.

Start the engine using the booster battery, then, with the engine running at idle speed, disconnect the jumper cables in the reverse order of connection.

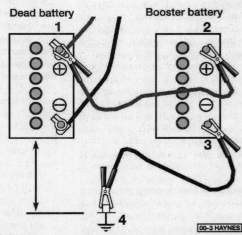

Make the booster battery cable connections in the numerical order shown (note that the negative cable of the booster battery is NOT attached to the negative terminal of the dead battery)

# Automotive chemicals and lubricants

A number of automotive chemicals and lubricants are available for use during vehicle maintenance and repair. They include a wide variety of products ranging from cleaning solvents and degreasers to lubricants and protective sprays for rubber, plastic and vinyl.

## Cleaners

**Carburetor cleaner and choke cleaner** is a strong solvent for gum, varnish and carbon. Most carburetor cleaners leave a dry-type lubricant film which will not harden or gum up. Because of this film it is not recommended for use on electrical components.

**Brake system cleaner** is used to remove grease and brake fluid from the brake system, where clean surfaces are absolutely necessary. It leaves no residue and often eliminates brake squeal caused by contaminants.

**Electrical cleaner** removes oxidation, corrosion and carbon deposits from electrical contacts, restoring full current flow. It can also be used to clean spark plugs, carburetor jets, voltage regulators and other parts where an oil-free surface is desired.

**Demoisturants** remove water and moisture from electrical components such as alternators, voltage regulators, electrical connectors and fuse blocks. They are non-conductive, non-corrosive and non-flammable.

**Degreasers** are heavy-duty solvents used to remove grease from the outside of the engine and from chassis components. They can be sprayed or brushed on and, depending on the type, are rinsed off either with water or solvent.

## Lubricants

**Motor oil** is the lubricant formulated for use in engines. It normally contains a wide variety of additives to prevent corrosion and reduce foaming and wear. Motor oil comes in various weights (viscosity ratings) from 5 to 80. The recommended weight of the oil depends on the season, temperature and the demands on the engine. Light oil is used in cold climates and under light load conditions. Heavy oil is used in hot climates and where high loads are encountered. Multi-viscosity oils are designed to have characteristics of both light and heavy oils and are available in a number of weights from 5W-20 to 20W-50.

**Gear oil** is designed to be used in differentials, manual transmissions and other areas where high-temperature lubrication is required.

**Chassis and wheel bearing grease** is a heavy grease used where increased loads and friction are encountered, such as for wheel bearings, balljoints, tie-rod ends and universal joints.

**High-temperature wheel bearing grease** is designed to withstand the extreme temperatures encountered by wheel bearings in disc brake equipped vehicles. It usually contains molybdenum disulfide (moly), which is a dry-type lubricant.

**White grease** is a heavy grease for metal-to-metal applications where water is a problem. White grease stays soft under both low and high temperatures (usually from -100 to +190-degrees F), and will not wash off or dilute in the presence of water.

**Assembly lube** is a special extreme pressure lubricant, usually containing moly, used to lubricate high-load parts (such as main and rod bearings and cam lobes) for initial start-up of a new engine. The assembly lube lubricates the parts without being squeezed out or washed away until the engine oiling system begins to function.

**Silicone lubricants** are used to protect rubber, plastic, vinyl and nylon parts.

**Graphite lubricants** are used where oils cannot be used due to contamination problems, such as in locks. The dry graphite will lubricate metal parts while remaining uncontaminated by dirt, water, oil or acids. It is electrically conductive and will not foul electrical contacts in locks such as the ignition switch.

**Moly penetrants** loosen and lubricate frozen, rusted and corroded fasteners and prevent future rusting or freezing.

**Heat-sink grease** is a special electrically non-conductive grease that is used for mounting electronic ignition modules where it is essential that heat is transferred away from the module.

## Sealants

**RTV sealant** is one of the most widely used gasket compounds. Made from silicone, RTV is air curing, it seals, bonds, waterproofs, fills surface irregularities, remains flexible, doesn't shrink, is relatively easy to remove, and is used as a supplementary sealer with almost all low and medium temperature gaskets.

**Anaerobic sealant** is much like RTV in that it can be used either to seal gaskets or to form gaskets by itself. It remains flexible, is solvent resistant and fills surface imperfections. The difference between an anaerobic sealant and an RTV-type sealant is in the curing. RTV cures when exposed to air, while an anaerobic sealant cures only in the absence of air. This means that an anaerobic sealant cures only after the assembly of parts, sealing them together.

**Thread and pipe sealant** is used for sealing hydraulic and pneumatic fittings and vacuum lines. It is usually made from a Teflon compound, and comes in a spray, a paint-on liquid and as a wrap-around tape.

## Chemicals

**Anti-seize compound** prevents seizing, galling, cold welding, rust and corrosion in fasteners. High-temperature anti-seize, usually made with copper and graphite lubricants, is used for exhaust system and exhaust manifold bolts.

**Anaerobic locking compounds** are used to keep fasteners from vibrating or working loose and cure only after installation, in the absence of air. Medium strength locking compound is used for small nuts, bolts and screws that may be removed later. High-strength locking compound is for large nuts, bolts and studs which aren't removed on a regular basis.

**Oil additives** range from viscosity index improvers to chemical treatments that claim to reduce internal engine friction. It should be noted that most oil manufacturers caution against using additives with their oils.

**Gas additives** perform several functions, depending on their chemical makeup. They usually contain solvents that help dissolve gum and varnish that build up on carburetor, fuel injection and intake parts. They also serve to break down carbon deposits that form on the inside surfaces of the combustion chambers. Some additives contain upper cylinder lubricants for valves and piston rings, and others contain chemicals to remove condensation from the gas tank.

## Miscellaneous

**Brake fluid** is specially formulated hydraulic fluid that can withstand the heat and pressure encountered in brake systems. Care must be taken so this fluid does not come in contact with painted surfaces or plastics. An opened container should always be resealed to prevent contamination by water or dirt.

**Weatherstrip adhesive** is used to bond weatherstripping around doors, windows and trunk lids. It is sometimes used to attach trim pieces.

**Undercoating** is a petroleum-based, tar-like substance that is designed to protect metal surfaces on the underside of the vehicle from corrosion. It also acts as a sound-deadening agent by insulating the bottom of the vehicle.

**Waxes and polishes** are used to help protect painted and plated surfaces from the weather. Different types of paint may require the use of different types of wax and polish. Some polishes utilize a chemical or abrasive cleaner to help remove the top layer of oxidized (dull) paint on older vehicles. In recent years many non-wax polishes that contain a wide variety of chemicals such as polymers and silicones have been introduced. These non-wax polishes are usually easier to apply and last longer than conventional waxes and polishes.

# Conversion factors

## Length (distance)

| | | | | | |
|---|---|---|---|---|---|
| Inches (in) | X | 25.4 | = Millimetres (mm) | X 0.0394 | = Inches (in) |
| Feet (ft) | X | 0.305 | = Metres (m) | X 3.281 | = Feet (ft) |
| Miles | X | 1.609 | = Kilometres (km) | X 0.621 | = Miles |

## Volume (capacity)

| | | | | | |
|---|---|---|---|---|---|
| Cubic inches (cu in; in$^3$) | X | 16.387 | = Cubic centimetres (cc; cm$^3$) | X 0.061 | = Cubic inches (cu in; in$^3$) |
| Imperial pints (Imp pt) | X | 0.568 | = Litres (l) | X 1.76 | = Imperial pints (Imp pt) |
| Imperial quarts (Imp qt) | X | 1.137 | = Litres (l) | X 0.88 | = Imperial quarts (Imp qt) |
| Imperial quarts (Imp qt) | X | 1.201 | = US quarts (US qt) | X 0.833 | = Imperial quarts (Imp qt) |
| US quarts (US qt) | X | 0.946 | = Litres (l) | X 1.057 | = US quarts (US qt) |
| Imperial gallons (Imp gal) | X | 4.546 | = Litres (l) | X 0.22 | = Imperial gallons (Imp gal) |
| Imperial gallons (Imp gal) | X | 1.201 | = US gallons (US gal) | X 0.833 | = Imperial gallons (Imp gal) |
| US gallons (US gal) | X | 3.785 | = Litres (l) | X 0.264 | = US gallons (US gal) |

## Mass (weight)

| | | | | | |
|---|---|---|---|---|---|
| Ounces (oz) | X | 28.35 | = Grams (g) | X 0.035 | = Ounces (oz) |
| Pounds (lb) | X | 0.454 | = Kilograms (kg) | X 2.205 | = Pounds (lb) |

## Force

| | | | | | |
|---|---|---|---|---|---|
| Ounces-force (ozf; oz) | X | 0.278 | = Newtons (N) | X 3.6 | = Ounces-force (ozf; oz) |
| Pounds-force (lbf; lb) | X | 4.448 | = Newtons (N) | X 0.225 | = Pounds-force (lbf; lb) |
| Newtons (N) | X | 0.1 | = Kilograms-force (kgf; kg) | X 9.81 | = Newtons (N) |

## Pressure

| | | | | | |
|---|---|---|---|---|---|
| Pounds-force per square inch (psi; lbf/in$^2$; lb/in$^2$) | X | 0.070 | = Kilograms-force per square centimetre (kgf/cm$^2$; kg/cm$^2$) | X 14.223 | = Pounds-force per square inch (psi; lbf/in$^2$; lb/in$^2$) |
| Pounds-force per square inch (psi; lbf/in$^2$; lb/in$^2$) | X | 0.068 | = Atmospheres (atm) | X 14.696 | = Pounds-force per square inch (psi; lbf/in$^2$; lb/in$^2$) |
| Pounds-force per square inch (psi; lbf/in$^2$; lb/in$^2$) | X | 0.069 | = Bars | X 14.5 | = Pounds-force per square inch (psi; lbf/in$^2$; lb/in$^2$) |
| Pounds-force per square inch (psi; lbf/in$^2$; lb/in$^2$) | X | 6.895 | = Kilopascals (kPa) | X 0.145 | = Pounds-force per square inch (psi; lbf/in$^2$; lb/in$^2$) |
| Kilopascals (kPa) | X | 0.01 | = Kilograms-force per square centimetre (kgf/cm$^2$; kg/cm$^2$) | X 98.1 | = Kilopascals (kPa) |

## Torque (moment of force)

| | | | | | |
|---|---|---|---|---|---|
| Pounds-force inches (lbf in; lb in) | X | 1.152 | = Kilograms-force centimetre (kgf cm; kg cm) | X 0.868 | = Pounds-force inches (lbf in; lb in) |
| Pounds-force inches (lbf in; lb in) | X | 0.113 | = Newton metres (Nm) | X 8.85 | = Pounds-force inches (lbf in; lb in) |
| Pounds-force inches (lbf in; lb in) | X | 0.083 | = Pounds-force feet (lbf ft; lb ft) | X 12 | = Pounds-force inches (lbf in; lb in) |
| Pounds-force feet (lbf ft; lb ft) | X | 0.138 | = Kilograms-force metres (kgf m; kg m) | X 7.233 | = Pounds-force feet (lbf ft; lb ft) |
| Pounds-force feet (lbf ft; lb ft) | X | 1.356 | = Newton metres (Nm) | X 0.738 | = Pounds-force feet (lbf ft; lb ft) |
| Newton metres (Nm) | X | 0.102 | = Kilograms-force metres (kgf m; kg m) | X 9.804 | = Newton metres (Nm) |

## Vacuum

| | | | | | |
|---|---|---|---|---|---|
| Inches mercury (in. Hg) | X | 3.377 | = Kilopascals (kPa) | X 0.2961 | = Inches mercury |
| Inches mercury (in. Hg) | X | 25.4 | = Millimeters mercury (mm Hg) | X 0.0394 | = Inches mercury |

## Power

| | | | | | |
|---|---|---|---|---|---|
| Horsepower (hp) | X | 745.7 | = Watts (W) | X 0.0013 | = Horsepower (hp) |

## Velocity (speed)

| | | | | | |
|---|---|---|---|---|---|
| Miles per hour (miles/hr; mph) | X | 1.609 | = Kilometres per hour (km/hr; kph) | X 0.621 | = Miles per hour (miles/hr; mph) |

## Fuel consumption*

| | | | | | |
|---|---|---|---|---|---|
| Miles per gallon, Imperial (mpg) | X | 0.354 | = Kilometres per litre (km/l) | X 2.825 | = Miles per gallon, Imperial (mpg) |
| Miles per gallon, US (mpg) | X | 0.425 | = Kilometres per litre (km/l) | X 2.352 | = Miles per gallon, US (mpg) |

## Temperature

Degrees Fahrenheit = (°C x 1.8) + 32     Degrees Celsius (Degrees Centigrade; °C) = (°F - 32) x 0.56

*It is common practice to convert from miles per gallon (mpg) to litres/100 kilometres (l/100km), where mpg (Imperial) x l/100 km = 282 and mpg (US) x l/100 km = 235

# Safety first!

Regardless of how enthusiastic you may be about getting on with the job at hand, take the time to ensure that your safety is not jeopardized. A moment's lack of attention can result in an accident, as can failure to observe certain simple safety precautions. The possibility of an accident will always exist, and the following points should not be considered a comprehensive list of all dangers. Rather, they are intended to make you aware of the risks and to encourage a safety conscious approach to all work you carry out on your vehicle.

## Essential DOs and DON'Ts

**DON'T** rely on a jack when working under the vehicle. Always use approved jackstands to support the weight of the vehicle and place them under the recommended lift or support points.

**DON'T** attempt to loosen extremely tight fasteners (i.e. wheel lug nuts) while the vehicle is on a jack - it may fall.

**DON'T** start the engine without first making sure that the transmission is in Neutral (or Park where applicable) and the parking brake is set.

**DON'T** remove the radiator cap from a hot cooling system - let it cool or cover it with a cloth and release the pressure gradually.

**DON'T** attempt to drain the engine oil until you are sure it has cooled to the point that it will not burn you.

**DON'T** touch any part of the engine or exhaust system until it has cooled sufficiently to avoid burns.

**DON'T** siphon toxic liquids such as gasoline, antifreeze and brake fluid by mouth, or allow them to remain on your skin.

**DON'T** inhale brake lining dust - it is potentially hazardous (see *Asbestos* below).

**DON'T** allow spilled oil or grease to remain on the floor - wipe it up before someone slips on it.

**DON'T** use loose fitting wrenches or other tools which may slip and cause injury.

**DON'T** push on wrenches when loosening or tightening nuts or bolts. Always try to pull the wrench toward you. If the situation calls for pushing the wrench away, push with an open hand to avoid scraped knuckles if the wrench should slip.

**DON'T** attempt to lift a heavy component alone - get someone to help you.

**DON'T** rush or take unsafe shortcuts to finish a job.

**DON'T** allow children or animals in or around the vehicle while you are working on it.

**DO** wear eye protection when using power tools such as a drill, sander, bench grinder, etc. and when working under a vehicle.

**DO** keep loose clothing and long hair well out of the way of moving parts.

**DO** make sure that any hoist used has a safe working load rating adequate for the job.

**DO** get someone to check on you periodically when working alone on a vehicle.

**DO** carry out work in a logical sequence and make sure that everything is correctly assembled and tightened.

**DO** keep chemicals and fluids tightly capped and out of the reach of children and pets.

**DO** remember that your vehicle's safety affects that of yourself and others. If in doubt on any point, get professional advice.

## Asbestos

Certain friction, insulating, sealing, and other products - such as brake linings, brake bands, clutch linings, torque converters, gaskets, etc. - may contain asbestos. Extreme care must be taken to avoid inhalation of dust from such products, since it is hazardous to health. If in doubt, assume that they do contain asbestos.

## Fire

Remember at all times that gasoline is highly flammable. Never smoke or have any kind of open flame around when working on a vehicle. But the risk does not end there. A spark caused by an electrical short circuit, by two metal surfaces contacting each other, or even by static electricity built up in your body under certain conditions, can ignite gasoline vapors, which in a confined space are highly explosive. Do not, under any circumstances, use gasoline for cleaning parts. Use an approved safety solvent.

Always disconnect the battery ground (-) cable at the battery before working on any part of the fuel system or electrical system. Never risk spilling fuel on a hot engine or exhaust component. It is strongly recommended that a fire extinguisher suitable for use on fuel and electrical fires be kept handy in the garage or workshop at all times. Never try to extinguish a fuel or electrical fire with water.

## Fumes

Certain fumes are highly toxic and can quickly cause unconsciousness and even death if inhaled to any extent. Gasoline vapor falls into this category, as do the vapors from some cleaning solvents. Any draining or pouring of such volatile fluids should be done in a well ventilated area.

When using cleaning fluids and solvents, read the instructions on the container carefully. Never use materials from unmarked containers.

Never run the engine in an enclosed space, such as a garage. Exhaust fumes contain carbon monoxide, which is extremely poisonous. If you need to run the engine, always do so in the open air, or at least have the rear of the vehicle outside the work area.

If you are fortunate enough to have the use of an inspection pit, never drain or pour gasoline and never run the engine while the vehicle is over the pit. The fumes, being heavier than air, will concentrate in the pit with possibly lethal results.

## The battery

Never create a spark or allow a bare light bulb near a battery. They normally give off a certain amount of hydrogen gas, which is highly explosive.

Always disconnect the battery ground (-) cable at the battery before working on the fuel or electrical systems.

If possible, loosen the filler caps or cover when charging the battery from an external source (this does not apply to sealed or maintenance-free batteries). Do not charge at an excessive rate or the battery may burst.

Take care when adding water to a non maintenance-free battery and when carrying a battery. The electrolyte, even when diluted, is very corrosive and should not be allowed to contact clothing or skin.

Always wear eye protection when cleaning the battery to prevent the caustic deposits from entering your eyes.

## Household current

When using an electric power tool, inspection light, etc., which operates on household current, always make sure that the tool is correctly connected to its plug and that, where necessary, it is properly grounded. Do not use such items in damp conditions and, again, do not create a spark or apply excessive heat in the vicinity of fuel or fuel vapor.

## Secondary ignition system voltage

A severe electric shock can result from touching certain parts of the ignition system (such as the spark plug wires) when the engine is running or being cranked, particularly if components are damp or the insulation is defective. In the case of an electronic ignition system, the secondary system voltage is much higher and could prove fatal.

# Troubleshooting

## Contents

This section provides an easy reference guide to the more common problems which may occur during the operation of your vehicle. These problems and their possible causes are grouped under headings denoting various components or systems, such as Engine, Cooling system, etc. They also refer you to the chapter and/or section which deals with the problem.

Remember that successful troubleshooting is not a mysterious black art practiced only by professional mechanics. It is simply the result of the right knowledge combined with an intelligent, systematic approach to the problem. Always work by a process of elimination, starting with the simplest solution and working through to the most complex - and never overlook the obvious. Anyone can run the gas tank dry or leave the lights on overnight, so don't assume that you are exempt from such oversights.

Finally, always establish a clear idea of why a problem has occurred and take steps to ensure that it doesn't happen again. If the electrical system fails because of a poor connection, check the other connections in the system to make sure that they don't fail as well. If a particular fuse continues to blow, find out why - don't just replace one fuse after another. Remember, failure of a small component can often be indicative of potential failure or incorrect functioning of a more important component or system.

## Engine

### 1 Engine will not rotate when attempting to start

1   Battery terminal connections loose or corroded (Chapter 1).
2   Battery discharged or faulty (Chapter 1).
3   Automatic transaxle not completely engaged in Park (Chapter 7) or clutch pedal not completely depressed (Chapter 8).
4   Broken, loose or disconnected wiring in the starting circuit (Chapters 5 and 12).
5   Starter motor pinion jammed in flywheel ring gear (Chapter 5).
6   Starter solenoid faulty (Chapter 5).
7   Starter motor faulty (Chapter 5).
8   Ignition switch faulty (Chapter 12).
9   Starter pinion or flywheel teeth worn or broken (Chapter 5).

### 2 Engine rotates but will not start

1   Fuel tank empty.
2   Battery discharged (engine rotates slowly) (Chapter 5).
3   Battery terminal connections loose or corroded (Chapter 1).
4   Leaking fuel injector(s), faulty fuel pump, pressure regulator, etc. (Chapter 4).
5   Broken or stripped timing belt or chain (Chapter 2A or 2B).
6   Ignition components damp or damaged (Chapter 5).
7   Worn, faulty or incorrectly gapped spark plugs (Chapter 1).
8   Broken, loose or disconnected wiring in the starting circuit (Chapter 5).
9   Loose distributor is changing ignition timing (Chapter 5).
10   Broken, loose or disconnected wires at the ignition coil or faulty coil (Chapter 5).

### 3 Engine hard to start when cold

1   Battery discharged or low (Chapter 1).

2   Malfunctioning fuel system (Chapter 4).
3   Faulty coolant temperature sensor or intake air temperature sensor (Chapter 6).
4   Injector(s) leaking (Chapter 4).
5   Faulty ignition system (Chapter 5).

### 4 Engine hard to start when hot

1   Air filter clogged (Chapter 1).
2   Fuel not reaching the fuel injection system (Chapter 4).
3   Corroded battery connections, especially ground (Chapter 1).
4   Faulty coolant temperature sensor or intake air temperature sensor (Chapter 6).

### 5 Starter motor noisy or excessively rough in engagement

1   Pinion or flywheel gear teeth worn or broken (Chapter 5).
2   Starter motor mounting bolts loose or missing (Chapter 5).

### 6 Engine starts but stops immediately

1   Loose or faulty electrical connections at distributor, coil or alternator (Chapter 5).
2   Insufficient fuel reaching the fuel injector(s) (Chapters 1 and 4).
3   Vacuum leak at the gasket between the intake manifold/plenum and throttle body (Chapter 4).
4   Idle speed incorrect (Chapter 1).

### 7 Oil leaks

1   Oil pan gasket and/or oil pan drain bolt washer leaking (Chapter 2A or 2B).
2   Oil pressure sending unit leaking (Chapter 2C).
3   Valve covers leaking (Chapter 2A or 2B).
4   Engine oil seals leaking (Chapter 2A or 2B).
5   Oil pump housing leaking (Chapter 2A or 2B).

### 8 Engine lopes while idling or idles erratically

1   Vacuum leakage (Chapters 2A, 2B and 4).
2   Leaking EGR valve (Chapter 6).
3   Air filter clogged (Chapter 1).
4   Malfunction in the fuel injection or engine control system (Chapters 4 and 6).
5   Leaking head gasket (Chapter 2A or 2B).
6   Timing belt and/or sprockets worn (Chapter 2A).
7   Camshaft lobes worn (Chapter 2A or 2B).

### 9 Engine misses at idle speed

1   Spark plugs worn or not gapped properly (Chapters 2A, 2B and 4).

2   Faulty spark plug wires (Chapter 1).
3   Vacuum leaks (Chapters 2A, 2B and 4).
4   Incorrect ignition timing (Chapter 1).
5   Uneven or low compression (Chapter 2C).
6   Problem with the fuel injection system (Chapter 4).

### 10 Engine misses throughout driving speed range

1   Fuel filter clogged and/or impurities in the fuel system (Chapter 1).
2   Low fuel output at the injector(s) (Chapter 4).
3   Faulty or incorrectly gapped spark plugs (Chapter 1).
4   Incorrect ignition timing (Chapter 5).
5   Cracked distributor cap, disconnected distributor wires or damaged distributor components (Chapters 1 and 5).
6   Leaking spark plug wires (Chapters 1 or 5).
7   Faulty emission system components (Chapter 6).
8   Low or uneven cylinder compression pressures (Chapter 2C).
9   Weak or faulty ignition system (Chapter 5).
10   Vacuum leak in fuel injection system, intake manifold, air control valve or vacuum hoses (Chapter 4).

### 11 Engine stumbles on acceleration

1   Spark plugs fouled (Chapter 1).
2   Problem with fuel injection or engine control system (Chapters 4 and 6).
3   Fuel filter clogged (Chapters 1 and 4).
4   Incorrect ignition timing (Chapter 5).
5   Intake manifold air leak (Chapters 2A, 2B and 4).
6   Problem with the emissions control system (Chapter 6).

### 12 Engine surges while holding accelerator steady

1   Intake air leak (Chapter 4).
2   Fuel pump or fuel pressure regulator faulty (Chapter 4).
3   Problem with fuel injection system (Chapter 4).
4   Problem with the emissions control system (Chapter 6).

### 13 Engine stalls

1   Idle speed incorrect (Chapter 1).
2   Fuel filter clogged and/or water and impurities in the fuel system (Chapters 1 and 4).
3   Distributor components damp or damaged (Chapter 5).
4   Faulty emissions system components (Chapter 6).
5   Faulty or incorrectly gapped spark plugs (Chapter 1).
6   Faulty spark plug wires (Chapter 1).
7   Vacuum leak in the fuel injection system, intake manifold or vacuum hoses (Chap-

ters 2A, 2B and 4).
8    Valve clearances incorrectly set (Chapter 1).

## 14    Engine lacks power

1    Incorrect ignition timing (Chapter 5).
2    Excessive play in distributor shaft (Chapter 5).
3    Worn rotor, distributor cap, spark plug wires or faulty coil (Chapters 1 and 5).
4    Faulty or incorrectly gapped spark plugs (Chapter 1).
5    Problem with the fuel injection system (Chapter 4).
6    Plugged air filter (Chapter 1).
7    Brakes binding (Chapter 9).
8    Automatic transaxle fluid level incorrect (Chapter 1).
9    Clutch slipping (Chapter 8).
10    Fuel filter clogged and/or impurities in the fuel system (Chapters 1 and 4).
11    Emission control system not functioning properly (Chapter 6).
12    Low or uneven cylinder compression pressures (Chapter 2C).
13    Obstructed exhaust system (Chapter 4).

## 15    Engine backfires

1    Emission control system not functioning properly (Chapter 6).
2    Ignition timing incorrect (Chapter 5).
3    Faulty secondary ignition system (cracked spark plug insulator, faulty plug wires, distributor cap and/or rotor) (Chapters 1 and 5).
4    Problem with the fuel injection system (Chapter 4).
5    Vacuum leak at fuel injector(s), intake manifold, air control valve or vacuum hoses (Chapters 2A, 2B and 4).
6    Valve clearances incorrectly set and/or valves sticking (Chapter 1).

## 16    Pinging or knocking engine sounds during acceleration or uphill

1    Incorrect grade of fuel.
2    Ignition timing incorrect (Chapter 5).
3    Fuel injection system faulty (Chapter 4).
4    Improper or damaged spark plugs or wires (Chapter 1).
5    Worn or damaged distributor components (Chapter 5).
6    EGR valve not functioning (Chapter 6).
7    Vacuum leak (Chapters 2A, 2B and 4).

## 17    Engine runs with oil pressure light on

1    Low oil level (Chapter 1).
2    Idle rpm below specification (Chapter 1).
3    Short in wiring circuit (Chapter 12).
4    Faulty oil pressure sender (Chapter 2C).
5    Worn engine bearings and/or oil pump (Chapters 2A, 2B or 2C).

## 18    Engine diesels (continues to run) after switching off

1    Idle speed too high (Chapter 1).
2    Excessive engine operating temperature (Chapter 3).
3    Ignition timing in need of adjustment (Chapter 5).

## Engine electrical system

## 19    Battery will not hold a charge

1    Alternator drivebelt defective or not adjusted properly (Chapter 1).
2    Battery electrolyte level low (Chapter 1).
3    Battery terminals loose or corroded (Chapter 1).
4    Alternator not charging properly (Chapter 5).
5    Loose, broken or faulty wiring in the charging circuit (Chapter 5).
6    Short in vehicle wiring (Chapter 12).
7    Internally defective battery (Chapters 1 and 5).

## 20    Alternator light fails to go out

1    Faulty alternator or charging circuit (Chapter 5).
2    Alternator drivebelt defective or out of adjustment (Chapter 1).
3    Alternator voltage regulator inoperative (Chapter 5).

## 21    Alternator light fails to come on when key is turned on

1    Warning light bulb defective (Chapter 12).
2    Fault in the printed circuit, dash wiring or bulb holder (Chapter 12).

## Fuel system

## 22    Excessive fuel consumption

1    Dirty or clogged air filter element (Chapter 1).
2    Incorrectly set ignition timing (Chapter 5).
3    Emissions system not functioning properly (Chapter 6).
4    Fuel injection system not functioning properly (Chapter 4).
5    Low tire pressure or incorrect tire size (Chapter 1).

## 23    Fuel leakage and/or fuel odor

1    Leaking fuel feed or return line (Chapters 1 and 4).
2    Tank overfilled.
3    Evaporative canister filter clogged (Chapters 1 and 6).
4    Problem with fuel injection system (Chapter 4).

## Cooling system

## 24    Overheating

1    Insufficient coolant in system (Chapter 1).
2    Water pump drivebelt defective or out of adjustment (Chapter 1).
3    Radiator core blocked or grille restricted (Chapter 3).
4    Thermostat faulty (Chapter 3).
5    Electric coolant fan inoperative or blades broken (Chapter 3).
6    Radiator cap not maintaining proper pressure (Chapter 3).
7    Ignition timing incorrect (Chapter 5).

## 25    Overcooling

1    Faulty thermostat (Chapter 3).
2    Inaccurate temperature gauge sending unit (Chapter 3).

## 26    External coolant leakage

1    Deteriorated/damaged hoses; loose clamps (Chapters 1 and 3).
2    Water pump defective (Chapter 3).
3    Leakage from radiator core or coolant reservoir bottle (Chapter 3).
4    Engine drain or water jacket core plugs leaking (Chapter 2).

## 27    Internal coolant leakage

1    Leaking cylinder head gasket (Chapter 2A or 2B).
2    Cracked cylinder bore or cylinder head (Chapter 2C).

## 28    Coolant loss

1    Too much coolant in reservoir (Chapter 1).
2    Coolant boiling away because of overheating (Chapter 3).
3    Internal or external leakage (Chapter 3).
4    Faulty radiator cap (Chapter 3).

## 29    Poor coolant circulation

1    Inoperative water pump (Chapter 3).
2    Restriction in cooling system (Chapters 1 and 3).
3    Water pump drivebelt defective/out of adjustment (Chapter 1).
4    Thermostat sticking (Chapter 3).

## Clutch

## 30    Pedal travels to floor - no pressure or very little resistance

1    Master or release cylinder faulty (Chapter 8).
2    Hose/pipe burst or leaking (Chapter 8).
3    Connections leaking (Chapter 8).
4    No fluid in reservoir (Chapter 8).
5    If fluid level in reservoir rises as pedal is

depressed, master cylinder center valve seal is faulty (Chapter 8).
6    If there is fluid on dust seal at master cylinder, piston primary seal is leaking (Chapter 8).
7    Broken release bearing or fork (Chapter 8).

### 31    Fluid in area of master cylinder dust cover and on pedal

Rear seal failure in master cylinder (Chapter 8).

### 32    Fluid on release cylinder

Release cylinder plunger seal faulty (Chapter 8).

### 33    Pedal feels spongy when depressed

Air in system (Chapter 8).

### 34    Unable to select gears

1    Faulty transaxle (Chapter 7).
2    Faulty clutch disc or pressure plate (Chapter 8).
3    Faulty release lever or release bearing (Chapter 8).
4    Faulty shift lever assembly or control cables (Chapter 8).
5    Faulty release cylinder.

### 35    Clutch slips (engine speed increases with no increase in vehicle speed)

1    Clutch plate worn (Chapter 8).
2    Clutch plate is oil soaked by leaking rear main seal (Chapter 8).
3    Clutch plate not seated (Chapter 8).
4    Warped pressure plate or flywheel (Chapter 8).
5    Weak diaphragm springs (Chapter 8).
6    Clutch plate overheated. Allow to cool.

### 36    Grabbing (chattering) as clutch is engaged

1    Oil on clutch plate lining, burned or glazed facings (Chapter 8).
2    Worn or loose engine or transaxle mounts (Chapters 2 and 7).
3    Worn splines on clutch plate hub (Chapter 8).
4    Warped pressure plate or flywheel (Chapter 8).
5    Burned or smeared resin on flywheel or pressure plate (Chapter 8).

### 37    Transaxle rattling (clicking)

1    Release lever loose (Chapter 8).
2    Clutch plate damper spring failure (Chapter 8).
3    Low engine idle speed (Chapter 1).

### 38    Noise in clutch area

1    Fork shaft improperly installed (Chapter 8).
2    Faulty bearing (Chapter 8).

### 39    Clutch pedal stays on floor

1    Clutch master cylinder piston binding in bore (Chapter 8).
2    Broken release bearing or fork (Chapter 8).
3    Broken pressure plate diaphragm spring (Chapter 8).

### 40    High pedal effort

1    Piston binding in bore (Chapter 8).
2    Pressure plate faulty (Chapter 8).
3    Incorrect size master or release cylinder (Chapter 8).

## Manual transaxle

### 41    Knocking noise at low speeds

1    Worn driveaxle constant velocity (CV) joints (Chapter 8).
2    Worn side gear shaft counterbore in differential case (Chapter 7A).*

### 42    Noise most pronounced when turning

Differential gear noise (Chapter 7A).*

### 43    Clunk on acceleration or deceleration

1    Loose engine or transaxle mounts (Chapters 2 and 7A).
2    Worn differential pinion shaft in case.*
3    Worn side gear shaft counterbore in differential case (Chapter 7A).*
4    Worn or damaged driveaxle inboard CV joints (Chapter 8).

### 44    Clicking noise in turns

Worn or damaged outboard CV joint (Chapter 8).

### 45    Vibration

1    Rough wheel bearing (Chapters 1 and 10).
2    Damaged driveaxle (Chapter 8).
3    Out-of-round tires (Chapter 1).
4    Tire out of balance (Chapters 1 and 10).
5    Worn CV joint (Chapter 8).

### 46    Noisy in neutral with engine running

1    Damaged input gear bearing (Chapter 7A).*

2    Damaged clutch release bearing (Chapter 8).

### 47    Noisy in one particular gear

1    Damaged or worn constant mesh gears (Chapter 7A).*
2    Damaged or worn synchronizers (Chapter 7A).*
3    Bent reverse fork (Chapter 7A).*
4    Damaged fourth speed gear or output gear (Chapter 7A).*
5    Worn or damaged reverse idler gear or idler bushing (Chapter 7A).*

### 48    Noisy in all gears

1    Insufficient lubricant (Chapter 7A).
2    Damaged or worn bearings (Chapter 7A).*
3    Worn or damaged input gear shaft and/or output gear shaft (Chapter 7A).*

### 49    Slips out of gear

1    Worn or improperly adjusted linkage (Chapter 7A).
2    Transaxle loose on engine (Chapter 7A).
3    Shift linkage does not work freely, binds (Chapter 7A).
4    Input gear bearing retainer broken or loose (Chapter 7A).*
5    Dirt between clutch cover and engine housing (Chapter 7A).
6    Worn shift fork (Chapter 7A).*

### 50    Leaks lubricant

1    Side gear shaft seals worn (Chapter 7).
2    Excessive amount of lubricant in transaxle (Chapters 1 and 7A).
3    Loose or broken input gear shaft bearing retainer (Chapter 7A).*
4    Input gear bearing retainer O-ring and/or lip seal damaged (Chapter 7A).*

### 51    Locked in gear

Lock pin or interlock pin missing (Chapter 7A).*

*Although the corrective action necessary to remedy the symptoms described is beyond the scope of this manual, the above information should be helpful in isolating the cause of the condition so that the owner can communicate clearly with a professional mechanic.*

## Automatic transaxle

**Note:** *Due to the complexity of the automatic transaxle, it is difficult for the home mechanic to properly diagnose and service this component. For problems other than the following, the vehicle should be taken to a dealer or transaxle shop.*

## 52  Fluid leakage

1   Automatic transaxle fluid is a deep red color. Fluid leaks should not be confused with engine oil, which can easily be blown onto the transaxle by air flow.
2   To pinpoint a leak, first remove all built-up dirt and grime from the transaxle housing with degreasing agents and/or steam cleaning. Then drive the vehicle at low speeds so air flow will not blow the leak far from its source. Raise the vehicle and determine where the leak is coming from. Common areas of leakage are:

   a) *Pan (Chapters 1 and 7B)*
   b) *Dipstick tube (Chapters 1 and 7B)*
   c) *Transaxle oil lines (Chapter 7B)*
   d) *Speed sensor (Chapter 7B)*
   e) *Driveaxle oil seals (Chapter 7B).*

## 53  Transaxle fluid brown or has a burned smell

Transaxle fluid overheated (Chapter 1).

## 54  General shift mechanism problems

1   Chapter 7, Part B, deals with checking and adjusting the shift linkage on automatic transaxles. Common problems which may be attributed to poorly adjusted linkage are:

   a) *Engine starting in gears other than Park or Neutral.*
   b) *Indicator on shifter pointing to a gear other than the one actually being used.*
   c) *Vehicle moves when in Park.*

2   Refer to Chapter 7B for the shift linkage adjustment procedure.

## 55  Transaxle will not downshift with accelerator pedal pressed to the floor

Throttle valve cable out of adjustment (Chapter 7B).

## 56  Engine will start in gears other than Park or Neutral

Park/Neutral position switch malfunctioning (Chapter 7B).

## 57  Transaxle slips, shifts roughly, is noisy or has no drive in forward or reverse gears

There are many probable causes for the above problems, but the home mechanic should be concerned with only one possibility - fluid level. Before taking the vehicle to a repair shop, check the level and condition of the fluid as described in Chapter 1. Correct the fluid level as necessary or change the fluid and filter if needed. If the problem persists, have a professional diagnose the cause.

## Driveaxles

## 58  Clicking noise in turns

Worn or damaged outboard CV joint (Chapter 8).

## 59  Shudder or vibration during acceleration

1   Excessive toe-in (Chapter 10).
2   Incorrect spring heights (Chapter 10).
3   Worn or damaged inboard or outboard CV joints (Chapter 8).
4   Sticking inboard CV joint assembly (Chapter 8).

## 60  Vibration at highway speeds

1   Out-of-balance front wheels and/or tires (Chapters 1 and 10).
2   Out-of-round front tires (Chapters 1 and 10).
3   Worn CV joint(s) (Chapter 8).

## Brakes

**Note:** *Before assuming that a brake problem exists, make sure that:*
   a) *The tires are in good condition and properly inflated (Chapter 1).*
   b) *The front end alignment is correct (Chapter 10).*
   c) *The vehicle is not loaded with weight in an unequal manner.*

## 61  Vehicle pulls to one side during braking

1   Incorrect tire pressures (Chapter 1).
2   Front end out of alignment (have the front end aligned).
3   Front, or rear, tire sizes not matched to one another.
4   Restricted brake lines or hoses (Chapter 9).
5   Malfunctioning drum brake or caliper assembly (Chapter 9).
6   Loose suspension parts (Chapter 10).
7   Loose calipers (Chapter 9).
8   Excessive wear of brake shoe or pad material or disc/drum on one side.

## 62  Noise (high-pitched squeal when the brakes are applied)

Front disc brake pads worn out. The noise comes from the wear sensor (or the pad backing plate) rubbing against the disc. Replace pads with new ones immediately (Chapter 9). Also inspect the brake discs for damage.

## 63  Brake roughness or chatter (pedal pulsates)

1   Excessive lateral disc runout (Chapter 9).
2   Uneven pad wear (Chapter 9).
3   Defective disc (Chapter 9).

## 64  Excessive brake pedal effort required to stop vehicle

1   Malfunctioning power brake booster (Chapter 9).
2   Partial system failure (Chapter 9).
3   Excessively worn pads or shoes (Chapter 9).
4   Piston in caliper or wheel cylinder stuck or sluggish (Chapter 9).
5   Brake pads or shoes contaminated with oil or grease (Chapter 9).
6   Brake disc grooved and/or glazed (Chapter 1).
7   New pads or shoes installed and not yet seated. It will take a while for the new material to seat against the disc or drum.

## 65  Excessive brake pedal travel

1   Partial brake system failure (Chapter 9).
2   Insufficient fluid in master cylinder (Chapters 1 and 9).
3   Air trapped in system (Chapters 1 and 9).

## 66  Dragging brakes

1   Incorrect adjustment of brake light switch (Chapter 9).
2   Master cylinder pistons not returning correctly (Chapter 9).
3   Restricted brakes lines or hoses (Chapters 1 and 9).
4   Incorrect parking brake adjustment (Chapter 9).

## 67  Grabbing or uneven braking action

1   Malfunction of proportioning valve (Chapter 9).
2   Brake pads or shoes worn out, or contaminated with grease, oil or brake fluid (Chapter 9).

## 68  Brake pedal feels spongy when depressed

1   Air in hydraulic lines (Chapter 9).
2   Master cylinder mounting bolts loose (Chapter 9).
3   Master cylinder defective (Chapter 9).

## 69   Brake pedal travels to the floor with little resistance

1   Little or no fluid in the master cylinder reservoir caused by leaking caliper piston(s) (Chapter 9).
2   Loose, damaged or disconnected brake lines (Chapter 9).

## 70   Parking brake does not hold

Parking brake linkage improperly adjusted (Chapters 1 and 9).

## Suspension and steering systems

**Note:** *Before attempting to diagnose the suspension and steering systems, perform the following preliminary checks:*

a) *Tires for wrong pressure and uneven wear.*
b) *Steering universal joints from the column to the rack and pinion for loose connectors or wear.*
c) *Front and rear suspension and the rack and pinion assembly for loose or damaged parts.*
d) *Out-of-round or out-of-balance tires, bent rims and loose and/or rough wheel bearings.*

## 71   Vehicle pulls to one side

1   Mismatched or uneven tires (Chapter 10).
2   Broken or sagging springs (Chapter 10).
3   Wheel alignment out-of-specifications (Chapter 10).
4   Front brake dragging (Chapter 9).

## 72   Abnormal or excessive tire wear

1   Wheel alignment out-of-specifications (Chapter 10).
2   Sagging or broken springs (Chapter 10).
3   Tire out-of-balance (Chapter 10).
4   Worn strut damper (Chapter 10).
5   Overloaded vehicle.
6   Tires not rotated regularly.

## 73   Wheel makes a thumping noise

1   Blister or bump on tire (Chapter 10).
2   Improper strut damper action (Chapter 10).

## 74   Shimmy, shake or vibration

1   Tire or wheel out-of-balance or out-of-round (Chapter 10).
2   Loose or worn wheel bearings (Chapters 1, 8 and 10).
3   Worn tie-rod ends (Chapter 10).
4   Worn balljoints (Chapters 1 and 10).

5   Excessive wheel runout (Chapter 10).
6   Blister or bump on tire (Chapter 10).

## 75   Hard steering

1   Lack of lubrication at balljoints, tie-rod ends and rack and pinion assembly (Chapter 10).
2   Front wheel alignment out-of-specifications (Chapter 10).
3   Low tire pressure(s) (Chapters 1 and 10).
4   Power steering fluid low (Chapter 1).
5   Defective power steering pump (Chapter 10).

## 76   Poor returnability of steering to center

1   Lack of lubrication at balljoints and tie-rod ends (Chapter 10).
2   Binding in balljoints (Chapter 10).
3   Binding in steering column (Chapter 10).
4   Lack of lubricant in steering gear assembly (Chapter 10).
5   Front wheel alignment out-of-specifications (Chapter 10).

## 77   Abnormal noise at the front end

1   Lack of lubrication at balljoints and tie-rod ends (Chapters 1 and 10).
2   Damaged strut mounting (Chapter 10).
3   Worn control arm bushings or tie-rod ends (Chapter 10).
4   Loose stabilizer bar (Chapter 10).
5   Loose wheel nuts (Chapters 1 and 10).
6   Loose suspension bolts (Chapter 10).

## 78   Wander or poor steering stability

1   Mismatched or uneven tires (Chapter 10).
2   Lack of lubrication at balljoints and tie-rod ends (Chapters 1 and 10).
3   Worn strut assemblies (Chapter 10).
4   Loose stabilizer bar (Chapter 10).
5   Broken or sagging springs (Chapter 10).
6   Wheels out of alignment (Chapter 10).

## 79   Erratic steering when braking

1   Wheel bearings worn (Chapter 10).
2   Broken or sagging springs (Chapter 10).
3   Leaking wheel cylinder or caliper (Chapter 10).
4   Warped discs or drums (Chapter 10).

## 80   Excessive pitching and/or rolling around corners or during braking

1   Loose stabilizer bar (Chapter 10).
2   Worn strut dampers or mountings (Chapter 10).

3   Broken or sagging springs (Chapter 10).
4   Overloaded vehicle.

## 81   Suspension bottoms

1   Overloaded vehicle.
2   Sagging springs (Chapter 10).

## 82   Cupped tires

1   Front wheel or rear wheel alignment out-of-specifications (Chapter 10).
2   Worn strut dampers (Chapter 10).
3   Wheel bearings worn (Chapter 10).
4   Excessive tire or wheel runout (Chapter 10).
5   Worn balljoints (Chapter 10).

## 83   Excessive tire wear on outside edge

1   Inflation pressures incorrect (Chapter 1).
2   Excessive speed in turns.
3   Front end alignment incorrect (excessive toe-in). Have professionally aligned.
4   Suspension arm bent or twisted (Chapter 10).

## 84   Excessive tire wear on inside edge

1   Inflation pressures incorrect (Chapter 1).
2   Front end alignment incorrect (toe-out). Have professionally aligned.
3   Loose or damaged steering components (Chapter 10).

## 85   Tire tread worn in one place

1   Tires out-of-balance.
2   Damaged or buckled wheel. Inspect and replace if necessary.
3   Defective tire (Chapter 1).

## 86   Excessive play or looseness in steering system

1   Wheel bearing(s) worn (Chapter 10).
2   Tie-rod end loose (Chapter 10).
3   Steering gear loose (Chapter 10).
4   Worn or loose steering intermediate shaft (Chapter 10).

## 87   Rattling or clicking noise in rack and pinion

1   Steering gear loose (Chapter 10).
2   Steering gear defective.

# Chapter 1
# Tune-up and routine maintenance

## Contents

## Specifications

### Recommended lubricants and fluids

**Note:** *Listed here are manufacturer recommendations at the time this manual was written. Manufacturers occasionally upgrade their fluid and lubricant specifications, so check with your local auto parts store for current recommendations.*

Engine oil
   Type ..................................................... API "certified for gasoline engines"
   Viscosity ............................................... See accompanying chart
Fuel....................................................... Unleaded gasoline, 87 octane or higher
Automatic transaxle
   Fluid..................................................... DEXRON III automatic transmission fluid
   Differential lubricant .................................. DEXRON III automatic transmission fluid
Manual transaxle lubricant ............................... API GL-5 75W-90 gear oil
Brake fluid................................................. DOT 3 brake fluid
Clutch fluid............................................... DOT 3 brake fluid
Power steering system .................................... DEXRON III automatic transmission fluid

### Capacities*

Engine oil (including filter)
   1.6L engine............................................. 3.0 qts
   1.8L engine............................................. 3.7 qts
Coolant..................................................... 6.0 qts
Transaxle
   Automatic (drain and refill)
      Three-speed........................................ 2.6 qts
         Differential ................................... 1.5 qts
      Four-speed ......................................... 3.3 qts
   Manual................................................. 2.7 qts

*All capacities approximate. Add as necessary to bring up to appropriate level.*

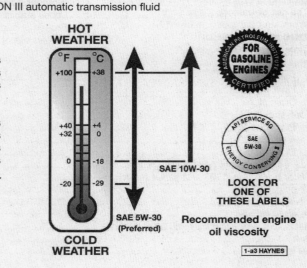

Recommended engine oil viscosity

## Ignition system

Spark plug type and gap

  Type

    1997 and earlier .......................................................... NGK BKR5EYA or equivalent

    1998 and 1999 ............................................................. NGK BKR5EKB11 or equivalent

    2000 and later .............................................................. NGK IFR5A11 or equivalent

  Gap

    1997 and earlier .......................................................... 0.031 inch

    1998 and later .............................................................. 0.043 inch

Spark plug wire resistance ................................................ 10,000 to 25,000 ohms

Engine firing order ............................................................. 1-3-4-2

## Valve clearances (engine cold)

Intake valve....................................................................... 0.006 to 0.010 inch

Exhaust valve..................................................................... 0.0010 to 0014 inch

## Cooling system

Thermostat rating

  Corolla

    Starts to open

      1997 and earlier.................................................... 176 to 183-degrees F

      1998 and later....................................................... 165 to 173-degrees F

    Fully open

      1997 and earlier.................................................... 203-degrees F

      1998 and later....................................................... 194-degrees

  Prizm

    Starts to open ............................................................. 176 to 183-degrees F

    Fully open ................................................................... 203-degrees F

Accessory drivebelt tension (with Burroughs or Nippondenso tension gauge) - 1997 and earlier models only

  Used belt

    Alternator ................................................................... 95 to 135 lbs

    Power steering pump................................................... 60 to 100 lbs

    Air conditioning compressor ....................................... 80 to 120 lbs

  New belt (a belt which has been used less than 5 minutes)

    Alternator ................................................................... 170 to 180 lbs

    Power steering pump................................................... 100 to 150 lbs

    Air conditioning compressor........................................ 140 to 180 lbs

## Clutch pedal

Freeplay ............................................................................ 3/16 to 5/8 inch

Height ............................................................................... 6 inches

## Brakes

Disc brake pad lining thickness (minimum) ........................ 3/16 inch

Drum brake shoe lining thickness (minimum)..................... 1/16 inch

Parking brake adjustment................................................... 4 to 7 clicks

## Suspension and steering

Steering wheel freeplay limit.............................................. 1-3/16 inch

Balljoint allowable movement ............................................. 0 inch

## Torque specifications                              **Ft-lbs** (unless otherwise indicated)

Automatic transaxle

  Three-speed

    Pan bolts.................................................................... 48 in-lbs

    Filter bolts .................................................................. 48 in-lbs

    Drain plug................................................................... 156 in-lbs

  Four-speed

    Pan bolts.................................................................... 48 in-lbs

    Filter bolts .................................................................. 84 in-lbs

    Drain plug................................................................... 36

Manual transaxle drain and filler plugs .............................. 29

Fuel filter banjo bolt (1997 and earlier models only)............ 22

Spark plugs........................................................................ 156 in-lbs

Seat bolts/nuts .................................................................. 27

Drivebelt tensioner

  Nut............................................................................. 21

  Bolt............................................................................. 51

Wheel lug nuts ................................................................... 76

**920361-specs HAYNES**

**Cylinder location and distributor rotation (1997 and earlier models)**

*The blackened terminal shown on the distributor cap indicates the Number One spark plug wire position*

FRONT OF VEHICLE

**1998 AND 1999**

**2000 AND LATER**

**92036-2B SPECS HAYNES**

**Cylinder numbering and coil terminal identification diagram**

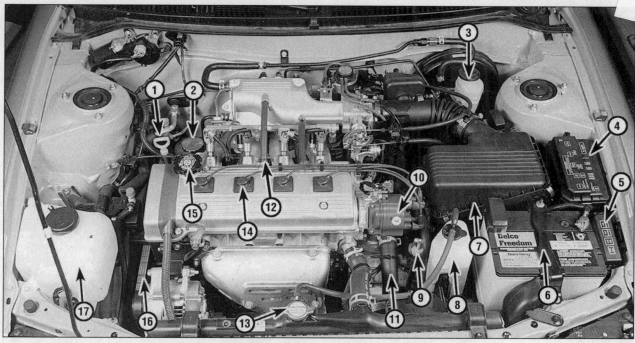

**Typical engine compartment components (1997 and earlier models)**

| | | | | | |
|---|---|---|---|---|---|
| 1 | Engine oil dipstick | 7 | Air cleaner assembly | 13 | Radiator cap |
| 2 | Power steering fluid reservoir | 8 | Coolant reservoir | 14 | Spark plug |
| 3 | Brake fluid reservoir | 9 | Automatic transaxle dipstick | 15 | Oil filler cap |
| 4 | Fuse/relay block | 10 | Distributor | 16 | Drivebelt |
| 5 | Relay block | 11 | Radiator hose | 17 | Windshield washer fluid reservoir |
| 6 | Battery | 12 | PCV valve | | |

**Typical engine compartment components (2000 model shown, most components on 1998 and later models similar)**

1  Engine oil dipstick
2  Drivebelt
3  Fuse/relay block
4  Windshield washer fluid reservoir
5  Power steering fluid reservoir
6  Ignition coils (2000 and later models only)
7  PCV valve
8  Oil filler cap
9  Brake fluid reservoir
10  Air cleaner assembly
11  Fuse/relay block
12  Fuse/relay block
13  Battery
14  Automatic transaxle dipstick
15  Coolant reservoir
16  Radiator cap
17  Radiator hoses

**1**

**Typical engine compartment underside components**

| 1 | Driveaxle boot | 4 | Steering gear boot | 7 | Radiator drain fitting |
|---|---|---|---|---|---|
| 2 | Automatic transaxle drain plug | 5 | Engine oil drain plug | 8 | Engine oil filter |
| 3 | Exhaust system catalytic converter | 6 | Front suspension strut unit | 9 | Front disc brake caliper |

**Typical rear underside components**

| 1 | Gas tank filler pipe | 3 | Suspension strut | 5 | Gas tank |
|---|---|---|---|---|---|
| 2 | Muffler | 4 | Exhaust system | 6 | Rear brake assembly |

# 1 Toyota Corolla and Geo/Chevrolet Prizm Maintenance schedule

The maintenance intervals in this manual are provided with the assumption that you, not the dealer, will be doing the work. These are the minimum maintenance intervals recommended by the factory for vehicles that are driven daily. If you wish to keep your vehicle in peak condition at all times, you may wish to perform some of these procedures even more often. Because frequent maintenance enhances the efficiency, performance and resale value of your car, we encourage you to do so. If you drive in dusty areas, tow a trailer, idle or drive at low speeds for extended periods or drive for short distances (less than four miles) in below freezing temperatures, shorter intervals are also recommended.

When your vehicle is new, it should be serviced by a factory authorized dealer service department to protect the factory warranty. In many cases, the initial maintenance check is done at no cost to the owner.

## Every 250 miles or weekly, whichever comes first

Check the engine oil level (Section 4)
Check the engine coolant level (Section 4)
Check the windshield washer fluid level (Section 4)
Check the brake fluid level (Section 4)
Check the power steering fluid level (Section 4)
Check the automatic transaxle fluid level (Section 4)
Check the tires and tire pressures (Section 5)

## Every 3000 miles or 3 months, whichever comes first

All items listed above plus:
Change the engine oil and oil filter (Section 6)

## Every 7500 miles or 6 months, whichever comes first

Inspect (and replace, if necessary) the windshield wiper blades (Section 7)
Check the clutch pedal for proper freeplay (Section 8)
Check and service the battery (Section 9)
Check and adjust if necessary the engine drivebelts (Section 10)
Inspect (and replace if, necessary) all underhood hoses (Section 11)
Check the cooling system (Section 12)
Rotate the tires (Section 13)
Check the seat belts (Section 14)

## Every 15,000 miles or 12 months, whichever comes first

All items listed above plus:
Inspect the brake system (Section 15)*

Replace the air filter (Section 16)
Inspect the fuel system (Section 17)
Check the three-speed automatic transaxle differential lubricant level (Section 18)
Check the manual transaxle lubricant level (Section 19)
Inspect the suspension and steering components (Section 20)*
Check the driveaxle boots (Section 21)

## Every 30,000 miles or 24 months, whichever comes first

All items listed above plus:
Change the brake fluid (Section 22)
Replace the fuel filter (1997 and earlier models) (Section 23)
Check (and replace, if necessary) the spark plugs (Section 24)
Inspect (and replace, if necessary) the spark plug wires, distributor cap and rotor (Section 25)
Service the cooling system (drain, flush and refill) (Section 26)
Inspect the evaporative emissions control system (Section 27)
Inspect the exhaust system (Section 28)
Change the automatic transaxle fluid and filter and differential lubricant (Section 29)**
Change the manual transaxle lubricant (Section 30)**
Check and replace if necessary the PCV valve (Section 31)

## Every 60,000 miles or 48 months, whichever comes first

Check and adjust the valve clearances (Section 32)
Replace the timing belt (Chapter 2A)

*This item is affected by "severe" operating conditions as described below. If your vehicle is operated under "severe" conditions, perform all maintenance indicated with an asterisk (*) at 3000 mile/3 month intervals. Severe conditions are indicated if you mainly operate your vehicle under one or more of the following conditions:

Operating in dusty areas
Towing a trailer
Idling for extended periods and/or low speed operation
Operating when outside temperatures remain below freezing and when most trips are less than 4 miles

** If operated under one or more of the following conditions, change the manual or automatic transaxle fluid and differential lubricant every 15,000 miles:

In heavy city traffic where the outside temperature regularly reaches 90-degrees F (32-degrees C) or higher
In hilly or mountainous terrain
Frequent trailer pulling

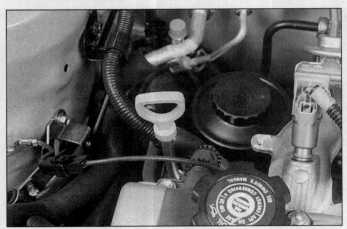

**4.2a  On 1997 and earlier models, the engine oil dipstick is located on the right rear side of the engine**

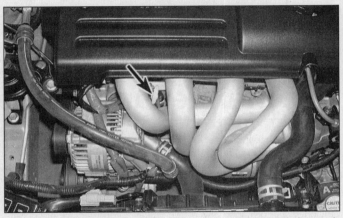

**4.2b  On 1998 and later models, the engine oil dipstick is located at the front of the engine, between the intake runners for cylinders 1 and 2**

## 2   Introduction

This chapter is designed to help the home mechanic maintain the Toyota Corolla or Geo/Chevrolet Prizm for peak performance, economy, safety and long life.

Included is a master maintenance schedule, followed by sections dealing specifically with each item on the schedule. Visual checks, adjustments, component replacement and other helpful items are included. Refer to the **accompanying illustrations** of the engine compartment and the underside of the vehicle for the location of various components.

Servicing your Corolla or Prizm in accordance with the mileage/time maintenance schedule and the following Sections will provide it with a planned maintenance program that should result in a long and reliable service life. This is a comprehensive plan, so maintaining some items but not others at the specified service intervals will not produce the same results.

As you service your Corolla or Prizm, you will discover that many of the procedures can - and should - be grouped together because of the nature of the particular procedure you're performing or because of the close proximity of two otherwise unrelated components to one another.

For example, if the vehicle is raised for any reason, you should inspect the exhaust, suspension, steering and fuel systems while you're under the vehicle. When you're rotating the tires, it makes good sense to check the brakes and wheel bearings since the wheels are already removed.

Finally, let's suppose you have to borrow or rent a torque wrench. Even if you only need to tighten the spark plugs, you might as well check the torque of as many critical fasteners as time allows.

The first step of this maintenance program is to prepare yourself before the actual work begins. Read through all sections pertinent to the procedures you're planning to do, then make a list of and gather together all the parts and tools you will need to do the job. If it looks as if you might run into problems during a particular segment of some procedure, seek advice from your local parts counterperson or dealer service department.

## 3   Tune-up general information

The term *tune-up* is used in this manual to represent a combination of individual operations rather than one specific procedure.

If, from the time the vehicle is new, the routine maintenance schedule is followed closely and frequent checks are made of fluid levels and high wear items, as suggested throughout this manual, the engine will be kept in relatively good running condition and the need for additional work will be minimized.

More likely than not, however, there will be times when the engine is running poorly due to lack of regular maintenance. This is even more likely if a used vehicle, which has not received regular and frequent maintenance checks, is purchased. In such cases, an engine tune-up will be needed outside of the regular routine maintenance intervals.

The first step in any tune-up or engine diagnosis to help correct a poor running engine would be a cylinder compression check. A check of the engine compression (Chapter 2 Part B) will give valuable information regarding the overall performance of many internal components and should be used as a basis for tune-up and repair procedures. If, for instance, a compression check indicates serious internal engine wear, a conventional tune-up will not help the running condition of the engine and would be a waste of time and money.

The following series of operations are those most often needed to bring a generally poor running engine back into a proper state of tune.

### Minor tune-up

Check all engine related fluids (Section 4)
Clean, inspect and test the battery
   (Section 9)

Check and adjust the drivebelts
   (Section 10)
Check all underhood hoses (Section 11)
Check the cooling system (Section 12)
Check the air filter (Section 16)
Replace the spark plugs (Section 24)
Inspect the distributor cap and rotor
   (Section 25)
Inspect the spark plug and coil wires
   (Section 25)

### Major tune-up

*All items listed under Minor tune-up, plus . . .*
Replace the air filter (Section 16)
Check the fuel system (Section 17)
Replace the distributor cap and rotor
   (Section 25)
Replace the spark plug wires (Section 25)
Check the charging system (Chapter 5)

## 4   Fluid level checks (every 250 miles or weekly)

1   Fluids are an essential part of the lubrication, cooling, brake, clutch and other systems. Because these fluids gradually become depleted and/or contaminated during normal operation of the vehicle, they must be periodically replenished. See *Recommended lubricants and fluids* and *Capacities* at the beginning of this Chapter before adding fluid to any of the following components. **Note:** *The vehicle must be on level ground before fluid levels can be checked.*

### Engine oil

*Refer to illustrations 4.2a, 4.2b, 4.4a, 4.4b and 4.6*

2   The engine oil level is checked with a dipstick located at the back side of the engine (1997 and earlier models) or at the front of the engine (1998 and later models) **(see illustrations)**. The dipstick extends through a metal tube from which it protrudes down into the engine oil pan.

3   The oil level should be checked before the vehicle has been driven, or about 5 min-

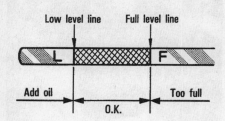

4.4a On 1997 and earlier models, the oil level should be at or near the F mark - if it isn't, add enough oil to bring the level to near the F mark (it takes one full quart to raise the level from the L to the F mark)

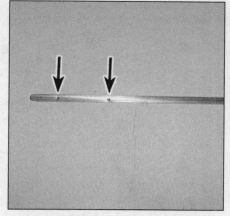

4.4b On 1998 and later models, the engine oil dipstick has two dimples - keep the oil level at or near the upper dimple

4.6 The threaded oil filler cap is located on the valve cover - always make sure the area around the opening is clean before unscrewing the cap to prevent dirt from contaminating the engine

**1**

utes after the engine has been shut off. If the oil is checked immediately after driving the vehicle, some of the oil will remain in the upper engine components, producing an inaccurate reading on the dipstick.

4    Pull the dipstick from the tube and wipe all the oil from the end with a clean rag or paper towel. Insert the clean dipstick all the way back into its metal tube and pull it out again. Observe the oil at the end of the dipstick. At its highest point, the level should be between the L and F marks (1997 and earlier models) or between the two dimples (1998 and later models) **(see illustrations)**.

5    It takes one quart of oil to raise the level from the L mark to the F mark (or the lower dimple and the upper dimple) on the dipstick. Do not allow the level to drop below the L mark (or the lower dimple) or oil starvation may cause engine damage. Conversely, overfilling the engine (adding oil above the F mark or upper dimple) may cause oil fouled spark plugs, oil leaks or oil seal failures.

6    Remove the threaded cap from the valve cover to add oil **(see illustration)**. Use a funnel to prevent spills. After adding the oil, install the filler cap hand tight. Start the engine and look carefully for any small leaks around the oil filter or drain plug. Stop the engine and check the oil level again after it has had sufficient time to drain from the upper block and cylinder head galleys.

7    Checking the oil level is an important preventive maintenance step. A continually dropping oil level indicates oil leakage through damaged seals, from loose connections, or past worn rings or valve guides. If the oil looks milky in color or has water droplets in it, a cylinder head gasket may be blown. The engine should be checked immediately. The condition of the oil should also be checked. Each time you check the oil level, slide your thumb and index finger up the dipstick before wiping off the oil. If you see small dirt or metal particles clinging to the dipstick, the oil should be changed (Section 6).

### Engine coolant

*Refer to illustrations 4.8a and 4.8b*
**Warning:** *Do not allow antifreeze to come in*

*contact with your skin or painted surfaces of the vehicle. Flush contaminated areas immediately with plenty of water. Don't store new coolant or leave old coolant lying around where it's accessible to children or pets - they're attracted by its sweet smell and may drink it. Ingestion of even a small amount of coolant can be fatal! Wipe up garage floor and drip pan spills immediately. Keep antifreeze containers covered and repair cooling system leaks as soon as they're noticed.*

8    All vehicles covered by this manual are equipped with a pressurized coolant recovery system. A white coolant reservoir located in the left front corner of the engine compartment is connected by a hose to the base of the radiator filler neck **(see illustrations)**. If the coolant heats up during engine operation, coolant can escape through a pressurized filler cap, then through a connecting hose into the reservoir. As the engine cools, the coolant is automatically drawn back into the cooling system to maintain the correct level.

9    The coolant level should be checked

regularly. It must be between the Full and Low lines on the tank. The level will vary with the temperature of the engine. When the engine is cold, the coolant level should be at or slightly above the Low mark on the tank. Once the engine has warmed up, the level should be at or near the Full mark. If it isn't, allow the fluid in the tank to cool, then remove the cap from the reservoir and add coolant to bring the level up to the Full line. Use only ethylene/glycol type coolant and water in the mixture ratio recommended by your owner's manual. Do not use supplemental inhibitor additives. If only a small amount of coolant is required to bring the system up to the proper level, water can be used. However, repeated additions of water will dilute the recommended antifreeze and water solution. In order to maintain the proper ratio of antifreeze and water, it is advisable to top up the coolant level with the correct mixture. Refer to your owner's manual for the recommended ratio.

4.8a On 1997 and earlier models, the coolant reservoir is located next to the battery - make sure the level is between Low and Full marks on the reservoir

4.8b On 1998 and later models the coolant reservoir is mounted below the battery, but the Low and Full marks are still visible from above

**4.14 The windshield washer fluid reservoir tank is located on the right front corner of the engine compartment**

**4.17 The brake fluid level should be kept between the MIN and MAX marks on the translucent plastic reservoir - lift up the cap to add fluid**

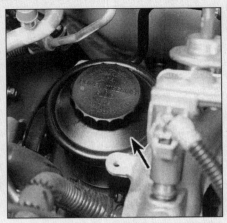

**4.25 The power steering fluid reservoir (arrow) is located in the right rear corner of the engine compartment on 1997 and earlier models**

10   If the coolant level drops within a short time after replenishment, there may be a leak in the system. Inspect the radiator, hoses, engine coolant filler cap, drain plugs, air bleeder plugs and water pump. If no leak is evident, have the radiator cap pressure tested by your dealer. **Warning:** *Never remove the radiator cap or the coolant recovery reservoir cap when the engine is running or has just been shut down, because the cooling system is hot. Escaping steam and scalding liquid could cause serious injury.*

11   If it is necessary to open the radiator cap, wait until the system has cooled completely, then wrap a thick cloth around the cap and turn it to the first stop. If any steam escapes, wait until the system has cooled further, then remove the cap.

12   When checking the coolant level, always note its condition. It should be relatively clear. If it is brown or rust colored, the system should be drained, flushed and refilled. Even if the coolant appears to be normal, the corrosion inhibitors wear out with use, so it must be replaced at the specified intervals.

13   Do not allow antifreeze to come in contact with your skin or painted surfaces of the vehicle. Flush contacted areas immediately with plenty of water.

### *Windshield washer fluid*

*Refer to illustration 4.14*

14   Fluid for the windshield washer system is stored in a plastic reservoir which is located on the right side of the engine compartment **(see illustration)**. In milder climates, plain water can be used to top up the reservoir, but the reservoir should be kept no more than two-thirds full to allow for expansion should the water freeze. In colder climates, the use of a specially designed windshield washer fluid, available at your dealer and any auto parts store, will help lower the freezing point of the fluid. Mix the solution with water in accordance with the manufacturer's directions on the container. Do not use regular antifreeze. It will damage the vehicle's paint.

### *Battery electrolyte*

15   On models not equipped with a sealed battery, unscrew the filler/vent cap and check the electrolyte level. It must be between the upper and lower levels. If the level is low, add distilled water. Install and securely retighten the cap. **Caution:** *Overfilling the cells may cause electrolyte to spill over during periods of heavy charging, causing corrosion or damage.*

### *Brake and clutch fluid*

*Refer to illustration 4.17*

16   The brake master cylinder is mounted on the front of the power booster unit in the engine compartment. The clutch master cylinder used on vehicles with manual transaxles is located next to the brake master cylinder.

17   To check either the fluid level of the brake master cylinder or clutch reservoir, simply look at the MAX and MIN marks on the reservoirs **(see illustration)**. The level should be at or near the maximum fill line for both reservoirs.

18   If the level is low for either reservoir, wipe the top of the reservoir cover with a clean rag to prevent contamination of the brake or clutch system before lifting the cover.

19   Add only the specified brake fluid to the brake or clutch reservoir (refer to *Recommended lubricants and fluids* at the front of this chapter or to your owner's manual). Mixing different types of brake fluid can damage the system. Fill the brake master cylinder reservoir only to the dotted line - this brings the fluid to the correct level when you put the cover back on. **Warning:** *Use caution when filling either reservoir - brake fluid can harm your eyes and damage painted surfaces. Do not use brake fluid that has been opened for more than one year (even if the cap has been on) or has been left open. Brake fluid absorbs moisture from the air. Excess moisture can cause a dangerous loss of braking.*

20   While the reservoir cap is removed,

inspect the master cylinder reservoir for contamination. If deposits, dirt particles or water droplets are present, the fluid in the brake system should be changed (see Section 22 for the brake fluid replacement procedure or chapter 8 for the clutch hydraulic system bleeding procedure).

21   After filling the reservoir to the proper level, make sure the lid is properly seated to prevent fluid leakage and/or system pressure loss.

22   The brake fluid in the master cylinder will drop slightly as the brake pads at each wheel wear down during normal operation. If the master cylinder requires repeated replenishing to keep it at the proper level, this is an indication of leakage in the brake system, which should be corrected immediately. Check all brake lines and connections, along with the wheel cylinders and booster (see Section 16 for more information).

23   If, upon checking the master cylinder fluid level, you discover one or both reservoirs empty or nearly empty, the brake system must be diagnosed immediately (see Chapter 9).

### *Power steering fluid*

*Refer to illustrations 4.25, 4.29a and 4.29b*

24   Unlike manual steering, the power steering system relies on fluid which may, over a period of time, require replenishing.

25   The fluid reservoir for the power steering pump is located on the power steering pump on 1997 and earlier models **(see illustration)**. On 1998 and later models it's mounted on the passenger's side inner fender panel, just behind the windshield washer fluid reservoir.

26   For the check, the front wheels should be pointed straight ahead and the engine should be off.

27   Use a clean rag to wipe off the reservoir cap and the area around the cap. This will help prevent any foreign matter from entering the reservoir during the check.

28   Twist off the cap and check the temperature of the fluid at the end of the dipstick

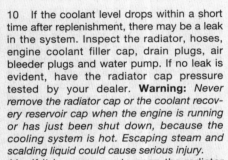

**4.29a On 1997 and earlier models, the power steering fluid is checked with a dipstick which is part of the cap - the fluid level varies with temperature, so the fluid can be checked hot or cold**

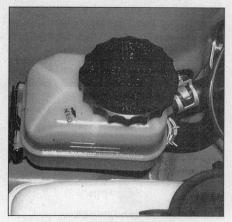

**4.29b On 1998 and later models, the power steering fluid level is checked by looking through the plastic reservoir**

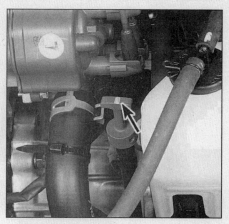

**4.35a The automatic transaxle dipstick (arrow) is located in a tube which extends forward from the transaxle toward the radiator**

**1**

with your finger.

29   On 1997 and earlier models, wipe the fluid off the dipstick with a clean rag, reinsert it, then withdraw it and read the fluid level. The level should be at the HOT mark if the fluid was hot to the touch. It should be at the COLD mark if the fluid was cool to the touch. Note that the marks (HOT and COLD) are on opposite sides of the dipstick **(see illustration)**. On 1998 and later models, view the level of power steering fluid through the translucent reservoir **(see illustration)**. At no time should the fluid level drop below the upper mark for each heat range.

30   If additional fluid is required, pour the specified type directly into the reservoir, using a funnel to prevent spills.

31   If the reservoir requires frequent fluid additions, all power steering hoses, hose connections, the power steering pump and the rack and pinion assembly should be carefully checked for leaks.

## *Automatic transaxle fluid*

*Refer to illustrations 4.35a and 4.35b*

32   The level of the automatic transaxle fluid should be carefully maintained. Low fluid level can lead to slipping or loss of drive, while overfilling can cause foaming, loss of fluid and transaxle damage.

33   The transaxle fluid level should only be checked when the transaxle is hot (at its normal operating temperature). If the vehicle has just been driven over 10 miles (15 miles in a frigid climate), and the fluid temperature is 160 to 175-degrees F, the transaxle is hot. **Caution:** *If the vehicle has just been driven for a long time at high speed or in city traffic in hot weather, or if it has been pulling a trailer, an accurate fluid level reading cannot be obtained. Allow the fluid to cool down for about 30 minutes.*

34   If the vehicle has not just been driven, park the vehicle on level ground, set the parking brake and start the engine. While the engine is idling, depress the brake pedal and

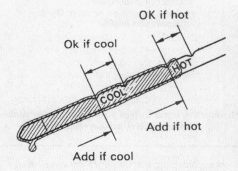

**4.35b If the automatic transaxle fluid is cold, the level should be between the two lower notches; if it's at operating temperature, the level should be between the two upper notches**

move the selector lever through all the gear ranges, beginning and ending in Park.

35   With the engine still idling, remove the dipstick from its tube **(see illustration)**. Check the level of the fluid on the dipstick **(see illustration)** and note its condition.

36   Wipe the fluid from the dipstick with a clean rag and reinsert it back into the filler tube until the cap seats.

37   Pull the dipstick out again and note the fluid level. If the transaxle is cold, the level should be in the COLD or COOL range on the dipstick. If it is hot, the fluid level should be in the HOT range. If the level is at the low side of either range, add the specified automatic transmission fluid through the dipstick tube with a funnel.

38   Add just enough of the recommended fluid to fill the transaxle to the proper level. It takes about one pint to raise the level from the low mark to the high mark when the fluid is hot, so add the fluid a little at a time and keep checking the level until it is correct.

39   The condition of the fluid should also be checked along with the level. If the fluid at the end of the dipstick is black or a dark reddish

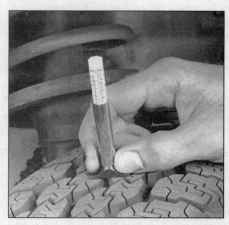

**5.2 A tire tread depth indicator should be used to monitor tire wear - they are available at auto parts stores and service stations and cost very little**

brown color, or if it emits a burned smell, the fluid should be changed (see Section 29). If you are in doubt about the condition of the fluid, purchase some new fluid and compare the two for color and smell.

---

## 5   Tire and tire pressure checks (every 250 miles or weekly)

*Refer to illustrations 5.2, 5.3, 5.4a, 5.4b and 5.8*

1   Periodic inspection of the tires may spare you from the inconvenience of being stranded with a flat tire. It can also provide you with vital information regarding possible problems in the steering and suspension systems before major damage occurs.

2   Normal tread wear can be monitored with a simple, inexpensive device known as a tread depth indicator **(see illustration)**. When the tread depth reaches the specified minimum, replace the tire(s).

**UNDERINFLATION**

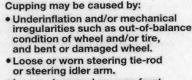

**CUPPING**

**OVERINFLATION**

**Cupping may be caused by:**
● Underinflation and/or mechanical irregularities such as out-of-balance condition of wheel and/or tire, and bent or damaged wheel.
● Loose or worn steering tie-rod or steering idler arm.
● Loose, damaged or worn front suspension parts.

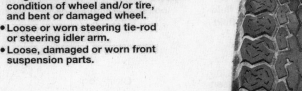

**INCORRECT TOE-IN OR EXTREME CAMBER**

**FEATHERING DUE TO MISALIGNMENT**

**5.3  This chart will help you determine the condition of your tires, the probable cause(s) of abnormal wear and the corrective action necessary**

3    Note any abnormal tread wear **(see illustration)**. Tread pattern irregularities such as cupping, flat spots and more wear on one side than the other are indications of front end alignment and/or balance problems. If any of these conditions are noted, take the vehicle to a tire shop or service station to correct the problem.

4    Look closely for cuts, punctures and embedded nails or tacks. Sometimes a tire will hold its air pressure for a short time or leak down very slowly even after a nail has embedded itself into the tread. If a slow leak persists, check the valve stem core to make sure it is tight **(see illustration)**. Examine the tread for an object that may have embedded itself into the tire or for a "plug" that may have begun to leak (radial tire punctures are repaired with a plug that is installed in a puncture). If a puncture is suspected, it can be easily verified by spraying a solution of soapy water onto the puncture area **(see illustration)**. The soapy solution will bubble if there is a leak. Unless the puncture is inordinately large, a tire shop or gas station can usually repair the punctured tire.

5    Carefully inspect the inner sidewall of each tire for evidence of brake fluid leakage. If you see any, inspect the brakes immediately.

6    Correct tire air pressure adds miles to the lifespan of the tires, improves mileage and enhances overall ride quality. Tire pressure cannot be accurately estimated by looking at a tire, particularly if it is a radial. A tire

pressure gauge is therefore essential. Keep an accurate gauge in the glovebox. The pressure gauges fitted to the nozzles of air hoses at gas stations are often inaccurate.

7    Always check tire pressure when the tires are cold. "Cold," in this case, means the vehicle has not been driven over a mile in the three hours preceding a tire pressure check. A pressure rise of four to eight pounds is not uncommon once the tires are warm.

8    Unscrew the valve cap protruding from the wheel or hubcap and push the gauge

**5.4a  If a tire loses air on a steady basis, check the valve core first to make sure it's snug (special inexpensive wrenches are commonly available at auto parts stores)**

firmly onto the valve **(see illustration)**. Note the reading on the gauge and compare this figure to the recommended tire pressure shown on the tire placard on the left door. Be sure to reinstall the valve cap to keep dirt and moisture out of the valve stem mechanism. Check all four tires and, if necessary, add enough air to bring them up to the recommended pressure levels.

9    Don't forget to keep the spare tire inflated to the specified pressure (consult your owner's manual). Note that the air pressure specified

**5.4b  If the valve core is tight, raise the corner of the vehicle with the low tire and spray a soapy water solution onto the tread as the tire is turned slowly - slow leaks will cause small bubbles to appear**

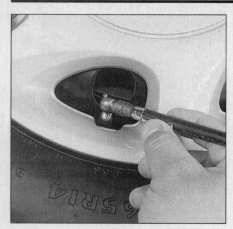

5.8  To extend the life of your tires, check the air pressure at least once a week with an accurate gauge (don't forget the spare!)

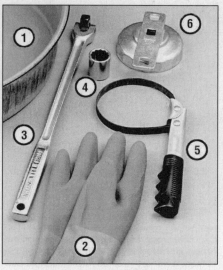

6.2  These tools are required when changing the engine oil and filter

1    **Drain pan** - *It should be fairly shallow in depth, but wide in order to prevent spills*
2    **Rubber gloves** - *When removing the drain plug and filter, it is inevitable that you will get oil on your hands (the gloves will prevent burns)*
3    **Breaker bar** - *Sometimes the oil drain plug is pretty tight and a long breaker bar is needed to loosen it*
4    **Socket** – *To be used with the breaker bar or a ratchet (must be the correct size to fit the drain plug)*
5    **Filter wrench** - *This is a metal band-type wrench, which requires clearance around the filter to be effective*
6    **Filter wrench** - *This type fits on the bottom of the filter and can be turned with a ratchet or beaker bar (different size wrenches are available for different types of filters)*

6.7  Use a proper size box-end wrench or socket to remove the oil drain plug and avoid rounding it off

**1**

6.13a  The oil filter is usually on very tight, so you'll need a special wrench for removal - DO NOT use the wrench to tighten the new filter (this is a 1997 or earlier model)

for the compact spare is significantly higher than the pressure of the regular tires.

---

## 6    Engine oil and oil filter change (every 3000 miles or 3 months)

*Refer to illustrations 6.2, 6.7, 6.13a, 6.13b and 6.15*

1    Frequent oil changes are the best preventive maintenance the home mechanic can give the engine, because aging oil becomes diluted and contaminated, which leads to premature engine wear.
2    Make sure that you have all the necessary tools before you begin this procedure **(see illustration)**. You should also have plenty of rags or newspapers handy for mopping up any spills.
3    Access to the underside of the vehicle is greatly improved if the vehicle can be lifted on a hoist, driven onto ramps or supported by jackstands. **Warning:** *Do not work under a vehicle which is supported only by a bumper, hydraulic or scissors-type jack.*
4    If this is your first oil change, get under the vehicle and familiarize yourself with the location of the oil drain plug. The engine and exhaust components will be warm during the actual work, so try to anticipate any potential problems before the engine and accessories are hot.
5    Park the vehicle on a level spot. Start the engine and allow it to reach its normal operating temperature (the needle on the temperature gauge should be at least above the bottom mark). Warm oil and sludge will flow out more easily. Turn off the engine when it's warmed up. Remove the filler cap in the rear cam cover.
6    Raise the vehicle and support it on jackstands. **Warning:** *To avoid personal injury, never get beneath the vehicle when it is supported by only by a jack. The jack provided with your vehicle is designed solely for raising the vehicle to remove and replace the wheels. Always use jackstands to support the vehicle when it becomes necessary to place your*

body underneath the vehicle.
7    Being careful not to touch the hot exhaust components, place the drain pan under the drain plug in the bottom of the pan and remove the plug **(see illustration)**. You may want to wear gloves while unscrewing the plug the final few turns if the engine is really hot.
8    Allow the old oil to drain into the pan. It may be necessary to move the pan farther under the engine as the oil flow slows to a trickle. Inspect the old oil for the presence of metal shavings and chips.
9    After all the oil has drained, wipe off the drain plug with a clean rag. Even minute metal particles clinging to the plug would immediately contaminate the new oil.
10   Clean the area around the drain plug opening, reinstall the plug and tighten it securely, but do not strip the threads.
11   Move the drain pan into position under the oil filter.
12   Remove all tools, rags, etc. from under the vehicle, being careful not to spill the oil in the drain pan, then lower the vehicle.

13   Loosen the oil filter **(see illustrations)** by turning it counterclockwise with the filter wrench. Any standard filter wrench should work. Once the filter is loose, use your hands to unscrew it from the block. Just as the filter is detached from the block, immediately tilt

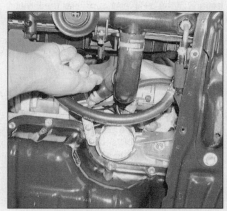

6.13b  On 1998 and later models the oil filter points straight down from the engine block

**6.15 Lubricate the oil filter gasket with clean engine oil before installing the filter on the engine**

**7.5 Push on the release lever and slide the wiper assembly down out of the hook in the end of the wiper arm**

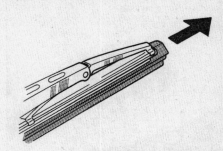

**7.6 After detaching the end of the element, slide it out of the end of the frame**

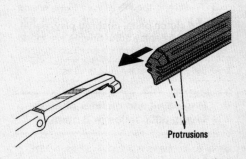

**7.7 Insert the end of the element with the protrusions in first**

the open end up to prevent the oil inside the filter from spilling out. **Warning:** *The engine exhaust manifold may still be hot, so be careful.*

14    With a clean rag, wipe off the mounting surface on the block. If a residue of old oil is allowed to remain, it will smoke when the block is heated up. It will also prevent the new filter from seating properly. Also make sure that the none of the old gasket remains stuck to the mounting surface. It can be removed with a scraper if necessary.

15    Compare the old filter with the new one to make sure they are the same type. Smear some engine oil on the rubber gasket of the new filter and screw it into place **(see illustration)**. Because overtightening the filter will damage the gasket, do not use a filter wrench to tighten the filter. Tighten it by hand until the gasket contacts the seating surface. Then seat the filter by giving it an additional 3/4-turn.

16    Add new oil to the engine through the oil filler cap in the valve cover. Use a spout or funnel to prevent oil from spilling onto the top of the engine. Pour three quarts of fresh oil into the engine. Wait a few minutes to allow the oil to drain into the pan, then check the level on the oil dipstick (see Section 4 if necessary). If the oil level is at or near the F mark, install the filler cap hand tight, start the engine and allow the new oil to circulate.

17    Allow the engine to run for about a minute. While the engine is running, look under the vehicle and check for leaks at the oil pan drain plug and around the oil filter. If either is leaking, stop the engine and tighten the plug or filter slightly.

18    Wait a few minutes to allow the oil to trickle down into the pan, then recheck the level on the dipstick and, if necessary, add enough oil to bring the level to the F mark.

19    During the first few trips after an oil change, make it a point to check frequently for leaks and proper oil level.

20    The old oil drained from the engine cannot be reused in its present state and should be discarded. Check with your local refuse disposal company, disposal facility or envi-

ronmental agency to see if they will accept the oil for recycling. Don't pour used oil into drains or onto the ground. After the oil has cooled, it can be drained into a suitable container (capped plastic jugs, topped bottles, milk cartons, etc.) for transport to one of these disposal sites.

## 7   Windshield wiper blade inspection and replacement (every 7500 miles or 6 months)

*Refer to illustrations 7.5, 7.6 and 7.7*

1    The windshield wiper and blade assembly should be inspected periodically for damage, loose components and cracked or worn blade elements.

2    Road film can build up on the wiper blades and affect their efficiency, so they should be washed regularly with a mild detergent solution.

3    The action of the wiping mechanism can loosen bolts, nuts and fasteners, so they should be checked and tightened, as necessary, at the same time the wiper blades are checked.

4    If the wiper blade elements are cracked, worn or warped, or no longer clean adequately, they should be replaced with new ones.

5    Remove the wiper blade assembly from the arm by pushing on the release lever, then sliding the assembly down and out of the hook in the end of the arm **(see illustration)**.

6    Detach the blade insert element and pull it out of the right end of the wiper frame **(see illustration)**.

7    Insert the new element end with the small protrusions into the right side of the wiper frame **(see illustration)**. Slide the element fully into place, then seat the protrusions in the end of the frames to secure it.

## 8   Clutch pedal freeplay check and adjustment (every 7500 miles or 6 months)

*Refer to illustrations 8.1 and 8.2*

1    Press down lightly on the clutch pedal

and, with a small steel ruler, measure the distance that it moves freely before the clutch resistance is felt **(see illustration)**. The freeplay should be within the specified limits. If it isn't, it must be adjusted.

2    Loosen the locknut on the pedal end of the clutch pushrod **(see illustration)**.

3    Turn the pushrod until pedal freeplay and pushrod freeplay are correct.

4    Tighten the locknut.

5    After adjusting the pedal freeplay, check the pedal height.

6    If pedal height is incorrect, loosen the locknut and turn the stopper bolt until the height is correct. Tighten the locknut.

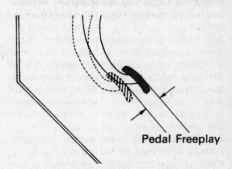

**8.1 To check clutch pedal freeplay, measure the distance between the natural resting place of the pedal and the point at which you encounter resistance**

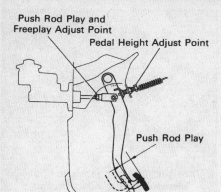

8.2  The clutch pedal pushrod play, pedal height and freeplay adjustments are made by loosening the locknut and turning the appropriate threaded adjuster

## 9    Battery check, maintenance and charging (every 7500 miles or 6 months)

Refer to illustrations 9.1, 9.6a, 9.6b, 9.7a, 9.7b and 9.8

**Warning:** *Certain precautions must be followed when checking and servicing the battery. Hydrogen gas, which is highly flammable, is always present in the battery cells, so keep lighted tobacco and all other open flames and sparks away from the battery. The electrolyte inside the battery is actually dilute sulfuric acid, which will cause injury if splashed on your skin or in your eyes. It will also ruin clothes and painted surfaces. When removing the battery cables, always detach the negative cable first and hook it up last!*

1    A routine preventive maintenance program for the battery in your vehicle is the only way to ensure quick and reliable starts. But before performing any battery maintenance, make sure that you have the proper equipment necessary to work safely around the battery **(see illustration).**

2    There are also several precautions that should be taken whenever battery maintenance is performed. Before servicing the battery, always turn the engine and all accessories off and disconnect the cable from the negative terminal of the battery.

3    The battery produces hydrogen gas, which is both flammable and explosive. Never create a spark, smoke or light a match around the battery. Always charge the battery in a ventilated area.

4    Electrolyte contains poisonous and corrosive sulfuric acid. Do not allow it to get in your eyes, on your skin on your clothes. Never ingest it. Wear protective safety glasses when working near the battery. Keep children away from the battery.

5    Note the external condition of the battery. If the positive terminal and cable clamp on your vehicle's battery is equipped with a

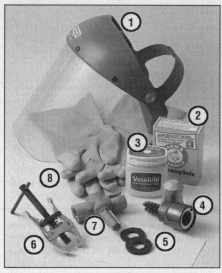

9.1  Tools and materials required for battery maintenance

1    **Face shield/safety goggles** - *When removing corrosion with a brush, the acidic particles can easily fly up into your eyes*

2    **Baking soda** - *A solution of baking soda and water can be used to neutralize corrosion*

3    **Petroleum jelly** - *A layer of this on the battery posts will help prevent corrosion*

4    **Battery post/cable cleaner** - *This wire brush cleaning tool will remove all traces of corrosion from the battery posts and cable clamps*

5    **Treated felt washers** - *Placing one of these on each post, directly under the cable clamps, will help prevent corrosion*

6    **Puller** - *Sometimes the cable clamps are very difficult to pull off the posts, even after the nut/bolt has been completely loosened. This tool pulls the clamp straight up and off the post without damage*

7    **Battery post/cable cleaner** - *Here is another cleaning tool which is a slightly different version of number 4 above, but it does the same thing*

8    **Rubber gloves** - *Another safety item to consider when servicing the battery; remember that's acid inside the battery*

rubber protector, make sure that it's not torn or damaged. It should completely cover the terminal. Look for any corroded or loose connections, cracks in the case or cover or loose hold-down clamps. Also check the entire length of each cable for cracks and frayed conductors.

6    If corrosion, which looks like white, fluffy deposits **(see illustration)** is evident, particularly around the terminals, the battery should be removed for cleaning. Loosen the cable clamp bolts with a wrench, being careful to remove the ground cable first, and slide them off the terminals **(see illustration).** Then disconnect the hold-down clamp bolt and nut,

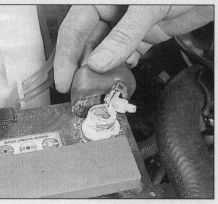

9.6a  Battery terminal corrosion usually appears as light, fluffy powder

9.6b  Removing a cable from the battery post with a wrench - sometimes a pair of special battery pliers are required for this procedure if corrosion has caused deterioration of the nut hex (always remove the ground (-) cable first and hook it up last!)

9.7a  When cleaning the cable clamps, all corrosion must be removed (the inside of the clamp is tapered to match the taper on the post, so don't remove too much material)

remove the clamp and lift the battery from the engine compartment.

7    Clean the cable clamps thoroughly with a battery brush or a terminal cleaner and a solution of warm water and baking soda **(see illustration).** Wash the terminals and the top

**1**

of the battery case with the same solution but make sure that the solution doesn't get into the battery. When cleaning the cables, terminals and battery top, wear safety goggles and rubber gloves to prevent any solution from coming in contact with your eyes or hands. Wear old clothes too - even diluted, sulfuric acid splashed onto clothes will burn holes in them. If the terminals have been extensively corroded, clean them up with a terminal cleaner **(see illustration)**. Thoroughly wash all cleaned areas with plain water.

8    Make sure that the battery tray is in good condition and the hold-down nut and bolt are tight **(see illustration)**. If the battery is removed from the tray, make sure no parts remain in the bottom of the tray when the battery is reinstalled. When reinstalling the hold-down clamp bolt or nut, do not overtighten it.

9    Information on removing and installing the battery can be found in Chapter 5. Information on jump starting can be found at the front of this manual. For more detailed battery checking procedures, refer to the *Haynes Automotive Electrical Manual*.

### Cleaning

10    Corrosion on the hold-down components, battery case and surrounding areas can be removed with a solution of water and baking soda. Thoroughly rinse all cleaned areas with plain water.

11    Any metal parts of the vehicle damaged by corrosion should be covered with a zinc-based primer, then painted.

### Charging

**Warning:** *When batteries are being charged, hydrogen gas, which is very explosive and flammable, is produced. Do not smoke or allow open flames near a charging or a recently charged battery. Wear eye protection when near the battery during charging. Also, make sure the charger is unplugged before connecting or disconnecting the battery from the charger.*

12    Slow-rate charging is the best way to restore a battery that's discharged to the point where it will not start the engine. It's also a good way to maintain the battery charge in a vehicle that's only driven a few miles between starts. Maintaining the battery charge is particularly important in the winter when the battery must work harder to start the engine and electrical accessories that drain the battery are in greater use.

13    It's best to use a one or two-amp battery charger (sometimes called a "trickle" charger). They are the safest and put the least strain on the battery. They are also the least expensive. For a faster charge, you can use a higher amperage charger, but don't use one rated more than 1/10th the amp/hour rating of the battery. Rapid boost charges that claim to restore the power of the battery in one to two hours are hardest on the battery and can damage batteries not in good condition. This type of charging should only be used in emergency situations.

**9.7b  Regardless of the type of tool used to clean the battery posts, a clean, shiny surface should be the result**

14    The average time necessary to charge a battery should be listed in the instructions that come with the charger. As a general rule, a trickle charger will charge a battery in 12 to 16 hours.

---

### 10   Drivebelt check, adjustment and replacement (every 7500 miles or 6 months)

---

### Check

*Refer to illustrations 10.3, 10.4 and 10.5*

1    The alternator, power steering pump and air conditioning compressor drivebelts, also referred to as simply "fan" belts, are located at the right end of the engine. The good condition and proper adjustment of the alternator belt is critical to the operation of the engine. Because of their composition and the high stresses to which they are sub-

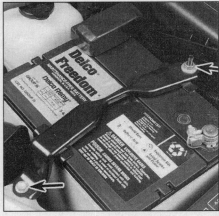

**9.8  Make sure the battery clamp nut and bolt (arrows) are tight**

jected, drivebelts stretch and deteriorate as they get older. They must therefore be periodically inspected.

2    On 1997 and earlier models, multiple belts are used; one belt transmits power from the crankshaft to the alternator, and one drives the air conditioning compressor, if equipped. If the vehicle is equipped with power steering, the pump is driven by it's own belt, powered by the water pump pulley. On 1998 and later models, a single "serpentine" drivebelt is used.

3    With the engine off, open the hood and locate the drivebelt(s). With a flashlight, check each belt for separation of the adhesive rubber on both sides of the core, core separation from the belt side, a severed core, separation of the ribs from the adhesive rubber, cracking or separation of the ribs, and torn or worn ribs or cracks in the inner ridges of the ribs **(see illustration)**. Also check for fraying and glazing, which gives the belt a shiny appearance. Both sides of the belt should be inspected, which means you will

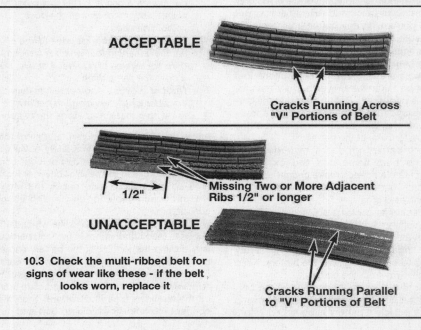

**10.3  Check the multi-ribbed belt for signs of wear like these - if the belt looks worn, replace it**

ACCEPTABLE

Cracks Running Across "V" Portions of Belt

1/2"

Missing Two or More Adjacent Ribs 1/2" or longer

UNACCEPTABLE

Cracks Running Parallel to "V" Portions of Belt

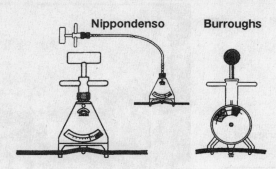

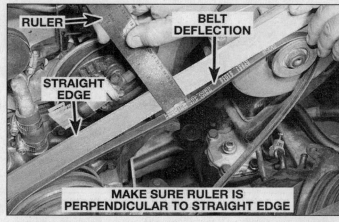

10.4 If you are able to borrow either a Nippondenso or Burroughs belt tension gauge, this is how it's installed on the belt - compare the reading on the scale with the specified drivebelt tension

10.5 Measuring drivebelt deflection with a straightedge and ruler

have to twist the belt to check the underside. Use your fingers to feel the belt where you can't see it. If any of the above conditions are evident, replace the belt (go to Step 8).

4    Belt tension on 1997 and earlier models must be manually adjusted, while on 1998 and later models the tension is adjusted automatically. To check the tension of each belt in accordance with factory specifications, install either a Nippondenso or Burroughs belt tension gauge on the belt **(see illustration)**. Measure the tension in accordance with the manufacturer's instructions and compare your measurement to the specified drivebelt tension for either a used or new belt. **Note:** *A "used" belt is defined as any belt which has been operated more than five minutes on the engine; a "new" belt is one that has been used for less than five minutes.*

5    If you don't have either of the above tools, and cannot borrow one, the following rule of thumb method is recommended: Push firmly on the belt with your thumb at a distance halfway between the pulleys and note how far the belt can be pushed (deflected). Measure this deflection with a ruler **(see illustration)**. The belt should deflect 1/4-inch if the distance from pulley center to pulley center is between 7 and 11 inches; the belt should deflect 1/2-inch if the distance from pulley center to pulley center is between 12 and 16 inches.

## *Adjustment*

### 1997 and earlier models

*Refer to illustration 10.7*

6    To adjust the alternator belt, loosen the alternator pivot bolt and the lock bolt at the alternator bracket. Now turn the adjusting bolt and measure the belt tension in accordance with one of the above methods. When the proper tension is achieved, tighten the lock bolt and pivot bolt securely.

7    To adjust the air conditioning compressor drivebelt, loosen the idler pulley lock nut and turn the adjusting bolt **(see illustration)** until the belt is properly tensioned. When the proper tension is achieved, tighten the lock bolt and pivot bolt securely.

8    Adjust the power steering pump belt by

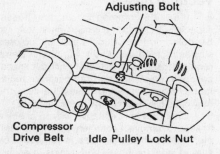

10.7 After loosening the idler pulley lock nut, turn the adjusting bolt to tension the air conditioning compressor drivebelt

loosening the pivot bolt and the lock bolt that secures the pump to the slotted bracket, then pivot the pump (away from the engine to tighten the belt, toward it to loosen it) until the proper tension is reached. When the proper tension is achieved, tighten the lock bolt and pivot bolt securely.

### 1998 and later models

9    Drivebelt tension on 1998 and later models is adjusted automatically by a spring-loaded tensioner.

## *Replacement*

### 1997 and earlier models

*Refer to illustration 10.12*

10    To replace a belt, follow the above procedures for drivebelt adjustment but slip the belt off the crankshaft pulley and remove it. If you are replacing the power steering pump belt, you will have to remove the air conditioning compressor belt first because of the way they are arranged on the crankshaft pulley. Because of this and because belts tend to wear out more or less together, it is a good idea to replace both belts at the same time. Mark each belt and its appropriate pulley groove so the replacement belts can be installed in their proper positions.

11    Take the old belts to the parts store in order to make a direct comparison for length,

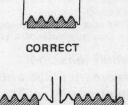

10.12 When installing a multi-ribbed belt, make sure that it is centered - it must not overlap either edge of the pulley

width and design.

12    After replacing the drivebelt, make sure that it fits properly in the ribbed grooves in the pulleys **(see illustration)**. It is essential that the belt be properly centered.

13    Adjust the belt(s) in accordance with the procedure outlined above.

### 1998 and later models

*Refer to illustrations 10.14 and 10.15*

14    Rotate the belt tensioner clockwise to release the tension, then slip the belt off the

10.14 To release the tension on the drivebelt, place a long wrench or socket and ratchet on the hex-shaped boss, then turn the tensioner clockwise (1998 and later models)

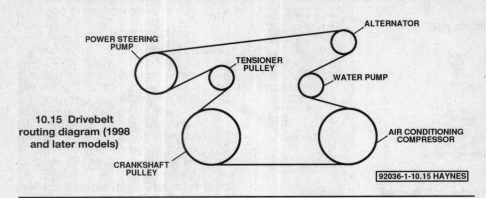

**10.15 Drivebelt routing diagram (1998 and later models)**

POWER STEERING PUMP
TENSIONER PULLEY
ALTERNATOR
WATER PUMP
AIR CONDITIONING COMPRESSOR
CRANKSHAFT PULLEY

92036-1-10.15 HAYNES

**10.17 The drivebelt tensioner on 1998 and later models is retained by a nut and a bolt**

pulleys **(see illustration)**. Slowly release the tensioner.

15   Route the new belt over the pulleys **(see illustration)**, again rotating the tensioner to allow the belt to be installed, then release the belt tensioner.

16   Make sure the belt is properly centered in the pulleys **(see illustration 10.12)**.

### Drivebelt tensioner replacement (1998 and later models)

*Refer to illustration 10.17*

17   To replace a tensioner that can't properly tension the belt, or one that exhibits binding or a worn-out bearing/pulley, remove the drivebelt (see Step 14) then unscrew the mounting bolt and nut **(see illustration)**.

18   Installation is the reverse of the removal procedure. Tighten the fasteners to the torque values listed in this Chapter's Specifications.

19   Install the drivebelt (see Steps 15 and 16).

### 11   Underhood hose check and replacement (every 7500 miles or 6 months)

**Caution:** *Replacement of air conditioning hoses must be left to a dealer service department or air conditioning shop that has the equipment to depressurize the system safely. Never remove air conditioning components or hoses until the system has been depressurized.*

### General

1   High temperatures in the engine compartment can cause the deterioration of the rubber and plastic hoses used for engine, accessory and emission systems operation. Periodic inspection should be made for cracks, loose clamps, material hardening and leaks.

2   Information specific to the cooling system hoses can be found in Section 12.

3   Some, but not all, hoses are secured to the fittings with clamps. Where clamps are used, check to be sure they haven't lost their tension, allowing the hose to leak. If clamps

aren't used, make sure the hose has not expanded and/or hardened where it slips over the fitting, allowing it to leak.

### Vacuum hoses

4   It's quite common for vacuum hoses, especially those in the emissions system, to be color coded or identified by colored stripes molded into them. Various systems require hoses with different wall thickness, collapse resistance and temperature resistance. When replacing hoses, be sure the new ones are made of the same material.

5   Often the only effective way to check a hose is to remove it completely from the vehicle. If more than one hose is removed, be sure to label the hoses and fittings to ensure correct installation.

6   When checking vacuum hoses, be sure to include any plastic T-fittings in the check. Inspect the fittings for cracks and the hose where it fits over the fitting for distortion, which could cause leakage.

7   A small piece of vacuum hose (1/4-inch inside diameter) can be used as a stethoscope to detect vacuum leaks. Hold one end of the hose to your ear and probe around vacuum hoses and fittings, listening for the "hissing" sound characteristic of a vacuum leak. **Warning:** *When probing with the vacuum hose stethoscope, be very careful not to come into contact with moving engine components such as the drivebelts, cooling fan, etc.*

### Fuel hose

**Warning:** *Gasoline is extremely flammable, so take extra precautions when you work on any part of the fuel system. Don't smoke or allow open flames or bare light bulbs near the work area, and don't work in a garage where a gas-type appliance (such as a water heater or a clothes dryer) is present. Since gasoline is carcinogenic, wear latex gloves when there's a possibility of being exposed to fuel, and, if you spill any fuel on your skin, rinse it off immediately with soap and water. Mop up any spills immediately and do not store fuel-soaked rags where they could ignite. The fuel system is under constant pressure, so, if any fuel lines are to be disconnected, the fuel pressure in the system must be relieved first. When you perform any kind of work on the fuel system, wear safety glasses and have a*

*Class B type fire extinguisher on hand.*

8   Check all rubber fuel lines for deterioration and chafing. Check especially for cracks in areas where the hose bends and just before fittings, such as where a hose attaches to the fuel filter.

9   High quality fuel line, specifically designed for fuel injection systems, must be used for fuel line replacement. **Warning:** *Never use anything other than the proper fuel line for fuel line replacement.*

10   Spring-type clamps are commonly used on fuel lines. These clamps often lose their tension over a period of time, and can be "sprung" during removal. Replace all spring-type clamps with screw clamps whenever a hose is replaced.

### Metal lines

11   Sections of metal line are often used for fuel line between the fuel pump and fuel injection unit. Check carefully to be sure the line has not been bent or crimped and that cracks have not started in the line.

12   If a section of metal fuel line must be replaced, only seamless steel tubing should be used, since copper and aluminum tubing don't have the strength necessary to withstand normal engine vibration.

13   Check the metal brake lines where they enter the master cylinder and brake proportioning unit (if used) for cracks in the lines or loose fittings. Any sign of brake fluid leakage calls for an immediate thorough inspection of the brake system.

### 12   Cooling system check (every 7500 miles or 6 months)

*Refer to illustration 12.4*

1   Many major engine failures can be attributed to a faulty cooling system. If the vehicle is equipped with an automatic transaxle, the cooling system also cools the transaxle fluid and thus plays an important role in prolonging transaxle life.

**Check for a chafed area that could fail prematurely.**

**Check for a soft area indicating the hose has deteriorated inside.**

**Overtightening the clamp on a hardened hose will damage the hose and cause a leak.**

**Check each hose for swelling and oil-soaked ends. Cracks and breaks can be located by squeezing the hose.**

**12.4  Hoses, like drivebelts, have a habit of failing at the worst possible time - to prevent the inconvenience of a blown radiator or heater hose, inspect them carefully as shown here**

2    The cooling system should be checked with the engine cold. Do this before the vehicle is driven for the day or after the engine has been shut off for at least three hours.

3    Remove the radiator cap by turning it to the left until it reaches a stop. If you hear a hissing sound (indicating there is still pressure in the system), wait until it stops. Now press down on the cap with the palm of your hand and continue turning to the left until the cap can be removed. Thoroughly clean the cap, inside and out, with clean water. Also clean the filler neck on the radiator. All traces of corrosion should be removed. The coolant inside the radiator should be relatively transparent. If it's rust colored, the system should be drained and refilled (see Section 26). If the coolant level isn't up to the top, add additional antifreeze/coolant mixture (see Section 4).

4    Carefully check the large upper and

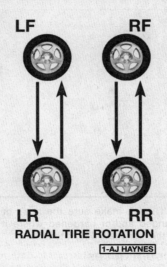

LF          RF

LR          RR

**RADIAL TIRE ROTATION**

1-AJ HAYNES

**13.2  The recommended tire rotation pattern for these vehicles**

lower radiator hoses along with the smaller diameter heater hoses which run from the engine to the firewall. Inspect each hose along its entire length, replacing any hose which is cracked, swollen or shows signs of deterioration. Cracks may become more apparent if the hose is squeezed **(see illustration)**. Regardless of condition, it's a good idea to replace hoses with new ones every two years.

5    Make sure that all hose connections are tight. A leak in the cooling system will usually show up as white or rust colored deposits on the areas adjoining the leak. If wire-type clamps are used at the ends of the hoses, it may be a good idea to replace them with more secure screw-type clamps.

6    Use compressed air or a soft brush to remove bugs, leaves, etc. from the front of the radiator or air conditioning condenser. Be careful not to damage the delicate cooling fins or cut yourself on them.

7    Every other inspection, or at the first indication of cooling system problems, have the cap and system pressure tested. If you don't have a pressure tester, most gas stations and repair shops will do this for a minimal charge.

## 13   Tire rotation (every 7500 miles or 6 months)

*Refer to illustration 13.2*

1    The tires should be rotated at the specified intervals and whenever uneven wear is noticed. Since the vehicle will be raised and the tires removed anyway, check the brakes (see Section 16) at this time.

2    Radial tires must be rotated in a specific pattern **(see illustration)**.

3    Refer to the information in *Jacking and towing* at the front of this manual for the proper procedures to follow when raising the

vehicle and changing a tire. If the brakes are to be checked, do not apply the parking brake as stated. Make sure the tires are blocked to prevent the vehicle from rolling.

4    Preferably, the entire vehicle should be raised at the same time. This can be done on a hoist or by jacking up each corner and then lowering the vehicle onto jackstands placed under the frame rails. Always use four jackstands and make sure the vehicle is firmly supported.

5    After rotation, check and adjust the tire pressures as necessary and be sure to check the lug nut tightness.

6    For further information on the wheels and tires, refer to Chapter 10.

## 14   Seat belt check (every 7500 miles or 6 months)

1    Check the seat belts, buckles, latch plates and guide loops for obvious damage and signs of wear. Seat belts that exhibit fraying along the edges should be replaced.

2    Where the seat belt receptacle bolts to the floor of the vehicle, check that the bolts are secure.

3    See if the seat belt reminder light comes on when the key is turned to the Run or Start position. A chime should also sound.

## 15   Brake check (every 15,000 miles or 12 months)

**Warning:** *Dust created by the brake system is harmful to your health. Never blow it out with compressed air and don't inhale any of it. An approved filtering mask should be worn when working on the brakes. Do not, under any circumstances, use petroleum-based solvents to clean brake parts. Use brake system cleaner only!*

**Note:** *For detailed photographs of the brake system, refer to Chapter 9.*

1    In addition to the specified intervals, the brakes should be inspected every time the wheels are removed or whenever a defect is suspected. Any of the following symptoms could indicate a potential brake system defect: The vehicle pulls to one side when the brake pedal is depressed; the brakes make squealing or dragging noises when applied; brake travel is excessive; the pedal pulsates; brake fluid leaks, usually onto the inside of the tire or wheel.

2    The disc brake pads have built-in wear indicators which should make a high-pitched squealing or scraping noise when they are worn to the replacement point. When you hear this noise, replace the pads immediately or expensive damage to the discs can result.

3    Loosen the wheel lug nuts.

4    Raise the vehicle and place it securely on jackstands.

5    Remove the wheels (see *Jacking and towing* at the front of this book, or your owner's manual, if necessary).

**1**

## Disc brakes

*Refer to illustration 15.6*

6    There are two pads - an outer and an inner - in each caliper. The pads are visible through small inspection holes in each caliper **(see illustration)**.

7    Check the pad thickness by looking at each end of the caliper and through the inspection hole in the caliper body. If the lining material is less than the thickness listed in this Chapter's Specifications, replace the pads. **Note:** *Keep in mind that the lining material is riveted or bonded to a metal backing plate and the metal portion is not included in this measurement.*

8    If it is difficult to determine the exact thickness of the remaining pad material by the above method, or if you are at all concerned about the condition of the pads, remove the caliper(s), then remove the pads from the calipers for further inspection (refer to Chapter 9).

9    Once the pads are removed from the calipers, clean them with brake cleaner and re-measure them.

10   Measure the disc thickness with a micrometer to make sure that it still has service life remaining. If any disc is thinner than the specified minimum thickness, replace it (refer to Chapter 9). Even if the disc has service life remaining, check its condition. Look for scoring, gouging and burned spots. If these conditions exist, remove the disc and have it resurfaced (see Chapter 9).

11   Before installing the wheels, check all brake lines and hoses for damage, wear, deformation, cracks, corrosion, leakage, bends and twists, particularly in the vicinity of the rubber hoses at the calipers.

12   Check the clamps for tightness and the connections for leakage. Make sure that all hoses and lines are clear of sharp edges, moving parts and the exhaust system. If any of the above conditions are noted, repair, reroute or replace the lines and/or fittings as necessary (see Chapter 9).

**15.6  You will find an inspection hole like this in each caliper - placing a ruler across the hole should enable you to determine the thickness of remaining pad material for both inner and outer pads**

## Rear drum brakes

*Refer to illustrations 15.13, 15.15 and 15.17*

13   To check the brake shoe lining thickness without removing the brake drums, remove the rubber plug from the backing plate and use a flashlight to inspect the linings **(see illustration)**. For a more thorough brake inspection, follow the procedure below.

14   Refer to Chapter 9 and remove the rear brake drums.

15   Note the thickness of the lining material on the rear brake shoes **(see illustration)** and look for signs of contamination by brake fluid and grease. If the lining material is within 1/16-inch of the recessed rivets or metal shoes, replace the brake shoes with new ones. The shoes should also be replaced if they are cracked, glazed (shiny lining surfaces) or contaminated with brake fluid or grease. See Chapter 9 for the replacement procedure.

16   Check the shoe return and hold-down

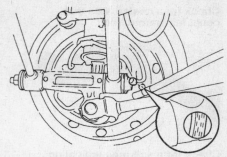

**15.13  A quick check of the remaining drum brake shoe lining material can be made by removing the rubber plug in the backing plate and looking through the inspection hole**

springs and the adjusting mechanism to make sure they're installed correctly and in good condition. Deteriorated or distorted springs, if not replaced, could allow the linings to drag and wear prematurely.

17   Check the wheel cylinders for leakage by carefully peeling back the rubber boots **(see illustration)**. If brake fluid is noted behind the boots, the wheel cylinders must be rebuilt or replaced (see Chapter 9).

18   Check the drums for cracks, score marks, deep scratches and hard spots, which will appear as small discolored areas. If imperfections cannot be removed with emery cloth, the drums must be resurfaced by an automotive machine shop (see Chapter 9 for more detailed information).

19   Install the brake drums.

20   Install the wheels and lug nuts.

21   Remove the jackstands and lower the vehicle.

22   Tighten the wheel lug nuts to the torque listed in this Chapter's Specifications.

## Brake booster check

23   Sit in the driver's seat and perform the following sequence of tests.

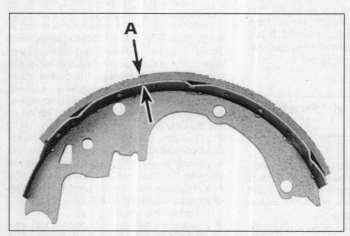

**15.15  If the lining is bonded to the brake shoe, measure the lining thickness from the outer surface to the metal shoe, as shown here; if the lining is riveted to the shoe, measure from the lining outer surface to the rivet head**

**15.17  Carefully peel back the wheel cylinder boot and check for leaking fluid indicating that the cylinder must be replaced or rebuilt**

**16.2a  Detach the clips and separate the cover from the air cleaner housing**

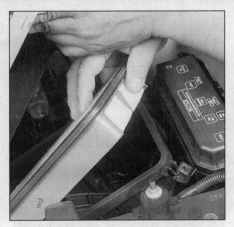

**16.2b  Hold the cover up out of the way and lift the element out**

**17.2  Use a small screwdriver to carefully pry out the old gasket - take care not to damage the cap**

1

24   With the brake fully depressed, start the engine - the pedal should move down a little when the engine starts.

25   With the engine running, depress the brake pedal several times - the travel distance should not change.

26   Depress the brake, stop the engine and hold the pedal in for about 30 seconds - the pedal should neither sink nor rise.

27   Restart the engine, run it for about a minute and turn it off. Then firmly depress the brake several times - the pedal travel should decrease with each application.

28   If your brakes do not operate as described above when the preceding tests are performed, the brake booster has failed. Refer to Chapter 9 for the replacement procedure.

## *Parking brake*

29   Slowly pull up on the parking brake and count the number of clicks you hear until the handle is up as far as it will go. The adjustment is correct if you hear the specified number of clicks. If you hear more or fewer clicks, it's time to adjust the parking brake (refer to Chapter 9).

30   An alternative method of checking the parking brake is to park the vehicle on a steep hill with the parking brake set and the transaxle in Neutral (be sure to stay in the vehicle for this procedure). If the parking brake cannot prevent the vehicle from rolling, it is in need of adjustment (see Chapter 9).

## 16   Air filter replacement (every 15,000 miles or 12 months)

*Refer to illustrations 16.2a and 16.2b*

1   The air filter is located inside a housing at the left (driver's) side of the engine compartment.

2   To remove the air filter, release the spring clips that keep the two halves of the air cleaner housing together, then lift the cover up and remove the air filter element **(see illustrations)**.

3   Inspect the outer surface of the filter element. If it is dirty, replace it. If it is only moderately dusty, it can be reused by blowing it clean from the back to the front surface with compressed air. Because it is a pleated paper type filter, it cannot be washed or oiled. If it cannot be cleaned satisfactorily with compressed air, discard and replace it. While the cover is off, be careful not to drop anything down into the housing. **Caution:** *Never drive the vehicle with the air cleaner removed. Excessive engine wear could result and backfiring could even cause a fire under the hood.*

4   Wipe out the inside of the air cleaner housing.

5   Place the new filter into the air cleaner housing, making sure it seats properly.

6   Installation of the cover is the reverse of removal.

## 17   Fuel system check (every 15,000 miles or 12 months)

*Refer to illustrations 17.2 and 17.5*

**Warning:** *Gasoline is extremely flammable, so take extra precautions when you work on any part of the fuel system. Don't smoke or allow open flames or bare light bulbs near the work area, and don't work in a garage where a gas-type appliance (such as a water heater or a clothes dryer) is present. Since gasoline is carcinogenic, wear latex gloves when there's a possibility of being exposed to fuel, and, if you spill any fuel on your skin, rinse it off immediately with soap and water. Mop up any spills immediately and do not store fuel-soaked rags where they could ignite. The fuel system is under constant pressure, so, if any fuel lines are to be disconnected, the fuel pressure in the system must be relieved first. When you perform any kind of work on the fuel system, wear safety glasses and have a Class B type fire extinguisher on hand.*

1   If you smell gasoline while driving or after the vehicle has been sitting in the sun, inspect the fuel system immediately.

2   Remove the gas filler cap and inspect if for damage and corrosion. The gasket should have an unbroken sealing imprint. If the gasket is damaged or corroded, remove it and install a new one **(see illustration)**.

3   Inspect the fuel feed and return lines for cracks. Make sure that the threaded flare-nut type connectors which secure the metal fuel lines to the fuel injection system and, on 1997 and earlier models, the banjo bolts which secure the banjo fittings to the in-line fuel filter are tight.

4   Since some components of the fuel system - the fuel tank and part of the fuel feed and return lines, for example - are underneath the vehicle, they can be inspected more easily with the vehicle raised on a hoist. If that's not possible, raise the vehicle and support it securely on jackstands.

5   With the vehicle raised and safely supported, inspect the gas tank and filler neck for punctures, cracks and other damage. The connection between the filler neck and the tank is particularly critical. Sometimes a rubber filler neck will leak because of loose clamps or deteriorated rubber **(see illustration)**. These are problems a home mechanic can usually rectify. **Warning:** *Do not, under*

**17.5  Inspect the filler hose for cracks and make sure the clamps are tight**

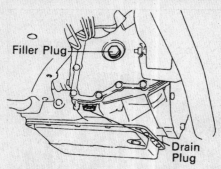

**18.2  Differential filler and drain plug details (three-speed automatic transaxle)**

any circumstances, try to repair a fuel tank (except rubber components). A welding torch or any open flame can easily cause fuel vapors inside the tank to explode.

6    Carefully check all rubber hoses and metal lines leading away from the fuel tank. Check for loose connections, deteriorated hoses, crimped lines and other damage. Carefully inspect the lines from the tank to the fuel injection system. Repair or replace damaged sections as necessary (see Chapter 4).

## 18  Automatic transaxle differential lubricant level check (three-speed only) (15,000 miles or 12 months)

*Refer to illustration 18.2*

1    The three-speed automatic transaxle differential has a separate lubricant supply with a check/fill plug which must be removed to check the level. If the vehicle is raised to gain access to the plug, be sure to support it safely on jackstands - DO NOT crawl under the vehicle when it's supported only by the jack.
2    Remove the filler plug from the front of the differential **(see illustration)**.
3    Use your little finger as a dipstick to make sure the lubricant level is even with the

bottom of the plug hole. If not, use a syringe or a gear oil pump to add the recommended lubricant (see this Chapter's Specifications) until it just starts to run out of the opening.
4    Install the plug and tighten it securely.

## 19  Manual transaxle lubricant level check (every 15,000 miles or 12 months)

*Refer to illustration 19.1*

1    The manual transaxle does not have a dipstick. To check the fluid level, raise the vehicle and support it securely on jackstands. On the lower front side of the transaxle housing, you will see a plug **(see illustration)**. Remove it. If the lubricant level is correct, it should be up to the lower edge of the hole.
2    If the transaxle needs more lubricant (if the level is not up to the hole), use a syringe or a gear oil pump to add more. Stop filling the transaxle when the lubricant begins to run out the hole.
3    Install the plug and tighten it securely. Drive the vehicle a short distance, then check for leaks.

## 20  Steering and suspension check (every 15,000 miles or 12 months)

*Refer to illustrations 20.1, 20.7 and 20.8*
**Note:** *For detailed illustrations of the steering and suspension components, refer to Chapter 10.*

### With the wheels on the ground

1    With the vehicle stopped and the front wheels pointed straight ahead, rock the steering wheel gently back and forth. If freeplay **(see illustration)** is excessive, a front wheel bearing, main shaft yoke, intermediate shaft yoke, lower arm balljoint or steering system joint is worn or the steering gear is out of adjustment or broken. Refer to

**19.1  Use your finger as a dipstick to check the manual transaxle lubricant level**

Chapter 10 for the appropriate repair procedure.
2    Other symptoms, such as excessive vehicle body movement over rough roads, swaying (leaning) around corners and binding as the steering wheel is turned, may indicate faulty steering and/or suspension components.
3    Check the shock absorbers by pushing down and releasing the vehicle several times at each corner. If the vehicle does not come back to a level position within one or two bounces, the shocks/struts are worn and must be replaced. When bouncing the vehicle up and down, listen for squeaks and noises from the suspension components. Additional information on suspension components can be found in Chapter 10.

### Under the vehicle

4    Raise the vehicle with a floor jack and support it securely on jackstands. See *Jacking and towing* at the front of this book for the proper jacking points.
5    Check the tires for irregular wear patterns and proper inflation. See Section 5 in this Chapter for information regarding tire wear and Chapter 10 for the wheel bearing replacement procedures.
6    Inspect the universal joint between the steering shaft and the steering gear housing.

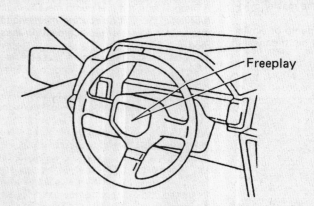

**20.1  Steering wheel freeplay is the amount of travel between an initial steering input and the point at which the front wheels begin to turn (indicated by a slight resistance)**

**20.7  To check a balljoint for wear, place a block of wood under the tire and lower the jack until there is about half a load on the coil spring, then move the control arm up and down with a prybar to make sure there is no play in the balljoint (if there is, replace it)**

**20.8  Push on the balljoint boot to check for damage**

**21.2  Flex the driveaxle boots by hand to check for cracks and/or leaking grease**

**23.4a  Using a backup wrench to prevent the filter from turning, remove the banjo bolt at the top and . . .**

Check the steering gear housing for grease leakage or oozing. Make sure that the dust seals and boots are not damaged and that the boot clamps are not loose. Check the steering linkage for looseness or damage. Check the tie-rod ends for excessive play. Look for loose bolts, broken or disconnected parts and deteriorated rubber bushings on all suspension and steering components. While an assistant turns the steering wheel from side to side, check the steering components for free movement, chafing and binding. If the steering components do not seem to be reacting with the movement of the steering wheel, try to determine where the slack is located.

7     Check the balljoints for wear by placing a wood block under each tire. Lower the jack until there is about half a load on the coil springs. Make sure that the front wheels are in a straight forward position and block the rear wheels with chocks. Move each control arm up and down with a pry bar **(see illustration)** to ensure that its balljoint has no play. If any balljoint does have play, replace it. See Chapter 10 for the balljoint replacement procedure.

8     Inspect the balljoint boots for damage and leaking grease **(see illustration)**. Replace the balljoints with new ones if they are damaged (see Chapter 10).

## 21  Driveaxle boot check (every 15,000 miles or 12 months)

*Refer to illustration 21.2*

1     The driveaxle boots are very important because they prevent dirt, water and foreign material from entering and damaging the constant velocity (CV) joints. Oil and grease can cause the boot material to deteriorate prematurely, so it's a good idea to wash the boots with soap and water. Because it constantly pivots back and forth following the steering action of the front hub, the outer CV boot wears out sooner and should be inspected regularly.

2     Inspect the boots for tears and cracks as well as loose clamps **(see illustration)**. If there is any evidence of cracks or leaking lubricant, they must be replaced as described in Chapter 8.

## 22  Brake fluid change (every 30,000 miles or 24 months)

**Warning:** *Brake fluid can harm your eyes and damage painted surfaces, so use extreme caution when handling or pouring it. Do not use brake fluid that has been standing open or is more than one year old. Brake fluid absorbs moisture from the air. Excess moisture can cause a dangerous loss of braking effectiveness.*

1     At the specified intervals, the brake fluid should be drained and replaced. Since the brake fluid may drip or splash when pouring it, place plenty of rags around the master cylinder to protect any surrounding painted surfaces.

2     Before beginning work, purchase the specified brake fluid (see *Recommended lubricants and fluids* at the beginning of this Chapter).

3     Remove the cap from the master cylinder reservoir.

4     Using a hand suction pump or similar device, withdraw the fluid from the master cylinder reservoir.

5     Add new fluid to the master cylinder until it rises to the line indicated on the reservoir.

6     Bleed the brake system as described in Chapter 9 at all four brakes until new and uncontaminated fluid is expelled from the bleeder screw. Be sure to maintain the fluid level in the master cylinder as you perform the bleeding process. If you allow the master cylinder to run dry, air will enter the system.

7     Refill the master cylinder with fluid and check the operation of the brakes. The pedal should feel solid when depressed, with no sponginess. **Warning:** *Do not operate the vehicle if you are in doubt about the effectiveness of the brake system.*

**23.4b  . . . loosen the fitting at the bottom of the filter**

## 23  Fuel filter replacement (1997 and earlier models) (every 30,000 miles or 24 months)

*Refer to illustrations 23.4a and 23.4b*

**Note:** *The fuel filter on 1998 and later models is integral with the fuel pump module in the fuel tank; there is no routine maintenance interval for fuel filter replacement on these models.*

1     Relieve the fuel system pressure (see Chapter 4). Disconnect the negative battery cable.

2     The canister filter is mounted in a bracket on the firewall near the brake master cylinder.

3     Remove any components that would interfere with access to the top of the filter.

4     Using a backup wrench to steady the filter, remove the banjo bolt at the top and, using a flare-nut wrench, if available, loosen the fitting at the bottom of the fuel filter **(see illustrations)**.

5     Remove both bracket bolts from the firewall and remove the old filter and the filter support bracket assembly.

6     Note that the inlet and outlet pipes are

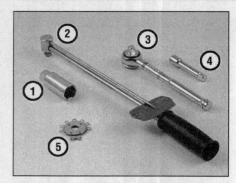

**24.1  Tools required for changing spark plugs**

1   *Spark plug socket - This will have special padding inside to protect the spark plug porcelain insulator*
2   *Torque wrench - Although not mandatory, use of this tool is the best way to ensure that the plugs are tightened properly*
3   *Ratchet - Standard hand tool to fit the plug socket*
4   *Extension - Depending on model and accessories, you may need special extensions and universal joints to reach one or more of the plugs*
5   *Spark plug gap gauge - This gauge for checking the gap comes in a variety of styles. Make sure the gap for your engine is included*

clearly labeled on their respective ends of the filter and that the flanged end of the filter faces down. Make sure the new filter is installed so that it's facing the proper direction as noted above. When correctly installed, the filter should be installed so that the outlet pipe faces up and the inlet pipe faces down.
7    Connect the threaded fitting to the bottom of the filter and tighten it securely. Using new sealing washers, connect the banjo fitting to the top of the filter and tighten the banjo bolt to the torque listed in this Chapter's Specifications.
8    The remainder of installation is the reverse of the removal procedure.

## 24   Spark plug check and replacement (every 30,000 miles or 24 months)

*Refer to illustrations 24.1, 24.4a, 24.4b, 24.4c, 24.6, 24.8, 24.10a and 24.10b*

1    Spark plug replacement requires a spark plug socket which fits onto a ratchet wrench. This socket is lined with a rubber grommet to protect the porcelain insulator of the spark plug and to hold the plug while you insert it into the spark plug hole. You will also need a wire-type feeler gauge to check and adjust the spark plug gap and a torque wrench to tighten the new plugs to the specified torque **(see illustration)**.
2    If you are replacing the plugs, purchase

**24.4a  Spark plug manufacturers recommend using a wire-type gauge when checking the gap - if the wire does not slide between the electrodes with a slight drag, adjustment is required**

**24.4c  To change the gap, bend the side electrode only, as indicated by the arrows, and be very careful not to crack or chip the porcelain insulator surrounding the center electrode**

the new plugs, adjust them to the proper gap and then replace each plug one at a time. **Note:** *When buying new spark plugs, it's essential that you obtain the correct plugs for your specific vehicle. This information can be found in the Specifications Section at the beginning of this Chapter, on the Vehicle Emissions Control Information (VECI) label located on the underside of the hood or in the owner's manual. If these sources specify different plugs, purchase the spark plug type specified on the VECI label because that information is provided specifically for your engine.*
3    Inspect each of the new plugs for defects. If there are any signs of cracks in the porcelain insulator of a plug, don't use it.
4    Check the electrode gaps of the new plugs. Check the gap by inserting the wire gauge of the proper thickness between the electrodes at the tip of the plug **(see illustrations)**. The gap between the electrodes should be identical to that listed in this Chap-

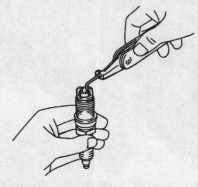

**24.4b  The spark plugs on 1998 and 1999 models have two ground electrodes - check the gap between each ground electrode and the center electrode**

**24.6  When removing the spark plug wires, pull only on the boot and use a twisting/pulling motion**

ter's Specifications or on the VECI label. If the gap is incorrect, use the notched adjuster on the feeler gauge body to bend the curved side electrode slightly **(see illustration)**. **Caution:** *The spark plugs on 2000 and later models use iridium-coated electrodes. Don't attempt to adjust the gap on a used iridium-coated plug.*
5    If the side electrode is not exactly over the center electrode, use the notched adjuster to align them. **Caution:** *If the gap of a new plug must be adjusted, bend only the base of the ground electrode - do not touch the tip.*

### Removal

6    To prevent the possibility of mixing up spark plug wires, work on one spark plug at a time. Remove the wire and boot from one spark plug (1999 and earlier models only). Grasp the boot - not the cable - as shown, give it a half twisting motion and pull straight up **(see illustration)**. On 2000 and later models, remove the ignition coils (see Chapter 5).
7    If compressed air is available, blow any dirt or foreign material away from the spark plug area before proceeding.

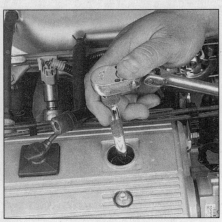

**24.8  Use a spark plug socket with a long extension to unscrew the spark plug**

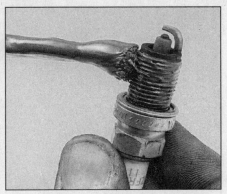

**24.10a  Apply a thin coat of anti-seize compound to the spark plug threads**

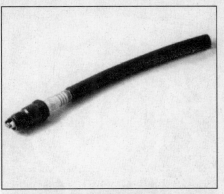

**24.10b  A length of snug-fitting rubber hose will save time and prevent damaged threads when installing the spark plugs**

1

8    Remove the spark plug (see illustration).

9    Whether you are replacing the plugs at this time or intend to reuse the old plugs, compare each old spark plug with the chart on the inside back cover of this manual to determine the overall running condition of the engine.

## Installation

10    Prior to installation, it's a good idea to coat the spark plug threads with anti-seize compound (see illustration). Also, it's often difficult to insert spark plugs into their holes without cross-threading them. To avoid this possibility, fit a length of snug-fitting rubber hose over the end of the spark plug (see illustration). The flexible hose acts as a universal joint to help align the plug with the plug hole. Should the plug begin to cross-thread, the hose will slip on the spark plug, preventing thread damage. Tighten the plug to the torque listed in this Chapter's Specifications.

11    On 1999 and earlier models, attach the plug wire to the new spark plug, again using a twisting motion on the boot until it is firmly seated on the end of the spark plug. On 2000 and later models, install the ignition coil.

12    Follow the above procedure for the remaining spark plugs, replacing them one at a time to prevent mixing up the spark plug wires.

## 25  Spark plug wire, distributor cap and rotor check and replacement (every 30,000 miles or 24 months)

Refer to illustrations 25.8, 25.11a, 25.11b, 25.12 and 25.13

**Note:** Only 1997 and earlier models are equipped with distributors. All 1999 and earlier models have spark plug wires, but 2000 and later models have "coil over plug" ignition systems, where the ignition coils fit directly over the spark plugs. Therefore, this procedure does not apply to 2000 and later models.

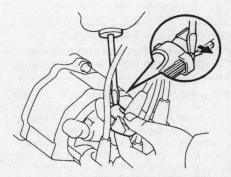

**25.8  Use a small screwdriver to lift the lock claw up when detaching the spark plug wire boot from the distributor**

1    The spark plug wires should be checked whenever new spark plugs are installed.

2    Begin this procedure by making a visual check of the spark plug wires while the engine is running. In a darkened garage (make sure there is ventilation) start the engine and observe each plug wire. Be careful not to come into contact with any moving engine parts. If there is a break in the wire, you will see arcing or a small spark at the damaged area. If arcing is noticed, make a note to obtain new wires, then allow the engine to cool and check the distributor cap and rotor.

3    The spark plug wires should be inspected one at a time to prevent mixing up the order, which is essential for proper engine operation. Each original plug wire should be numbered to help identify its location. If the number is illegible, a piece of tape can be marked with the correct number and wrapped around the plug wire.

4    Disconnect the plug wire from the spark plug. A removal tool can be used for this purpose or you can grasp the rubber boot, twist the boot half a turn and pull the boot free. Do not pull on the wire itself.

5    Check inside the boot for corrosion, which will look like a white crusty powder.

6    Push the wire and boot back onto the end of the spark plug. It should fit tightly onto

**25.11a  Remove the screws (arrows) and detach the distributor cap (third screw not visible in this photo)**

the end of the plug. If it doesn't, remove the wire and use pliers to carefully crimp the metal connector inside the wire boot until the fit is snug.

7    Using a clean rag, wipe the entire length of the wire to remove built-up dirt and grease. Once the wire is clean, check for burns, cracks and other damage. Do not bend the wire sharply, because the conductor might break.

8    Disconnect the wire from the distributor cap (1997 and earlier models) or coil pack (1998 and 1999 models), using a small screwdriver to lift up on the lock claw, if equipped (see illustration). Check for corrosion and a tight fit. Reconnect the wire.

9    Inspect the remaining spark plug wires, making sure that each one is securely fastened at the distributor or coil pack and spark plug when the check is complete.

10    If new spark plug wires are required, purchase a set for your specific engine model. Pre-cut wire sets with the boots already installed are available. Remove and replace the wires one at a time to avoid mix-ups in the firing order.

11    On 1997 and earlier models, detach the distributor cap by removing the retaining screws (see illustration). Look inside it for

**25.11b   Inspect the distributor cap for carbon tracks, charred or eroded terminals and other damage (if in doubt about its condition, install a new one)**

**25.12   Check the rotor for damage, wear and corrosion (if in doubt about its condition, buy a new one)**

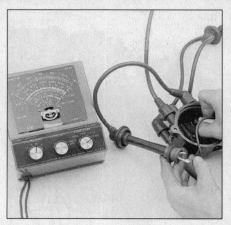

**25.13   Measure the resistance value of the distributor cap and the spark plug wires - if it exceeds the specified maximum value, replace either the cap, or the wires, or both**

cracks, carbon tracks and worn, burned or loose contacts **(see illustration)**.

12    Pull the rotor off the distributor shaft and examine it for cracks and carbon tracks **(see illustration)**. Replace the cap and rotor if any damage or defects are noted.

13    It is common practice to install a new cap and rotor whenever new spark plug wires are installed, but if you wish to continue using the old cap, check the resistance between the spark plug wires and the cap first **(see illustration)**. If the indicated resistance is more than the maximum value listed in this Chapter's Specifications, replace the cap and/or wires.

14    When installing a new cap, remove the wires from the old cap one at a time and attach them to the new cap in the exact same location - do not simultaneously remove all the wires from the old cap or firing order mix-ups may occur.

## 26   Cooling system servicing (draining, flushing and refilling) (every 30,000 miles or 24 months)

**Warning:** *Do not allow engine coolant (antifreeze) to come in contact with your skin or painted surfaces of the vehicle. Rinse off spills immediately with plenty of water. Antifreeze is highly toxic if ingested. Never leave antifreeze lying around in an open container or in puddles on the floor; children and pets are attracted by it's sweet smell and may drink it. Check with local authorities on disposing of used antifreeze. Many communities have collection centers which will see that antifreeze is disposed of safely. Antifreeze is flammable under certain conditions - be sure to read the precautions on the container.*
**Note:** *Non-toxic antifreeze is available at most auto parts stores. Although the antifreeze is non-toxic when fresh, proper disposal is still required.*

1    Periodically, the cooling system should be drained, flushed and refilled to replenish the antifreeze mixture and prevent formation

of rust and corrosion, which can impair the performance of the cooling system and cause engine damage. When the cooling system is serviced, all hoses and the radiator cap should be checked and replaced if necessary.

### Draining

Refer to illustrations 26.4, 26.5a and 26.5b

2    Apply the parking brake and block the wheels. If the vehicle has just been driven, wait several hours to allow the engine to cool down before beginning this procedure.

3    Once the engine is completely cool, remove the radiator cap.

4    Move a large container under the radiator drain to catch the coolant. Attach a 3/8-inch inner diameter hose to the drain fitting to direct the coolant into the container (some models are already equipped with a hose), then open the drain fitting (a pair of pliers may be required to turn it) **(see illustration)**.

5    After the coolant stops flowing out of the radiator, move the container under the engine block drain plug **(see illustrations)**.

**26.4   On most models you will have to remove a cover for access to the radiator drain fitting located at the bottom of the radiator - before opening the valve, push a short section of 3/8-inch ID hose onto the plastic fitting to prevent the coolant from splashing**

**26.5a   After draining the radiator, remove the block drain plug (arrow) located on the side of the engine block - 1997 or earlier model shown**

**26.5b   On 1998 and later models, attach a length of tubing to the spigot on the engine block drain, then loosen the drain bolt and allow the block to drain**

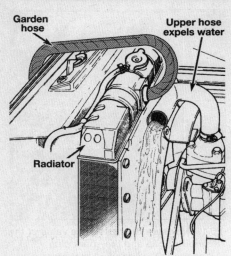

**26.9 With the thermostat removed, disconnect the upper radiator hose and flush the radiator and engine block with a garden hose**

**27.2a Check the evaporative emissions control canister for damage and the hose connections for cracks and damage (arrows) - 1997 and earlier models shown**

**27.2b On 1998 and later models, the charcoal canister is located above the rear suspension crossmember**

Loosen the plug and allow the coolant in the block to drain.

6    While the coolant is draining, check the condition of the radiator hoses, heater hoses and clamps (refer to Section 12 if necessary). Replace any damaged clamps or hoses.

## Flushing

*Refer to illustration 26.9*

7    Once the system is completely drained, remove the thermostat from the engine (see Chapter 3). Then reinstall the thermostat housing without the thermostat. This will allow the system to be thoroughly flushed.

8    Tighten the radiator drain plug. Turn your heating system controls to Hot, so that the heater core will be flushed at the same time as the rest of the cooling system.

9    Disconnect the upper radiator hose from the radiator, then place a garden hose in the upper radiator inlet and flush the system until the water runs clear at the upper radiator hose **(see illustration)**.

10   In severe cases of contamination or clogging of the radiator, remove the radiator (see Chapter 3) and have a radiator repair facility clean and repair it if necessary.

11   Many deposits can be removed by the chemical action of a cleaner available at auto parts stores. Follow the procedure outlined in the manufacturer's instructions. **Note:** *When the coolant is regularly drained and the system refilled with the correct antifreeze/water mixture, there should be no need to use chemical cleaners or descalers.*

12   Remove the overflow hose from the coolant recovery reservoir. Drain the reservoir and flush it with clean water, then reconnect the hose.

## Refilling

13   Close and tighten the radiator drain. Install and/or tighten the block drain plug. Reinstall the thermostat (see Chapter 3).

14   Place the heater temperature control in the maximum heat position.

15   Slowly add new coolant (a 50/50 mixture of water and antifreeze) to the radiator until it's full. Add coolant to the reservoir up to the lower mark.

16   Leave the radiator cap off and run the engine in a well-ventilated area until the thermostat opens (coolant will begin flowing through the radiator and the upper radiator hose will become hot).

17   Turn the engine off and let it cool. Add more coolant mixture to bring the level back up to the lip on the radiator filler neck.

18   Squeeze the upper radiator hose to expel air, then add more coolant mixture if necessary. Replace the radiator cap.

19   Start the engine, allow it to reach normal operating temperature and check for leaks.

## 27    Evaporative emissions control system check (every 30,000 miles or 24 months)

*Refer to illustrations 27.2a and 27.2b*

1    The function of the evaporative emissions control system is to draw fuel vapors from the gas tank and fuel system, store them in a charcoal canister and then burn them during normal engine operation.

2    The most common symptom of a fault in the evaporative emissions control system is a strong fuel odor in the engine compartment (1997 and earlier models) or from the rear of the vehicle (1998 and later models). If a fuel odor is detected, inspect the charcoal canister. On 1997 and earlier models it's located in the engine compartment, below the air intake duct **(see illustration)**. On 1998 and later models it's located under the rear of the vehicle, above the rear suspension crossmember **(see illustration)**. Check the canister and all hoses for damage and deterioration.

3    The evaporative emissions control system is explained in more detail in Chapter 6.

## 28    Exhaust system check (every 30,000 miles or 24 months)

*Refer to illustration 28.4*

1    With the engine cold (at least three hours after the vehicle has been driven), check the complete exhaust system from its starting point at the engine to the end of the tailpipe. This should be done on a hoist where unrestricted access is available.

2    Check the pipes and connections for evidence of leaks, severe corrosion or damage. Make sure that all brackets and hangers are in good condition and tight.

3    At the same time, inspect the underside of the body for holes, corrosion, open seams, etc. which may allow exhaust gases to enter the passenger compartment. Seal all body openings with silicone or body putty.

4    Rattles and other noises can often be traced to the exhaust system, especially the mounts and hangers. Try to move the pipes, muffler and catalytic converter. If the components can come in contact with the body or suspension parts, secure the exhaust system with new mounts **(see illustration)**.

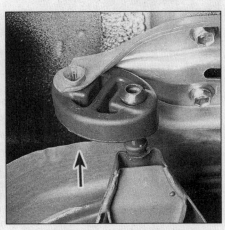

**28.4 Be sure to check each exhaust system rubber hanger (arrow) for damage**

29.7  On some models you'll need an Allen wrench to remove the transaxle drain plug

29.8a  After loosening the front bolts, remove the rear transaxle pan bolts . . .

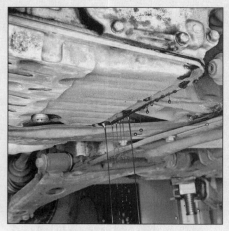

29.8b  . . . and allow the remaining fluid to drain out

5    Check the running condition of the engine by inspecting inside the end of the tailpipe. The exhaust deposits here are an indication of engine state-of-tune. If the pipe is black and sooty or coated with white deposits, the engine is in need of a tune-up, including a thorough fuel engine control system inspection.

## 29  Automatic transaxle/differential fluid and filter change (every 30,000 miles or 24 months)

*Refer to illustrations 29.7, 29.8a, 29.8b, 29.9, 29.11, 29.13a and 29.13b*

1    At the specified time intervals, the automatic transaxle and differential fluid should be drained and replaced.
2    Before beginning work, purchase the specified transmission fluid (see *Recommended fluids and lubricants* at the front of this Chapter).
3    Other tools necessary for this job include jackstands to support the vehicle in a

raised position, wrenches, drain pan capable of holding at least four quarts, newspapers and clean rags.
4    The fluid should be drained immediately after the vehicle has been driven. Hot fluid is more effective than cold fluid at removing built up sediment. **Warning:** *Fluid temperature can exceed 350-degrees F in a hot transaxle. Wear protective gloves.*
5    After the vehicle has been driven to warm up the fluid, raise it and support it on jackstands for access to the transaxle and differential drain plugs.
6    Move the necessary equipment under the vehicle, being careful not to touch any of the hot exhaust components.
7    Place the drain pan under the drain plug in the transaxle pan and remove the drain plug with the Allen wrench **(see illustration)**. Be sure the drain pan is in position, as fluid will come out with some force. Once the fluid is drained, reinstall the drain plug securely.
8    Remove the front transaxle pan bolts, then loosen the rear bolts and carefully pry the pan loose with a screwdriver and allow

the remaining fluid to drain **(see illustrations)**. Once the fluid has drained, remove the bolts and lower the pan.
9    Remove the filter retaining bolts, disconnect the clip (some models) and lower the filter from the transaxle **(see illustration)**. Be careful when lowering the filter as it contains residual fluid.
10   Place the new filter in position, connect the clip (if equipped) and install the bolts. Tighten the bolts to the torque listed in the Specifications Section at the beginning of this Chapter.
11   Carefully clean the gasket surfaces of the fluid pan, removing all traces of old gasket material. Noting their location, remove the magnets, wash the pan in clean solvent and dry it. Be sure to clean and reinstall any magnets **(see illustration)**.
12   Install a new gasket, place the fluid pan in position and install the bolts in their original positions. Tighten the bolts to the torque listed in this Chapter's Specifications.
13   On three-speed models, locate the differential drain plug. Place the drain pan

29.9  Remove the filter bolts and lower the filter (be careful, there will be some residual fluid) - note that here one of the pan magnets is stuck to the filter (arrow); be sure to clean any magnets and return them to the pan

29.11  Noting their locations, remove any magnets and wash them and the pan in solvent before reinstalling them

1

29.13a  Use an Allen wrench to remove the differential drain plug (three-speed models)

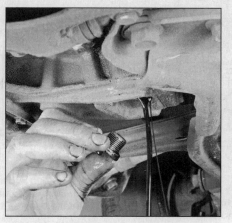

29.13b  Be careful when removing the plug because the fluid usually comes out with some force

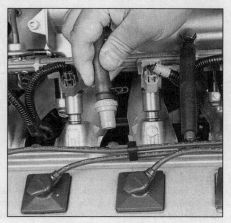

31.1a  On 1997 and earlier models, the PCV valve is located in the center of the valve cover

underneath the plug, remove it with the Allen wrench and drain the fluid **(see illustrations)**. When the differential fluid has drained, reinstall the plug securely.

14    Referring to Section 18, add new fluid to the differential until it begins to run out of the filler hole (see *Recommended lubricants and fluids* at the beginning of this Chapter for the specified fluid type and capacity). **Caution:** *Do not overfill. The automatic transaxle and the differential are separate units.*

15    Lower the vehicle.

16    With the engine off, add new fluid to the transaxle through the dipstick tube (see *Recommended fluids and lubricants* for the recommended fluid type and capacity). Use a funnel to prevent spills. It is best to add a little fluid at a time, continually checking the level with the dipstick (see Section 4). Allow the fluid time to drain into the pan.

17    Start the engine and shift the selector into all positions from P through L, then shift into P and apply the parking brake.

18    With the engine idling, check the fluid level. Add fluid up to the Cool level on the dipstick.

## 30   Manual transaxle lubricant change (every 30,000 miles or 24 months)

1    Raise the vehicle and support it securely on jackstands.
2    Remove the drain plug(s) and drain the fluid.
3    Reinstall the drain plug(s) and tighten to the torque listed in this Chapter's Specifications.
4    Add new fluid until it is even with the lower edge of the filler hole (Section 19). See *Recommended lubricants and fluids* for the specified lubricant type.

## 31   Positive Crankcase Ventilation (PCV) valve and hose check and replacement (every 30,000 miles or 24 months)

*Refer to illustrations 31.1a, 31.1b, 31.1c and 31.4*

1    The PCV valve and hose is located in

the valve cover. On 1997 and earlier models it's located in the center of the valve cover, on the firewall side **(see illustration)**. On 1998 and 1999 models it's mounted in a grommet on the end of the valve cover, near the ignition coil pack **(see illustration)**. On 2000 and later models it's threaded into the left (driver's) end of the valve cover **(see illustration)**.

2    Remove the PCV valve from the cover. On 1999 and earlier models, grasp the valve firmly and pull it out of its grommet. On 2000 and later models, detach the hose, unscrew the valve from the cover, then reattach the hose.

3    With the engine idling at normal operating temperature, place your finger over the end of the valve. If there's no vacuum at the valve, check for a plugged hose or valve. Replace any plugged or deteriorated hoses.

4    Turn off the engine. Remove the PCV valve from the hose. Connect a clean piece of hose and blow through the valve from the valve cover (cylinder head) end. If air will not pass through the valve in this direction, replace it with a new one **(see illustration)**.

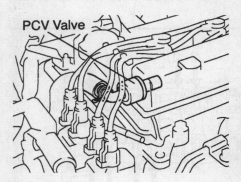

PCV Valve

31.1b  The PCV valve on 1998 and 1999 models is also mounted in a rubber grommet, but on the end of the valve cover near the ignition coil pack

31.1c  On 2000 and later models the PCV valve is threaded into the left end of the valve cover

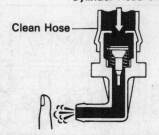
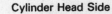

Cylinder Head Side

Clean Hose

31.4  To check the PVC valve, first attach a clean section of hose to the cylinder head side of the valve and blow through it - air should pass through easily - then blow through the intake manifold side of the valve and verify that air passes through with difficulty

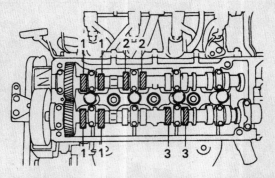

**32.6a  1997 and earlier models: When the no. 1 piston is at TDC on the compression stroke, the valve clearance for the no. 1 and no. 3 cylinder exhaust valves and the no. 1 and no. 2 cylinder intake valves can be measured**

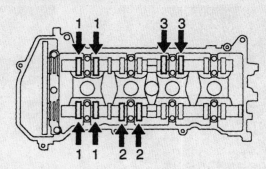

**32.6b  1998 and later models: The same valve clearances are checked as on earlier models, but on these later models the intake camshaft is at the front of the cylinder head and the exhaust camshaft is at the rear**

**32.6c  Check the clearance for each valve with a feeler gauge of the specified thickness - if the clearance is correct, you should feel a slight drag on the gauge as you pull it out**

5   When purchasing a replacement PCV valve, make sure it's for your particular vehicle and engine size. Compare the old valve with the new one to make sure they're the same.

## 32  Valve clearance check and adjustment (every 60,000 miles or 48 months)

**Warning:** *These models are equipped with airbags. Always disable the airbag system before working in the vicinity of any airbag system component to avoid the possibility of accidental deployment of the airbag(s), which could cause personal injury (see Chapter 12).*
**Note:** *If you're working on a 1997 or earlier model, the following procedure requires the use of a special valve lifter tool. It is impossible to perform this task without it.*

### Check
*Refer to illustrations 32.6a, 32.6b, 32.6c, 32.7a and 32.7b*

1   Disconnect the negative cable from the battery. **Caution:** *If the stereo in your vehicle is equipped with an anti-theft system, make sure you have the correct activation code before disconnecting the battery.*
2   Disconnect the spark plug wires (1999 and earlier) (Section 25) or ignition coils (2000 and later models - see Chapter 5) and remove any other components that will interfere with valve cover removal.
3   Blow out the recessed area around the

spark plug openings with compressed air, if available, to remove any debris that might fall into the cylinders, then remove the spark plugs (see Section 24).
4   Remove the valve cover (refer to Chapter 2).
5   Refer to Chapter 2 and position the number 1 piston at TDC on the compression stroke.
6   Measure the clearances of the indicated valves with feeler gauges **(see illustrations)**. Record the measurements which are out of specification. They will be used later to determine the required replacement shims (1997 and earlier models) or lifters (1998 and later models).
7   Turn the crankshaft one complete revolution and realign the timing marks. Measure the remaining valves **(see illustrations)**.

### Adjustment
**Note:** *1997 and earlier models use replaceable adjustment shims that ride on top of the lifters. 1998 and later models don't have shims; instead, they use lifters of different thickness to provide proper valve clearance.*

#### 1997 and earlier models
*Refer to illustrations 32.9a, 32.9b and 32.9c*
8   If any of the valve clearances were out

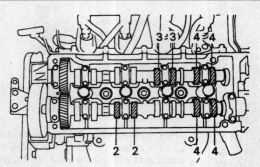

**32.7a  When the no. 4 piston is at TDC on the compression stroke, the valve clearance for the no. 2 and no. 4 exhaust valves and the no. 3 and no. 4 intake valves can be measured - this is on a 1997 or earlier engine . . .**

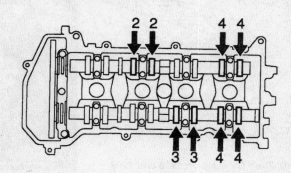

**32.7b  . . . and this is on a 1998 or later engine**

**32.9a  Install the valve lifter tool as shown and squeeze the
handles together to depress the valve lifter, then hold the lifter
down with the smaller tool so the shim can be removed**

**32.9b  Keep pressure on the lifter with the smaller tool and
remove the shim with a small screwdriver . . .**

of specification, turn the crankshaft pulley until the camshaft lobe above the first valve which you intend to adjust is pointing upward, away from the shim.

9    Position the notch in the valve lifter toward the spark plug. Then depress the valve lifter with the special valve lifter tools **(see illustration)**. Place the special valve lifter tool in position as shown, with the longer jaw of the tool gripping the lower edge of the cast lifter boss and the upper, shorter jaw gripping the upper edge of the lifter itself. Depress the valve lifter by squeezing the handles of the valve lifter tool together, then hold the lifter down with the smaller tool and remove the larger one. Remove the adjusting shim with a small screwdriver or a pair of tweezers **(see illustrations)**. Note that the wire hook on the end of some valve lifter tool handles can be used to clamp both handles together to keep the lifter depressed while the shim is removed.

### 1998 and later models

10   If any of the valve clearances were out

of adjustment, remove the camshaft(s) from over the lifter(s) that was/were out of the specified clearance range (see Chapter 2, Part B).

### All models

*Refer to illustrations 32.11a, 32.11b, 32.12a and 32.12b*

11   Measure the thickness of the shim (1997 and earlier models) or lifter (1998 and later models) with a micrometer **(see illustrations)**. To calculate the correct thickness of a replacement shim or lifter that will place the valve clearance within the specified value, use the following formula:

$N = T + (A - V)$

$T = thickness\ of\ the\ old\ shim\ or\ lifter$
$A = valve\ clearance\ measured$
$N = thickness\ of\ the\ new\ shim\ or\ lifter$
$V = desired\ valve\ clearance$ (see this Chapter's Specifications)

12   Select a shim (1997 and earlier models) or lifter (1998 and later models) with a thickness as close as possible to the valve clearance calculated. The shims on 1997 and ear-

**32.9c  . . . a pair of tweezers or a magnet
(as shown here)**

lier models, which are available in 16 sizes in increments of 0.0020-inch (0.050 mm), range

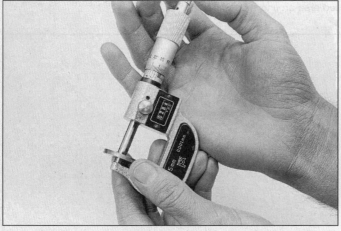

**32.11a  Measure the shim thickness with a micrometer (1997 and
earlier models)**

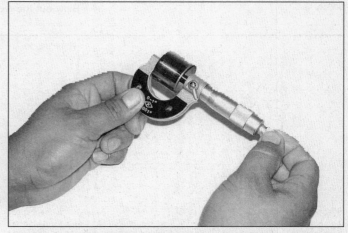

**32.11b  On 1998 and later models, measure the thickness of the
lifter head with a micrometer**

New shim thickness      mm (in.)

| Shim No. | Thickness | Shim No. | Thickness |
|---|---|---|---|
| 1 | 2.55 (0.1004) | 9 | 2.95 (0.1161) |
| 2 | 2.60 (0.1024) | 10 | 3.00 (0.1181) |
| 3 | 2.65 (0.1043) | 11 | 3.05 (0.1201) |
| 4 | 2.70 (0.1063) | 12 | 3.10 (0.1220) |
| 5 | 2.75 (0.1083) | 13 | 3.15 (0.1240) |
| 6 | 2.80 (0.1102) | 14 | 3.20 (0.1260) |
| 7 | 2.85 (0.1122) | 15 | 3.25 (0.1280) |
| 8 | 2.90 (0.1142) | 16 | 3.30 (0.1299) |

**32.12a  Valve adjusting shim thickness chart (1997 and earlier models)**

New lifter thickness      mm (in.)

| Lifter No. | Thickness | Lifter No. | Thickness | Lifter No. | Thickness |
|---|---|---|---|---|---|
| 06 | 5.060 (0.1992) | 30 | 5.300 (0.2087) | 54 | 5.540 (0.2181) |
| 08 | 5.080 (0.2000) | 32 | 5.320 (0.2094) | 56 | 5.560 (0.2189) |
| 10 | 5.100 (0.2008) | 34 | 5.340 (0.2102) | 58 | 5.580 (0.2197) |
| 12 | 5.120 (0.2016) | 36 | 5.360 (0.2110) | 60 | 5.600 (0.2205) |
| 14 | 5.140 (0.2024) | 38 | 5.380 (0.2118) | 62 | 5.620 (0.2213) |
| 16 | 5.160 (0.2031) | 40 | 5.400 (0.2126) | 64 | 5.640 (0.2220) |
| 18 | 5.180 (0.2039) | 42 | 5.420 (0.2134) | 66 | 5.660 (0.2228) |
| 20 | 5.200 (0.2047) | 44 | 5.440 (0.2142) | 68 | 5.680 (0.2236) |
| 22 | 5.220 (0.2055) | 46 | 5.460 (0.2150) | 70 | 5.700 (0.2244) |
| 24 | 5.240 (0.2063) | 48 | 5.480 (0.2157) | 72 | 5.720 (0.2252) |
| 26 | 5.260 (0.2071) | 50 | 5.500 (0.2165) | 74 | 5.740 (0.2260) |
| 28 | 5.280 (0.2079) | 52 | 5.520 (0.2173) | | |

**32.12b  Valve lifter thickness chart (1998 and later models)**

in size from 1.0039-inch (2.550 mm) to 0.1299-inch (3.300 mm) **(see illustration)**. The lifters on 1998 and later models are available in 35 sizes in increments of 0.0008-inch (0.020 mm), range in size from 0.1992-inch (5.060 mm) to 0.2260-inch (5.740 mm) **(see illustration)**. **Note:** *Through careful analysis of the shim or lifter sizes needed to bring the out-of-specification valve clearance within specification, it is often possible to simply move a shim or lifter that has to come out anyway to another location requiring a shim or lifter of that particular size, thereby reducing the number of new shims or lifters that must be purchased.*

## 1997 and earlier models

13    Place the special valve lifter tool in position **as shown in illustration 32.9a**, with the longer jaw of the tool gripping the lower edge of the cast lifter boss and the upper, shorter jaw gripping the upper edge of the lifter itself, press down the valve lifter by squeezing the handles of the valve lifter tool together and install the new adjusting shim (note that the wire hook on the end of one valve lifter tool handle can be used to clamp the handles together to keep the lifter depressed while the shim is inserted. Measure the clearance with a feeler gauge to make sure that your calculations are correct.

14    Repeat this procedure until all the valves which are out of clearance have been corrected.

## 1998 and later models

15    Install the proper thickness lifter(s) in position, making sure to lubricate them with camshaft installation lube first. **Note:** *Apply the lubricant to the underside of the lifter where it contacts the valve stem, the walls of the lifter and the face of the lifter.*

16    Install the camshaft(s) (see Chapter 2B).

17    Recheck the valve clearances.

## All models

18    Installation of the spark plugs, valve cover, spark plug wires and boots (or ignition coils), etc. is the reverse of removal.

# Chapter 2 Part A
# Engines - 1997 and earlier

## Contents

**2A**

## Specifications

### General

| | |
|---|---|
| Engine type | DOHC, inline four-cylinder, four valves per cylinder |
| Cylinder numbers (drivebelt end-to-transaxle end) | 1-2-3-4 |
| Firing order | 1-3-4-2 |
| Displacement | |
|   4A-FE (VIN 6) | 1.6L (97 cu. in.) |
|   7A-FE (VIN 8) | 1.8L (108 cu. in.) |

### Timing belt

| | |
|---|---|
| Tensioner spring free length | |
|   1.6L | 1.717 inches |
|   1.8L | 1.480 inches |
| Timing belt deflection | 13/64 to 1/4 inch |

### Camshaft

| | |
|---|---|
| Journal diameter | |
|   Exhaust camshaft, no. 1 journal | 0.9822 to 0.9829 inch |
|   All others | 0.9035 to 0.9041 inches |
| Bearing oil clearance | |
|   Standard | 0.0014 to 0.0028 inch |
|   Service limit | 0.0039 inch |
| Runout limit | 0.0016 inch |
| Lobe height | |
|   Intake camshaft | |
|     1993 and 1994 | |
|       Standard | 1.650 to 1.6539 inches |
|       Service limit (minimum) | 1.6338 inches |
|     1995 | |
|       Corolla | |
|         1.6L | |
|           Standard | 1.650 to 1.6539 inches |
|           Service limit (minimum) | 1.6338 inches |
|         1.8L | |
|           Standard | 1.6421 to 1.6461 inches |
|           Service limit (minimum) | 1.6260 inches |
|       Prizm | |
|         Standard | 1.650 to 1.6539 inches |
|         Service limit (minimum) | 1.6338 inches |

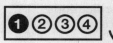 

**Cylinder location and distributor rotation**

*The blackened terminal shown on the distributor cap indicates the Number One spark plug wire position*

## Camshaft (continued)

Lobe height
  Intake camshaft
    1996 and 1997

| | |
|---|---|
| Standard | 1.6421 to 1.6461 inches |
| Service limit (minimum) | 1.6260 inches |

  Exhaust camshaft

| | |
|---|---|
| Standard | 1.6520 to 1.6560 inches |
| Service limit (minimum) | 1.6358 inches |

Camshaft thrust clearance (endplay)
  Intake camshaft

| | |
|---|---|
| Standard | 0.0012 to 0.0033 inch |
| Service limit (maximum) | 0.0043 inch |

  Exhaust camshaft

| | |
|---|---|
| Standard | 0.0014 to 0.0035 inch |
| Service limit (maximum) | 0.0043 inch |
| Camshaft gear spring free length | 0.669 to 0.693 inch |

Camshaft gear backlash

| | |
|---|---|
| Standard | 0.0008 to 0.0079 inch |
| Service limit | 0.0188 inch |

Valve lifter

| | |
|---|---|
| Diameter | 1.2191 to 1.2195 inches |
| Bore diameter | 1.2205 to 1.2215 inches |

Lifter oil clearance

| | |
|---|---|
| Standard | 0.0009 to 0.0023 inch |
| Service limit | 0.0028 inch |

## Oil pump

Rotor-to-body clearance

| | |
|---|---|
| Standard | 0.0031 to 0.0071 inch |
| Service limit | 0.0079 inch |

Rotor tip clearance

| | |
|---|---|
| Standard | 0.0010 to 0.0033 inch |
| Service limit | 0.0079 inch |

Rotor-to-cover clearance

| | |
|---|---|
| Standard | 0.0010 to 0.0033 inch |
| Service limit | 0.0039 inch |

## Torque specifications

**Ft-lbs** (unless otherwise indicated)

Intake manifold bolts

| | |
|---|---|
| At EGR port | 108 in-lbs |
| Others | 168 in-lbs |

Intake manifold brace bolts

| | |
|---|---|
| At manifold | 168 in-lbs |
| At block | 29 |
| Intake air chamber cover | 168 in-lbs |
| Exhaust manifold nuts/bolts | 25 |
| Exhaust manifold brace bolts | 46 |
| Crankshaft pulley-to-crankshaft bolt | 87 |

Flywheel/driveplate bolts

| | |
|---|---|
| Flywheel (manual transaxle) | 58 |
| Driveplate (automatic transaxle) | 47 |
| Idler pulley bolts | 27 |

Cylinder head bolts

| | |
|---|---|
| Step 1 | 22 |
| Step 2 | Tighten an additional 90-degrees |
| Step 3 | Tighten an additional 90-degrees |
| Camshaft bearing cap bolts | 108 in-lbs |
| Camshaft sprocket bolt | 43 |
| Oil pump bolts | 16 |
| Oil pick-up/strainer nuts/bolts | 82 in-lbs |

Oil pan bolts

| | |
|---|---|
| 1.6L | 43 in-lbs |

  1.8L

| | |
|---|---|
| Reinforcement section-to-block | 144 in-lbs |
| Pan-to-reinforcement section | 43 in-lbs |
| Center chassis brace bolts | 45 |
| Engine/transaxle stiffener bolts (1.6L) | 17 |
| Rear main oil seal retainer bolts | 82 in-lbs |

## 1 General information

This Part of Chapter 2 is devoted to in-vehicle repair procedures for the four-cylinder engines. All information concerning engine removal and installation and engine block and cylinder head overhaul can be found in Part C of this Chapter.

The following repair procedures are based on the assumption that the engine is installed in the vehicle. If the engine has been removed from the vehicle and mounted on a stand, many of the steps outlined in this Part of Chapter 2 will not apply.

The Specifications included in this Part of Chapter 2 apply only to the procedures contained in this Part. Part C of Chapter 2 contains the Specifications necessary for cylinder head and engine block rebuilding.

On 1997 and earlier models, the four-cylinder engines in the Corolla are designated the 4A-FE (1.6L) and the 7A-FE (1.8L). The two engines are almost identical, and incorporate dual overhead camshafts (DOHC) and four valves per cylinder. In the Geo Prizm models, the 1.6L engine is designated VIN 6, and the 1.8L is the VIN 8.

## 2 Repair operations possible with the engine in the vehicle

Many major repair operations can be accomplished without removing the engine from the vehicle.

Clean the engine compartment and the exterior of the engine with some type of degreaser before any work is done. It will make the job easier and help keep dirt out of the internal areas of the engine.

Depending on the components involved, it may be helpful to remove the hood to improve access to the engine as repairs are performed (refer to Chapter 11 if necessary). Cover the fenders to prevent damage to the paint. Special pads are available, but an old bedspread or blanket will also work.

If vacuum, exhaust, oil or coolant leaks develop, indicating a need for gasket or seal replacement, the repairs can generally be made with the engine in the vehicle. The intake and exhaust manifold gaskets, oil pan gasket, crankshaft oil seals and cylinder head gasket are all accessible with the engine in place.

Exterior engine components, such as the intake and exhaust manifolds, the oil pan, the oil pump, the water pump, the starter motor, the alternator, the distributor and the fuel system components can be removed for repair with the engine in place, although when replacing the oil pump an engine hoist is required to support the engine from above, while the right engine mount is removed.

Since the cylinder head can be removed without pulling the engine, camshaft and valve component servicing can also be accomplished with the engine in the vehicle. Replacement of the timing belt and pulleys is

also possible with the engine in the vehicle.

In extreme cases caused by a lack of necessary equipment, repair or replacement of piston rings, pistons, connecting rods and rod bearings is possible with the engine in the vehicle. However, this practice is not recommended because of the cleaning and preparation work that must be done to the components involved.

## 3 Top Dead Center (TDC) for number one piston - locating

*Refer to illustration 3.8*
**Note:** *The following procedure is based on the assumption that the distributor is correctly installed. If you are trying to locate TDC to install the distributor correctly, piston position must be determined by feeling for compression at the number one spark plug hole, then aligning the ignition timing marks as described in Step 8.*

1 Top Dead Center (TDC) is the highest point in the cylinder that each piston reaches as it travels up the cylinder bore. Each piston reaches TDC on the compression stroke and again on the exhaust stroke, but TDC generally refers to piston position on the compression stroke.

2 Positioning the piston(s) at TDC is an essential part of many procedures such as camshaft and timing belt/pulley removal and distributor removal.

3 Before beginning this procedure, be sure to place the transmission in Neutral and apply the parking brake or block the rear wheels. Also, disable the ignition system by detaching the coil wire from the center terminal of the distributor cap and grounding it on the block with a jumper wire. Remove the spark plugs (see Chapter 1).

4 In order to bring any piston to TDC, the crankshaft must be turned using one of the methods outlined below. When looking at the timing belt end of the engine, normal crankshaft rotation is clockwise.

a) *The preferred method is to turn the crankshaft with a socket and ratchet attached to the bolt threaded into the front of the crankshaft. Turn the bolt in a clockwise direction only. Never turn the bolt counterclockwise.*

b) *A remote starter switch, which may save some time, can also be used. Follow the instructions included with the switch. Once the piston is close to TDC, use a socket and ratchet as described in the previous paragraph.*

c) *If an assistant is available to turn the ignition switch to the Start position in short bursts, you can get the piston close to TDC without a remote starter switch. Make sure your assistant is out of the vehicle, away from the ignition switch, then use a socket and ratchet as described in Paragraph a) to complete the procedure.*

5 Note the position of the terminal for the

**3.8 Align the crankshaft drivebelt pulley notch (arrow) with the 0 (zero) on the timing plate**

2A

number one spark plug wire on the distributor cap. If the terminal isn't marked, follow the plug wire from the number one cylinder spark plug to the cap.

6 Use a felt-tip pen or chalk to make a mark on the distributor body directly under the number one terminal.

7 Detach the cap from the distributor and set it aside (see Chapter 1 if necessary).

8 Turn the crankshaft until the notch in the crankshaft pulley is aligned with the 0 on the timing plate located at the front of the engine **(see illustration)**.

9 Look at the distributor rotor - it should be pointing directly at the mark you made on the distributor body. If so, you are at TDC for number 1.

10 If the rotor is 180-degrees off, the number one piston is at TDC on the exhaust stroke.

11 To get the piston to TDC on the compression stroke, turn the crankshaft one complete turn (360-degrees) clockwise. The rotor should now be pointing at the mark on the distributor. When the rotor is pointing at the number one spark plug wire terminal in the distributor cap and the ignition timing marks are aligned, the number one piston is at TDC on the compression stroke. **Note:** *If it's impossible to align the ignition timing marks when the rotor is pointing at the mark on the distributor body, the timing belt may have jumped the teeth on the pulleys or may have been installed incorrectly.*

12 After the number one piston has been positioned at TDC on the compression stroke, TDC for any of the remaining pistons can be located by turning the crankshaft and following the firing order. Mark the remaining spark plug wire terminal locations on the distributor body just like you did for the number one terminal, then number the marks to correspond with the cylinder numbers. As you turn the crankshaft, the rotor will also turn. When it's pointing directly at one of the marks on the distributor, the piston for that particular cylinder is at TDC on the compression stroke.

**4.4 Remove the two harness-cover mounting bolts at the right end of the valve cover, allowing the wiring harness to be pulled up and clear of the valve cover**

**4.5 The valve cover is held in place by four nuts (arrows)**

**4.6 The valve cover gasket (left arrow) is a rubber O-ring-like seal - it can be reused if it hasn't hardened - make sure the spark plug tube seals (right arrow) are in place before replacing the valve cover**

## 4   Valve cover - removal and installation

### Removal

*Refer to illustrations 4.4 and 4.5*

1    Disconnect the negative cable from the battery.
**Caution:** *If the stereo in your vehicle is equipped with an anti-theft system, make sure you have the correct activation code before disconnecting the battery.*
2    Detach the PCV hoses from the valve cover.
3    Remove the spark plug wires from the spark plugs, handling them by the boots, not pulling on the wires.
4    Remove two bolts at the timing belt end of the valve cover, then pull up the wiring harness cover and wiring **(see illustration)**.
5    Remove the valve cover mounting nuts, then detach the valve cover and gasket from the cylinder head **(see illustration)**. If the valve cover is stuck to the cylinder head, bump the end with a wood block and a hammer to jar it loose. If that doesn't work, try to slip a flexible putty knife between the cylinder head and valve cover to break the seal. **Caution:** *Don't pry at the valve cover-to-cylinder head joint or damage to the sealing surfaces may occur, leading to oil leaks after the valve cover is reinstalled.*

### Installation

*Refer to illustrations 4.6 and 4.7*

6    The mating surfaces of the housing or cylinder head and valve cover must be clean when the valve cover is installed. The rubber sealing gasket can be reused unless it has high mileage and the rubber has hardened or cracked, then pull out the rubber seal and clean the mating surfaces with lacquer thinner or acetone. Install a new rubber gasket, pressing it evenly into the groove around the underside of the valve cover. If there's residue or oil on the mating surfaces when

the valve cover is installed, oil leaks may develop. **Note:** *Make sure that the spark plug tube gaskets are in place on the underside of the valve cover before reinstalling it* **(see illustration)**.
7    Apply RTV sealant near the rubber plug, cam seal cap and distributor **(see illustration)**.
8    Position a new gasket on the cylinder head, then install the valve cover and nuts.
9    Tighten the nuts to the torque listed in this Chapter's Specifications in three or four equal steps.
10   Reinstall the remaining parts, run the engine and check for oil leaks.

## 5   Intake manifold - removal and installation

### Removal

*Refer to illustrations 5.3, 5.6a, 5.6b, 5.7a and 5.7b*
**Note:** *If the intake manifold is to be unbolted only for removal of the cylinder head, then the intake manifold can simply be unbolted from the cylinder head and pushed toward the firewall, without disconnecting any hoses, wires or linkage. The following procedure is for complete removal of the manifold from the vehicle.*

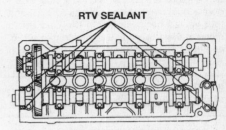

**RTV SEALANT**

**4.7 Apply sealant to the five points indicated by the shaded areas before installing the valve cover**

1    Disconnect the negative cable from the battery. **Caution:** *If the stereo in your vehicle is equipped with an anti-theft system, make sure you have the correct activation code before disconnecting the battery.*
2    Refer to Chapter 4 to remove the throttle body and linkages, and safely relieve the fuel system pressure.
3    Label and detach the PCV and vacuum hoses connected to the intake manifold, including those from the MAP sensor, brake booster and the air conditioning idle-up actuator **(see illustration)**.

**5.3 The various hoses should be marked to insure correct reinstallation**

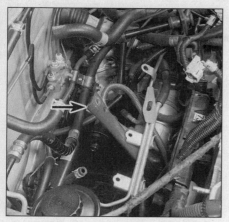

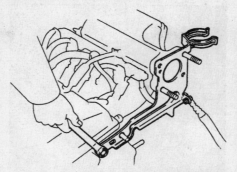

5.6b On 1993 through 1995 models, unbolt this intake manifold brace from the cylinder head

5.6a From underneath the vehicle, remove the bolt retaining this brace (arrow) to the intake manifold (intake manifold removed for clarity)

5.7a From underneath the vehicle, unbolt the wiring harness (right arrow) from the water neck and cylinder head, then remove the lower manifold bolts (left arrow indicates one, of the four bolts)

2A

4    The intake manifold can be removed with the injectors and fuel rail in place. If the injectors are to be removed from the intake manifold, refer to Chapter 4.

5    Disconnect the electrical connector from the EGR valve, label and disconnect the vacuum hose, and unbolt the EGR pipe from the intake manifold (see Chapter 4). Set the EGR assembly aside.

6    Unbolt the upper end of the intake manifold-to-block brace (accessible from under the vehicle), and on 1993 through 1995 models, remove the intake manifold-to-cylinder head brace from the cylinder head (see illustrations).

7    Remove the ground strap and mounting nuts/bolts, then detach the manifold from the engine (see illustrations). Note: *From under the vehicle, it will be necessary to unbolt the wiring harness, the lower intake manifold bolts, and the two nuts securing the pair of steel lines to the underside of the intake manifold.*

## Installation

8    Clean the mating surfaces of the intake manifold and the cylinder head mounting sur-face with lacquer thinner or acetone. If the gasket shows signs of leaking, have the manifold checked for warpage at an automotive machine shop and resurfaced if necessary.

9    The manifold is a two piece design, and a new gasket set may include the gasket for the air chamber cover. Unless the engine has high mileage or you suspect a vacuum leak at the mating surfaces, don't unbolt the air chamber cover. If it's necessary to replace the gasket, unbolt the cover, clean the surfaces, position the new gasket and reinstall the cover, tightening it to Specifications (see Chapter 4).

10    Install a new gasket, then position the manifold on the cylinder head and install the nuts/bolts.

11    Tighten the nuts/bolts in three or four equal steps to the torque listed in this Chapter's Specifications. Work from the center out towards the ends to avoid warping the manifold.

12    Install the remaining parts in the reverse order of removal.

13    Before starting the engine, check the throttle linkage for smooth operation.

14    Run the engine and check for coolant and vacuum leaks.

15    Road test the vehicle and check for proper operation of all accessories, including the cruise control system, if equipped.

## 6    Exhaust manifold - removal and installation

**Warning:** *The engine must be completely cool before beginning this procedure.*

## Removal

*Refer to illustrations 6.2, 6.4 and 6.5*

1    Disconnect the negative cable from the battery. **Caution:** *If the stereo in your vehicle is equipped with an anti-theft system, make sure you have the correct activation code before disconnecting the battery.*

2    Remove the upper heat insulator from the manifold (see illustration). Note: *There is also a lower heat insulator, but is attached to the manifold from underneath and does not need to be removed.*

3    Apply penetrating oil to the exhaust manifold mounting nuts/bolts, and the nuts retaining the exhaust pipe to the manifold. After the nuts have soaked, remove the nuts retaining the exhaust pipe to the manifold (see Chapter 4).

5.7b Remove the intake manifold bolts/nuts and remove the intake manifold

6.2 Remove the five upper heat insulator bolts (arrows)

6.4  Remove the upper bolt retaining the exhaust brace to the manifold (top arrow), and loosen the two bolts retaining the brace to the block (lower arrows)

4    Unbolt the exhaust manifold brace (see illustration).
5    Remove the nuts/bolts and detach the manifold and gasket (see illustration).

## Installation

6    Use a scraper to remove all traces of old gasket material and carbon deposits from the manifold and cylinder head mating surfaces. If the gasket was leaking, have the manifold checked for warpage at an automotive machine shop and resurfaced if necessary.
7    Position a new gasket over the cylinder head studs. Note: The marks on the gasket should face out (away from the cylinder head) and the arrow should point toward the rear (transaxle end) of the engine.
8    Install the manifold and thread the mounting nuts/bolts into place.
9    Working from the center out, tighten the nuts/bolts to the torque listed in this Chapter's Specifications in three or four equal steps.
10   Reinstall the remaining parts in the reverse order of removal.
11   Run the engine and check for exhaust leaks.

6.5  Remove the bolts and nuts and remove the exhaust manifold

## 7    Timing belt and sprockets - removal, inspection and installation

### Removal

*Refer to illustrations 7.7, 7.12, 7.13, 7.14, 7.17a, 7.17b, 7.17c and 7.19*

1    Disconnect the negative cable from the battery. Caution: *If the stereo in your vehicle is equipped with an anti-theft system, make sure you have the correct activation code before disconnecting the battery.*
2    Block the rear wheels and set the parking brake.
3    Loosen the lug nuts on the right front wheel and raise the vehicle. Support the front of the vehicle securely on jackstands.
4    Remove the right front wheel and fender apron seal.
5    Remove the coolant overflow tank (see Chapter 3).
6    Remove the spark plugs and drivebelts (see Chapter 1).
7    Remove the air conditioning compressor belt idler pulley (see illustration).
8    Unbolt the cruise control actuator (if

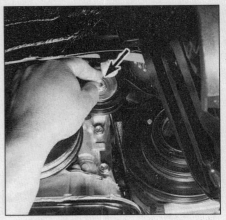

7.7  With the belts removed, unbolt and remove the air conditioning idler pulley (arrow, shown from below through the right fenderwell)

equipped) and set it aside.
9    Refer to Section 4 and remove the valve cover.
10   Support the engine from underneath with a jack (use a wood block on the jack but don't place the block under the oil pan drain plug. Note: *If you're planning on removing the oil pan in addition to the timing belt, support the engine with a hoist from above (see Chapter 2, Part C).*
11   Position the number one piston at TDC on the compression stroke (see Section 3).
12   Remove the starter (see Chapter 5) and wedge a large screwdriver into the flywheel teeth to hold the engine while an assistant loosens the crankshaft pulley bolt (see illustration).
13   The crankshaft pulley should slide off the crankshaft without a puller (see illustration).
14   Remove the three timing belt covers (see illustration).
15   Remove the timing belt guide, noting that the cupped side faces out and the smooth side is next to the belt.

7.12  Use a pry bar wedged into the flywheel (arrow) to keep the engine from rotating while loosening the crankshaft pulley bolt

7.13  After removing the center bolt, remove the crankshaft pulley

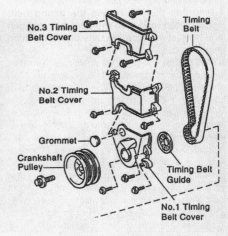

7.14  Timing belt cover details

**7.17a Loosen the center bolt on the belt tensioner, pry it to the rear of the car, then retighten to keep tension off the belt for removal/installation**

**7.17b With the engine supported by a jack, remove these two nuts (arrows) from the right engine mount**

**7.17c Remove the upper engine mount bolt (arrow) and lower the engine, then pull the upper part of the mount up (the rubber will allow movement) until the belt can be slipped out between the upper and lower sections of the mount - this saves removing the mount completely**

**2A**

**7.19 The crankshaft sprocket should slide off the crankshaft easily - if it's stuck, protect the oil pump case with rags and pry the sprocket off with two screwdrivers. While the engine is at TDC, apply marks (arrows) on the crankshaft sprocket and oil pump case so you can find TDC without the crankshaft pulley in place**

**7.21 Check the timing belt for cracked and missing teeth**

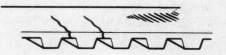

**7.22 If the face of the belt is cracked or worn, check the idler pulleys for nicks or burrs**

**7.23 Wear on one side of the belt indicates sprocket misalignment problems**

16    If you plan to re-use the timing belt, apply match marks on the sprocket and belt and an arrow indicating direction of rotation on the belt.

17    Loosen the belt tensioner adjustment bolt, pry the tensioner to the left and retighten the bolt in this position **(see illustration)**. Slip the timing belt off the sprocket. If you're removing the belt for camshaft seal replacement or cylinder head removal, it isn't necessary to detach the belt from the crankshaft sprocket, or unbolt the right engine mount. If you are removing the belt completely, support the engine with a jack and remove the two nuts from the right engine mount (from underneath) and the bolt from above **(see illustrations)**. Lower the engine slightly and remove the engine mount bracket.

18    If the camshaft sprocket is worn or damaged, hold the front (exhaust) camshaft with a

large wrench and remove the bolt, then detach the sprocket (see Section 9).

19    Slip the timing belt off the crankshaft sprocket and remove it. If the sprocket is worn or damaged, or if you need to replace the crankshaft front oil seal, remove the sprocket from the crankshaft. Before removing it, make matching marks on the crankshaft sprocket and the oil pump case, so that you can find the TDC position without the crankshaft pulley in place **(see illustration)**.

### Inspection

*Refer to illustrations 7.21, 7.22, 7.23 and 7.25*
**Caution:** *Do not bend, twist or turn the timing belt inside out. Do not allow it to come in contact with oil, coolant or fuel. Do not utilize timing belt tension to keep the camshaft or crankshaft from turning when installing the sprocket bolts. Do not turn the crankshaft or camshaft more than a few degrees (if necessary for tooth alignment) while the timing belt is removed.*

20    If the timing belt broke during engine operation, the belt may have been contaminated or over-tightened. **Caution:** *If the timing belt broke during engine operation, the valves may have come in contact with the pistons, causing damage. Check the valve clear-*

ances (see Chapter 1) - *bent valves usually will have excessive clearance, indicating damage that will require cylinder head removal to repair.*

21    If the belt teeth are cracked or missing **(see illustration)**, the distributor, oil pump or camshafts may have seized.

22    If there is noticeable wear or cracks on the face of the belt, check to see if there are nicks or burrs on the idler pulleys **(see illustration)**.

23    If there is wear or damage on only one side of the belt, check the belt guide and the alignment of the sprockets **(see illustration)**.

24    Replace the timing belt with a new one if obvious wear or damage is noted or if it is the least bit questionable. Correct any problems which contributed to belt failure prior to belt installation. **Note:** *Professionals recommend*

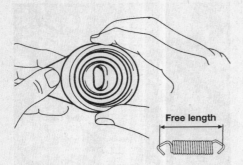

**7.25  Check the idler pulley bearing for smooth operation and measure the free length of the tension spring for comparison to this Chapter's Specifications**

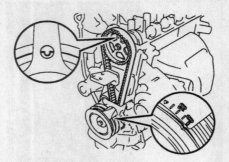

**7.27  The camshaft sprocket is at TDC when the hole in the sprocket lines up with the notch in the front bearing cap**

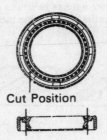

**8.2b  The seal may be removed more easily by carefully cutting the lip as indicated**

**8.2a  Wrap tape around the screwdriver tip and carefully work the crankshaft front oil seal out of the bore - DO NOT nick or scratch the crankshaft in the process!**

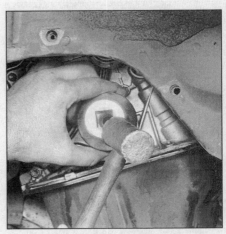

**8.4  Gently drive the new seal into place with the spring side installed toward the engine**

*replacing the belt whenever it is removed, since belt failure can lead to expensive engine damage. The manufacturer recommends replacing the belt at 60,000-mile intervals.*

25  Release the bolt on the belt tensioner, then remove the tensioner and its spring. Check the idler for free rotation and measure the spring's free length **(see illustration)**. Replace the spring if it doesn't meet Specifications. reinstall the tensioner and spring.

## Installation

*Refer to illustration 7.27*

26  Remove all dirt, oil and grease from the timing belt area at the front of the engine.

27  Recheck the camshaft and crankshaft timing marks to be sure they are properly aligned **(see illustration)**. If the camshaft sprocket had been removed, reinstall it with the timing mark hole in the sprocket aligned with the notch in the front bearing cap of the exhaust camshaft. Make sure the crankshaft sprocket is still aligned at the TDC marks you made in Step 19.

28  Install the timing belt on the crankshaft and camshaft sprockets. If the original belt is being reinstalled, align the marks made during removal.

29  Slip the belt guide onto the crankshaft with the cupped side facing out.

30  Reinstall the lower timing belt cover and crankshaft pulley and recheck the TDC marks.

31  Keeping tension on the side of the belt nearest the front of the vehicle, loosen the idler pulley bolt 1/2-turn, allowing the spring to apply pressure to the idler pulley.

32  Slowly turn the crankshaft clockwise two complete revolutions (720-degrees), then tighten the idler pulley mounting bolt to the torque listed in this Chapter's Specifications. Measure the deflection of the belt on the radiator side, halfway between the camshaft and crankshaft sprockets, and compare it to this Chapter's Specifications.

33  Recheck the timing marks. With the crankshaft at TDC for number one cylinder, the hole in the camshaft sprocket must align

with the timing mark **(see illustration 7.27)**. If the marks are not aligned exactly, repeat the belt installation procedure. **Caution:** *DO NOT start the engine until you're absolutely certain that the timing belt is installed correctly. Serious and costly engine damage could occur if the belt is installed out-of-time.*

34  Reinstall the remaining parts in the reverse order of removal.

35  Run the engine and check for proper operation.

## 8   Crankshaft front oil seal - replacement

*Refer to illustrations 8.2a, 8.2b and 8.4*

1  Remove the timing belt and crankshaft sprocket (see Section 7).

2  Note how far the seal is recessed in the bore, then carefully pry it out of the oil pump housing with a screwdriver or seal removal tool **(see illustration)**. Don't scratch the housing bore or damage the crankshaft in the process (if the crankshaft is damaged, the new seal will end up leaking). **Note:** *The seal may be easier to remove if the old seal lip is cut with a sharp utility knife first* **(see illustration).**

3  Clean the bore in the housing and coat the outer edge of the new seal with engine oil or multi-purpose grease. Apply moly-base grease to the seal lip.

4  Using a socket with an outside diameter slightly smaller than the outside diameter of the seal, carefully drive the new seal into place with a seal driver or large socket **(see illustration)**. Make sure it's installed squarely and driven in to the same depth as the original. If a socket isn't available, a short section of large diameter pipe will also work. Check the seal after installation to make sure the spring didn't pop out of place.

5  Reinstall the crankshaft sprocket and timing belt (see Section 7).

6  Run the engine and check for oil leaks at the front seal.

## 9   Camshaft oil seal - replacement

*Refer to illustrations 9.2, 9.3 and 9.5*

1  Refer to Section 4 and remove the valve cover, then remove the timing belt.

2  Use a large wrench to hold the exhaust camshaft at its hex portion, while removing

9.2 Use an open-end wrench to hold the camshaft while removing the camshaft sprocket bolt

9.3 Carefully pry the camshaft seal out of the bore - DO NOT nick or scratch the camshaft journal

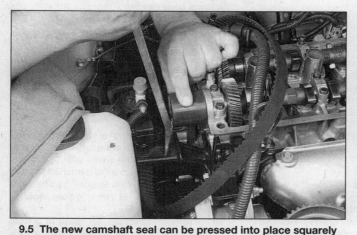

9.5 The new camshaft seal can be pressed into place squarely with an appropriate-size socket and a pry bar working against the engine mount

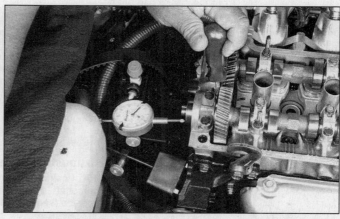

10.4 Mount a dial indicator as shown to measure camshaft endplay - with the dial zeroed, pry the camshaft forward and back and read the endplay on the dial

the bolt from the camshaft sprocket (see illustration), then pull off the sprocket.

3    Note how far the seal is seated in the bore, then carefully pry it out with a small screwdriver (see illustration). Don't scratch the bore or damage the camshaft in the process (if the camshaft is damaged, the new seal will end up leaking).

4    Clean the bore and coat the outer edge of the new seal with engine oil or multi-purpose grease. Apply multi-purpose grease to the seal lip.

5    Using a socket with an outside diameter slightly smaller than the outside diameter of the seal, carefully drive the new seal into place with a seal installer or large socket. Make sure it's installed squarely and driven in to the same depth as the original. If a socket isn't available, a short section of pipe will also work. Note: There isn't much room for a hammer, so you can also pry between the engine mount and the socket to press the seal in (see illustration).

6    Reinstall the camshaft sprocket and timing belt (see Section 7). Refer to Section 4 and reinstall the valve cover.

7    Run the engine and check for oil leaks at the camshaft seal.

## 10   Camshafts and valve lifters - removal, inspection and installation

Note: Before beginning this procedure, obtain two 6 x 1.0 mm bolts 16 to 20 mm long. They will be referred to as service bolts in the text.

### Removal

Refer to illustrations 10.4, 10.5, 10.6, 10.7, 10.10, 10.14a, 10.14b, 10.15, 10.16a and 10.16b

1    Remove the valve cover as described in Section 4.

2    Refer to Section 3 and place the engine on TDC for number 1 cylinder. Apply a dab of paint to the camshaft gear TDC alignment marks (the marks where the gears mesh just at the valve cover/head interface line). Note: There are two sets of marks on the camshaft gears. The marks that align at TDC are for TDC reference only, the other two marks are used to align the camshafts during installation.

3    Remove the distributor (see Chapter 5). Remove the timing belt and camshaft sprocket (see Sections 7 and 9).

4    Measure the camshaft thrust clearance (endplay) with a dial indicator (see illustration). If the clearance is greater than the service limit, replace the camshaft and/or the cylinder head.

### Intake camshaft

5    Position the knock pin in the EXHAUST camshaft just above the top of the cylinder head (see illustration). This will position the intake camshaft lobes so the camshaft will be

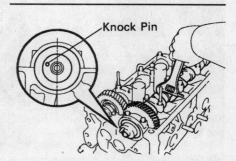

10.5 Place the EXHAUST camshaft knock pin between the 9 o'clock and 10 o'clock position

**10.6  Remove the two bolts and the front camshaft bearing cap from the intake camshaft**

**10.7  Install a service bolt through the sub-gear and thread it into the main gear (arrow)**

pushed up evenly by valve spring pressure. **Caution:** *This positioning is important to avoid damaging the cylinder head or camshaft as the camshaft is removed.*

6    Remove the two bolts and the front bearing cap from the intake camshaft **(see illustration)**.

7    Secure the intake camshaft sub-gear to the main gear by installing one of the service

**10.10  Turn the exhaust camshaft until the knock pin is pointed at about the 5 o'clock position**

bolts into the threaded hole **(see illustration)**.

8    Following the reverse of the tightening sequence **(see illustration 10.40)**, loosen the remaining intake camshaft bearing cap bolts in 1/4-turn increments until the bolts can be removed by hand. Lift the bearing caps straight up and off.

9    Lift the camshaft straight up and out of the cylinder head.

**Exhaust camshaft**

10    Position the knock pin in the exhaust camshaft at approximately the 5 o'clock position **(see illustration)**.

11    Remove the front (timing belt end) exhaust camshaft bearing cap bolts and detach the bearing cap and oil seal. **Caution:** *Do not pry the cap off. If it doesn't come loose easily, leave it in place without bolts.*

12    Following the reverse of the tightening sequence **(see illustration 10.30)**, loosen the remaining exhaust camshaft bearing cap bolts in 1/4-turn increments until the bolts can be removed by hand. Lift off the bearing caps. **Caution:** *As the center bearing cap*

*bolts are being loosened, make sure the camshaft is moving up evenly. If one end or the other stops moving and the cam gets cocked, start over by reinstalling the bearing caps and resetting the knock pin. DO NOT try to pry or force the camshaft out.*

13    Lift the camshaft straight up and out of the cylinder head.

14    Clean the oil off the valve lifter shims, mark them with a felt-tip marker and remove the lifters, keeping the shims with their lifters **(see illustration)**. Store the camshaft bearing caps, lifters and shims so they can be reinstalled without mixing them up **(see illustration)**.

15    Position the intake camshaft in a vise, clamping it on the hex portion. Using a two-pin spanner, rotate the sub-gear clockwise and remove the service bolt from the threaded hole, then allow the subgear to rotate back until all tension is relieved **(see illustration)**.

16    Remove the sub-gear snap-ring **(see illustration)**. The wave washer, sub-gear and camshaft gear spring can now be removed from the camshaft **(see illustration)**.

**10.14a  Mark the lifters/shims (I for intake, E for exhaust, and number their location) and remove them with a magnetic retrieval tool**

**10.14b  Mark up a cardboard box to store the lifters/shims and bearing caps**

10.15  With the hex portion of the intake camshaft clamped in a vise, use a pin spanner to relieve the tension on the service bolt, remove the bolt, then release the tension on the subgear

## Inspection

*Refer to illustrations 10.17, 10.18, 10.21, 10.22, 10.23a, 10.23b and 10.24*

17   Measure the free length (distance between the ends) of the camshaft gear spring **(see illustration)** and compare it to

10.16a  Remove the snap-ring with a pair of snap-ring pliers

this Chapter's Specifications. If not as specified, replace the spring.

18   Inspect each lifter for scuffing and score marks **(see illustration)**.

19   Measure the outside diameter of each lifter and the corresponding lifter bore inside diameter. Subtract the lifter diameter from the lifter bore diameter to determine the oil clearance. Compare it to this Chapter's Specifications. If the oil clearance is exces-

sive, a new cylinder head and/or new lifters will be required.

20   Visually examine the cam lobes and bearing journals for score marks, pitting, galling and evidence of overheating (blue, discolored areas). Look for flaking away of the hardened surface layer of each lobe.

21   Using a micrometer, measure the height of each camshaft lobe **(see illustration)**. Compare your measurements with this Chapter's Specifications. If the height for any one lobe is less than the specified minimum, replace the camshaft.

22   Using a micrometer, measure the diameter of each journal at several points **(see illustration)**. Compare your measurements with this Chapter's Specifications. If the diameter of any one journal is less than specified, replace the camshaft.

23   Check the oil clearance for each camshaft journal as follows:

a)  Clean the bearing caps and the camshaft journals with lacquer thinner or acetone.

b)  Carefully lay the camshaft(s) in place in the cylinder head. Don't install the lifters or intake camshaft subgear and don't use any lubrication.

c)  Lay a strip of Plastigage on each journal.

**2A**

10.16b  Remove the wave washer (1), the camshaft subgear (2) and the gear spring (3)

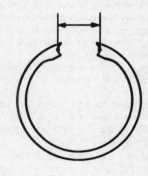

10.17  Measure the distance between the ends of the camshaft  gear spring

10.18  Wipe off the oil and inspect each lifter for wear and scuffing

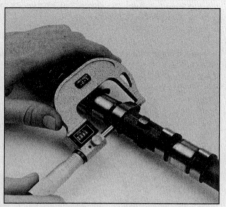

10.21  Measure the lobe heights on each camshaft - if any lobe height is less than the specified allowable minimum, replace that camshaft

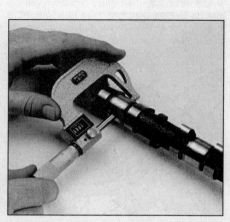

10.22  Measure each journal diameter with a micrometer (if any journal measures less than the specified limit, replace the camshaft)

**10.23a  The camshaft bearing caps are numbered and have an arrow that should face the timing belt end of the engine**

**10.23b  Compare the width of the crushed Plastigage to the scale on the envelope to determine the oil clearance**

**10.24  Position a dial indicator as shown here to measure gear backlash - hold one camshaft steady with a wrench while moving the other camshaft with another wrench**

d)  *Install the bearing caps with the arrows pointing toward the front (timing belt end) of the engine* **(see illustration)**.

e)  *Tighten the bolts to the torque listed in this Chapter's Specifications in 1/4-turn increments.* **Note:** *Don't turn the camshaft while the Plastigage is in place.*

f)  *Remove the bolts and detach the caps.*

g)  *Compare the width of the crushed Plastigage (at its widest point) to the scale on the Plastigage envelope* **(see illustration)**.

h)  *If the clearance is greater than specified, replace the camshaft and/or cylinder head.*

i)  *Scrape off the Plastigage with your fingernail or the edge of a credit card - don't scratch or nick the journals or bearing caps.*

24   With the caps reinstalled temporarily, use a dial indicator to measure the backlash between the two camshaft gears. Hold one camshaft from turning (using a wrench on the hex portion) while measuring the movement in the other camshaft gear, and compare the results to Specifications **(see illustration)**. If the backlash is beyond Specifications, replace both camshafts.

## Installation

*Refer to illustrations 10.28, 10.30, 10.37 and 10.40*

### Exhaust camshaft

25   Apply moly-base grease or engine assembly lube to the lifters, then install them in their original locations. Make sure the valve adjustment shims are on place on the lifters.

26   Apply moly-base grease or engine assembly lube to the camshaft lobes and bearing journals.

27   Position the exhaust camshaft in the cylinder head with the knock pin at approximately the 5 o'clock position **(see illustration 10.10)**.

28   Apply a thin coat of RTV sealant to the outer edge of the front bearing-cap-to-cylinder-head mating surface **(see illustration)**. **Note:** *The cap must be installed immediately or the sealer will dry prematurely.*

29   Install the bearing caps in numerical order with the arrows pointing toward the timing belt end of the engine.

30   Following the recommended tightening sequence **(see illustration)**, tighten the bearing cap bolts in 1/4-turn increments to the torque listed in this Chapter's Specifications.

31   Refer to Section 9 and install a new camshaft oil seal.

### Intake camshaft

32   Reassemble the intake camshaft sub-gear. Install the camshaft gear spring, sub-gear and wave washer. Secure them with the snap-ring **(see illustrations 10.16a and 10.16b)**.

33   Refer to Step 15 and install a service bolt.

34   Apply moly-base grease or engine assembly lube to the lifters, then install them in their original locations. Make sure the valve adjustment shims are in place on the lifters.

35   Apply moly-base grease or engine assembly lube to the camshaft lobes and bearing journals.

36   Rotate the EXHAUST camshaft until the knock pin is positioned between the 9 o'clock and 10 o'clock positions **(see illustration 10.5)**.

37   Align the intake camshaft gear with the exhaust camshaft gear by matching up the *"installation"* marks on the gears **(see illustration)**. **Caution:** *There are two sets of marks. Do not use the TDC marks you applied paint to in Step 2. The camshaft installation marks are the only marks that are duplicated on both front and rear sides of the two camshaft gears.*

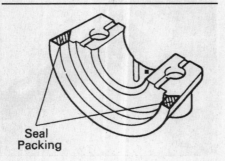

**10.28  Apply sealer to the shaded areas of the front exhaust camshaft bearing cap**

Seal
Packing

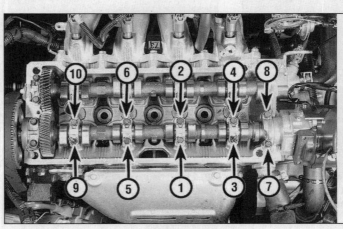

**10.30  Exhaust camshaft bearing cap bolt TIGHTENING sequence**

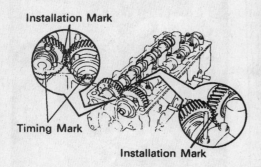

**10.37 Align the camshaft gears as shown here - use the installation marks, NOT the TDC marks**

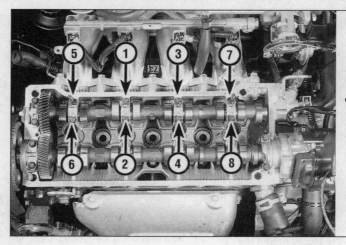

**10.40 INTAKE camshaft bearing cap bolt tightening sequence**

38    Roll the intake camshaft down into position. Turn the exhaust camshaft back and forth a little until the intake camshaft sits in the bearings evenly.

39    Install the intake camshaft bearing caps in numerical order with the arrows pointing toward the timing belt end of the engine.

40    Following the recommended sequence **(see illustration)**, tighten the bearing cap bolts in 1/4-turn increments to the torque listed in this Chapter's Specifications.

41    Remove the service bolt from the intake camshaft gear.

42    Install the timing belt pulley on the exhaust camshaft and tighten the bolt to the torque listed in this Chapter's Specifications. Prevent the camshaft from turning by holding it with a wrench on the large hex.

43    Install the timing belt (see Section 7).

44    The remainder of installation is the reverse of the removal procedure.

## 11  Cylinder head - removal and installation

**Note:** *The engine must be completely cool before beginning this procedure.*

### Removal

*Refer to illustration 11.14*

1    Disconnect the negative cable from the battery. **Caution:** *If the stereo in your vehicle is equipped with an anti-theft system, make sure you have the correct activation code before disconnecting the battery.*

2    Drain the coolant from the engine block and radiator (see Chapter 1).

3    Drain the engine oil and remove the oil filter (see Chapter 1).

4    Remove the throttle body, fuel injectors and fuel rail (see Chapter 4).

5    Remove the intake manifold (see Section 5).

6    Remove the exhaust manifold (see Section 6). **Note:** *It is possible to leave the intake and exhaust manifolds attached to the cylinder head, to be removed along with the cylinder head for disassembly on the bench.*

7    Remove the timing belt and camshaft sprocket (see Sections 7 and 9).

8    Remove the camshafts and lifters (see Section 10).

9    Remove the alternator and distributor (see Chapter 5).

10    Unbolt the upper bracket of the power steering pump and set the pump aside without disconnecting the hoses.

11    Label and remove any remaining items, such as coolant fittings, tubes, cables, hoses or wires.

12    Refer to Chapter 3 and detach the water necks from each end of the cylinder head. At this point the cylinder head should be ready for removal.

13    Using an 8 mm hex-head socket bit and a breaker bar, loosen the cylinder head bolts in 1/4-turn increments until they can be removed by hand. Loosen the cylinder head bolts opposite of the recommended tightening sequence **(see illustration 11.25)** to avoid warping or cracking the cylinder head.

14    Lift the cylinder head off the engine block. If it's stuck, very carefully pry up at the transaxle end, beyond the gasket surface **(see illustration)**.

15    Remove all external components from the cylinder head to allow for thorough cleaning and inspection. See Chapter 2, Part C, for cylinder head servicing procedures.

### Installation

*Refer to illustrations 11.17, 11.22 and 11.25*

16    The mating surfaces of the cylinder head and block must be perfectly clean when the cylinder head is installed.

17    Use a gasket scraper to remove all traces of carbon and old gasket material **(see**

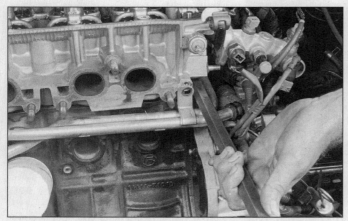

**11.14 If the cylinder head is stuck, pry only at the overhang, not between the mating surfaces**

**11.17 Remove all traces of old gasket material - the cylinder head and block mating surfaces must be perfectly clean to ensure a good gasket seal**

**11.22 Place the new cylinder head gasket over the dowels in the block, noting the markings for UP on the gasket**

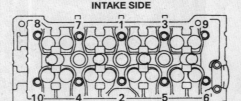

INTAKE SIDE

EXHAUST SIDE

**11.25 Cylinder head bolt TIGHTENING sequence**

**12.8 Remove the front bolts of the chassis brace (arrows) - for better access for oil pan removal, you'll have to insert a large pry bar between the brace and the chassis to pry the brace down**

illustration), then clean the mating surfaces with lacquer thinner or acetone. If there's oil on the mating surfaces when the cylinder head is installed, the gasket may not seal correctly and leaks could develop. When working on the block, stuff the cylinders with clean shop rags to keep out debris. Use a vacuum cleaner to remove material that falls into the cylinders.

18    Check the block and cylinder head mating surfaces for nicks, deep scratches and other damage. If damage is slight, it can be removed with a file; if it's excessive, machining may be the only alternative.

19    Use a tap of the correct size to chase the threads in the cylinder head bolt holes, then clean the holes with compressed air - make sure that nothing remains in the holes. **Warning:** *Wear eye protection when using compressed air!*

20    Mount each bolt in a vise and run a die down the threads to remove corrosion and restore the threads. Dirt, corrosion, sealant and damaged threads will affect torque readings.

21    Install the components that were removed from the cylinder head.

22    Position the new gasket over the dowel pins in the block **(see illustration).**

23    Carefully set the cylinder head on the block without disturbing the gasket.

24    Before installing the cylinder head bolts, apply a small amount of clean engine oil to the threads and under the bolt heads.

25    Install the bolts in their original locations and tighten them finger tight. Install the shorter bolts along the intake side of the cylinder head and the longer bolts along the exhaust side. Following the recommended sequence, tighten the bolts in three steps to the torque listed in this Chapter's Specifications **(see illustration).** Steps 2 and 3 of the tightening sequence each require the bolts to be tightened an additional 90 degrees. If you don't have an angle-torque attachment for your torque wrench, simply apply a paint mark at one edge of each cylinder head bolt and tighten the bolt until that mark is 90

degrees from where you started (Step 2). After Step 3, the marks will be 180 degrees from where they started.

26    The remaining installation steps are the reverse of removal. **Note:** *If the semi-circular rubber plug had been removed from the cylinder head (ahead of the intake camshaft), reinstall it with some RTV sealant.*

27    Check and adjust the valves as necessary (see Chapter 1).

28    Refill the cooling system, install a new oil filter and add oil to the engine (see Chapter 1).

29    Run the engine and check for leaks. Set the ignition timing (see Chapter 5) and road test the vehicle.

## 12   Oil pan - removal and installation

### Removal

*Refer to illustration 12.8*

1    Disconnect the negative cable from the battery. **Caution:** *If the stereo in your vehicle is equipped with an anti-theft system, make sure you have the correct activation code before disconnecting the battery.*

2    Set the parking brake and block the rear wheels.

3    Raise the front of the vehicle and support it securely on jackstands.

4    Remove the two plastic splash shields

under the engine.

5    Drain the engine oil and remove the oil filter (see Chapter 1). Remove the oil dipstick.

6    Disconnect the electrical connector to the oxygen sensor (see Chapter 6).

7    Remove the two nuts retaining the front exhaust pipe to the exhaust manifold, then the two bolts/nuts connecting the pipe to the rear of the exhaust system (see Chapter 4).

8    Remove the two front bolts of the longitudinal chassis brace and the two nuts securing the radiator-side engine mount **(see illustration).**

### 1.6L engine

*Refer to illustration 12.10*

9    Unbolt and remove the block-to-transaxle brace.

10    Remove the bolts and detach the oil pan. If it's stuck, pry it loose very carefully with a small screwdriver or putty knife **(see illustration).** Don't damage the mating surfaces of the pan and block or oil leaks could develop.

**12.10 Carefully pry the oil pan away from the block - if the mating surfaces are damaged, oil leaks could develop**

**12.12 Remove the bolts and nuts securing the oil pan to the reinforcement section, then break its seal with a putty knife**

**12.13 Remove the oil baffle plate and the pick-up tube/strainer**

**12.15a Unbolt the six fasteners inside the area concealed by the oil pan - one is an Allen-head bolt (arrow)**

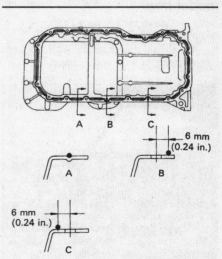

**12.15b To access the bolts between the reinforcement section and the transaxle (arrows), you'll need to bend the exhaust pipe down at its flex joint and pry the longitudinal chassis brace down**

### 1.8L engine

*Refer to illustrations 12.12, 12.13, 12.15a and 12.15b*

11   The 1.8L engine uses a steel oil pan mounted to a large cast aluminum reinforcement section that stiffens the block/transaxle mounting. There is no separate engine/transaxle brace as on the 1.6L engine. **Note:** *The manufacturer uses a very tough sealant on the oil pan flange which make the steel and aluminum components difficult to separate. Take your time and use a putty knife rather than a screwdriver to break the seal.*

12   Remove the bolts and stud nuts around the perimeter of the oil pan, then carefully pry between the steel and aluminum sections to loosen the oil pan **(see illustration)**. Be careful not to gouge the softer aluminum or distort the flange of the steel pan.

13   Remove the two bolts and two nuts retaining the oil baffle and remove the baffle **(see illustration)**.

14   Unbolt the pick-up tube/oil strainer assembly and remove it for cleaning.

15   If necessary, remove the bolts and nuts retaining the aluminum reinforcement section to the block and transaxle **(see illustrations)**.

Carefully pry the aluminum section down without gouging the sealing surface. Six of the fasteners are inside the area originally covered by the oil pan. **Caution:** *Do not pry on the reinforcement section until you are certain all fasteners are removed. There are two Allen bolts near the right front which will require using a ball-hex wrench to remove. If you don't have this type wrench (which allows you to turn an Allen bolt from a slight angle), you will have to remove the air conditioning compressor and bracket to allow using a L-shaped Allen wrench.*

## Installation

*Refer to illustrations 12.20a, 12.20b and 12.20c*

16   Use a scraper to remove all traces of old gasket material and sealant from the block and oil pan. Clean the mating surfaces with lacquer thinner or acetone.

17   Make sure the threaded bolt holes in the block are clean.

18   Check the oil pan flange for distortion, particularly around the bolt holes. If necessary, place the oil pan on a wood block and use a hammer to flatten and restore the gas-

ket surface.

19   Inspect the oil pump pick-up tube assembly for cracks and a blocked strainer. On 1.6L models, if the pick-up was removed, clean it thoroughly and install it now, using a new O-ring or gasket. Tighten the nuts/bolts to the torque listed in this Chapter's Specifications.

20   Apply a 5 mm wide bead of RTV sealant to the oil pan flange (1.6L engine) or reinforcement section (1.8L engine) **(see illustrations)**. **Note:** *Installation must be completed within 5 minutes once the sealer has been applied.*

21   Carefully position the oil pan or reinforcement section on the engine block and install the bolts. Working from the center out. Tighten the bolts to the torque listed in this Chapter's Specifications in three or four steps. On 1.8L engines, reinstall the oil baffle plate and the pick-up tube/strainer. Be sure to use a new gasket on the pick-up tube. On 1.8L engines, apply RTV to the oil pan flange

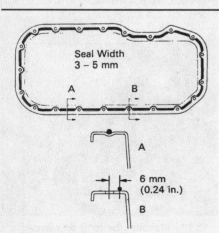

**12.20a RTV sealant application details - 1.6L engine oil pan**

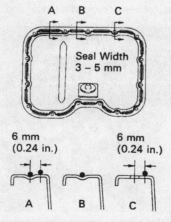

**12.20b RTV sealant application details - 1.8L engine oil pan**

**12.20c RTV sealant application details - 1.8L engine aluminum reinforcement section**

and install it after the aluminum reinforcement section has been sealed and tightened to specifications.

22   The remainder of installation is the reverse of removal. Be sure to add oil and install a new oil filter. Use new gasket/seals on the front exhaust pipe.

23   Run the engine and check for oil pressure and leaks.

## 13   Oil pump - removal, inspection and installation

### *Removal*

*Refer to illustrations 13.3, 13.4 and 13.6*

1   Remove the oil dipstick, oil pan, baffle plate (1.8L only) and the oil pick-up/strainer assembly (see Section 12).

2   Support the engine securely from above and remove the timing belt, crankshaft pulley and sprocket (see Section 7).

3   On 1996 and some 1995 models, remove the crankshaft position sensor **(see illustration)**. Unbolt and remove the oil dipstick pipe (see Chapter 3).

4   Remove the seven bolts and detach the oil pump body from the engine **(see illustration)**. You may have to pry carefully between the front main bearing cap and the pump body with a screwdriver, or tap behind the oil pump with a soft-faced hammer.

5   Use a scraper to remove all traces of sealant and old gasket material from the pump body and engine block, then clean the mating surfaces with lacquer thinner or acetone.

6   Remove the five Torx screws and separate the pump cover from the body. Lift out the drive and driven rotors **(see illustration)**.

7   Remove the oil pressure relief valve snap-ring retainer, spring and piston. **Warning:** *The spring is tightly compressed - be careful and wear eye protection.*

### *Inspection*

*Refer to illustrations 13.10a, 13.10b and 13.10c*

8   Clean all components with solvent, inspect them for wear and damage.

**13.3  Remove the bolt (arrow) and withdraw the crankshaft position sensor from the oil pump body**

**13.4  Remove the oil pump body-to-block bolts (arrows)**

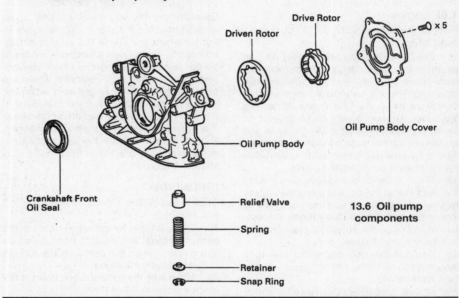

**13.6  Oil pump components**

Driven Rotor

Drive Rotor

x 5

Oil Pump Body Cover

Oil Pump Body

Crankshaft Front Oil Seal

Relief Valve

Spring

Retainer

Snap Ring

9   Check the oil pressure relief valve piston sliding surface and valve spring. If either the spring or the valve is damaged, they must be replaced as a set.

10   Check the driven rotor-to-body clearance, rotor-to-cover clearance and drive rotor tip clearance with a feeler gauge **(see illustrations)** and compare the results to this Chapter's Specifications. If any clearance is excessive, replace the rotors as a set. If necessary, replace the oil pump body.

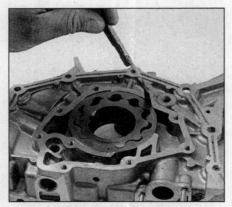

**13.10a  Measure the driven rotor-to-body clearance with a feeler gauge**

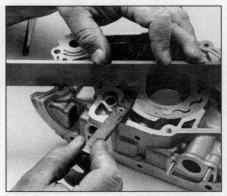

**13.10b  Using a straightedge and feeler gauge, measure the rotor-to-cover clearance**

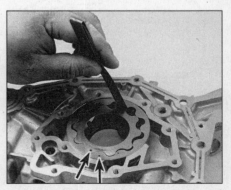

**13.10c  Measure the rotor tip clearance with a feeler gauge - install the rotors with the marks facing out, against the cover (arrows)**

13.15 When installing the oil pump, align the flats in the pump rotor with the flats on the crankshaft (arrows)

14.3 Mark the flywheel/driveplate and the crankshaft so they can be reassembled in the same relative positions

14.5 On vehicles equipped with a spacer plate, note the position of the locating pin (arrow)

## Installation

*Refer to illustration 13.15*

11    Lubricate the drive and driven rotors with clean engine oil and place them in the pump body with the marks facing out (see illustration 13.10c).

12    Pack the pump cavity with petroleum jelly and attach the pump cover, tighten the screws to the torque listed in this Chapter's Specifications.

13    Lubricate the oil pressure relief valve piston with clean engine oil and reinstall the valve components in the pump body.

14    Place a new gasket on the engine block (the dowel pins should hold it in place).

15    Position the pump assembly against the block and install the mounting bolts. Make sure that the flats on the oil pump drive rotor aligns with the flats on the crankshaft (see illustration).

16    Tighten the bolts to the torque listed in this Chapter's Specifications in three or four steps. Follow a criss-cross pattern to avoid warping the body.

17    Reinstall the remaining parts in the reverse order of removal.

18    Using a new gasket, install the oil pick-up tube/strainer assembly and baffle plate (see Section 12). Tighten the fasteners to the torque listed in this Chapter's Specifications. When reinstalling the dipstick tube, use a new O-ring on the oil-pump end.

19    Add oil, start the engine and check for oil pressure and leaks.

20    Recheck the engine oil level.

## 14    Flywheel/driveplate - removal and installation

## Removal

*Refer to illustrations 14.3 and 14.5*

1    Raise the vehicle and support it securely on jackstands, then refer to Chapter 7 and remove the transaxle.

2    Remove the pressure plate and clutch disc (Chapter 8) (manual transaxle equipped

vehicles).

3    Use a center punch or paint to make alignment marks on the flywheel/driveplate and crankshaft to ensure correct alignment during reinstallation (see illustration).

4    Remove the bolts that secure the flywheel/driveplate to the crankshaft. If the crankshaft turns, wedge a screwdriver in the ring gear teeth to jam the flywheel (see illustration 7.12).

5    Remove the flywheel/driveplate from the crankshaft. Since the flywheel is fairly heavy, be sure to support it while removing the last bolt. Automatic transaxle equipped vehicles have spacers on both sides of the driveplate (see illustration). Keep them with the drive-plate. **Warning:** *The ring-gear teeth may be sharp, wear gloves to protect your hands.*

## Installation

6    Clean the flywheel to remove grease and oil. Inspect the surface for cracks, rivet grooves, burned areas and score marks. Light scoring can be removed with emery cloth. Check for cracked and broken ring gear teeth. Lay the flywheel on a flat surface and use a straightedge to check for warpage.

7    Clean and inspect the mating surfaces of the flywheel/driveplate and the crankshaft. If the crankshaft rear seal is leaking, replace it before reinstalling the flywheel/driveplate (see Section 15).

8    Position the flywheel/driveplate against the crankshaft. Be sure to align the marks made during removal. Note that some engines have an alignment dowel or staggered bolt holes to ensure correct installation. Before installing the bolts, apply thread-locking compound to the threads.

9    Wedge a screwdriver in the ring gear teeth to keep the flywheel/driveplate from turning and tighten the bolts to the torque listed in this Chapter's Specifications. Follow a criss-cross pattern and work up to the final torque in three or four steps.

10    The remainder of installation is the reverse of the removal procedure.

## 15    Rear main oil seal - replacement

*Refer to illustrations 15.2, 15.5 and 15.6*

1    Remove the transaxle (see Chapter 7). Remove the rear end plate.

2    The seal can be replaced without removing the oil pan or seal retainer. However, this method is not recommended because the lip of the seal is quite stiff and it's possible to cock the seal in the retainer bore or damage it during installation. If you want to take the chance, pry out the old seal with a screwdriver (see illustration). Apply multi-purpose grease to the crankshaft seal journal and the lip of the new seal and carefully push the new seal into place. The lip is stiff so carefully work it onto the seal journal of the crankshaft with a smooth object like the end of an extension as you tap the seal into place. Don't rush it or you may damage the seal.

3    The following method is recommended but requires resealing the rear of the oil pan

15.2 The quick way to replace the rear crankshaft oil seal is to simply pry the old one out with a screwdriver, lubricate the crankshaft journal and the lip of the new seal with moly-base grease and push the new seal into place - the seal lip is very stiff and can be easily damaged during installation if you're not careful

(see Section 12) and removing the seal retainer.

4    After removing the two rearmost oil pan bolts that go into the seal retainer, break the seal between the rear of the oil pan and the bottom of the seal retainer with a putty knife. Remove the bolts, detach the seal retainer and remove all the old gasket material. remove the sealant from the top of the oil pan flange. **Note:** *Cover the open area of the oil pan with clean rags to keep debris out while bracing the pan flange.*

5    Position the seal and retainer assembly between two wood blocks on a workbench and drive the old seal out from the back side with a screwdriver **(see illustration)**.

6    Drive the new seal into the retainer with a wood block **(see illustration)** or a section of pipe slightly smaller in diameter than the outside diameter of the seal.

7    Lubricate the crankshaft seal journal and the lip of the new seal with multi-purpose grease. Position a new gasket on the engine block. Apply a bead of RTV sealant to the exposed portion of oil pan flange and particularly at the pan-to-block mating surface.

8    Slowly and carefully push the seal and retainer onto the crankshaft. The seal lip is stiff, so work it onto the crankshaft with a smooth object such as the end of an extension as you push the retainer against the block.

9    Install and tighten the retainer bolts to the torque listed in this Chapter's Specifications.

10   The remainder of installation is the reverse of removal.

## 16   Engine mounts - check and replacement

1    Engine mounts seldom require attention, but broken or deteriorated mounts should be replaced immediately or the added strain placed on the driveline components may cause damage or wear.

### Check

*Refer to illustration 16.3*

**Note:** *During the check, the engine must be raised slightly to remove the weight from the mounts.*

2    Raise the vehicle and support it securely on jackstands, then position a jack under the engine oil pan. Place a large wood block between the jack head and the oil pan, then carefully raise the engine just enough to take the weight off the mounts. Do not position the wood block under the drain plug. **Warning:** *DO NOT place any part of your body under the engine when it's supported only by a jack!*

3    Check the mounts to see if the rubber is cracked, hardened or separated from the metal casing, which would indicate the need for replacement **(see illustration)**.

4    Check for relative movement between the inner and outer portions of the mount (use a large screwdriver or pry bar to attempt to move the mounts). If movement is noted, replace the mount.

**15.5  After removing the retainer assembly from the block, support it between two wooden blocks and drive out the old seal with a screwdriver and hammer**

**15.6  Drive the new seal into the retainer with a wood block or a section of pipe - make sure that you don't cock the seal in the retainer bore**

5    Check the mount fasteners to make sure they are tight.

6    Rubber preservative should be applied to the mounts to slow deterioration.

### Replacement

*Refer to illustrations 18.8a, 18.8b, 18.9, 18.10a and 18.10b*

7    Disconnect the cable from the negative terminal of the battery, then raise the vehicle and support it securely on jackstands (if not already done). Support the engine as described in Step 3. **Caution:** *If the stereo in your vehicle is equipped with an anti-theft system, make sure you have the correct activation code before disconnecting the battery.*

8    To remove the right (passenger side) engine mount, remove the three bolts securing the mount to the body, then remove the three bolts and two nuts securing the mount bracket to the engine bracket **(see illustra-**tions)**. If the vehicle is equipped with air conditioning, detach the refrigerant line bracket retaining nut and remove the mount from the engine compartment. Be sure to remove the mount bracket from the old mount and reinstall it on the new mount.

9    Remove the mount-to-chassis nuts and detach the mount.

10   To remove the rear engine mount, pull the rubber plugs from the longitudinal chassis brace to access the two nuts.

11   To remove the front engine mount, remove the two nuts retaining the insulator to the longitudinal chassis brace **(see illustration 12.8)**, then the three nuts retaining the insulator to the chassis.

12   Installation is the reverse of removal. Use thread locking compound on the mount bolts/nuts and be sure to tighten them securely.

13   See Chapter 7 for transaxle mount replacement.

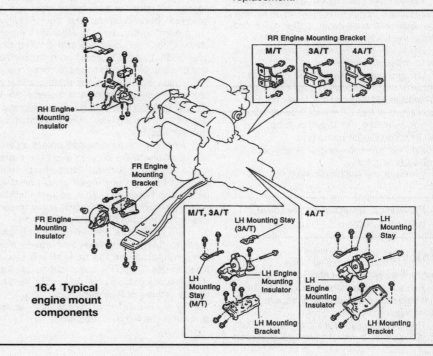

**16.4  Typical engine mount components**

RR Engine Mounting Bracket

| M/T | 3A/T | 4A/T |

RH Engine Mounting Insulator

FR Engine Mounting Bracket

FR Engine Mounting Insulator

M/T, 3A/T

LH Mounting Stay (3A/T)

LH Mounting Stay (M/T)

LH Engine Mounting Insulator

LH Mounting Bracket

4A/T

LH Mounting Stay

LH Engine Mounting Insulator

LH Mounting Bracket

# Chapter 2  Part B
# Engines - 1998 and later

## Contents

## Specifications

### General

| | |
|---|---|
| Engine type | DOHC, inline four-cylinder, four valves per cylinder |
| Cylinder numbers (drivebelt end-to-transaxle end) | 1-2-3-4 |
| Firing order | 1-3-4-2 |
| Displacement | |
| 1ZZ-FE | 1.8 liters (108 cu. in.) |

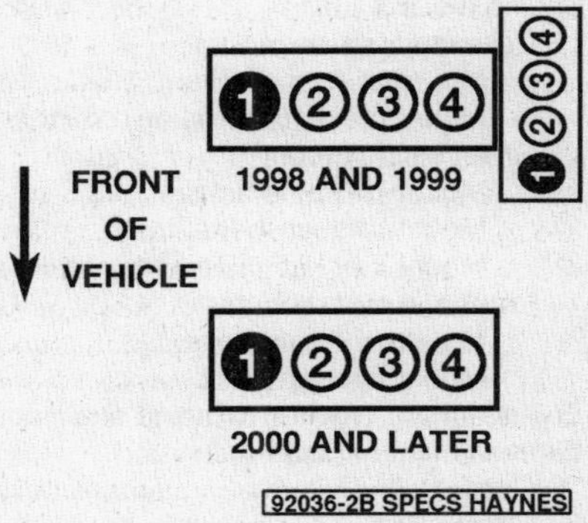

### Timing chain

| | |
|---|---|
| Timing chain sprocket wear limit | |
| Camshaft sprocket(s) (w/chain) | 3.831 inches |
| Crankshaft sprocket (w/chain) | 2.031 inches |
| Timing chain stretch limit | |
| 8 links (16 pins) | 4.827 inches |
| Timing chain guide wear limit | 0.039 inch |

FRONT OF VEHICLE

1998 AND 1999

2000 AND LATER

92036-2B SPECS HAYNES

**Cylinder numbering and coil terminal
identification diagram**

## Camshaft and lifters

Journal diameter
   No. 1 journal ........................................................................... 1.3563 to 1.3569 inch
   All others ................................................................................ 0.9035 to 0.9041 inches
Bearing oil clearance
   Standard ................................................................................. 0.0014 to 0.0028 inch
   Service limit (maximum) ......................................................... 0.0039 inch
Runout limit................................................................................. 0.0012 inch
Thrust clearance (endplay)
   Standard ................................................................................. 0.0016 to 0.0037 inch
   Service limit (maximum) ......................................................... 0.0043 inch
Lobe height
   Intake camshaft
      Standard ........................................................................ 1.7454 to 1.7493 inches
      Service limit (minimum) ................................................... 1.7394 inches
   Exhaust camshaft
      Standard ........................................................................ 1.7229 to 1.7268 inches
      Service limit (minimum) ................................................... 1.7169 inches
Valve lifter
   Diameter ................................................................................. 1.2191 to 1.2195 inches
   Bore diameter ......................................................................... 1.2205 to 1.2215 inches
Lifter oil clearance
   Standard ................................................................................. 0.0009 to 0.0023 inch
   Service limit ............................................................................ 0.0031 inch

## Oil pump

Rotor-to-body clearance
   Standard.................................................................................. 0.0039 to 0.0071 inch
   Service limit ............................................................................ 0.0118 inch
Rotor tip clearance
   Standard ................................................................................. 0.0024 to 0.0071 inch
   Service limit ............................................................................ 0.0138 inch
Rotor side clearance
   Standard ................................................................................. 0.0010 to 0.0030 inch
   Service limit ............................................................................ 0.0059 inch

## Torque specifications                    **Ft-lbs** (unless otherwise indicated)

Camshaft bearing cap bolts
   Journal No.1 ............................................................................ 17
   All others ................................................................................ 120 in-lbs
Camshaft sprocket bolts
   1998 ....................................................................................... 33
   1999 and later ........................................................................ 40
Crankshaft pulley/vibration damper bolt ..................................... 102
Cylinder head bolts
   Step 1 ..................................................................................... 18
   Step 2 ..................................................................................... 36
   Step 3 ..................................................................................... Tighten an additional 90-degrees
Drivebelt tensioner
   Bolt ......................................................................................... 51
   Nut.......................................................................................... 21
Engine mounts
   Passenger's side mount
      Mount-to-frame bolts.................................................... 40
      Mount bracket-to engine mount bracket nuts/bolts..... 40
   Driver's side mount
      Mount-to-frame bolts.................................................... 44
      Mount through-bolt........................................................ 44
      Mount bracket (manual transaxle) ............................... 15
   Front and rear mounts
      Mount-to-frame bolts.................................................... 44
      Mount through-bolt........................................................ 44
Engine mount bracket-to timing chain cover bolts ...................... 35
Exhaust manifold nuts/bolts
   Corolla .................................................................................... 27
   Prizm ...................................................................................... 36
Exhaust manifold brace bolts
   Corolla .................................................................................... 37
   Prizm ...................................................................................... 24

Exhaust manifold heat shield bolts
    Corolla ........................................................................ 108 in-lbs
    Prizm .......................................................................... 132 in-lbs
Exhaust pipe-to-exhaust manifold bolts ............................ 46
Flywheel/driveplate bolts
    Manual transaxle
        Step 1 ................................................................ 36
        Step 2 ................................................................ Tighten an additional 90-degrees
    Automatic transaxle .................................................. 61
Intake manifold nuts/bracket bolts ................................... 164 in-lbs
Oil pump bolts
    Pump cover-to-body screws ...................................... 93 in-lbs
    Relief valve plug ....................................................... 27
    Oil pump-to-engine block .......................................... 80 in-lbs
Oil pick-up/strainer nuts/bolts ........................................ 80 in-lbs
Oil pan bolts ................................................................. 80 in-lbs
Timing chain guide bolts (stationary) ............................... 80 in-lbs
Timing chain tensioner pivot arm bolt .............................. 164 in-lbs
Timing chain cover bolts **(see illustrations 7.14a and 7.14b)**
    10 mm head ............................................................. 89 in-lbs
    12 mm head ............................................................. 164 in-lbs
Timing chain tensioner nuts/bolts ................................... 80 in-lbs
Valve cover .................................................................. 89 in-lbs
Water pump mounting bolts
    10 mm head ............................................................. 89 in-lbs
    12 mm head ............................................................. 164 in-lbs

**2B**

## 1  General information

This Part of Chapter 2 is devoted to in-vehicle repair procedures for the 1998 and later 1.8L four-cylinder engine. All information concerning engine removal and installation and engine block and cylinder head overhaul can be found in Part C of this Chapter.

The following repair procedures are based on the assumption that the engine is installed in the vehicle. If the engine has been removed from the vehicle and mounted on a stand, many of the steps outlined in this Part of Chapter 2 will not apply.

The Specifications included in this Part of Chapter 2 apply only to the procedures contained in this Part. Part C of Chapter 2 contains the Specifications necessary for cylinder head and engine block rebuilding.

The 1998 and later four-cylinder engines in the Corolla and the Geo Prizm models are designated 1ZZ-FE (1.8L). This engine is an all new design from the previous model years; it incorporates an aluminum cylinder block with a bedplate to strengthen the lower half of the block. Although the cylinder head utilizes the usual dual overhead camshafts (DOHC) with four valves per cylinder as in previous models years, it is of new design also. The camshafts are driven from a single timing chain off the crankshaft, and 2000 and later models are equipped with Variable Valve Timing (VVT) to increase horsepower and decrease emissions.

## 2  Repair operations possible with the engine in the vehicle

Many major repair operations can be accomplished without removing the engine from the vehicle.

Clean the engine compartment and the exterior of the engine with some type of degreaser before any work is done. It will make the job easier and help keep dirt out of the internal areas of the engine.

Depending on the components involved, it may be helpful to remove the hood to improve access to the engine as repairs are performed (refer to Chapter 11 if necessary). Cover the fenders to prevent damage to the paint. Special pads are available, but an old bedspread or blanket will also work.

If vacuum, exhaust, oil or coolant leaks develop, indicating a need for gasket or seal replacement, the repairs can generally be made with the engine in the vehicle. The intake and exhaust manifold gaskets, oil pan gasket, crankshaft oil seals and cylinder head gasket are all accessible with the engine in place.

Exterior engine components, such as the intake and exhaust manifolds, the oil pan, the oil pump, the water pump, the starter motor, the alternator and the fuel system components can be removed for repair with the engine in place.

Since the cylinder head can be removed without pulling the engine, camshaft and valve component servicing can also be accomplished with the engine in the vehicle. Replacement of the timing chain and sprockets is also possible with the engine in the vehicle.

In extreme cases caused by a lack of necessary equipment, repair or replacement of piston rings, pistons, connecting rods and rod bearings is possible with the engine in the vehicle. However, this practice is not recommended because of the cleaning and preparation work that must be done to the components involved.

## 3  Top Dead Center (TDC) for number one piston - locating

*Refer to illustrations 3.5 and 3.8*

1   Top Dead Center (TDC) is the highest point in the cylinder that each piston reaches as it travels up the cylinder bore. Each piston reaches TDC on the compression stroke and again on the exhaust stroke, but TDC generally refers to piston position on the compression stroke.

2   Positioning the piston(s) at TDC is an essential part of many procedures such as valve timing, camshaft and timing chain/sprocket removal.

3   Before beginning this procedure, be sure to place the transmission in Neutral and apply the parking brake or block the rear wheels. Also, disable the ignition system by disconnecting the primary electrical connectors at the ignition coils and remove the spark

**3.5  A compression gauge can be used in the number one plug hole to assist in finding TDC**

**3.8  Align the groove in the damper with the "0" mark on the front timing chain cover**

9    After the number one piston has been positioned at TDC on the compression stroke, TDC for any of the remaining cylinders can be located by turning the crankshaft 180 degrees and following the firing order (refer to the Specifications). Rotating the engine 180 degrees past TDC #1 will put the engine at TDC compression for cylinder #3.

## 4    Valve cover - removal and installation

### Removal

*Refer to illustrations 4.2a, 4.2b and 4.6*

1    Disconnect the cable from the negative terminal of the battery. **Caution:** *If the stereo in your vehicle is equipped with an anti-theft system, make sure you have the correct activation code before disconnecting the battery.*
2    Remove the engine cover or the fuel injector wire harness cover **(see illustrations).**
3    On 1998 and 1999 models, remove the spark plug wires from the spark plugs and position them aside. Handle the spark plug wires by the boots only - do not pull on the wires.
4    On 2000 and later models, disconnect the electrical connectors from the ignition coils, remove the nuts securing the wiring harness to the valve cover and position the ignition coil wiring harness aside. Then remove the ignition coil pack from each of the spark plugs (see Chapter 5).
5    Detach the PCV hoses from the valve cover.
6    Remove the valve cover mounting nuts, then detach the valve cover and gasket from the cylinder head **(see illustration).** If the valve cover is stuck to the cylinder head, bump the end with a wood block and a hammer to jar it loose. If that doesn't work, try to slip a flexible putty knife between the cylinder head and valve cover to break the seal. **Caution:** *Don't pry at the valve cover-to-cylinder head joint or damage to the sealing surfaces*

plugs (see Chapter 1).
4    In order to bring any piston to TDC, the crankshaft must be turned using one of the methods outlined below. When looking at the front of the engine, normal crankshaft rotation is clockwise.

  a) *The preferred method is to turn the crankshaft with a socket and ratchet attached to the bolt threaded into the front of the crankshaft. Turn the bolt in a clockwise direction only.*
  b) *A remote starter switch, which may save some time, can also be used. Follow the instructions included with the switch. Once the piston is close to TDC, use a socket and ratchet as described in the previous paragraph.*
  c) *If an assistant is available to turn the ignition switch to the Start position in short bursts, you can get the piston close to TDC without a remote starter switch. Make sure your assistant is out of the vehicle, away from the ignition switch, then use a socket and ratchet as described in Paragraph (a) to complete the procedure.*

5    Install a compression gauge in the num-

ber one spark plug hole **(see illustration).** It should be a gauge with a screw-in fitting and a hose at least six inches long.
6    Rotate the crankshaft using one of the methods described above while observing for pressure on the compression gauge. The moment the gauge shows pressure indicates that the number one cylinder has begun the compression stroke.
7    Once the compression stroke has begun, TDC for the compression stroke is reached by bringing the piston to the top of the cylinder.
8    Continue turning the crankshaft until the notch in the crankshaft damper is aligned with the "TDC" or the "0" mark on the timing chain cover **(see illustration).** At this point, the number one cylinder is at TDC on the compression stroke. If the marks are aligned but there was no compression, the piston was on the exhaust stroke; continue rotating the crankshaft 360-degrees (1-turn). **Note:** *If a compression gauge is not available, you can simply place a blunt object over the spark plug hole and listen for compression as the engine is rotated. Once compression at the No.1 spark plug hole is noted the remainder of the Step is the same.*

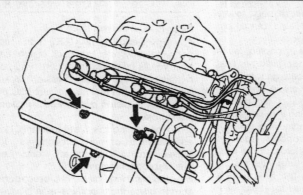

**4.2a  On 1998 and 1999 models, remove the fuel injector wiring harness cover -  arrows indicate the fasteners which secure the cover to the valve cover and the intake manifold**

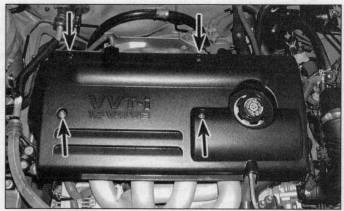

**4.2b  On 2000 and later models, remove the engine cover - upper arrows indicate plastic retaining clips, lower arrows indicate two retaining nuts**

4.6 Valve cover fastener locations

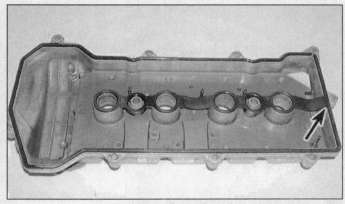

4.7 The valve cover gasket and the spark plug tube seals are incorporated into a single rubber O-ring-like seal (arrow) - press the gasket evenly into the grooves around the underside of the valve cover and the spark plug openings

*may occur, leading to oil leaks after the valve cover is reinstalled.*

## Installation

*Refer to illustrations 4.7 and 4.8*

7    Remove the valve cover gasket from the valve cover and clean the mating surfaces with lacquer thinner or acetone. Install a new rubber gasket, pressing it evenly into the grooves around the underside of the valve cover. **Note:** *Make sure the spark plug tube seals are in place on the underside of the valve cover before reinstalling it* **(see illustration).** The mating surfaces of the timing chain cover, the cylinder head and valve cover must be perfectly clean when the valve cover is installed. If there's residue or oil on the mating surfaces when the valve cover is installed, oil leaks may develop.

8    Apply RTV sealant at the timing chain cover-to-cylinder head joint, then install the valve cover and fasteners **(see illustration).**

9    Tighten the nuts/bolts to the torque listed in this Chapter's Specifications in three or four equal steps.

10   Reinstall the remaining parts, run the engine and check for oil leaks.

## 5    Variable Valve Timing (VVT) system - description and component check

1    The VVT system was introduced on all 2000 and later Corolla and Prizm models. The differences between the 1998 and 1999 1.8L engine and the 2000 and later models is strictly in the VVT system components and operation of the valve train. The engine short block, oiling and cooling systems are identical, as are all attached components.

2    The VVT system varies intake camshaft timing by directing oil pressure to advance or retard the intake camshaft sprocket/actuator assembly. Changing the intake camshaft timing during certain engine conditions increases engine power output, fuel economy

and reduces emissions.

3    System components include the Powertrain Control Module (PCM), the VVT oil control valve (OCV), the VVT oil control valve filter and the intake camshaft sprocket/actuator assembly.

4    The PCM uses inputs from the following sensors to turn the oil control valve ON or OFF:

a) *Vehicle Speed Sensor (VSS)*
b) *Throttle Position Sensor (TPS)*
c) *Mass Air Flow (MAF) sensor*
d) *Engine Coolant Temperature (ECT) sensor*

5    Once the VVT oil control valve is actuated by the PCM it directs the specified amount of oil pressure from the engine to advance or retard the intake camshaft sprocket/actuator assembly.

6    The intake camshaft sprocket/actuator assembly is equipped with an inner hub that is attached to the camshaft. The inner hub consists of a series of fixed vanes that use oil pressure as a wedge against the vanes to rotate the camshaft. The higher the oil pressure (or flow) the more the actuator assembly will rotate, thereby advancing or retarding the camshaft.

7    When oil is applied to the advance side of the vanes, the actuator can advance the camshaft up to 21 degrees in a clockwise direction. When oil is applied to the retard side of the vanes, the actuator will start to rotate the camshaft counterclockwise back to 0 degrees which is the normal position of the actuator during engine operation under no load or at idle. The PCM can also send a signal to the oil control valve to stop oil flow to both (advance and retard) passages to hold camshaft advance in its current position.

8    Under light engine loads, the VVT system will retard the camshaft timing to decrease valve overlap and stabilize engine output. Under medium engine loads, the VVT system will advance the camshaft timing to increase valve overlap, thereby increasing fuel economy and decreasing exhaust emissions. Under heavy engine loads at low RPM,

4.8 Apply sealant at the timing chain cover-to-cylinder head joint before installing the valve cover

the VVT system will advance the camshaft timing to help close the intake valve faster, which improves low to midrange torque. Under heavy engine loads at high RPM, the VVT system will retard the camshaft timing to slow the closing of the intake valve to improve engine horsepower.

## Component checks

### VVT oil control valve and filter

*Refer to illustrations 5.9a, 5.9b, 5.10 and 5.11*

**Note 1:** *A problem in the VVT oil control valve circuit will set a diagnostic* **trouble code** *and turn on the Check Engine light on the dash. Refer to Chapter 6 for accessing trouble codes.*

**Note 2:** *Most problems in the VVT system are with the oil control valve and its filter. Regular engine oil and filter changes are necessary for trouble-free operation of the valve.*

**Note 3:** *Some checks and inspections of the VVT system require removal of the valve cover, the intake camshaft sprocket/actuator assembly and the intake camshaft.*

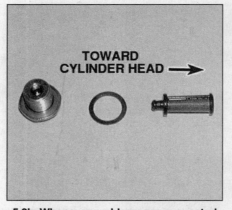

5.9a  The Variable Valve Timing (VVT) oil control valve and filter are located at the front of the cylinder head

5.9b  Whenever problems are suspected in the VVT system, check the oil control valve filter for clogging - note that the big end of the filter must be installed in the direction shown

5.10  Measure the resistance between the terminals of the VVT oil control valve

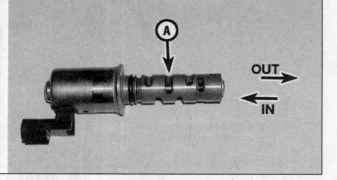

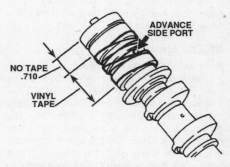

5.11  The oil control valve plunger (A) can be seen through the slots in the housing - the plunger should move outward when voltage is applied and inward when voltage is released

5.13  To test the actuator assembly, it will be necessary to block off all of the oil control orifices except the advance side port

9    The first check of the VVT system components begins with removing the oil control valve filter from the cylinder head and check the filter/O-ring for clogging (see illustrations). Clean and reinstall with a new O-ring. A clogged filter screen is often the cause of system problems.

10    The second and third check of the VVT system involve checking the operation of the oil control valve. Disconnect the electrical connector from the oil control valve and measure the resistance between the terminals of the control valve (see illustration). There should be 6.9 to 7.9 ohms; if not, replace the VVT oil control valve.

11    If the resistance figures check out OK, remove the oil control valve from the cylinder head and check the operation of the control valve plunger. Using a pair of fused jumper wires, apply battery voltage to terminal No. 1 on the control valve, then apply ground to terminal No. 2 on the oil control valve. With battery voltage applied to the oil control valve, check for free movement of the plunger (see illustration). The plunger should move out when voltage is applied and move back inward when voltage is not applied. If the oil control valve does not operate as described, replace the VVT oil control valve. Always use a new O-ring when reinstalling the control valve.

### VVT camshaft sprocket/actuator assembly

*Refer to illustrations 5.13 and 5.15*

12    The fourth and final check of the VVT system requires removing the valve cover, the intake camshaft and the camshaft sprocket/actuator assembly from the engine. Refer to Section 8 and perform Steps 1 through 10, removing the intake camshaft and sprocket from the engine only. Do not remove the exhaust camshaft or sprocket from the engine.

13    Clean the snout of the intake camshaft with lacquer thinner or acetone to remove all traces of oil from the front journals and the VVT oil control orifices. Apply vinyl tape over all the oil control orifices at the front of the camshaft, except the advance side oil port (see illustration). Do not apply tape over the front of the camshaft were the sprocket fits.

14    Install the intake camshaft sprocket/actuator assembly onto the intake camshaft and tighten the bolt to 40 ft-lbs. Check that the actuator assembly will not rotate from the locked position. The locked position is a "neutral position" in which the actuator is placed during idle and no load conditions or when the VVT system is not activated by the PCM.

15    Apply 14 psi of air pressure to the advance side oil port and try to rotate the

actuator assembly by hand (see illustration). The actuator should rotate approximately 30 degrees in a counterclockwise direction from the locked position. Also check that it rotates freely with no obvious binding. **Note:** *It is critical to have an airtight seal between the air gun nozzle and the advance oil port hole to accomplish this task, as the lock pin in the actuator may not be forced out of its locating hole. If leakage at the air gun nozzle occurs, it may be necessary to apply a greater amount of air pressure to the advance side oil port in order*

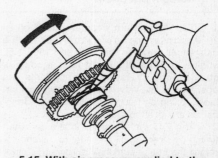

5.15  With air pressure applied to the advance side oil port, try to rotate the actuator in the direction shown

**6.4 You'll need an air hose adapter this long to reach down into the spark plug tubes - they're commonly available from auto parts stores**

**6.8b ... and lift them out with a magnet or needle-nose pliers**

**6.10 A pair of pliers will be required to remove the valve seal from the valve guide**

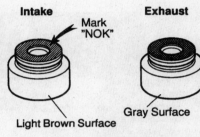

**6.8a Compress the valve spring enough to release the keepers . . .**

**6.15a Valve seal installation details**

Intake — Mark "NOK" — Light Brown Surface
Exhaust — Gray Surface

**2B**

to force the lock pin from the locating hole.

16    If the actuator does not rotate freely as described, replace the intake camshaft sprocket/actuator assembly.

17    Reassembly is the reverse of removal.

## 6    Valve springs, retainers and seals - replacement

*Refer to illustrations 6.4, 6.8a, 6.8b, 6.10, 6.15a, 6.15b and 6.17*

**Note:** *Broken valve springs and defective valve stem seals can be replaced without removing the cylinder head. Two special tools and a compressed air source are normally required to perform this operation, so read through this Section carefully. The universal shaft-type valve spring compressor required for the tight valve spring pockets of this vehicle may not be available at all tool rental yards, so check on the availability before beginning the job.*

1    Remove the valve cover (see Section 4). Refer to Section 7 and remove the timing chain, then refer to Section 8 and remove the camshafts and lifters from the cylinder head.

2    Remove the spark plug from the cylinder which has the defective component. If all of the valve stem seals are being replaced, all of the spark plugs should be removed.

3    Turn the crankshaft until the piston in the affected cylinder is at top dead center (TDC) on the compression stroke (see Section 3 for instructions). If you're replacing all of the valve stem seals, begin with cylinder number one and work on the valves for one cylinder at a time. Move from cylinder-to-

cylinder following the firing order sequence (see the Specifications listed at the beginning of this Chapter).

4    Thread an adapter into the spark plug hole **(see illustration)** and connect an air hose from a compressed air source to it. Most auto parts stores can supply the air hose adapter. **Note:** *Many cylinder compression gauges utilize a screw-in fitting that may work with your air hose quick-disconnect fitting.*

5    Apply compressed air to the cylinder. The valves should be held in place by the air pressure. **Warning:** *If the cylinder isn't exactly at TDC, air pressure may force the piston down, causing the engine to quickly rotate. DO NOT leave a wrench on the crankshaft drive sprocket bolt or you may be injured by the tool.*

6    Stuff shop rags into the cylinder head holes around the valves to prevent parts and tools from falling into the engine.

7    Using a socket and a hammer, gently tap on the top of each valve spring retainer several times. This will break the bond between the valve keepers and the spring retainer and allow the keepers to separate from the valve spring retainer as the valve spring is compressed.

8    Use a valve spring compressor to compress the spring, then remove the keepers with small needle-nose pliers or a magnet **(see illustrations)**. **Note:** *Several different types of tools are available for compressing the valve springs with the head in place. Be sure to purchase or rent the "import type" that bolts to the top of the cylinder head. This type uses a support bar across the cylinder*

*head for leverage as the valve spring is compressed. The lack of clearance surrounding the valve springs on these engines prohibits typical types of valve spring compressors from being used.*

9    Remove the valve spring and retainer. **Note:** *If air pressure fails to retain the valve in the closed position during this operation, the valve face or seat may be damaged. If so, the cylinder head will have to be removed for repair.*

10    Remove the old valve stem seals, noting differences between the intake and exhaust seals **(see illustration)**.

11    Wrap a rubber band or tape around the top of the valve stem so the valve won't fall into the combustion chamber, then release the air pressure.

12    Inspect the valve stem for damage. Rotate the valve in the guide and check the end for eccentric movement, which would indicate that the valve is bent.

13    Move the valve up-and-down in the guide and make sure it doesn't bind. If the valve stem binds, either the valve is bent or the guide is damaged. In either case, the head will have to be removed for repair.

14    Reapply air pressure to the cylinder to retain the valve in the closed position, then remove the tape or rubber band from the valve stem.

15    Lubricate the valve stem with engine oil and install a new seal on the valve guide **(see illustrations)**.

16    Install the valve spring and the spring

**6.15b Using a deep socket and hammer, gently tap the new seals onto the valve guide only until seated**

**6.17 Apply a small dab of grease to each keeper as shown here before installation - it will hold them in place on the valve stem as the spring is released**

**7.7 Verify the engine is at TDC by observing the position of the camshaft sprocket marks (upper arrows) - they must be in a straight line and parallel with the top of the timing chain cover (lower arrow)**

**7.9 If an engine support fixture is not available, the engine can be supported from below using a floor jack and a block of wood**

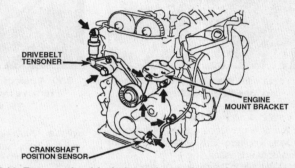

**7.11 Remove the drivebelt tensioner, the engine mount bracket and the crankshaft position sensor from the timing chain cover - mounting bolt locations are indicated by arrows**

retainer in position over the valve.

17    Compress the valve spring and carefully position the keepers in the groove. Apply a small dab of grease to the inside of each keeper to hold it in place **(see illustration)**.

18    Remove the pressure from the spring tool and make sure the keepers are seated.

19    Disconnect the air hose and remove the adapter from the spark plug hole.

20    Install the camshaft, lifters, timing chain and the valve cover by referring to the appropriate Sections.

21    Install the spark plug(s) and the spark plug wires (on 1998 and 1999 models) or the ignition coils (on 2000 and later models).

22    Start and run the engine, then check for oil leaks and unusual sounds coming from the valve cover area.

## 7    Timing chain and sprockets - removal, inspection and installation

**Note:** *Special tools are required for this procedure. Read through the entire procedure and acquire the necessary tools and equipment before beginning work.*

### Removal

*Refer to illustrations 7.7, 7.9, 7.11, 7.13, 7.14a, 7.14b, 7.14c, 7.15, 7.16 and 7.19*

1    Detach the cable from the negative terminal of the battery. **Caution:** *On models equipped with the Theftlock audio system, be sure the lockout feature is turned off before performing any procedure which requires disconnecting the battery (see the front of this manual).*

2    Remove the drivebelt (see Chapter 1) and the alternator (see Chapter 5).

3    Remove the windshield washer reservoir and the valve cover (see Section 4).

4    With the parking brake applied and the shifter in Park (automatic) or in gear (manual), loosen the lug nuts from the right front wheel, then raise the front of the vehicle and support it securely on jackstands. Remove the right front wheel and the right splash shield from the wheelwell.

5    Drain the cooling system (see Chapter 1).

6    While the coolant is draining, refer to Chapter 10 and remove the power steering pump from the engine without disconnecting the fluid lines. Tie the power steering pump to the body with a piece of wire and position it

out of the way.

7    Position the number one piston at TDC on the compression stroke (see Section 3). Visually confirm the engine is at TDC on the compression stroke by verifying that the timing mark on the crankshaft pulley/vibration damper is aligned with the "0" mark on the timing chain cover and the camshaft sprocket marks are aligned and parallel with the top of the timing chain cover **(see illustration 3.8 and the accompanying illustration)**. **Note:** *There are two sets of marks on the camshaft sprockets. The marks that align at TDC are for TDC reference only; the other two marks are used to align the sprockets with the timing chain during installation.*

8    Remove the crankshaft pulley/vibration damper, being careful not to rotate the engine from TDC (see Section 12). If the engine rotates off TDC during this step, reposition the engine back to TDC before proceeding. The engine should be left at TDC for the No. 1 piston during this entire procedure.

9    Support the engine from above, using an engine support fixture (available at rental yards), or from below using a floor jack. Use a wood block between the floor jack and the engine to prevent damage **(see illustration)**.

10    Remove the passenger side engine

7.13  Timing chain tensioner mounting nuts (arrows)

7.14a  Timing chain cover upper fasteners - this stud (A) will have to be removed with a torx bit and tightened to 82 in-lbs during installation

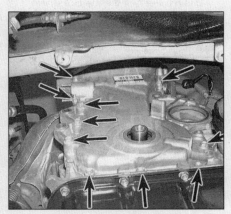

7.14b  Timing chain cover lower fasteners - make a note of the fastener sizes, locations and lengths as your removing them as they must be installed back in the original position

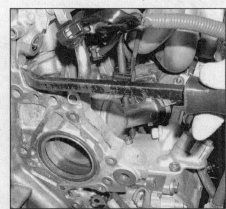

7.14c  Pry the timing chain cover off by the casting protrusion

7.15  Slide the crankshaft position sensor reluctor ring off the crankshaft - note the "F" mark (arrow) on the front (it must be facing outward upon installation)

2B

mount (see Section 18).

11  Remove the drivebelt tensioner, the engine mount bracket and the crankshaft position sensor from the timing chain cover **(see illustration)**. Also remove the bolt securing the crankshaft position sensor wiring harness to the timing chain cover.

12  Remove the water pump (see Chapter 3).

13  Remove the timing chain tensioner from the rear side of the timing chain cover **(see illustration)**.

14  Remove the timing chain cover fasteners and pry the cover off the engine **(see illustrations)**.

15  Slide the crankshaft position sensor reluctor ring off the crankshaft **(see illustration)**.

16  Remove the timing chain tensioner pivot arm/chain guide **(see illustration)**.

17  Lift the timing chain off the camshaft sprockets and remove the timing chain and the crankshaft sprocket as an assembly from the engine. The crankshaft sprocket should slip off the crankshaft by hand. If not, use several flat bladed screwdrivers to evenly pry the sprocket off the crankshaft. **Note:** *If you*

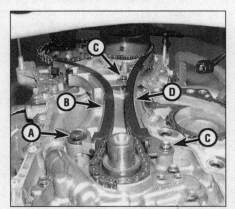

7.16  Timing chain guide mounting details

A  Pivot bolt
B  Tensioner pivot arm/chain guide
C  Stationary chain guide mounting bolts
D  Stationary chain guide

*intend to reuse the timing chain, use white paint or chalk to make a mark indicating the front of the chain. If a used timing chain is reinstalled with the wear pattern in the opposite direction noise and increased wear may occur.*

18  Remove the stationary timing chain guide **(see illustration 7.16)**.

19  To remove the camshaft sprockets, loosen the bolts while holding the lug on the

7.19  Hold the lug on the camshaft with a wrench to keep it from rotating as the sprocket bolts are loosened - 2000 and later models are equipped with a variable valve timing actuator on the intake camshaft sprocket - loosen the center bolt (arrow) securing the sprocket to the camshaft only

camshaft with a wrench **(see illustration)**. Note the identification marks on the camshaft sprockets before removal, then remove the bolts. Pull on the sprockets by hand until they

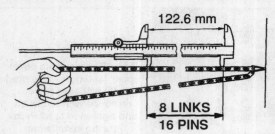

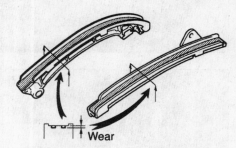

**7.20a  Timing chain stretch is measured by checking the length of the chain between 8 links (16 pins) at 3 or more places (selected randomly) around the chain - if chain stretch exceeds the specifications between any 8 links, the chain must be replaced**

**7.20b  Wrap the chain around each of the timing sprockets and measure the diameter of the sprockets across the chain rollers - if the measurement exceeds the minimum sprocket diameter, the chain and the timing sprockets must be replaced**

slip off the dowels. If necessary, use a small puller, with the legs inserted in the relief holes, to pull the sprockets off. **Note 1:** *All 2000 and later models are equipped with variable valve timing, which consists of an actuator assembly attached to the intake camshaft sprocket. When removing the intake camshaft sprocket on these models only loosen and remove the center bolt, which fastens the sprocket to the camshaft. Do not loosen the outer four bolts that secure the actuator to the sprocket.*
**Note 2:** *1998 and 1999 models are not equipped with variable valve timing and have two very similar looking camshaft sprockets. On these models, the camshaft sprockets are marked "EX" for exhaust and "IN" for intake. When looking at the engine from the front of the vehicle, the forward facing cam is the intake camshaft and the cam nearest the firewall is the exhaust camshaft. It is very important that the sprockets are returned to their original location during installation.*

### Inspection

*Refer to illustrations 7.20a, 7.20b and 7.21*
20   Visually inspect all parts for wear and damage. Check the timing chain for loose pins, cracks, worn rollers and side plates. Check the sprockets for hook-shaped, chipped and broken teeth. Also check the timing chain for stretching and the diameter of the timing sprockets for wear with the chain assembled on the sprockets **(see illustrations)**. Be sure to measure across the chain rollers when checking the sprocket diameter and to measure chain stretch at three or more places around the chain. Maximum chain elongation and minimum sprocket diameter (with chain) should not exceed the amount listed in this Chapter's Specifications. Replace the timing chain and sprockets as a set if the engine has high mileage or fails inspection.
21   Check the chain guides for excessive wear **(see illustration)**. Replace the chain guides if scoring or wear exceeds the amount listed in this Chapter's Specifications. Note that some scoring of the timing chain guide shoes is normal. If excessive wear is indicated, it will also be necessary to inspect the chain guide oil hole on the front of the oil pump for clogging (see Section 15).

### Installation

*Refer to illustrations 7.27, 7.28, 7.29, 7.32a, 7.32b, 7.34, 7.35 and 7.37*
22   Remove all traces of old sealant from

**7.21  Timing chain guide wear is measured from the top of the chain contact surface to the bottom of the wear grooves**

the timing chain cover and the mating surfaces of the engine block and cylinder head.
23   Make sure the camshafts are positioned with the dowel pins at the top in the 12 o'clock position, then install both camshaft sprockets in their original locations by aligning the dowel pin hole on the rear of the sprockets with dowel pin on the camshaft. Apply medium strength thread locking compound to the camshaft sprocket bolt threads and make sure the washers are in place. Hold the camshaft from turning as described in Step 19 and tighten the bolts to the torque listed in this Chapter's Specifications.
24   Rotate the camshafts as necessary to align the TDC marks on the camshaft sprock-

ets **(see illustration 7.7)**.
25   If the crankshaft has been rotated off TDC during this procedure, it will be necessary to rotate the crankshaft until the keyway is pointing straight up in the 12 o'clock position with the centerline of the cylinder bores.
26   Install the stationary timing chain guide **(see illustration 7.16)**.
27   Loop the timing chain around the crankshaft sprocket and align the No.1 colored link with the mark on the crankshaft sprocket. Install the chain and crankshaft sprocket as an assembly on the engine **(see illustration)**. **Note:** *There are three colored links on the timing chain. The No.1 colored link is the link farthest away from the two colored links that are closest together.*
28   Slip the timing chain into the lip of the stationary timing chain guide and over the intake camshaft sprocket, then around the exhaust camshaft sprocket making sure to align the remaining two colored links with the marks on the camshaft sprockets **(see illus-**

**7.27  Loop the timing chain around the crankshaft sprocket and align the No.1 colored link with the mark on the crankshaft sprocket, then install the chain and crankshaft sprocket as an assembly on the engine**

**7.28  Loop the timing chain up over the intake camshaft and around the exhaust camshaft, while aligning the remaining two colored links with the marks on the camshaft sprockets**

**7.29  After the tensioner pivot arm is installed, make sure the tab (arrow) on the pivot arm can't move past the stopper on the cylinder head**

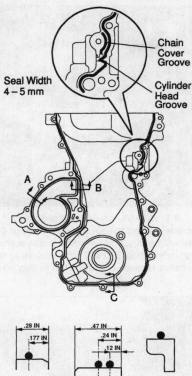

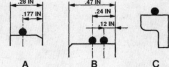

**7.32a  Timing chain cover sealant installation details**

**7.32b  Also apply a bead of sealant on each side of the parting line between the cylinder head and the engine block**

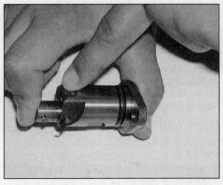

**7.34  Raise the ratchet pawl and push the plunger inward until the hook on the tensioner body can be engaged with pin on the plunger to lock the plunger in place**

2B

tration). Make sure to remove all slack from the right side of the chain when doing so.

29   Use one hand to remove the slack from the left side of the chain and install the timing chain tensioner pivot arm/chain guide. Tighten the pivot bolt to torque listed in this Chapter's Specifications. After installation, make sure the tab on the pivot arm can't move past the stopper on the cylinder head **(see illustration)**.

30   Reconfirm that the number one piston is still at TDC on the compression stroke and that the timing marks on the crankshaft and camshaft sprockets are aligned with the colored links on the chain.

31   Install the crankshaft position sensor reluctor ring with the "F" mark facing outward.

32   Apply a bead of RTV sealant to the timing chain cover sealing surfaces **(see illustrations)**. Place the timing chain cover in position on the engine and install the bolts in their original locations. Tighten the bolts evenly in several steps to the torque listed in this Chapter's Specifications **(see illustration)**. Be sure to follow the sealant manufacturer's recommendations for assembly and sealant curing times.

33   Install the water pump O-ring, the water pump and the engine mount bracket all within the sealant manufacturer's specified drying times. Before installing the engine mount bracket, apply a small amount of RTV sealant to the threads of the engine mount bracket bolts.

34   Reload and lock the timing chain tensioner to its "zero" position as follows:

  a) *Raise the ratchet pawl and push the plunger inward until it bottoms out* **(see illustration)**.

  b) *Engage the hook on the tensioner body with the pin on the tensioner plunger to lock the plunger in place.*

35   Lubricate the tensioner O-ring with a small amount of oil and install the tensioner into the timing chain cover with the hook facing up **(see illustration)**.

36   Install the crankshaft pulley/vibration

damper (see Section 12).

37   Rotate the engine clockwise slightly to set the chain tension. As the engine is rotated, the hook on the tensioner body should release itself from the pin on the plunger and allow the plunger to spring out and apply tension to the timing chain. If the plunger does not spring outward and apply tension to the timing chain, press downward on the pivot arm and release the hook with a screwdriver **(see illustration)**.

38   Rotate the engine clockwise several turns and reposition the number one piston at TDC on the compression stroke (see Sec-

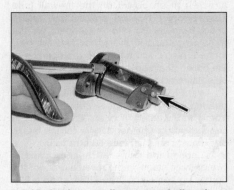

**7.35  Apply a small amount of oil to the tensioner O-ring and insert the tensioner into the timing chain cover with the hook facing upward**

tion 3). Visually confirm that the timing mark on the crankshaft pulley/vibration damper is aligned with the "0" mark on the timing chain cover and the camshaft sprocket marks are aligned and parallel with the top of the timing chain cover as shown in **illustration 7.7**.

39   The remainder of the installation is the reverse of removal.

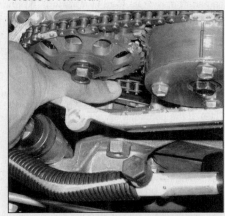

**7.37  Depress the tensioner pivot arm and disengage the hook from the plunger pin to set tension on the timing chain**

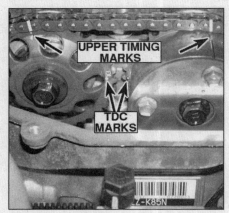

**8.4  With the TDC marks aligned, apply a dab of paint to the timing chain links where they meet the upper timing marks on the camshaft sprockets**

**8.7  With the camshaft sprockets removed, hang the timing chain out of the way with a piece of wire and place a shop rag in the timing chain cover opening to prevent foreign objects from falling into the engine**

**8.8  The camshaft bearing caps are numbered and have an arrow that should face the timing chain end of the engine**

## 8    Camshafts and lifters - removal, inspection and installation

**Note:** *The camshafts should always be thoroughly inspected before installation and camshaft endplay should always be checked prior to camshaft removal (see Step 13).*

### Removal

*Refer to illustrations 8.4, 8.7, 8.8, 8.11a and 8.11b*

1    Disconnect the cable from the negative terminal of the battery.

2    Remove the valve cover (see Section 4).

3    Refer to Section 3 and place the engine on TDC for number 1 cylinder. Visually confirm the engine is at TDC on the compression stroke by verifying that the timing mark on the crankshaft pulley/vibration damper is aligned with the "0" mark on the timing chain cover and the camshaft sprocket TDC marks are aligned and parallel with the top of the timing chain cover **(see illustrations 3.8 and 7.7)**.

4    With the TDC marks aligned, apply a dab of paint to the timing chain links where they meet the upper timing marks on the camshaft sprockets **(see illustration)**. **Note:** *There are two sets of marks on the camshaft sprockets. The marks that align at TDC are for TDC reference only, the other two marks are used to align the sprockets with the timing chain during installation.*

5    Using a wrench to hold the camshaft sprockets from turning, loosen the camshaft sprocket bolts several turns **(see illustration 7.19)**. If the camshaft sprockets have rotated during the bolt loosening process, rotate the engine clockwise until the "TDC" marks on the cam sprockets are realigned.

6    Remove the timing chain tensioner from the timing cover **(see illustration 7.13)** and the camshaft position sensor from the cylinder head (see Chapter 6).

7    Remove the camshaft sprocket retaining bolts. Disengage the timing chain from the sprockets and remove the camshaft sprockets from the engine. Make sure to note that

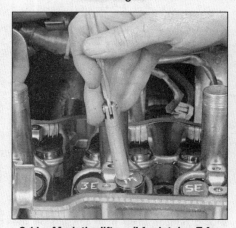

**8.11a  Mark the lifters (I for intake, E for exhaust, and number their location) and remove them with a magnetic retrieval tool**

the sprockets are marked "IN" for intake or "EX" for exhaust and cannot be interchanged. After removing the sprockets, hang the timing chain up with a piece of wire and attach it to an object on the firewall **(see illustration)**. This will prevent the timing chain from falling into the engine as the remaining steps in this procedure are performed. Also place a rag in the opening of the timing chain cover to prevent any foreign objects from falling into the engine.

8    Verify the markings on the camshaft bearing caps. The caps should be marked from 1 to 5 with an "I" or an "E" mark on the cap indicating whether they're for the intake or exhaust camshaft **(see illustration)**.

9    Loosen the camshaft bearing caps in two or three steps, in the reverse order of the tightening sequence **(see illustration 8.22)**. **Caution:** *Keep the caps in order. They must go back in the same location they were removed from.*

10    Detach the bearing caps, then remove the camshaft(s) from the cylinder head. Mark

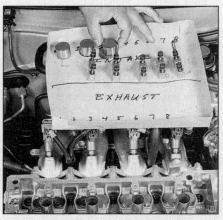

**8.11b  Mark up a cardboard box to store the lifters and bearing caps in order**

the camshaft(s) "Intake" or "Exhaust" to avoid mixing them up. **Note:** *When looking at the engine from the front of the vehicle, the forward facing cam is the intake camshaft and the cam nearest the firewall is the exhaust camshaft. It is very important that the camshafts are returned to their original locations during installation.*

11    Remove the lifters from the cylinder head, keeping them in order with their respective valve and cylinder **(see illustrations)**. **Caution:** *Keep the lifters in order. They must go back in the position from which they were removed.*

12    Inspect the camshafts and lifters as described below. Also inspect the camshaft sprockets for wear on the teeth. Inspect the chains for cracks or excessive wear of the rollers, and for stretching (see Section 7). If any of the components show signs of excessive wear they must be replaced.

### Inspection

*Refer to illustrations 8.13, 8.14, 8.15, 8.16, 8.18 and 8.19*

13    Before the camshafts are removed from the engine, check the camshaft endplay by placing a dial indicator with the stem in line with the camshaft and touching the snout

**8.13 Mount a dial indicator as shown to measure camshaft endplay - with the dial zeroed, pry the camshaft forward and back and read the endplay on the dial**

**8.14 Inspect the cam bearing surfaces in the cylinder head for pits, score marks and abnormal wear - if wear or damage is noted, the cylinder head must be replaced**

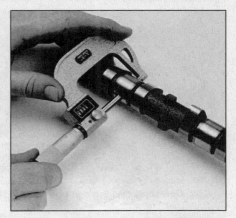

**8.15 Measure each journal diameter with a micrometer - if any journal measures less than the specified limit, replace the camshaft**

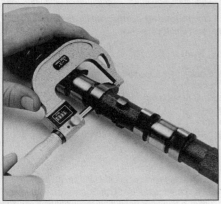

**8.16 Measure the lobe heights on each camshaft - if any lobe height is less than the specified allowable minimum, replace that camshaft**

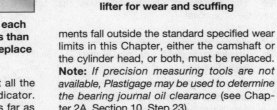

**8.18 Wipe off the oil and inspect each lifter for wear and scuffing**

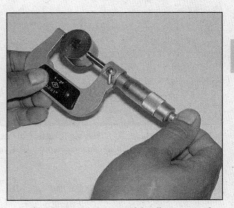

**8.19 Measure the outside diameter of each lifter and the inside diameter of each lifter bore to determine the oil clearance measurement**

(see illustration). Push the camshaft all the way to the rear and zero the dial indicator. Next, pry the camshaft to the front as far as possible and check the reading on the dial indicator. The distance it moves is the endplay. If it's greater than the Specifications listed in this Chapter, check the bearing caps for wear. If the bearing caps are worn, the cylinder head must be replaced.

14   With the camshafts removed, visually check the camshaft bearing surfaces in the cylinder head for pitting, score marks, galling and abnormal wear. If the bearing surfaces are damaged, the cylinder head may have to be replaced **(see illustration)**.

15   Measure the outside diameter of each camshaft bearing journal and record your measurements **(see illustration)**. Compare them to the journal outside diameter specified in this Chapter, then measure the inside diameter of each corresponding camshaft bearing and record the measurements. Subtract each cam journal outside diameter from its respective cam bearing bore inside diameter to determine the oil clearance for each bearing. Compare the results to the specified journal-to-bearing clearance. If any of the measure-

ments fall outside the standard specified wear limits in this Chapter, either the camshaft or the cylinder head, or both, must be replaced. **Note:** *If precision measuring tools are not available, Plastigage may be used to determine the bearing journal oil clearance (see Chapter 2A, Section 10, Step 23).*

16   Using a micrometer, measure the height of each camshaft lobe **(see illustration)**. Compare your measurements with this Chapter's Specifications. If the height for any one lobe is less than the specified minimum, replace the camshaft.

17   Check the camshaft runout by placing the camshaft back into the cylinder head and set up a dial indicator on the center journal. Zero the dial indicator. Turn the camshaft slowly and note the dial indicator readings. Runout should not exceed .0012 inch. If the measured runout exceeds the specified runout, replace the camshaft.

18   Inspect each lifter for scuffing and score marks **(see illustration)**.

19   Measure the outside diameter of each lifter **(see illustration)** and the corresponding lifter bore inside diameter. Subtract the lifter diameter from the lifter bore diameter to determine the oil clearance. Compare it to this Chapter's Specifications. If the oil clearance is excessive, a new cylinder head and/or new lifters will be required.

### Installation

*Refer to illustrations 8.20 and 8.22*

20   Apply moly-based engine assembly lubricant to the camshaft lobes and journals

**8.20 Place the camshafts in the cylinder head with the No.1 cylinder lobes pointing inward at approximately a 30 degree angle**

**2B**

8.22  Camshaft bearing cap bolt TIGHTENING sequence

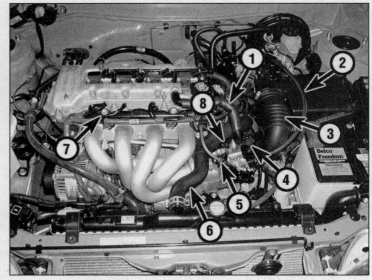

9.4  Label and disconnect the following components required for intake manifold removal

1    PCV hose
2    Throttle cable
3    Air intake duct and upper air cleaner cover
4    Throttle body
5    Throttle valve cable (if equipped)
6    Upper radiator hose
7    Fuel rail and injectors
8    Vacuum hoses at rear of manifold (not visible)

9.6  Intake manifold fastener locations - note that the three lower bracket bolts (A) are removed first and tightened last

and install the camshaft into the cylinder head with the No.1 cylinder camshaft lobes pointing inward toward each other and the dowel pins facing upward **(see illustration)**. If the old camshafts are being used, make sure they're installed in the exact location from which they came.
21   Install the bearing caps and bolts and tighten them hand tight.
22   Tighten the bearing cap bolts in several equal steps, to the torque listed in this Chapter's Specifications, using the proper tightening sequence **(see illustration)**.
23   Engage the camshaft sprocket teeth with the timing chain links so that the match marks made during removal align with the upper timing marks on the sprockets, then position the sprockets over the dowels on the camshaft hubs and install the camshaft sprocket bolts finger tight. At this point the mark on the crankshaft pulley should be aligned with the "0" mark on the timing chain cover, the camshaft sprocket TDC marks should be aligned and parallel with the top of the timing chain cover and the timing chain match marks should be aligned with the upper timing sprocket marks with all of the slack in the chain positioned towards the tensioner side of the engine **(see illustration 8.4)**.
24   Double check that the timing sprockets are returned to the proper camshaft and tighten the camshaft sprocket bolts to the torque listed in this Chapter's Specifications.
25   Install the timing chain tensioner as

described in Section 7, Steps 34 and 35.
26   The remainder of installation is the reverse of removal.

## 9    Intake manifold - removal and installation

**Warning:** *Wait until the engine is completely cool before beginning this procedure.*

### Removal

*Refer to illustrations 9.4 and 9.6*
1    Relieve the fuel system pressure (see Chapter 4), then disconnect the negative cable from the battery. **Caution:** *If the stereo in your vehicle is equipped with an anti-theft system, make sure you have the correct activation code before disconnecting the battery.*
2    On 1998 and 1999 models, remove the fuel injector wiring harness cover **(see illustration 4.2a)**. On 2000 and later models, remove the engine cover **(see illustration)**.
3    Drain the cooling system (see Chapter 1) and remove the upper radiator hose (see Chapter 3).
4    Remove the fuel rail and injectors as an assembly (see chapter 4). Also remove the throttle linkage and the throttle body from the intake manifold **(see illustration)**.
5    Label and detach the PCV and vacuum hoses connected to the rear of the intake manifold.

6    Remove the intake manifold mounting nuts, bolts and support bracket bolts and remove the manifold and the gasket from the engine **(see illustration)**.

### Installation

7    Clean the mating surfaces of the intake manifold and the cylinder head mounting surface with lacquer thinner or acetone. If the gasket shows signs of leaking, have the manifold checked for warpage at an automotive machine shop and resurfaced if necessary.
8    Install a new gasket over the manifold studs, then position the manifold on the cylinder head and install the nuts/bolts and brackets.
9    Tighten the manifold-to-cylinder head nuts/bolts in three or four equal steps to the torque listed in this Chapter's Specifications. Work from the center out towards the ends to avoid warping the manifold. After the manifold-to-cylinder head bolts have been tightened to the proper torque, tighten the lower bracket bolts **(see illustration 9.6)**.
10   Install the remaining parts in the reverse order of removal. Refill the cooling system (see Chapter 1).
11   Before starting the engine, check the throttle linkage for smooth operation.
12   Run the engine and check for coolant and vacuum leaks.
13   Road test the vehicle and check for proper operation of all accessories, including the cruise control system, if equipped.

10.3 Working below the vehicle, remove the exhaust pipe-to-manifold mounting bolts (A) and lower the front exhaust pipe - being careful not to damage (C) the oxygen sensor - (B) indicates the mounting bolts for the lower exhaust manifold brace

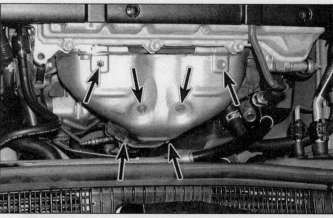

10.5 Working from the engine compartment, remove the upper heat shield mounting bolts (arrows) . . .

2B

## 10 Exhaust manifold - removal and installation

**Warning:** *The engine must be completely cool before beginning this procedure.*

### Removal

*Refer to illustrations 10.3, 10.5 and 10.6*

1 Disconnect the negative cable from the battery. **Caution:** *If the stereo in your vehicle is equipped with an anti-theft system, make sure you have the correct activation code before disconnecting the battery.*

2 Raise the front of the vehicle and support it securely on jackstands.

3 Working below the vehicle, apply penetrating oil to the bolts and springs retaining the exhaust pipe to the manifold. After the bolts have soaked, remove the bolts retaining the exhaust pipe to the manifold. Separate the front exhaust pipe from the manifold, being careful not to damage the oxygen sensor **(see illustration)**. **Note:** *It may be necessary to remove the inner heat shield from above the right driveaxle to access one of the exhaust pipe-to-manifold bolts.*

4 Unbolt the lower exhaust manifold brace and remove it from the engine.

5 Working in the engine compartment, remove the upper heat shield from the manifold **(see illustration)**. **Note:** *There is also a lower heat shield, but it is attached to the manifold from underneath and does not need to be removed.*

6 Remove the nuts/bolts and detach the manifold and gasket **(see illustration)**.

### Installation

7 Use a scraper to remove all traces of old gasket material and carbon deposits from the manifold and cylinder head mating surfaces. If the gasket was leaking, have the manifold checked for warpage at an automotive machine shop and resurfaced if necessary. **Note:** *If the manifold is being replaced with a new one it will be necessary to remove the lower heat shield and fasten it to the new manifold.*

8 Position a new gasket over the cylinder head studs noting any directional marks or arrows on the gasket if equipped.

9 Install the manifold and thread the mounting nuts into place.

10 Working from the center out, tighten the nuts/bolts to the torque listed in this Chapter's Specifications in three or four equal steps.

11 Reinstall the remaining parts in the reverse order of removal.

12 Run the engine and check for exhaust leaks.

## 11 Cylinder head - removal and installation

**Warning:** *The engine must be completely cool before beginning this procedure.*

### Removal

*Refer to illustrations 11.11 and 11.13*

1 Relieve the fuel system pressure (see Chapter 4), then disconnect the cable from the negative terminal of the battery. **Caution:** *If the stereo in your vehicle is equipped with*

10.6 . . . and the exhaust manifold retaining nuts, then pull the manifold off the studs on the cylinder head and remove from above

*an anti-theft system, make sure you have the correct activation code before disconnecting the battery.*

2 Drain the coolant from the engine block and radiator (see Chapter 1).

3 Remove the drivebelt and the alternator (see Chapter 5).

4 Remove the valve cover (see Section 4).

5 Remove the throttle body, fuel injectors and fuel rail (see Chapter 4).

6 Remove the intake manifold (see Section 9).

7 Remove the exhaust manifold (see Section 10).

8 Remove the timing chain and camshaft sprockets (see Section 7).

9 Remove the camshafts and lifters (see Section 8).

10 On 2000 and later models, refer to Section 5 and remove the variable valve timing control valve and filter.

11 Label and remove the coolant hoses, tubes and electrical connections from the cylinder head **(see illustration)**.

12 Using a 10 mm hex-head socket bit and a breaker bar, loosen the cylinder head bolts in 1/4-turn increments until they can be removed by hand. Loosen the cylinder head

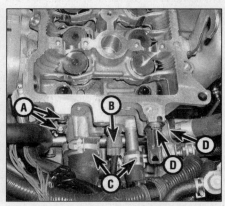

11.11 Detach the ground straps (A), the ECT sensor connector (B), the coolant hoses (C) and the bracket retaining bolts (D)

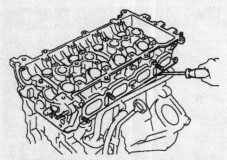

**11.13 Pry the cylinder head off the engine block at the casting protrusion - be careful not to damage the mating surfaces**

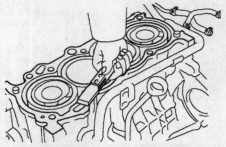

**11.16 Remove all traces of old gasket material - the cylinder head and block mating surfaces must be perfectly clean to ensure a good gasket seal**

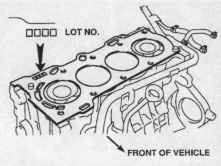

**11.21 Place the head gasket over the dowels in the block with the marks facing UP**

bolts in the reverse order of the recommended tightening sequence **(see illustration 11.24)** to avoid warping or cracking the cylinder head.

13   Lift the cylinder head off the engine block. If it's stuck, very carefully pry up at the transaxle end, beyond the gasket surface **(see illustration)**.

14   Remove any remaining external components from the cylinder head to allow for thorough cleaning and inspection. See Chapter 2C for cylinder head servicing procedures.

## Installation

*Refer to illustrations 11.16, 11.21 and 11.24*

15   The mating surfaces of the cylinder head and block must be perfectly clean when the cylinder head is installed.

16   Use a gasket scraper to remove all traces of carbon and old gasket material **(see illustration)**, then clean the mating surfaces with lacquer thinner or acetone. If there's oil on the mating surfaces when the cylinder head is installed, the gasket may not seal correctly and leaks could develop. When working on the block, stuff the cylinders with clean shop rags to keep out debris. Use a vacuum cleaner to remove material that falls into the cylinders.

17   Check the block and cylinder head mating surfaces for nicks, deep scratches and other damage. If damage is slight, it can be removed with a file; if it's excessive, machining may be the only alternative.

18   Use a tap of the correct size to chase the threads in the cylinder head bolt holes, then clean the holes with compressed air - make sure that nothing remains in the holes. **Warning:** *Wear eye protection when using compressed air!*

19   Using a wire brush, clean the threads on each bolt to remove corrosion and restore the threads. Dirt, corrosion, sealant and damaged threads will affect torque readings. Also check the cylinder head bolts for stretching. Measure the entire length of each bolt from the top of head to the tip of the threads. If the length of any bolt exceeds 6.280 inches, it must be replaced.

20   Install the components that were removed from the cylinder head.

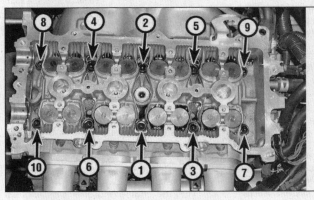

**11.24  Cylinder head bolt TIGHTENING sequence**

21   Position the new gasket over the dowel pins in the block **(see illustration)**.

22   Carefully set the cylinder head on the block without disturbing the gasket.

23   Before installing the cylinder head bolts, apply a small amount of clean engine oil to the threads and under the bolt heads.

24   Install the bolts in their original locations and tighten them finger tight. Following the recommended sequence, tighten the bolts in three steps to the torque listed in this Chapter's Specifications **(see illustration)**. Step 3 of the tightening sequence requires each bolt to be tightened an additional 90 degrees. If you don't have an angle-torque attachment for your torque wrench, simply apply a paint mark at one edge of each cylinder head bolt and tighten the bolt until that mark is 90 degrees (1/4-turn) from where you started at the beginning of Step 3.

25   The remaining installation steps are the reverse of removal.

26   Check and adjust the valves as necessary (see Chapter 1).

27   Change the engine oil and filter (see Chapter 1).

28   Refill the cooling system (see Chapter 1), run the engine and check for leaks.

## 12   Crankshaft pulley/vibration damper - removal and installation

*Refer to illustrations 12.4, 12.5 and 12.6*

1   Detach the cable from the negative ter-

minal of the battery. **Caution:** *If the stereo in your vehicle is equipped with an anti-theft system, make sure you have the correct activation code before disconnecting the battery.*

2   Remove the drivebelt (see Chapter 1).

3   With the parking brake applied and the shifter in Park (automatic) or in gear (manual), loosen the lug nuts from the right front wheel, then raise the front of the vehicle and support it securely on jackstands. Remove the right

**12.4 A puller base and several spacers can be mounted to the center hub of the pulley to keep the crankshaft from turning as the pulley retaining bolt is loosened - install the socket over the crankshaft bolt head before installing the puller, then insert the extension through the center hole of the puller**

**12.5 Reinstall the puller back onto the crankshaft pulley with the center bolt attached and remove the crankshaft pulley**

**12.6 Align the keyway in the crankshaft pulley hub with the Woodruff key in the crankshaft (arrow)**

**13.2 Carefully pry the old seal out of the timing chain cover - don't damage the crankshaft in the process**

2B

front wheel and the right splash shield from the wheelwell.

4   Remove the bolt from the front of the crankshaft. A breaker bar will probably be necessary, since the bolt is very tight **(see illustration)**.

5   Using a puller that bolts to the crankshaft hub, remove the crankshaft pulley from the crankshaft **(see illustration)**. **Note:** *Depending on the type of puller you have it may be necessary to support the engine from above, remove the right side engine mount and lower to engine to gain sufficient clearance to use the puller.*

6   To install the crankshaft pulley, slide the pulley onto the crankshaft as far as it will slide on, then use a vibration damper installation tool to press the pulley onto the crankshaft. Note that the slot (keyway) in the hub must be aligned with the Woodruff key in the end of the crankshaft **(see illustration)** and that the crankshaft bolt can also be used to press the crankshaft pulley into position.

7   Tighten the crankshaft bolt to the torque and angle of rotation listed in this Chapter's Specifications.

8   The remaining installation steps are the reverse of removal.

## 13   Crankshaft front oil seal - replacement

*Refer to illustrations 13.2, 13.3 and 13.4*

1   Remove the crankshaft pulley (see Section 12).

2   Note how the seal is installed - the new one must be installed to the same depth and facing the same way. Carefully pry the oil seal out of the cover with a seal puller or a large screwdriver **(see illustration)**. Be very careful not to distort the cover or scratch the crankshaft! Wrap electrician's tape around the tip of the screwdriver to avoid damage to the crankshaft.

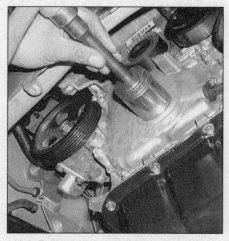

**13.3 Drive the new seal into place with a large socket and hammer**

3   Apply clean engine oil or multi-purpose grease to the outer edge of the new seal, then install it in the cover with the lip (spring side) facing IN. Drive the seal into place with a seal driver or a large socket and a hammer **(see illustration)**. Make sure the seal enters the bore squarely and stop when the front face is at the proper depth.

4   Check the surface on the pulley hub that the oil seal rides on. if the surface has been grooved from long-time contact with the seal, a press-on sleeve may be available to renew the sealing surface **(see illustration)**. This sleeve is pressed into place with a hammer and a block of wood and is commonly available at auto parts stores for various applications.

5   Lubricate the pulley hub with clean engine oil and reinstall the crankshaft pulley (see Section 12).

6   Install the crankshaft pulley retaining bolt and tighten it to the torque listed in this Chapter's Specifications.

7   The remainder of installation is the reverse of the removal.

**13.4 If the sealing surface of the pulley hub has a wear groove from contact with the seal, repair sleeves are available at most auto parts stores**

## 14   Oil pan - removal and installation

### *Removal*

*Refer to illustrations 14.6a and 14.6b*

1   Disconnect the cable from the negative terminal of the battery. **Caution:** *If the stereo in your vehicle is equipped with an anti-theft system, make sure you have the correct activation code before disconnecting the battery.*

2   Set the parking brake and block the rear wheels.

3   Raise the front of the vehicle and support it securely on jackstands.

4   Remove the two plastic splash shields under the engine, if equipped.

5   Drain the engine oil and remove the oil filter (see Chapter 1). Remove the oil dipstick.

6   Remove the bolts and detach the oil pan. If it's stuck, pry it loose very carefully with a small screwdriver or putty knife **(see**

**14.6a  Oil pan mounting bolts**

**14.6b  Pry the oil pan loose with a screwdriver or putty knife - be careful not to damage the mating surfaces of the pan and block or oil leaks may develop**

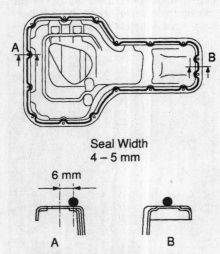

**Seal Width 4 – 5 mm**

**6 mm**

**14.11  Oil pan sealant installation details**

**15.2  Oil pump mounting bolts (arrows)**

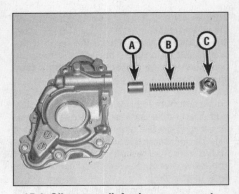

**15.4  Oil pump relief valve components**

A    Relief valve        C    Plug
B    Spring

illustrations). Don't damage the mating surfaces of the pan and block or oil leaks could develop.

## Installation

*Refer to illustration 14.11*

7    Use a scraper to remove all traces of old sealant from the block and oil pan. Clean the mating surfaces with lacquer thinner or acetone.
8    Make sure the threaded bolt holes in the block are clean.
9    Check the oil pan flange for distortion, particularly around the bolt holes. Remove any nicks or burrs as necessary.
10   Inspect the oil pump pick-up tube assembly for cracks and a blocked strainer. If the pick-up was removed, clean it thoroughly and install it now, using a new gasket. Tighten the nuts/bolts to the torque listed in this Chapter's Specifications.
11   Apply a 3/16-inch wide bead of RTV sealant to the mating surface of the oil pan **(see illustration)**. **Note:** *Be sure follow the sealant manufacturers recommendations for assembly and sealant curing times.*
12   Carefully position the oil pan on the

engine block and install the oil pan-to-engine block bolts loosely.
13   Working from the center out, tighten the oil pan-to-engine block bolts to the torque listed in this Chapter's Specifications in three or four steps.
14   The remainder of installation is the reverse of removal. Be sure to wait at least one hour before adding oil to allow the sealant to properly cure.
15   Run the engine and check for oil pressure and leaks.

## 15   Oil pump - removal, inspection and installation

## Removal

*Refer to illustrations 15.2 and 15.4*

1    Refer to Section 7, Steps 1 through 17 and remove the timing chain and the crankshaft sprocket.
2    Remove the five bolts and detach the oil pump body from the engine **(see illustration)**. You may have to pry carefully between the front of the block and the pump body with a screwdriver to remove it.
3    Use a scraper to remove all traces of sealant and old gasket material from the pump

body and engine block, then clean the mating surfaces with lacquer thinner or acetone.
4    Remove the oil pressure relief valve from the pump body **(see illustration)**.
5    Remove the three screws and separate the pump cover from the body. Lift out the drive and driven rotors.

## Inspection

*Refer to illustrations 15.8a, 15.8b, 15.8c and 15.9*

6    Clean all components with solvent, then inspect them for wear and damage.
7    Check the oil pressure relief valve piston sliding surface and valve spring. If either the spring or the valve is damaged, they must be replaced as a set.
8    Check the driven rotor-to-body clearance, rotor-to-cover clearance and drive rotor tip clearance with a feeler gauge **(see illustrations)** and compare the results to this Chapter's Specifications. If any clearance is excessive, replace the rotors as a set. If necessary, replace the oil pump body.
9    Check the timing chain guide oil jet for blockage **(see illustration)**.

15.8a  Measure the driven rotor-to-body clearance with a feeler gauge

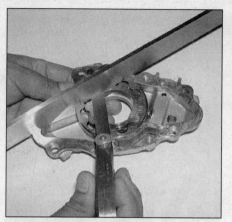

15.8b  Using a straightedge and feeler gauge, measure the rotor-to-cover clearance

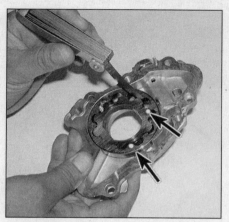

15.8c  Measure the rotor tip clearance with a feeler gauge - install the rotors with the marks facing out, against the cover (arrows)

**2B**

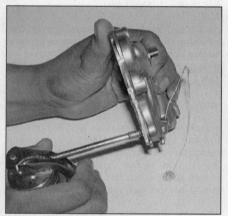

15.9  Check that the oil jet is free of debris - a blockage here will lead to an excessively worn timing chain and guides

15.14  Place the gasket over the dowels (lower arrows) on the engine block - when installing the oil pump, align the flats in the pump rotor with the flats on the crankshaft (upper arrow)

17.3 Carefully pry out the old seal with a screwdriver - it may be helpful to cut the seal lip with a razor blade first to make removal of the seal easier

## Installation

*Refer to illustration 15.14*

10   Lubricate the drive and driven rotors with clean engine oil and place them in the pump body with the marks facing out **(see illustration 15.8c)**.

11   Pack the pump cavity with petroleum jelly and attach the pump cover, tighten the screws to the torque listed in this Chapter's Specifications.

12   Lubricate the oil pressure relief valve piston with clean engine oil and reinstall the valve components in the pump body.

13   Place a new oil pump gasket on the engine block.

14   Position the pump assembly against the block and install the mounting bolts. Make sure that the flats on the oil pump drive rotor align with the flats on the crankshaft **(see illustration)**.

15   Tighten the bolts to the torque listed in this Chapter's Specifications in three or four steps. Follow a criss-cross pattern to avoid warping the body.

16   Reinstall the remaining parts in the reverse order of removal.

17   Add oil if necessary, start the engine and check for oil pressure and leaks.

## 16   Flywheel/driveplate - removal and installation

The flywheel/driveplate replacement for 1998 and later engines is identical to the flywheel/driveplate replacement procedure for the 1997 and earlier engines. Refer to Chapter 2 Part A for the procedure and use the torque figures in this Chapter's Specifications.

## 17   Rear main oil seal - replacement

*Refer to illustrations 17.3 and 17.4*

1   Remove the transmission (see Chapter 7A or 7B).

2   Remove the flywheel/driveplate (see Section 16).

3   Pry the oil seal from the rear of the engine with a screwdriver **(see illustration)**. Be careful not to nick or scratch the

crankshaft or the seal bore. Be sure to note how far it's recessed into the bore before removal so the new seal can be installed to the same depth. Thoroughly clean the seal bore in the block with a shop towel. Remove all traces of oil and dirt.

4   Lubricate the outside diameter of the seal and install the seal over the end of the crankshaft. Make sure the lip of the seal

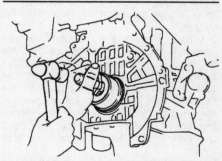

17.4  The rear oil seal can be pressed into place with a seal installation tool, a section of pipe or a blunt object shown here - in any case be sure the seal is installed squarely into the seal bore and flush with the rear of the engine

**18.8a To remove the passenger's side mount, detach the three mount-to-body bolts (lower arrows), the air conditioning refrigerant line clamp (upper arrow, if equipped) . . .**

**18.8b . . . and the mount bracket-to-engine mount bracket nuts and bolts (arrows)**

**18.9 Driver's side engine mount details**

- A   *Mount-to-body bolts*
- B   *Through-bolt*

points toward the engine. Preferably, a seal installation tool (available at most auto parts store) is needed to press the new seal back into place. If the proper seal installation tool is unavailable, use a large socket, section of pipe or a blunt tool and carefully drive the new seal squarely into the seal bore and flush with the edge of the engine block **(see illustration).**

5   Install the flywheel/driveplate (see Section 16).

6   Install the transmission (see Chapter 7A or 7B).

## 18   Powertrain mounts - check and replacement

1   Engine mounts seldom require attention, but broken or deteriorated mounts should be replaced immediately or the added strain placed on the driveline components may cause damage or wear.

### Check

**Note:** *During the check, the engine must be raised slightly to remove the weight from the mounts.*

2   Raise the vehicle and support it securely on jackstands, then position a jack under the engine oil pan. Place a large wood block between the jack head and the oil pan, then carefully raise the engine just enough to take the weight off the mounts. Do not position the wood block under the drain plug. **Warning:** *DO NOT place any part of your body under the engine when it's supported only by a jack!*

3   Check the mounts to see if the rubber is cracked, hardened or separated from the metal casing, which would indicate the need for replacement.

4   Check for relative movement between the inner and outer portions of the mount (use a large screwdriver or pry bar to attempt

to move the mounts). If movement is noted, replace the mount.

5   Check the mount fasteners to make sure they are tight.

6   Rubber preservative should be applied to the mounts to slow deterioration.

### Replacement

*Refer to illustrations 18.8a, 18.8b, 18.9, 18.10a and 18.10b*

7   Disconnect the cable from the negative terminal of the battery, then raise the vehicle and support it securely on jackstands (if not already done). Support the engine as described in Step 2. **Caution:** *If the stereo in your vehicle is equipped with an anti-theft system, make sure you have the correct activation code before disconnecting the battery.*

8   To remove the right (passenger side) engine mount, remove the three bolts securing the mount to the body, then remove the three bolts and two nuts securing the mount bracket to the engine bracket **(see illustrations).** If the vehicle is equipped with air con-

ditioning, detach the refrigerant line bracket retaining nut and remove the mount from the engine compartment. Be sure to remove the mount bracket from the old mount and reinstall it on the new mount.

9   To remove the left (driver's side) engine mount, remove the four bolts securing the mount to the body, then remove the through-bolt securing the mount to the transaxle bracket **(see illustration).** On manual transaxle equipped vehicles, it will be necessary to remove two bolts and an additional mount support bracket before removing the mount-to-body bolts and the mount through bolt.

10   To remove the front and rear engine mounts, remove the bolts securing the mount to the lower frame crossmember, then remove the through-bolt securing the mount to the engine bracket **(see illustrations).**

11   Installation is the reverse of removal. Use thread locking compound on the mount bolts/nuts and be sure to tighten them securely.

**18.10a Front engine mount details**

- A   *Mount-to-lower frame crossmember*
- B   *Through-bolt*

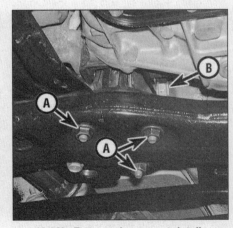

**18.10b Rear engine mount details**

- A   *Mount-to-frame crossmember*
- B   *Through-bolt*

# Chapter 2 Part C
# General engine overhaul procedures

## Contents

**2C**

## Specifications

### 1997 and earlier

#### General

| | |
|---|---|
| Displacement | |
| 1.6L ....................................................................... | 97 cubic inches |
| 1.8L ....................................................................... | 108 cubic inches |
| Cylinder compression pressure @ 250 rpm | |
| Standard............................................................... | 191 psi |
| Minimum............................................................... | 142 psi |
| Oil pressure (engine warm) | |
| At idle ................................................................... | 4.3 psi minimum |
| At 3000 rpm........................................................... | 36 to 71 psi |

#### Cylinder head

| | |
|---|---|
| Warpage limits | |
| Block surface ......................................................... | 0.0020 inch |
| Manifold surfaces ................................................... | 0.0039 inch |

#### Valves and related components

| | |
|---|---|
| Valve margin width | |
| Standard................................................................ | 0.031 to 0.047 inch |
| Minimum................................................................ | 0.020 inch |
| Valve stem diameter | |
| Intake ................................................................... | 0.2350 to 0.2356 inch |
| Exhaust................................................................. | 0.2348 to 0.2354 inch |
| Valve stem-to-guide clearance | |
| Intake | |
| Standard ........................................................... | 0.0010 to 0.0024 inch |
| Service limit....................................................... | 0.0031 inch |
| Exhaust | |
| Standard ........................................................... | 0.0012 to 0.0026 inch |
| Service limit....................................................... | 0.0039 inch |

## 1997 and earlier (continued)

### Valves and related components

Valve spring
    Out-of-square limit ................................................................. 0.079 inch
    Free length ............................................................................ 1.669 inches
    Installed height ..................................................................... 1.248 inches
Valve lifter
    Diameter ............................................................................... 1.2191 to 1.2195 inches
    Lifter bore diameter ............................................................. 1.2205 to 1.2215 inches
    Lifter-to-bore clearance
        Standard .......................................................................... 0.0009 to 0.0023 inch
        Service limit ..................................................................... 0.0028 inch
Valve clearance .......................................................................... See Chapter 1

### Crankshaft and connecting rods

Connecting rod journal
    Diameter
        1.6L ................................................................................. 1.5742 to 1.5748 inches
        1.8L ................................................................................. 1.8893 to 1.8898 inches
    Taper and out-of-round limits ............................................... 0.0002 inch
Connecting rod bearing oil clearance
    Standard
        1.6L ................................................................................. 0.0008 to 0.0020 inch
        1.8L ................................................................................. 0.0008 to 0.0017 inch
    Service limit ......................................................................... 0.0031 inch
Connecting rod side clearance (endplay)
    Standard ............................................................................... 0.0059 to 0.0098 inch
    Service limit ......................................................................... 0.0118 inch
Connecting rod bolts
    Minimum diameter
        1.6L (at middle of threads) ............................................. 0.339 inch
        1.8L (at middle of bolt) .................................................... 0.276 inch
Main bearing journal
    Diameter ............................................................................... 1.8891 to 1.8898 inches
    Taper and out-of-round limits ............................................... 0.0002 inch
    Runout limit .......................................................................... 0.0012 inch
Main bearing oil clearance
    Standard ............................................................................... 0.0006 to 0.0013 inch
    Service limit ......................................................................... 0.0039 inch
Crankshaft endplay
    Standard ............................................................................... 0.0008 to 0.0087 inch
    Service limit ......................................................................... 0.0118 inch
    Thrust washer thickness ....................................................... 0.0961 to 0.0980 inch

### Engine block

Deck warpage limit ..................................................................... 0.0020 inch
Cylinder bore diameter
    Standard
        Mark 1 ............................................................................. 3.1890 to 3.1894 inches
        Mark 2 ............................................................................. 3.1894 to 3.1898 inches
        Mark 3 ............................................................................. 3.1898 to 3.1902 inches
    Service limit ......................................................................... 3.1982 inches
Taper and out-of-round limits ..................................................... 0.0008 inch

### Pistons and rings

Piston diameter
    Mark 1 .................................................................................. 3.1852 to 3.1856 inches
    Mark 2 .................................................................................. 3.1856 to 3.1860 inches
    Mark 3 .................................................................................. 3.1860 to 3.1864 inches
Piston-to-bore clearance
    Standard ............................................................................... 0.0033 to 0.0041 inch
    Service limit ......................................................................... 0.009 inch
Piston ring end gap
    No. 1 (top) compression ring
        Standard .......................................................................... 0.0098 to 0.0138 inch
        Service limit ..................................................................... 0.0413 inch
    No. 2 (middle) compression ring
        Standard .......................................................................... 0.0138 to 0.0197 inch
        Service limit ..................................................................... 0.0472 inch

Oil ring
Standard ............................................................................... 0.0039 to 0.0157 inch
Service limit ......................................................................... 0.0413 inch
Piston ring groove clearance
No. 1 (top) compression ring .................................................... 0.0018 to 0.0033 inch
No. 2 (middle) compression ring ............................................... 0.0012 to 0.0028 inch

## Torque specifications*
**Ft-lbs** (unless otherwise indicated)
Main bearing cap bolts ................................................................ 44
Connecting rod cap nuts/bolts
Step 1
1.6L ....................................................................................... 22
1.8L ....................................................................................... 18
Step 2 ...................................................................................... Turn an additional 90-degrees (1/4-turn)

*Note: Refer to Part A for additional torque specifications.*

## 1998 and later

### General
Displacement
1.8L ........................................................................................ 108 cubic inches
Cylinder compression pressure @ 250 rpm
Standard ................................................................................. 218 psi
Minimum ................................................................................. 145 psi
Maximum variation between cylinders ....................................... 15 psi
Oil pressure (engine warm)
At idle (minimum) ................................................................... 4.3 psi
At 3000 rpm ........................................................................... 43 to 78 psi

### Cylinder head
Warpage limits
Block surface .......................................................................... 0.0020 inch
Manifold surfaces ................................................................... 0.0039 inch
Cylinder head bolts
Minimum diameter (at middle of bolt) ....................................... 0.354 inch
Maximum overall length ......................................................... 6.280 inches

### Valves and related components
Valve margin width
Standard ................................................................................. 0.039 inch
Minimum ................................................................................. 0.028 inch
Valve stem diameter
Intake ..................................................................................... 0.2154 to 0.2159 inch
Exhaust .................................................................................. 0.2152 to 0.2157 inch
Valve stem-to-guide clearance
Intake
Standard ............................................................................. 0.0010 to 0.0024 inch
Service limit ......................................................................... 0.0031 inch
Exhaust
Standard ............................................................................. 0.0012 to 0.0026 inch
Service limit ......................................................................... 0.0039 inch
Valve spring
Out-of-square limit ................................................................. 0.063 inch
Free length ............................................................................. 1.807 inches
Installed height ....................................................................... 1.323 inches
Valve lifter
Diameter ................................................................................. 1.2191 to 1.2195 inches
Lifter bore diameter ................................................................ 1.2205 to 1.2215 inches
Lifter-to-bore clearance
Standard ............................................................................. 0.0009 to 0.0023 inch
Service limit ......................................................................... 0.0031 inch
Valve lash ................................................................................. See Chapter 1

### Crankshaft and connecting rods
Connecting rod journal
Diameter ................................................................................. 1.7320 to 1.7328 inches
Taper and out-of-round limits .................................................. 0.0008 inch
Connecting rod bearing oil clearance
Standard ................................................................................. 0.0011 to 0.0024 inch
Service limit ............................................................................ 0.0031 inch

**2C**

## 1998 and later (continued)

### Crankshaft and connecting rods

Connecting rod side clearance (endplay)
  Standard ....................................................................................... 0.0063 to 0.0135 inch
  Service limit ................................................................................. 0.0157 inch
Connecting rod bolts
  Minimum diameter (at middle of bolt) ....................................... 0.252 inch
Main bearing journal
  Diameter ...................................................................................... 1.8893 to 1.8898 inches
  Taper and out-of-round limits ................................................... 0.0008 inch
  Runout limit ................................................................................. 0.0012 inch
Main bearing oil clearance
  Standard ....................................................................................... 0.0006 to 0.0013 inch
  Service limit ................................................................................. 0.0020 inch
Crankshaft endplay
  Standard ....................................................................................... 0.0016 to 0.0094 inch
  Service limit ................................................................................. 0.0118 inch
  Thrust washer thickness ............................................................ 0.0957 to 0.0976 inch
Main bearing cap sub-assembly bolts **(bolts 1 through 10 in illustration 23.12b)**
  Minimum diameter (at middle of bolt) ....................................... 0.287 inch

### Engine block

Deck warpage limit ........................................................................... 0.0020 inch
Cylinder bore diameter
  Standard ....................................................................................... 3.1102 to 3.1107 inches
  Service limit ................................................................................. 3.1107 inches
Taper and out-of-round limits .......................................................... 0.0008 inch

### Pistons and rings

Piston diameter ................................................................................. 3.1073 to 3.1077 inches
Piston-to-bore clearance
  Standard ....................................................................................... 0.0026 to 0.0035 inch
  Service limit ................................................................................. 0.0039 inch
Piston ring end gap
  No. 1 (top) compression ring
    Standard ..................................................................................... 0.0098 to 0.0138 inch
    Service limit ............................................................................... 0.0413 inch
  No. 2 (middle) compression ring
    Standard ..................................................................................... 0.0138 to 0.0197 inch
    Service limit ............................................................................... 0.0472 inch
  Oil ring
    Standard ..................................................................................... 0.0059 to 0.0157 inch
    Service limit ............................................................................... 0.0413 inch
Piston ring groove clearance .......................................................... 0.0012 to 0.0028 inch

### Torque specifications*                                                  **Ft-lbs** (unless otherwise indicated)

Main bearing cap sub-assembly/bedplate bolts
  Bolts 1 through 10
    Step 1 ......................................................................................... 16
    Step 2 ......................................................................................... 32
    Step 3 ......................................................................................... Turn an additional 45 degrees
    Step 4 ......................................................................................... Turn an additional 45 degrees
  Bolts 11 through 20
    Step 5 ......................................................................................... 168 in-lbs
Connecting rod cap bolts
  Step 1 ........................................................................................... 15
  Step 2 ........................................................................................... Turn an additional 90-degrees

*Note: Refer to Part B for additional torque specifications.*

## 1   General information

Included in this portion of Chapter 2 are the general overhaul procedures for the cylinder head(s) and internal engine components.

The information ranges from advice concerning preparation for an overhaul and the purchase of replacement parts to detailed, step-by-step procedures covering removal and installation of internal engine components and the inspection of parts.

The following Sections have been written based on the assumption that the engine has been removed from the vehicle. For information concerning in-vehicle engine repair, as well as removal and installation of the external components necessary for the overhaul, see Chapter 2A or 2B and Section 8 of this Chapter.

The Specifications included in this Part are only those necessary for the inspection and overhaul procedures which follow. Refer to Part A or Part B for additional Specifications.

2.4a  The oil pressure can be checked by removing the sending unit and installing a pressure gauge in its place

2.4b  The oil pressure sending unit (arrow) on 1997 and earlier engines is located on the front of the engine block, just to the right of the oil filter

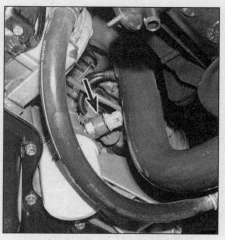

2.4c  The oil pressure sending unit (arrow) on 1998 and later engines is located on the front of the engine block, just above the oil filter

2C

## 2  Engine overhaul - general information

*Refer to illustrations 2.4a, 2.4b and 2.4c*

It's not always easy to determine when, or if, an engine should be completely over-hauled, as a number of factors must be considered.

High mileage is not necessarily an indication that an overhaul is needed, while low mileage doesn't preclude the need for an overhaul. Frequency of servicing is probably the most important consideration. An engine that's had regular and frequent oil and filter changes, as well as other required maintenance, will most likely give many thousands of miles of reliable service. Conversely, a neglected engine may require an overhaul very early in its life.

Excessive oil consumption is an indication that piston rings, valve seals and/or valve guides are in need of attention. Make sure that oil leaks aren't responsible before deciding that the rings and/or guides are bad. Perform a cylinder compression check to determine the extent of the work required (see Section 4). Also check the vacuum readings under various conditions (see Section 3).

Check the oil pressure with a gauge installed in place of the oil pressure sending unit **(see illustrations)** and compare it to this Chapter's Specifications. If it's extremely low, the bearings and/or oil pump are probably worn out.

Loss of power, rough running, knocking or metallic engine noises, excessive valve train noise and high fuel consumption rates may also point to the need for an overhaul, especially if they're all present at the same time. If a complete tune-up doesn't remedy the situation, major mechanical work is the only solution.

An engine overhaul involves restoring the internal parts to the specifications of a new engine. During an overhaul, the piston rings are replaced and the cylinder walls are reconditioned (rebored and/or honed). If a rebore is done by an automotive machine shop, new oversize pistons will also be installed. The main bearings, connecting rod bearings and camshaft bearings are generally replaced with new ones and, if necessary, the crankshaft may be reground to restore the journals. Generally, the valves are serviced as well, since they're usually in less-than-perfect condition at this point. While the engine is being overhauled, other components, such as the distributor, starter and alternator, can be rebuilt as well. The end result should be a like new engine that will give many trouble free miles. **Note:** *Critical cooling system components such as the hoses, drivebelts, thermostat and water pump should be replaced with new parts when an engine is overhauled. The radiator should be checked carefully to ensure that it isn't clogged or leaking (see Chapter 3). If you purchase a rebuilt engine or short block, some rebuilders will not warranty their engines unless the radiator has been professionally flushed. Also, we don't recommend overhauling the oil pump - always install a new one when an engine is rebuilt.*

Before beginning the engine overhaul, read through the entire procedure to familiarize yourself with the scope and requirements of the job. Overhauling an engine isn't difficult, but it is time-consuming. Plan on the vehicle being tied up for a minimum of two weeks, especially if parts must be taken to an automotive machine shop for repair or reconditioning. Check on availability of parts and make sure that any necessary special tools and equipment are obtained in advance. Most work can be done with typical hand tools, although a number of precision measuring tools are required for inspecting parts to determine if they must be replaced. Often an automotive machine shop will handle the inspection of parts and offer advice concerning reconditioning and replacement. **Note:**

*Always wait until the engine has been completely disassembled and all components, especially the engine block, have been inspected before deciding what service and repair operations must be performed by an automotive machine shop.* Since the block's condition will be the major factor to consider when determining whether to overhaul the original engine or buy a rebuilt one, never purchase parts or have machine work done on other components until the block has been thoroughly inspected. As a general rule, time is the primary cost of an overhaul, so it doesn't pay to install worn or substandard parts.

As a final note, to ensure maximum life and minimum trouble from a rebuilt engine, everything must be assembled with care in a spotlessly-clean environment.

## 3  Vacuum gauge diagnostic checks

*Refer to illustrations 3.4 and 3.6*

1    A vacuum gauge provides valuable information about what is going on in the engine at a low-cost. You can check for worn rings or cylinder walls, leaking head or intake manifold gaskets, incorrect carburetor adjustments, restricted exhaust, stuck or burned valves, weak valve springs, improper ignition or valve timing and ignition problems.
2    Unfortunately, vacuum gauge readings are easy to misinterpret, so they should be used in conjunction with other tests to confirm the diagnosis.
3    Both the absolute readings and the rate of needle movement are important for accurate interpretation. Most gauges measure vacuum in inches of mercury (in-Hg). As vacuum increases (or atmospheric pressure decreases), the reading will increase. Also, for every 1,000 foot increase in elevation above sea level, the gauge readings will decrease about one inch of mercury.

**3.4 The vacuum gauge is easily attached to this unused port (arrow) on the intake manifold**

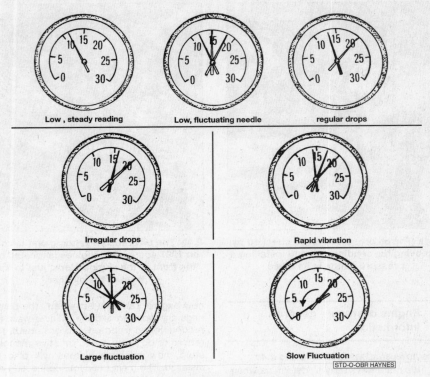

Low , steady reading    Low, fluctuating needle    regular drops

Irregular drops    Rapid vibration

Large fluctuation    Slow Fluctuation

STD-O-OBR HAYNES

**3.6 Typical vacuum gauge readings**

4    Connect the vacuum gauge directly to intake manifold vacuum, not to ported (above the throttle plate) vacuum **(see illustration)**. Be sure no hoses are left disconnected during the test or false readings will result.

5    Before you begin the test, allow the engine to warm up completely. Block the wheels and set the parking brake. With the transmission in neutral (or Park, on automatics), start the engine and allow it to run at normal idle speed. **Warning:** *Carefully inspect the fan blades for cracks or damage before starting the engine. Keep your hands and the vacuum tester clear of the fan and do not stand in front of the vehicle or in line with the fan when the engine is running.*

6    Read the vacuum gauge; an average, healthy engine should normally produce between 17 and 22 inches of vacuum with a fairly steady needle. Refer to the following vacuum gauge readings and what they indicate about the engines condition **(see illustration)**.

7    A low steady reading usually indicates a leaking gasket between the intake manifold and carburetor or throttle body, a leaky vacuum hose, late ignition timing or incorrect camshaft timing. Check ignition timing with a timing light and eliminate all other possible causes, utilizing the tests provided in this Chapter before you remove the timing belt cover to check the timing marks.

8    If the reading is three to eight inches below normal and it fluctuates at that low reading, suspect an intake manifold gasket leak at an intake port or a faulty injector.

9    If the needle has regular drops of about two to four inches at a steady rate the valves are probably leaking. Perform a compression or leak-down test to confirm this.

10    An irregular drop or down-flick of the needle can be caused by a sticking valve or an ignition misfire. Perform a compression or leak-down test and read the spark plugs.

11    A rapid vibration of about four in-Hg vibration at idle combined with exhaust smoke indicates worn valve guides. Perform a leak-down test to confirm this. If the rapid

vibration occurs with an increase in engine speed, check for a leaking intake manifold gasket or head gasket, weak valve springs, burned valves or ignition misfire.

12    A slight fluctuation, say one inch up and down, may mean ignition problems. Check all the usual tune-up items and, if necessary, run the engine on an ignition analyzer.

13    If there is a large fluctuation, perform a compression or leak-down test to look for a weak or dead cylinder or a blown head gasket.

14    If the needle moves slowly through a wide range, check for a clogged PCV system, incorrect idle fuel mixture, throttle body or intake manifold gasket leaks.

15    Check for a slow return after revving the engine by quickly snapping the throttle open until the engine reaches about 2,500 rpm and let it shut. Normally the reading should drop to near zero, rise above normal idle reading (about 5 in.-Hg over) and then return to the previous idle reading. If the vacuum returns slowly and doesn't peak when the throttle is snapped shut, the rings may be worn. If there is a long delay, look for a restricted exhaust system (often the muffler or catalytic converter). An easy way to check this is to temporarily disconnect the exhaust ahead of the suspected part and redo the test.

## 4    Cylinder compression check

*Refer to illustration 4.6*

1    A compression check will tell you what

mechanical condition the upper end (pistons, rings, valves, head gasket[s]) of your engine is in. Specifically, it can tell you if the compression is down due to leakage caused by worn piston rings, defective valves and seats or a blown head gasket. **Note:** *The engine must be at normal operating temperature and the battery must be fully charged for this check.*

2    Begin by cleaning the area around the spark plugs before you remove them (com-

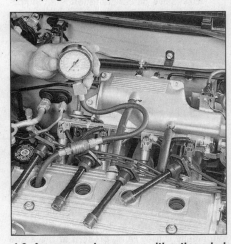

**4.6 A compression gauge with a threaded fitting for the spark plug hole is preferred over the type that requires hand pressure to maintain the seal - be sure to block open the throttle valve as far as possible during the compression check!**

pressed air should be used, if available). The idea is to prevent dirt from getting into the cylinders as the compression check is being done.

3    Remove all of the spark plugs from the engine (see Chapter 1).

4    Block the throttle wide open.

5    Disable the fuel system during the compression test (see Chapter 4, Section 2). On 1997 and earlier models, detach the coil wire from the center of the distributor cap and ground it on the engine block. Use a jumper wire with alligator clips on each end to ensure a good ground. On 1998 and 1999 models, disconnect the main electrical connectors from the two coil packs. On 2000 and later models, make sure the ignition coils have been disconnected and removed from the engine during spark plug removal.

6    Install the compression gauge in the spark plug hole (see illustration).

7    Crank the engine over at least seven compression strokes and watch the gauge. The compression should build up quickly in a healthy engine. Low compression on the first stroke, followed by gradually increasing pressure on successive strokes, indicates worn piston rings. A low compression reading on the first stroke, which doesn't build up during successive strokes, indicates leaking valves or a blown head gasket (a cracked head could also be the cause). Deposits on the undersides of the valve heads can also cause low compression. Record the highest gauge reading obtained.

8    Repeat the procedure for the remaining cylinders and compare the results to this Chapter's Specifications.

9    Add some engine oil (about three squirts from a plunger-type oil can) to each cylinder, through the spark plug hole, and repeat the test.

10    If the compression increases after the oil is added, the piston rings are definitely worn. If the compression doesn't increase significantly, the leakage is occurring at the valves or head gasket. Leakage past the valves may be caused by burned valve seats and/or faces or warped, cracked or bent valves.

11    If two adjacent cylinders have equally low compression, there's a strong possibility that the head gasket between them is blown. The appearance of coolant in the combustion chambers or the crankcase would verify this condition.

12    If one cylinder is 20-percent lower than the others, and the engine has a slightly rough idle, a worn exhaust lobe on the camshaft could be the cause.

13    If the compression is unusually high, the combustion chambers are probably coated with carbon deposits. If that's the case, the cylinder head(s) should be removed and decarbonized.

14    If compression is way down or varies greatly between cylinders, it would be a good idea to have a leak-down test performed by an automotive repair shop. This test will pinpoint exactly where the leakage is occurring and how severe it is.

## 5    Engine removal - methods and precautions

If you've decided that an engine must be removed for overhaul or major repair work, several preliminary steps should be taken.

Locating a suitable place to work is extremely important. Adequate work space, along with storage space for the vehicle, will be needed. If a shop or garage isn't available, at the very least a flat, level, clean work surface made of concrete or asphalt is required.

Cleaning the engine compartment and engine before beginning the removal procedure will help keep tools clean and organized.

An engine hoist or A-frame will also be necessary. Make sure the equipment is rated in excess of the combined weight of the engine and transaxle. Safety is of primary importance, considering the potential hazards involved in lifting the engine out of the vehicle.

If the engine is being removed by a novice, a helper should be available. Advice and aid from someone more experienced would also be helpful. There are many instances when one person cannot simultaneously perform all of the operations required when lifting the engine out of the vehicle.

Plan the operation ahead of time. Arrange for or obtain all of the tools and equipment you'll need prior to beginning the job. Some of the equipment necessary to perform engine removal and installation safely and with relative ease are (in addition to an engine hoist) a heavy duty floor jack, complete sets of wrenches and sockets as described in the front of this manual, wooden blocks and plenty of rags and cleaning solvent for mopping up spilled oil, coolant and gasoline. If the hoist must be rented, make sure that you arrange for it in advance and perform all of the operations possible without it beforehand. This will save you money and time.

Plan for the vehicle to be out of use for quite a while. A machine shop will be required to perform some of the work which the do-it-yourselfer can't accomplish without special equipment. These shops often have a busy schedule, so it would be a good idea to consult them before removing the engine in order to accurately estimate the amount of time required to rebuild or repair components that may need work.

Always be extremely careful when removing and installing the engine. Serious injury can result from careless actions. Plan ahead, take your time and a job of this nature, although major, can be accomplished successfully.

## 6    Engine - removal and installation

Refer to illustrations 6.7, 6.19, 6.24 and 6.25
Note: Read through the entire Section before beginning this procedure. The factory recom-

mends removing the engine and transaxle from the top as a unit, then separating the engine from the transaxle on the shop floor. If the transaxle is not being serviced, it is possible to leave the transaxle in the vehicle and remove the engine from the top by itself, by removing the crankshaft pulley and tilting up the timing belt or timing chain end of the engine for clearance.

### Removal

Warning: These models are equipped with airbags. The airbag is armed and can deploy (inflate) anytime the battery is connected. To prevent accidental deployment (and possible injury), turn the ignition key to LOCK and disconnect the negative battery cable whenever working near airbag components. After the battery is disconnected, wait at least two minutes before beginning work (the system has a back-up capacitor that must fully discharge). For more information see Chapter 12.

1    Relieve the fuel system pressure (see Chapter 4).

2    Disconnect the negative cable from the battery. Caution: If the stereo in your vehicle is equipped with an anti-theft system, make sure you have the correct activation code before disconnecting the battery.

3    Remove the battery and battery tray.

4    Place protective covers on the fenders and cowl and remove the hood (see Chapter 11).

5    Remove the air cleaner assembly (see Chapter 4).

6    Raise the vehicle and support it securely on jackstands. Drain the cooling system and engine oil and remove the drivebelts (see Chapter 1).

7    Clearly label, then disconnect all vacuum lines, coolant and emissions hoses, wiring harness connectors, ground straps and fuel lines. Masking tape and/or a touch up paint applicator work well for marking items (see illustration). Take instant photos or sketch the locations of components and brackets.

8    Remove the battery and battery tray (see Chapter 5).

9    Remove the windshield washer tank.

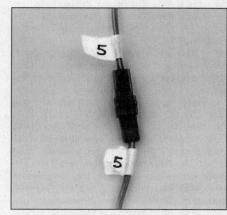

6.7  Label both ends of each wire and hose  before disconnecting it

6.19 Engine mount-to-lower crossmember bolts (A) and lower crossmember-to-chassis bolts (B)

6.24 Lift the engine high enough to clear the vehicle, then move it away and lower the hoist - this engine is being removed with the transaxle still attached

10    Remove the cooling fan(s), the radiator and the coolant reservoir tank (see Chapter 3).

11    Disconnect the heater hoses.

12    Release the residual fuel pressure in the tank by removing the gas cap, then detach the fuel lines connecting the engine to the chassis (see Chapter 4). Plug or cap all open fittings.

13    Disconnect the throttle linkage, transmission Throttle Valve (TV) linkage and speed control cable, if equipped, from the engine (see Chapter 4). If equipped with cruise control, remove the actuator from the engine compartment.

14    Refer to Part A or Part B of this Chapter and remove the drivebelt(s), the crankshaft pulley, the intake and exhaust manifold. Also remove the water pump pulley on 1997 and earlier models.

15    On power steering-equipped vehicles, unbolt the power steering pump. If clearance allows, tie the pump aside without disconnecting the hoses. If necessary, remove the pump (see Chapter 10).

16    On air-conditioned models, unbolt the compressor and set it aside. Do not disconnect the refrigerant hoses. **Note:** *Wire the compressor out of the way with a coat hanger, don't let the compressor hang on the hoses.*

17    Attach a lifting sling to the engine. Position a hoist and connect the sling to it. Take up the slack until there is slight tension on the hoist.

18    Refer to Chapter 8 and remove the driveaxles.

19    Remove the through-bolts holding the front (radiator side) and rear (firewall side) engine mount to its engine bracket, then detach the bolts securing the mounts to the lower engine crossmember and remove the front and rear mounts from the vehicle. Detach the remaining engine crossmember-to-chassis bolts and slide the lower engine crossmember out of the vehicle **(see illustration).**

20    On automatic transaxle-equipped models, pry out the torque converter dust shield from the lower bellhousing. Remove the torque converter-to-driveplate fasteners (see Chapter 7B) and push the converter back slightly into the bellhousing.

21    Recheck to be sure nothing except the right and left engine mounts and the belhousing bolts are still connecting the engine to the vehicle or to the transaxle. Disconnect and label anything still remaining.

22    Support the transaxle with a floor jack. Place a block of wood on the jack head to prevent damage to the transaxle. Remove the bolts from the right (passenger side) engine mount, leaving the left (driver's side) engine mount attached to the transaxle and the chassis. **Warning:** *Do not place any part of your body under the engine/transaxle when it's supported only by a hoist or other lifting device.*

23    Remove the engine-to-transaxle bolts and separate the engine from the transaxle (see Chapter 7A). The torque converter should remain in the transaxle. **Note:** *If the transaxle is to be removed at the same time, the driver's side engine mount should be disconnected, along with any wires, cables or hoses connected to the transaxle. The engine-to-transaxle bolts should also remain in place when removing the transaxle with the engine.*

24    Slowly lift the engine (or engine/transaxle) out of the vehicle **(see illustration)**. It may be necessary to pry the mounts away from the frame brackets. **Note:** *When removing the engine from a stick-shift vehicle and the transaxle is to remain in the vehicle, you may have to use the jack supporting the transaxle to tilt the transaxle enough to allow the engine to be angled forward and up out of the vehicle.*

25    Move the engine away from the vehicle and carefully lower the hoist until the engine can be set on the floor; or remove the flywheel/driveplate and mount the engine on an engine stand **(see illustration). Note:** *On*

*automatic transaxle-equipped models, mark the front and rear spacer plates and keep them with the driveplate.*

## Installation

26    Check the engine/transaxle mounts. If they're worn or damaged, replace them.

27    On manual transaxle-equipped models, inspect the clutch components (see Chapter 8) and on automatic models inspect the converter seal and bushing.

28    On automatic transaxle-equipped models, apply a dab of grease to the nose of the converter.

29    Carefully guide the transaxle into place, following the procedure outlined in Chapter 7. **Caution:** *Do not use the bolts to force the engine and transaxle into alignment. It may crack or damage major components.*

30    Install the engine-to-transaxle bolts and tighten them to the torque listed in the Chapter 7 Specifications.

31    Attach the hoist to the engine and carefully lower the engine/transaxle assembly into the engine compartment. **Note:** *If the engine*

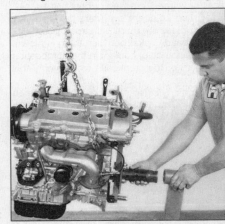

6.25 Lower the engine outside of the vehicle, remove the driveplate, and attach the engine to a suitable work stand

was removed with the transaxle remaining in the car, lower the engine into the car until an assistant can help you line up the dowel pins on the block with the transaxle. Some twisting and angling of the engine and/or the transaxle will be necessary to secure proper alignment of the two.

32  Install the lower crossmember and the engine mount bolts and tighten them securely.

33  Reinstall the remaining components and fasteners in the reverse order of removal.

34  Add coolant, oil, power steering and transmission fluids as needed (see Chapter 1).

35  Run the engine and check for proper operation and leaks. Shut off the engine and recheck the fluid levels.

## 7  Engine rebuilding alternatives

The do-it-yourselfer is faced with a number of options when performing an engine overhaul. The decision to replace the engine block, piston/connecting rod assemblies and crankshaft depends on a number of factors, with the number one consideration being the condition of the block. Other considerations are cost, access to machine shop facilities, parts availability, time required to complete the project and the extent of prior mechanical experience on the part of the do-it-yourselfer.

Some of the rebuilding alternatives include:

**Individual parts** - If the inspection procedures reveal that the engine block and most engine components are in reusable condition, purchasing individual parts may be the most economical alternative. The block, crankshaft and piston/connecting rod assemblies should all be inspected carefully. Even if the block shows little wear, the cylinder bores should be surface honed.

**Short block** - A short block consists of an engine block with a crankshaft and piston/connecting rod assemblies already installed. All new bearings are incorporated and all clearances will be correct. The existing camshafts, valve train components, cylinder head and external parts can be bolted to the short block with little or no machine shop work necessary.

**Long block** - A long block consists of a short block plus an oil pump, oil pan, cylinder head, valve cover, camshaft and valve train components, timing sprockets and chain or gears and timing cover. All components are installed with new bearings, seals and gaskets incorporated throughout. The installation of manifolds and external parts is all that's necessary.

Give careful thought to which alternative is best for you and discuss the situation with local automotive machine shops, auto parts dealers and experienced rebuilders before ordering or purchasing replacement parts.

## 8  Engine overhaul - disassembly sequence

*Refer to illustrations 8.5a and 8.5b*

1  It's much easier to disassemble and work on the engine if it's mounted on a portable engine stand. A stand can often be rented quite cheaply from an equipment rental yard. Before the engine is mounted on a stand, the flywheel/driveplate and rear oil seal retainer should be removed from the engine.

2  If a stand isn't available, it's possible to disassemble the engine with it blocked up on the floor. Be extra careful not to tip or drop the engine when working without a stand.

3  If you're going to obtain a rebuilt engine, all external components must come off first, to be transferred to the replacement engine, just as they will if you're doing a complete engine overhaul yourself. These include:

Alternator and brackets
Emissions control components
Distributor, spark plug wires and spark plugs (1997 and earlier)
Ignition coils (1998 and later)
Thermostat and housing cover
Water pump
EFI components
Intake/exhaust manifolds
Oil filter
Engine mounts
Clutch and flywheel/driveplate
Engine rear plate

**Note:** *When removing the external components from the engine, pay close attention to details that may be helpful or important during installation. Note the installed position of gaskets, seals, spacers, pins, brackets, washers, bolts and other small items.*

4  If you're obtaining a short block, which consists of the engine block, crankshaft, pistons and connecting rods all assembled, then the cylinder head, oil pan and oil pump will have to be removed as well from your engine so that your short block can be turned in to the rebuilder as a core. See *Engine rebuilding alternatives* for additional information regarding the different possibilities to be considered.

5  If you're planning a complete overhaul, the engine must be disassembled and the internal components removed in the following order **(see illustrations)**.

Intake and exhaust manifolds
Valve cover
Timing belt and sprockets (1997 and earlier)
Timing chain and sprockets (1998 and later)
Cylinder head
Oil pan
Oil pump
Piston/connecting rod assemblies
Crankshaft and main bearings
Rear main (crankshaft) oil seal and retainer (if equipped)

6  Before beginning the disassembly and

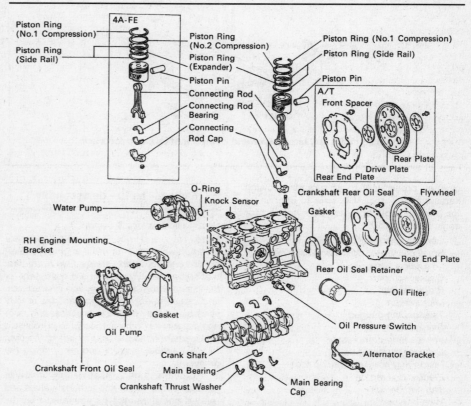

8.5a  1997 and earlier lower end components - exploded view

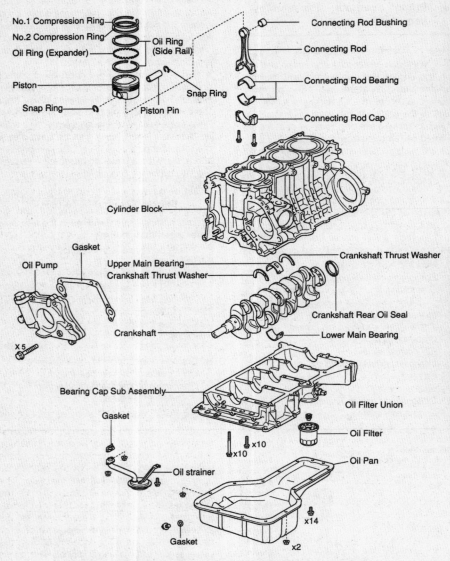

No.1 Compression Ring
No.2 Compression Ring
Oil Ring (Expander)
Oil Ring (Side Rail)
Piston
Snap Ring
Snap Ring
Piston Pin

Connecting Rod Bushing
Connecting Rod
Connecting Rod Bearing
Connecting Rod Cap

Cylinder Block

Gasket
Oil Pump
Upper Main Bearing
Crankshaft Thrust Washer
Crankshaft Thrust Washer
Crankshaft Rear Oil Seal
Crankshaft
Lower Main Bearing

X 5

Bearing Cap Sub Assembly

Oil Filter Union
Gasket
Oil Filter
x10
x10
Oil Pan
Oil strainer
Gasket
x14
x2

**8.5b  1998 and later lower end components - exploded view**

**9.2  A small plastic bag, with an appropriate label, can be used to store the valve train components so they can be kept together and reinstalled in the correct guide**

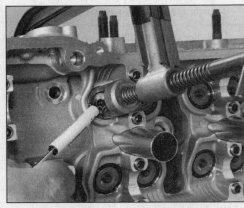

**9.3  Compress the spring until the keepers can be removed with a small magnetic screwdriver or needle-nose pliers**

overhaul procedures, make sure the following items are available. Also, refer to Section 21 for a list of tools and materials needed for engine reassembly.

*Common hand tools*
*Small cardboard boxes or plastic bags for storing parts*
*Gasket scraper*
*Ridge reamer*
*Micrometers*
*Telescoping gauges*
*Dial indicator set*
*Valve spring compressor*
*Cylinder surfacing hone*
*Piston ring groove-cleaning tool*
*Electric drill motor*
*Tap and die set*
*Wire brushes*
*Oil gallery brushes*
*Cleaning solvent*

## 9   Cylinder head - disassembly

*Refer to illustrations 9.2 and 9.3*
**Note:** *New and rebuilt cylinder heads are commonly available for most engines at dealerships and auto parts stores. Due to the fact that some specialized tools are necessary for the disassembly and inspection procedures, and replacement parts may not be readily available, it may be more practical and economical for the home mechanic to purchase a replacement head rather than taking the time to disassemble, inspect and recondition the original.*

1    Cylinder head disassembly involves removal of the intake and exhaust valves and related components. It's assumed that the lifters and camshafts have already been removed (see Chapter 2A or 2B as needed).

2    Before the valves are removed, arrange to label and store them, along with their related components, so they can be kept separate and reinstalled in the same guides they are removed from **(see illustration)**.
3    Compress the springs on the first valve with a spring compressor and remove the keepers **(see illustration)**. Carefully release the valve spring compressor and remove the retainer, the spring and the spring seat (if used). **Caution:** *Be very careful not to nick or otherwise damage the lifter bores when compressing the valve springs.* **Note:** *If your spring compressor does not have an end (such as the one shown) with cutouts on the side, an adapter is available to use with a standard spring compressor.*
4    Pull the valve out of the head, then remove the oil seal from the guide. If the valve binds in the guide (won't pull through), push it back into the head and deburr the area around the keeper groove with a fine file or whetstone.
5    Repeat the procedure for the remaining valves. Remember to keep all the parts for each valve together so they can be reinstalled in the same locations.
6    Once the valves and related components have been removed and stored in an

**10.12 Check the cylinder head gasket surfaces for warpage by trying to slip a feeler gauge under the precision straightedge (see the Specifications for the maximum warpage allowed and use a feeler gauge of that thickness)**

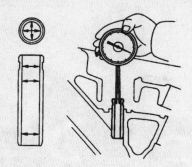

**10.14 Use a small dial bore gauge to determine the inside diameter of the valve guides**

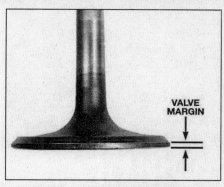

**10.16 The margin width on each valve must be as specified (if no margin exists, the valve cannot be re-used)**

organized manner, the head should be thoroughly cleaned and inspected. If a complete engine overhaul is being done, finish the engine disassembly procedures before beginning the cylinder head cleaning and inspection process.

## 10  Cylinder head - cleaning and inspection

*Refer to illustrations 10.12, 10.14, 10.16, 10.17 and 10.18*

1   Thorough cleaning of the cylinder head(s) and related valve train components, followed by a detailed inspection, will enable you to decide how much valve service work must be done during the engine overhaul. **Note:** *If the engine was severely overheated, the cylinder head is probably warped (see Step 12).*

### *Cleaning*

2   Scrape all traces of old gasket material and sealing compound off the head gasket, intake manifold and exhaust manifold sealing surfaces. Be very careful not to gouge the cylinder head. Special gasket-removal solvents that soften gaskets and make removal much easier are available at auto parts stores.
3   Remove all built up scale from the coolant passages.
4   Run a stiff wire brush through the various holes to remove deposits that may have formed in them. If there are heavy rust deposits in the water passages, the bare head should be professionally cleaned at a machine shop.
5   Run an appropriate-size tap into each of the threaded holes to remove corrosion and thread sealant that may be present. If compressed air is available, use it to clear the holes of debris produced by this operation.

**Warning:** *Wear eye protection when using compressed air!*
6   Clean the exhaust and intake manifold stud threads with a wire brush.
7   Clean the cylinder head with solvent and dry it thoroughly. Compressed air will speed the drying process and ensure that all holes and recessed areas are clean. **Note:** *Decarbonizing chemicals are available and may prove very useful when cleaning cylinder heads and valve train components. They are very caustic and should be used with caution. Be sure to follow the instructions on the container.*
8   Clean the lifters with solvent and dry them thoroughly. Compressed air will speed the drying process and can be used to clean out the oil passages. Don't mix them up during the cleaning process; keep them in a box with numbered compartments.
9   Clean all the valve springs, spring seats, keepers and retainers with solvent and dry them thoroughly. Work on the components from one valve at a time to avoid mixing up the parts.
10   Scrape off any heavy deposits that may have formed on the valves, then use a motorized wire brush to remove deposits from the valve heads and stems. Again, make sure the valves don't get mixed up.

### *Inspection*

**Note:** *Be sure to perform all of the following inspection procedures before concluding that machine shop work is required. Make a list of the items that need attention. The inspection procedures for the lifters and camshafts, can be found in Chapter 2A or 2B.*

#### Cylinder head

11   Inspect the head very carefully for cracks, evidence of coolant leakage and other damage. If cracks are found, check with an automotive machine shop concerning repair. If repair isn't possible, a new cylinder head should be obtained. On 1998 and later engines, inspect the cylinder head bolts for stretching. Measure the diameter of the each bolt at the middle past the threads and measure the overall length. Compare the measurements of each bolt to the Specifications

listed in this Chapter. If the bolts have stretched past their limit, they must be replaced.
12   Using a straightedge and feeler gauge, check the head gasket mating surface for warpage **(see illustration)**. If the warpage exceeds the limit found in this Chapter's Specifications, it can be resurfaced at an automotive machine shop.
13   Examine the valve seats in each of the combustion chambers. If they're pitted, cracked or burned, the head will require valve service that's beyond the scope of the home mechanic.
14   Check the valve stem-to-guide clearance with a small hole gauge and micrometer, or a small dial bore gauge **(see illustration)**. Also, check the valve stem deflection with a dial indicator attached securely to the head. The valve must be in the guide and approximately 1/16-inch off the seat. The total valve stem movement indicated by the gauge needle must be noted, then divided by two to obtain the actual clearance value. If it exceeds the stem-to-guide clearance limit found in this Chapter's Specifications, the valve guides should be replaced. After this is done, if there's still some doubt regarding the condition of the valve guides they should be checked by an automotive machine shop (the cost should be minimal). **Note:** *Most home mechanics will not have a precision small bore gauge, but your local machine shop can measure the guides for you.*

#### Valves

15   Carefully inspect each valve face for uneven wear, deformation, cracks, pits and burned areas. Check the valve stem for scuffing and galling and the neck for cracks. Rotate the valve and check for any obvious indication that it's bent. Look for pits and excessive wear on the end of the stem. The presence of any of these conditions indicates the need for valve service by an automotive machine shop.
16   Measure the margin width on each valve **(see illustration)**. Any valve with a margin narrower than that listed in this Chapter's Specifications will have to be replaced with a new one.

10.17  Measure the free length of each valve spring with a dial or vernier caliper

10.18  Check each valve spring for squareness

12.3  Gently tap the valve seals (arrow) into place with a deep socket and hammer - the intake seals are painted brown, while the exhaust seals are painted black or gray (don't mix them up)

## Valve components

17   Check each valve spring for wear (on the ends) and pits. Measure the free length and compare it to this Chapter's Specifications (see illustration). Any springs that are shorter than specified have sagged and should not be re-used. The tension of all springs should be pressure checked with a special fixture before deciding that they're suitable for use in a rebuilt engine (take the springs to an automotive machine shop for this check).

18   Stand each spring on a flat surface and check it for squareness (see illustration). If any of the springs are distorted or sagged, replace all of them with new parts.

19   Check the spring retainers and keepers for obvious wear and cracks. Any questionable parts should be replaced with new ones, as extensive damage will occur if they fail during engine operation.

20   Any damaged or excessively worn parts must be replaced with new ones.

21   If the inspection process indicates that the valve components are in generally poor condition and worn beyond the limits specified, which is usually the case in an engine that's being overhauled, reassemble the valves in the cylinder head and refer to Section 11 for valve servicing recommendations.

## 11   Valves - servicing

1   Because of the complex nature of the job and the special tools and equipment needed, servicing of the valves, the valve seats and the valve guides, commonly known as a valve job, should be done by a professional.

2   The home mechanic can remove and disassemble the head(s), do the initial cleaning and inspection, then reassemble and deliver them to a dealer service department or an automotive machine shop for the actual service work. Doing the inspection will enable you to see what condition the head(s) and valvetrain components are in and will ensure

that you know what work and new parts are required when dealing with an automotive machine shop.

3   The dealer service department, or automotive machine shop, will remove the valves and springs, recondition or replace the valves and valve seats, recondition the valve guides, check and replace the valve springs, spring retainers and keepers (as necessary), replace the valve seals with new ones, reassemble the valve components and make sure the installed spring height is correct. The cylinder head gasket surface will also be resurfaced if it's warped.

4   After the valve job has been performed by a professional, the head(s) will be in like new condition. When the heads are returned, be sure to clean them again before installation on the engine to remove any metal particles and abrasive grit that may still be present from the valve service or head resurfacing operations. Use compressed air, if available, to blow out all the oil holes and passages.

## 12   Cylinder head - reassembly

*Refer to illustrations 12.3 and 12.6*

1   Regardless of whether or not the head was sent to an automotive machine shop for valve servicing, make sure it's clean before beginning reassembly. Note that there are several small core plugs (also called freeze plugs or expansion plugs) in the head. These should be replaced whenever the engine is overhauled or the cylinder head is reconditioned (see Section 15 for replacement procedure).

2   If the head was sent out for valve servicing, the valves and related components will already be in place. Begin the reassembly procedure with Step 8.

3   Install new seals on each of the valve guides. **Note:** *Intake and exhaust valves require different seals - DO NOT mix them up!* Gently tap each intake valve seal into place until it's seated on the guide (see illustra-

tion). **Caution:** *Don't hammer on the valve seals once they're seated or you may damage them. Don't twist or cock the seals during installation or they won't seat properly on the valve stems.*

4   Beginning at one end of the head, lubricate and install the first valve. Apply moly-base grease or clean engine oil to the valve stem and tip.

5   Drop the spring seat or shim(s) over the valve guide and set the valve spring and retainer in place.

6   Compress the springs with a valve spring compressor and carefully install the keepers in the upper groove, then slowly release the compressor and make sure the keepers seat properly. Apply a small dab of grease to each keeper to hold it in place if necessary (see illustration).

7   Repeat the procedure for the remaining valves. Be sure to return the components to their original locations - don't mix them up!

8   Check the valve spring installed height with a dial or vernier caliper.

12.6  The small valve stem keepers are easier to position when coated with grease

**13.1  A ridge reamer is required to remove the ridge from the top of each cylinder - do this before removing the pistons!**

**13.3  Check the connecting rod side clearance with a feeler gauge as shown here**

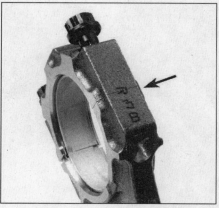

**13.4  The connecting rods and caps should be marked to indicate which cylinder they're installed in - if they aren't, mark them with a center punch to avoid confusion during reassembly - do not confuse the markings shown here as rod numbers; these are bearing size identifications**

2C

## 13  Pistons/connecting rods - removal

Refer to illustrations 13.1, 13.3, 13.4 and 13.6
**Note:** *Prior to removing the piston/connecting rod assemblies, remove the cylinder head(s), the oil pan and the oil pump pick-up tube by referring to the appropriate Sections in Chapter 2A or 2B.*

1   Use your fingernail to feel if a ridge has formed at the upper limit of ring travel (about 1/4-inch down from the top of each cylinder). If carbon deposits or cylinder wear have produced ridges, they must be completely removed with a special tool **(see illustration)**. Follow the manufacturer's instructions provided with the tool. Failure to remove the ridges before attempting to remove the piston/connecting rod assemblies may result in piston damage.
2   After the cylinder ridges have been removed, turn the engine upside-down so the crankshaft is facing up.
3   Before the connecting rods are removed, check the endplay with feeler gauges. Slide them between the first connecting rod and the crankshaft throw until the play is removed **(see illustration)**. The endplay is equal to the thickness of the feeler gauge(s). If the endplay exceeds the specified service limit, new connecting rods will be required. If new rods (or a new crankshaft) are installed, the endplay may fall under the service limit (if it does, the rods will have to be machined to restore it - consult an automotive machine shop for advice if necessary). Repeat the procedure for the remaining connecting rods.
4   Check the connecting rods and caps for identification marks. If they aren't plainly marked, use a small center punch to make the appropriate number of indentations on each rod and cap (1, 2, 3, etc., depending on the engine type and cylinder they're associated with) **(see illustration)**.
5   Loosen each of the connecting rod cap nuts 1/2-turn at a time until they can be removed by hand. Remove the number one

connecting rod cap and bearing insert. Don't drop the bearing insert out of the cap. **Note:** *The 1.6L engines use conventional rod bolts (pressed into the rod) and nuts, while the 1.8L engine uses cap bolts that go through the rod cap and screw into the rod.*
6   Slip a short length of plastic or rubber hose over each connecting rod cap bolt to protect the crankshaft journal and cylinder wall as the piston is removed **(see illustration)**.
7   Remove the bearing insert and push the connecting rod/piston assembly out through the top of the engine. Use a wooden hammer handle to push on the upper bearing surface in the connecting rod. If resistance is felt, double-check to make sure that all of the ridge was removed from the cylinder.
8   Repeat the procedure for the remaining cylinders. **Note:** *Turn the crankshaft as needed to put the rod to be removed close to parallel with the cylinder bore, i.e. don't try to drive it out while at a large angle to the bore.*
9   After removal, reassemble the connecting rod caps and bearing inserts in their respective connecting rods and install the cap nuts/bolts finger tight. Leaving the old bearing inserts in place until reassembly will help prevent the connecting rod bearing surfaces from being accidentally nicked or gouged.
10   Don't separate the pistons from the connecting rods (see Section 18 for additional information).

## 14  Crankshaft - removal

Refer to illustrations 14.1, 14.3 and 14.5
**Note:** *The crankshaft can be removed only after the engine has been removed from the vehicle. It's assumed that the flywheel or driveplate, crankshaft sprocket, timing belt or chain, oil pan, oil pick-up tube, oil pump and piston/connecting rod assemblies have already been removed. On 1997 and earlier models, the rear main oil seal retainer must also be removed from the block before proceeding with crankshaft removal.*

**13.6  On 1.6 liter engines, slip sections of hose over the rod bolts before removing the pistons to prevent damage to the crankshaft journals and cylinder walls**

**14.1  Checking crankshaft endplay with a dial indicator**

1   Before the crankshaft is removed, check the endplay. Mount a dial indicator with the stem in line with the crankshaft and touching the nose of the crank **(see illustration)**.

**14.3  Checking crankshaft endplay with a feeler gauge**

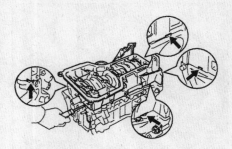

**14.5  On 1998 and later engines, pry the bearing cap sub-assembly/bedplate off the engine block at the points shown - be careful not to damage the mating surfaces of the cylinder block and bearing cap sub-assembly**

**15.1a  A hammer and a large punch can be used to knock the core plugs sideways in their bores**

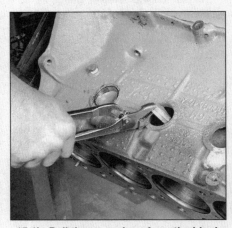

**15.1b  Pull the core plugs from the block with pliers**

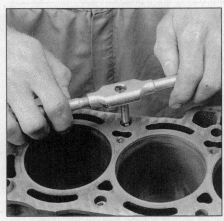

**15.8  All bolt holes in the block - particularly the main bearing cap and head bolt holes - should be cleaned and restored with a tap (be sure to remove debris from the holes after this is done)**

2    Push the crankshaft all the way to the rear and zero the dial indicator. Next, pry the crankshaft to the front as far as possible and check the reading on the dial indicator. The distance that it moves is the endplay. If it's greater than specified, check the crankshaft thrust surfaces for wear. If no wear is evident, new thrust washers should correct the endplay.

3    If a dial indicator isn't available, feeler gauges can be used. Gently pry or push the crankshaft all the way to the front of the engine. Slip feeler gauges between the crankshaft and the front face of the number 3 (thrust) main bearing to determine the clearance **(see illustration)**.

4    On 1997 and earlier engines, check the main bearing caps to see if they're marked to indicate their locations. They should be numbered consecutively from the front of the engine to the rear. If they aren't, mark them with number stamping dies or a center punch. Main bearing caps generally have a cast-in arrow, which points to the front of the engine. Loosen the main bearing cap bolts (1997 and earlier) or the bearing cap sub-assembly (1998 and later) bolts a 1/4-turn at a time, in the reverse order of the recommended tightening sequence **(see illustration 23.12a and 23.12b)**, until they can be removed by hand.

5    On 1997 and earlier engines, gently tap the caps with a soft-face hammer, then separate them from the engine block. If necessary, use the bolts as levers to remove the caps. Try not to drop the bearing inserts if they come out with the caps. On 1998 and later engines, gently pry the bearing cap sub-assembly/bedplate off the engine and place it on a workbench **(see illustration)**.

6    Carefully lift the crankshaft out of the engine. It may be a good idea to have an assistant available, since the crankshaft is quite heavy. With the bearing inserts in place in the engine block and in the bearing caps , return the caps or bearing cap sub-assembly to their respective locations on the engine block and tighten the bolts finger tight.

---

## 15   Engine block - cleaning

*Refer to illustrations 15.1a, 15.1b, 15.8 and 15.10*

**Caution:** *The core plugs (also known as freeze or soft plugs) may be difficult or impossible to retrieve if they're driven completely into the block coolant passages.*

1    Using the blunt end of a punch, tap on in the outer edge of the core plug to turn the plug sideways in the bore. Then using pliers, pull the core plug from the engine block **(see illustrations)**.

2    Using a gasket scraper, remove all traces of gasket material from the engine block. Be very careful not to nick or gouge the gasket sealing surfaces.

3    Remove the main bearing caps or cap assembly and separate the bearing inserts from the caps and the engine block. Tag the bearings, indicating which cylinder they were removed from and whether they were in the cap or the block, then set them aside.

4    Remove all of the threaded oil gallery plugs from the block. The plugs are usually very tight - they may have to be drilled out and the holes retapped. Use new plugs when the engine is reassembled.

5    If the engine is extremely dirty, it should

be taken to an automotive machine shop to be steam cleaned or hot tanked.

6    After the block is returned, clean all oil holes and oil galleries one more time. Brushes specifically designed for this purpose are available at most auto parts stores. Flush the passages with warm water until the water runs clear, dry the block thoroughly and wipe all machined surfaces with a light, rust preventive oil. If you have access to compressed air, use it to speed the drying process and to blow out all the oil holes and galleries. **Warning:** *Wear eye protection when using compressed air!*

7    If the block isn't extremely dirty or sludged up, you can do an adequate cleaning job with hot soapy water and a stiff brush. Take plenty of time and do a thorough job. Regardless of the cleaning method used, be sure to clean all oil holes and galleries very thoroughly, dry the block completely and coat all machined surfaces with light oil.

8    The threaded holes in the block must be clean to ensure accurate torque readings during reassembly. Run the proper size tap

**15.10  A large socket on an extension can be used to drive the new core plugs into the bores**

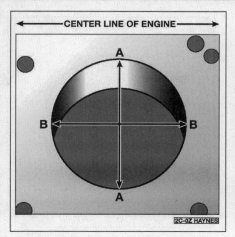

**16.4a  Measure the diameter of each cylinder at a right angle to engine centerline (A), and parallel to engine centerline (B) - out-of-round is the difference between A and B; taper is the difference between A and B at the top of the cylinder and A and B at the bottom of the cylinder**

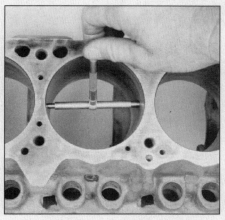

**16.4b  The ability to "feel" when the telescoping gauge is at the correct point will be developed over time, so work slowly and repeat the check until you're satisfied that the bore measurement is accurate**

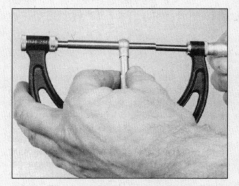

**16.4c  The gauge is then measured with a micrometer to determine the bore size**

**16.9  Check the block deck for distortion with a precision straightedge and feeler gauges**

2C

into each of the holes to remove rust, corrosion, thread sealant or sludge and restore damaged threads **(see illustration)**. If possible, use compressed air to clear the holes of debris produced by this operation. Now is a good time to clean the threads on the head bolts and the main bearing cap bolts as well.

9    Reinstall the main bearing caps and tighten the bolts finger tight.

10    After coating the sealing surfaces of the new core plugs with Permatex no. 2 sealant, install them in the engine block **(see illustration)**. Make sure they're driven in straight and seated properly or leakage could result. Special tools are available for this purpose, but a large socket, with an outside diameter that will just slip into the core plug, a 1/2-inch drive extension and a hammer will work just as well.

11    Apply non-hardening sealant (such as Permatex no. 2 or Teflon pipe sealant) to the new oil gallery plugs and thread them into the holes in the block. Make sure they're tightened securely.

12    If the engine isn't going to be reassembled right away, cover it with a large plastic trash bag to keep it clean.

## 16   Engine block - inspection

*Refer to illustrations 16.4a, 16.4b, 16.4c and 16.9*

1    Before the block is inspected, it should be cleaned as described in Section 15.

2    Visually check the block for cracks, rust and corrosion. Look for stripped threads in the threaded holes. On 1998 and later engines, check the main bearing cap bolts for stretching. Measure the diameter of the each bolt at the middle past the threads. Compare the measurements of each bolt to the Specifications listed in this Chapter. If the bolts have stretched past their limit, they must be replaced. It's a good idea to have the block checked for hidden cracks by an automotive machine shop that has the special equipment

to do this type of work, especially if the vehicle had a history of overheating or using coolant. If defects are found, have the block repaired, if possible, or replaced.

3    Check the cylinder bores for scuffing and scoring.

4    Check the cylinders for taper and out-of-round conditions as follows **(see illustrations)**:

5    Measure the diameter of each cylinder at the top (just under the ridge area), center and bottom of the cylinder bore, parallel to the crankshaft axis.

6    Next, measure each cylinder's diameter at the same three locations perpendicular to the crankshaft axis.

7    The taper of each cylinder is the difference between the bore diameter at the top of the cylinder and the diameter at the bottom. The out-of-round specification of the cylinder bore is the difference between the parallel and perpendicular readings. Compare your results to this Chapter's Specifications.

8    If the cylinder walls are badly scuffed or scored, or if they're out-of-round or tapered

beyond the limits given in this Chapter's Specifications, have the engine block rebored and honed at an automotive machine shop. If a rebore is done, oversize pistons and rings will be required.

9    Using a precision straightedge and feeler gauge, check the block deck (the surface that mates with the cylinder head) for distortion **(see illustration)**. If it's distorted beyond the specified limit, it can be resurfaced by an automotive machine shop.

10    If the cylinders are in reasonably good condition and not worn to the outside of the limits, and if the piston-to-cylinder clearances can be maintained properly, then they don't have to be rebored. Honing is all that's necessary (refer to Section 17).

## 17   Cylinder honing

*Refer to illustrations 17.3a and 17.3b*

1    Prior to engine reassembly, the cylinder bores must be honed so the new piston rings

**17.3a A "bottle brush" hone is the easiest type of hone to use**

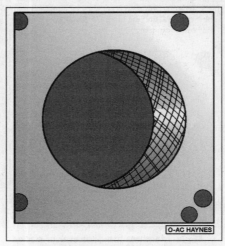

**17.3b  The cylinder hone should leave a smooth, crosshatch pattern with the lines intersecting at approximately a 60-degree angle**

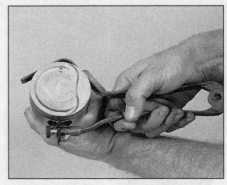

**18.4a  The piston ring grooves can be cleaned with a special tool, as shown here . . .**

**18.4b . . . or with a section of a broken ring**

will seat correctly and provide the best possible combustion chamber seal. **Note:** *If you don't have the tools or don't want to tackle the honing operation, most automotive machine shops will do it for a reasonable fee.*

2    Before honing the cylinders, install the main bearing caps or cap assembly (without bearing inserts) and tighten the bolts to the specified torque.

3    Two types of cylinder hones are commonly available - the flex hone or "bottle brush" type and the more traditional surfacing hone with spring-loaded stones. Both will do the job, but for the less-experienced mechanic the "bottle brush" hone will probably be easier to use. You'll also need some kerosene or honing oil, rags and an electric drill motor. The drill motor should be operated at a steady, slow speed. Use a large 1/2-inch drill or a 3/8-inch variable-speed drill. Proceed as follows:

a) *Mount the hone in the drill motor, compress the stones and slip it into the first cylinder* (see illustration). **Warning:** *Be sure to wear safety goggles or a face shield!*

b) *Lubricate the cylinder with plenty of honing oil, turn on the drill and move the hone up-and-down in the cylinder at a pace that will produce a fine crosshatch pattern on the cylinder walls. Ideally, the crosshatch lines should intersect at approximately a 60-degree angle* (see illustration). *Be sure to use plenty of lubricant and don't take off any more material than is absolutely necessary to produce the desired finish.* **Note:** *Piston ring manufacturers may specify a smaller crosshatch angle than the traditional 60-degrees - read and follow any instructions included with the new rings.*

c) *Don't withdraw the hone from the cylinder while it's running. Instead, shut off the drill and continue moving the hone up-and-down in the cylinder until it comes to a complete stop, then compress the stones and withdraw the hone. If you're using a "bottle brush" type*

*hone, stop the drill motor, then turn the chuck in the normal direction of rotation while withdrawing the hone from the cylinder.*

d) *Wipe the oil out of the cylinder and repeat the procedure for the remaining cylinders.*

4    After the honing job is complete, chamfer the top edges of the cylinder bores with a small file so the rings won't catch when the pistons are installed. Be very careful not to nick the cylinder walls with the end of the file.

5    The entire engine block must be washed again very thoroughly with warm, soapy water to remove all traces of the abrasive grit produced during the honing operation. **Note:** *The bores can be considered clean when a lint-free white cloth - dampened with clean engine oil - used to wipe them out doesn't pick up any more honing residue, which will show up as gray areas on the cloth. Be sure to run a brush through all oil holes and galleries and flush them with running water.*

6    After rinsing, dry the block and apply a coat of light rust preventive oil to all machined surfaces. Wrap the block in a plastic trash bag to keep it clean and set it aside until reassembly.

## 18    Pistons/connecting rods - inspection

*Refer to illustrations 18.4a, 18.4b, 18.10 and 18.11*

1    Before the inspection process can be carried out, the piston/connecting rod assemblies must be cleaned and the original piston rings removed from the pistons. **Note:** *Always use new piston rings when the engine is reassembled.*

2    Using a piston ring installation tool, carefully remove the rings from the pistons. Be careful not to nick or gouge the pistons in

the process.

3    Scrape all traces of carbon from the top of the piston. A hand-held wire brush or a piece of fine emery cloth can be used once the majority of the deposits have been scraped away. Do not, under any circumstances, use a wire brush mounted in a drill motor to remove deposits from the pistons. The piston material is soft and may be eroded away by the wire brush.

4    Use a piston ring groove-cleaning tool to remove carbon deposits from the ring grooves. If a tool isn't available, a piece broken off the old ring will do the job. Be very careful to remove only the carbon deposits - don't remove any metal and do not nick or scratch the sides of the ring grooves **(see illustrations)**.

5    Once the deposits have been removed, clean the piston/rod assemblies with solvent and dry them with compressed air (if available). Make sure the oil return holes in the back sides of the ring grooves and the oil hole in the lower end of each rod are clear.

6    If the pistons and cylinder walls aren't damaged or worn excessively, and if the engine block is not rebored, new pistons won't be necessary. Normal piston wear appears as even vertical wear on the piston thrust surfaces and slight looseness of the top ring in its groove. New piston rings, however, should always be used when an engine

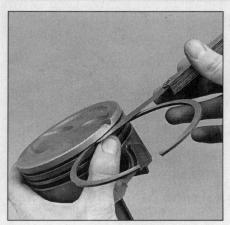

**18.10  Check the ring groove clearance with a feeler gauge at several points around the groove**

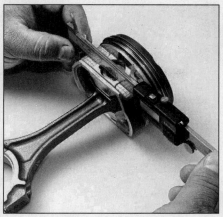

**18.11  Measure the piston diameter at a 90-degree angle to the piston pin, at the bottom of the piston pin area - a precision caliper may be used if a micrometer isn't available**

**19.6  Measure the diameter of each crankshaft journal at several points to detect taper and out-of-round conditions**

is rebuilt.

7    Carefully inspect each piston for cracks around the skirt, at the pin bosses and at the ring lands.

8    Look for scoring and scuffing on the thrust faces of the skirt, holes in the piston crown and burned areas at the edge of the crown. If the skirt is scored or scuffed, the engine may have been suffering from over-heating and/or abnormal combustion, which caused excessively high operating temperatures. The cooling and lubrication systems should be checked thoroughly. A hole in the piston crown is an indication that abnormal combustion (preignition) was occurring. Burned areas at the edge of the piston crown are usually evidence of spark knock (detonation). If any of the above problems exist, the causes must be corrected or the damage will occur again. The causes may include intake air leaks, incorrect air/fuel mixture, incorrect ignition timing and EGR system malfunctions.

9    Corrosion of the piston, in the form of small pits, indicates that coolant is leaking into the combustion chamber and/or the crankcase. Again, the cause must be corrected or the problem may persist in the rebuilt engine.

10    Measure the piston ring groove clearance by laying a new piston ring in each ring groove and slipping a feeler gauge in beside it **(see illustration)**. Check the clearance at three or four locations around each groove. Be sure to use the correct ring for each groove - they are different. If the clearance is greater than that listed in this Chapter's Specifications, new pistons will have to be used.

11    Check the piston-to-bore clearance by measuring the bore (see Section 16) and the piston diameter. Make sure the pistons and bores are correctly matched. Measure the piston across the skirt, at a 90-degree angle to the piston pin **(see illustration)**. Subtract the piston diameter from the bore diameter to obtain the clearance. If it's greater than specified, the block will have to be rebored and new pistons and rings installed.

12    Check the piston-to-rod clearance by twisting the piston and rod in opposite directions. Any noticeable play indicates excessive wear, which must be corrected.

13    If the pistons must be removed from the connecting rods for any reason, the rods should be taken to an automotive machine shop, to be checked for bend and twist, since automotive machine shops have special equipment for this purpose.

14    Check the connecting rods for cracks and other damage. Temporarily remove the rod caps, lift out the old bearing inserts, wipe the rod and cap bearing surfaces clean and inspect them for nicks, gouges and scratches. Also check the connecting rod bolts for stretching. Measure the diameter of each bolt at the middle past the threads. Compare the measurements of each bolt to the Specifications listed in this Chapter. If the bolts have stretched past their limit, they must be replaced. After checking the rods, replace the old bearings, slip the caps into place and tighten the nuts finger tight. If necessary install new rod bolts. **Note:** *If the engine is being rebuilt because of a connecting rod knock, be sure to install new rods.*

## 19   Crankshaft - inspection

*Refer to illustration 19.6*

1    Clean the crankshaft with solvent and dry it with compressed air (if available).

2    Check the main and connecting rod bearing journals for uneven wear, scoring, pits and cracks.

3    Remove all burrs from the crankshaft oil holes with a stone, file or scraper.

4    Clean the oil holes with a stiff brush and flush them with solvent.

5    Check the rest of the crankshaft for cracks and other damage. It should be magnafluxed to reveal hidden cracks - an automotive machine shop will handle the procedure.

6    Using a micrometer, measure the diameter of the main and connecting rod journals and compare the results to this Chapter's Specifications **(see illustration)**. By measuring the diameter at a number of points around each journal's circumference, you'll be able to determine whether or not the journal is out-of-round. Take the measurement at each end of the journal, near the crank throws, to determine if the journal is tapered. Crankshaft runout should be checked also, but large V-blocks and a dial indicator are needed to do it correctly. If you don't have the equipment, have a machine shop check the runout.

7    If the crankshaft journals are damaged, tapered, out-of-round or worn beyond the limits given in the Specifications, have the crankshaft reground by an automotive machine shop. Be sure to use the correct size bearing inserts if the crankshaft is reconditioned.

8    Check the oil seal journals at each end of the crankshaft for wear and damage. If the seal has worn a groove in the journal, or if it's nicked or scratched, the new seal may leak when the engine is reassembled. In some cases, an automotive machine shop may be able to repair the journal by pressing on a thin sleeve. If repair isn't feasible, a new or different crankshaft should be installed.

9    Refer to Section 20 and examine the main and rod bearing inserts.

## 20   Main and connecting rod bearings - inspection and selection

### Inspection

*Refer to illustration 20.1*

1    Even though the main and connecting rod bearings should be replaced with new ones during the engine overhaul, the old bearings should be retained for close examination, as they may reveal valuable informa-

**2C**

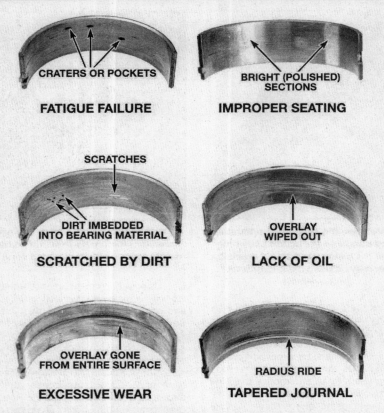

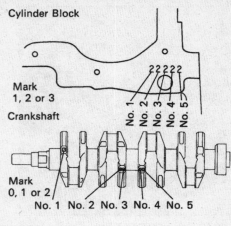

**20.10a  Crankshaft bearing grade markings for STANDARD size bearings on 1997 and earlier vehicles**

**20.1  When inspecting the main and connecting rod bearings, look for these problems**

tion about the condition of the engine **(see illustration)**.

2    Bearing failure occurs because of lack of lubrication, the presence of dirt or other foreign particles, overloading the engine and corrosion. Regardless of the cause of bearing failure, it must be corrected before the engine is reassembled to prevent it from happening again.

3    When examining the bearings, remove them from the engine block, the main bearing caps, the connecting rods and the rod caps and lay them out on a clean surface in the same general position as their location in the engine. This will enable you to match any bearing problems with the corresponding crankshaft journal.

4    Dirt and other foreign particles get into the engine in a variety of ways. It may be left in the engine during assembly, or it may pass through filters or the PCV system. It may get into the oil, and from there into the bearings. Metal chips from machining operations and normal engine wear are often present. Abrasives are sometimes left in engine components after reconditioning, especially when parts are not thoroughly cleaned using the proper cleaning methods. Whatever the source, these foreign objects often end up embedded in the soft bearing material and are easily recognized. Large particles will not embed in the bearing and will score or gouge the bearing and journal. The best prevention for this cause of bearing failure is to clean all

parts thoroughly and keep everything spotlessly clean during engine assembly. Frequent and regular engine oil and filter changes are also recommended.

5    Lack of lubrication (or lubrication breakdown) has a number of interrelated causes. Excessive heat (which thins the oil), overloading (which squeezes the oil from the bearing face) and oil leakage or throw off (from excessive bearing clearances, worn oil pump or high engine speeds) all contribute to lubrication breakdown. Blocked oil passages, which usually are the result of misaligned oil holes in a bearing shell, will also oil starve a bearing and destroy it. When lack of lubrication is the cause of bearing failure, the bearing material is wiped or extruded from the steel backing of the bearing. Temperatures may increase to the point where the steel backing turns blue from overheating.

6    Driving habits can have a definite effect on bearing life. Low speed operation in too high a gear (lugging the engine) puts very high loads on bearings, which tends to squeeze out the oil film. These loads cause the bearings to flex, which produces fine cracks in the bearing face (fatigue failure). Eventually the bearing material will loosen in pieces and tear away from the steel backing. Short trip driving leads to corrosion of bearings because insufficient engine heat is produced to drive off the condensed water and corrosive gases. These products collect in the engine oil, forming acid and sludge. As the oil is carried to the engine

bearings, the acid attacks and corrodes the bearing material.

7    Incorrect bearing installation during engine assembly will lead to bearing failure as well. Tight-fitting bearings leave insufficient bearing oil clearance and will result in oil starvation. Dirt or foreign particles trapped behind a bearing insert result in high spots on the bearing which lead to failure.

### Selection

*Refer to illustrations 20.10a and 20.10b*

8    If the original bearings are worn or damaged, or if the oil clearances are incorrect (see Section 23 or 25), the following procedures should be used to select the correct new bearings for engine reassembly. However, if the crankshaft has been reground, new undersize bearings must be installed - the following procedure should not be used if undersize bearings are required! The automotive machine shop that reconditions the crankshaft will provide or help you select the correct size bearings. Regardless of how the bearing sizes are determined, use the oil clearance, measured with Plastigage, as a guide to ensure the bearings are the right size.

### Main bearings

9    If you need to use a STANDARD size main bearing, install one that has the same number as the original bearing **(see illustration 20.10a or 20.10b for the bearing number locations)**. There are five sizes of main bearings.

10    If the number on the original main bearing has been obscured, locate the main journal grade numbers stamped into the oil pan mating surface on the engine block and the crankshaft counterweights **(see illustrations)**.

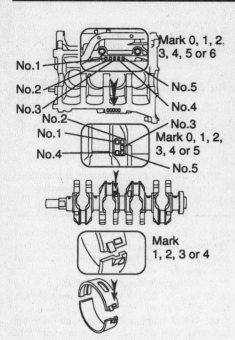

**22.3  When checking piston ring end gap, the ring must be square in the cylinder bore (this is done by pushing the ring down with the top of a piston as shown)**

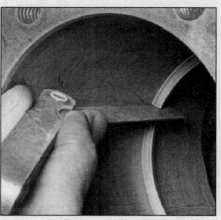

**22.4  With the ring square in the cylinder, measure the end gap with a feeler gauge**

2C

**20.10b  Crankshaft bearing grade markings for STANDARD size bearings on 1998 and later vehicles**

11   Adding the block number to the crank number for a particular journal will give the recommended bearing size.

### Connecting rod bearings

12   If you need to use a STANDARD size rod bearing, install one that has the same number as the number stamped into the connecting rod cap **(see illustration 13.4)**.

### All bearings

13   Remember, the oil clearance is the final judge when selecting new bearing sizes. If you have any questions or are unsure which bearings to use, get help from a dealer parts or service department.

### 21   Engine overhaul - reassembly sequence

1   Before beginning engine reassembly, make sure you have all the necessary new parts, gaskets and seals as well as the following items on hand:

*Common hand tools*
*A 1/2-inch drive torque wrench*
*Piston ring installation tool*
*Piston ring compressor*
*Short lengths of rubber or plastic hose to fit over connecting rod bolts*
*Plastigage*
*Feeler gauges*
*A fine-tooth file*
*New engine oil*
*Engine assembly lube or moly-base grease*

*Gasket sealant*
*Thread locking compound*

2   In order to save time and avoid problems, engine reassembly must be done in the following general order:

*Piston rings*
*Crankshaft and main bearings*
*Piston/connecting rod assemblies*
*Rear main (crankshaft) oil seal*
*Cylinder head and lifters*
*Camshafts*
*Oil pump*
*Timing belt and sprockets*
*(1997 and earlier)*
*Timing chain and sprockets*
*(1998 and later)*
*Timing covers*
*Oil pick-up*
*Oil pan*
*Intake and exhaust manifolds*
*Valve cover*
*Flywheel/driveplate*

### 22   Piston rings - installation

*Refer to illustrations 22.3, 22.4, 22.9a, 22.9b and 22.12*

1   Before installing the new piston rings, the ring end gaps must be checked. It's assumed that the piston ring groove clearance has been checked and verified correct (see Section 18).

2   Lay out the piston/connecting rod assemblies and the new ring sets so the ring sets will be matched with the same piston and cylinder during the end gap measurement and engine assembly.

3   Insert the top (number one) ring into the first cylinder and square it up with the cylinder walls by pushing it in with the top of the piston **(see illustration)**. The ring should be near the bottom of the cylinder, at the lower limit of ring travel.

4   To measure the end gap, slip feeler gauges between the ends of the ring until a

**22.9a  Install the spacer/expander in the oil control ring groove**

gauge equal to the gap width is found **(see illustration)**. The feeler gauge should slide between the ring ends with a slight amount of drag. Compare the measurement to that found in this Chapter's Specifications. If the gap is larger or smaller than specified, double-check to make sure you have the correct rings before proceeding.

5   If the gap is too small, replace the rings - DO NOT file the ends to increase the clearance.

6   Excess end gap isn't critical unless it's greater than the service limit listed in this Chapter's Specifications. Again, double-check to make sure you have the correct rings for your engine.

7   Repeat the procedure for each ring that will be installed in the first cylinder and for each ring in the remaining cylinders. Remember to keep rings, pistons and cylinders matched up.

8   Once the ring end gaps have been checked/corrected, the rings can be installed on the pistons.

9   The oil control ring (lowest one on the piston) is usually installed first. It's composed of three separate components. Slip the spacer/expander into the groove **(see illustration)**. If an anti-rotation tang is used, make

sure it's inserted into the drilled hole in the ring groove. Next, install the lower side rail. Don't use a piston ring installation tool on the oil ring side rails, as they may be damaged. Instead, place one end of the side rail into the groove between the spacer/expander and the ring land, hold it firmly in place and slide a finger around the piston while pushing the rail into the groove **(see illustration)**. Next, install the upper side rail in the same manner.

10   After the three oil ring components have been installed, check to make sure that both the upper and lower side rails can be turned smoothly in the ring groove.

11   The number two (middle) ring is installed next. It's usually stamped with a mark which must face up, toward the top of the piston. **Note:** *Always follow the instructions printed on the ring package or box - different manufacturers may require different approaches. Do not mix up the top and middle rings, as they have different cross sections.*

12   Use a piston ring installation tool and make sure the ring's identification mark is facing the top of the piston, then slip the ring into the middle groove on the piston **(see illustration)**. Don't expand the ring any more than necessary to slide it over the piston.

13   Install the number one (top) ring in the same manner. Make sure the mark is facing up. Be careful not to confuse the number one and number two rings.

14   Repeat the procedure for the remaining pistons and rings.

## 23   Crankshaft - installation and main bearing oil clearance check

*Refer to illustrations 23.10, 23.12a, 23.12b, 23.14, 23.19a, 23.19b and 23.20*

1   Crankshaft installation is the first major step in engine reassembly. It's assumed at this point that the engine block and crankshaft have been cleaned, inspected and repaired or reconditioned.

2   Position the engine with the bottom facing up.

**23.10  Lay the Plastigage strips (arrow) on the main bearing journals, parallel to the crankshaft centerline**

**22.9b  DO NOT use a piston ring installation tool when installing the oil ring side rails**

3   Remove the main bearing cap bolts and lift out the caps. Lay the caps out in the proper order.

4   If they're still in place, remove the old bearing inserts from the block and the main bearing caps. Wipe the main bearing surfaces of the block and caps with a clean, lint free cloth. They must be kept spotlessly clean!

## *Main bearing oil clearance check*

5   Clean the back sides of the new main bearing inserts and lay the upper bearing halves with the oil grooves in the corresponding main bearing saddle in the block. Lay each of the lower bearing halves from the bearing set in the corresponding main bearing caps (1997 and earlier) or the bearing cap sub-assembly (1998 and later). Make sure the tab on each bearing insert fits into the recess in the block or cap. Also, the oil holes in the block must line up with the oil holes in the bearing insert. **Caution:** *Do not hammer the bearings into place and don't nick or gouge the bearing faces. No lubrication should be used at this time.*

6   The thrust bearings (washers) must be installed on the number three main journal. **Note:** *On 1997 and earlier engines, four thrust washers are used , two on each side of the cap and two on each side of the saddle in the engine block. On 1998 and later engines,*

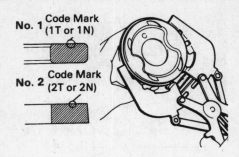

**22.12  Using a ring expander, install the compression rings with the marks facing up**

*only two thrust washers are used on the number three saddle of the engine block. There are no thrust washers on the bearing cap sub-assembly.*

7   Clean the faces of the bearings in the block and the crankshaft main bearing journals with a clean, lint free cloth. Check or clean the oil holes in the crankshaft, as any dirt here can go only one way - straight through the new bearings.

8   Once you're certain the crankshaft is clean, carefully lay it in position in the main bearings.

9   Before the crankshaft can be permanently installed, the main bearing oil clearance must be checked.

10   Trim several pieces of the appropriate size Plastigage (they must be slightly shorter than the width of the main bearings) and place one piece on each crankshaft main bearing journal, parallel with the journal axis **(see illustration)**.

11   Clean the faces of the bearings in the caps. On 1997 and earlier models, install the caps in their respective positions (don't mix them up) with the arrows pointing toward the front of the engine. On 1998 and later models, place the bearing cap sub-assembly/bedplate over the crankshaft and carefully tap it down into position on the engine block. Don't disturb the Plastigage. Apply a light coat of oil to the bolt threads and the undersides of the bolt heads, then install them.

12   Following the recommended sequence **(see illustrations)**, tighten the main bearing cap bolts, in the recommended steps, to the torque listed in this Chapter's Specifications.

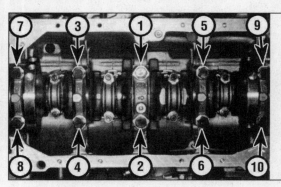

**23.12a  Main bearing cap TIGHTENING sequence - 1997 and earlier**

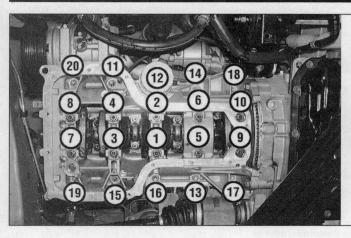

**23.12b  Main bearing cap sub-assembly/bedplate TIGHTENING sequence - 1998 and later**

**23.14  Compare the width of the crushed Plastigage to the scale on the envelope to determine the main bearing oil clearance (always take the measurement at the widest point of the Plastigage) - be sure to use the correct scale; standard and metric scales are included**

Don't rotate the crankshaft at any time during this operation! **Note:** *On 1998 and later engines, tighten the center bolts (1 through 10) in four steps to the recommended torque, then tighten the outer bolts (11 through 20) last.*

13    Remove the bolts and carefully lift off the main bearing caps or the bearing cap sub-assembly/bedplate. Keep them in order. Don't disturb the Plastigage or rotate the crankshaft. If any of the main bearing caps are difficult to remove, tap them gently from side-to-side with a soft-face hammer to loosen them.

14    Compare the width of the crushed Plastigage on each journal to the scale printed on the Plastigage envelope to obtain the main bearing oil clearance **(see illustration)**. Check the Specifications to make sure it's correct.

15    If the clearance is not as specified, the bearing inserts may be the wrong size (which means different ones will be required - see Section 20). Before deciding that different inserts are needed, make sure that no dirt or oil was between the bearing inserts and the caps or block when the clearance was measured. If the Plastigage is noticeably wider at one end than the other, the journal may be tapered (see Section 19).

16    Carefully scrape all traces of the Plasti-

gage material off the main bearing journals and/or the bearing faces. Don't nick or scratch the bearing faces.

## Final crankshaft installation

17    Carefully lift the crankshaft out of the engine. Clean the bearing faces in the block, then apply a thin, uniform layer of clean moly-base grease or engine assembly lube to each of the bearing surfaces. Coat the thrust washers as well.

18    Lubricate the crankshaft surfaces that contact the oil seals with moly-base grease, engine assembly lube or clean engine oil.

19    Make sure the crankshaft journals are clean, then lay the crankshaft back in place in the block. Install the thrust washers in the number 3 main journal. **Note 1:** *The upper (block side) thrust washers can be rotated into position around the crank with the crank in the block, with the thrust washer grooves facing OUT* **(see illustration)**. **Note 2:** *On 1997 and earlier models, the tanged lower thrust washers should be placed on the caps with their grooves OUT and the tangs fitting into the cap slots* **(see illustrations)**.

20    Clean the faces of the lower bearings, then apply lubricant to them. On 1997 and earlier models, install the caps in their respective positions (don't mix them up) with the arrows pointing toward the front of the

engine. On 1998 and later models, apply a bead of RTV sealant **(see illustration)** to the mating surface of the bearing cap sub-assembly, then place the bearing cap sub-assembly/bedplate over the crankshaft and carefully tap it down into position on the engine block.

21    Apply a light coat of oil to the bolt threads and the undersides of the bolt heads, then install them. Tighten all main bearing cap bolts to the torque listed in this Chapter's Specifications, following the recommended sequence **(see illustrations 23.12a and 23.12b)**.

22    Rotate the crankshaft a number of times by hand to check for any obvious binding.

23    Check the crankshaft endplay with a feeler gauge or a dial indicator as described

**23.19a  Rotate the upper thrust washer into position with the oil grooves facing OUT**

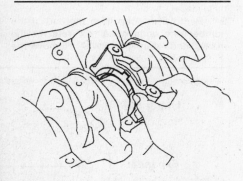

**23.19b  On 1997 and earlier engines, lower thrust washers must be installed the in the number three cap with the oil grooves facing OUT**

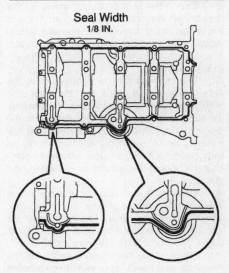

**23.20  Main bearing cap sub-assembly sealant installation details - 1998 and earlier**

**2C**

**24.3  After removing the retainer from the block, support it on a couple of wood blocks and drive out the old seal with a punch or screwdriver and hammer**

**24.5  Drive the new seal into the retainer with a wood block or a section of pipe, if you have one large enough - make sure you don't cock the seal in the retainer bore**

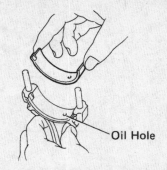

**25.3  Align the oil hole in the bearing with the oil hole in the rod**

in Section 14. The endplay should be correct if the crankshaft thrust faces aren't worn or damaged and new thrust washers have been installed.

24    Install a new rear main oil seal (see Section 24).

## 24   Rear main oil seal installation

*Refer to illustrations 24.3 and 24.5*

**Note:** *This procedure applies to 1997 and earlier engines only. If you're working on a 1998 or later engine, refer to Chapter 2B for this procedure (disregard the steps that do not apply since the engine is already removed from the vehicle).*

1    The crankshaft must be installed first and the main bearing caps bolted in place, then the new seal should be installed in the retainer and the retainer bolted to the block.

2    Check the seal contact surface on the crankshaft very carefully for scratches and nicks that could damage the new seal lip and cause oil leaks. If the crankshaft is damaged, the only alternative is a new or different crankshaft.

3    The old seal can be removed from the retainer by driving it out from the back side with a hammer and punch **(see illustration)**. Be sure to note how far it's recessed into the bore before removing it; the new seal will have to be recessed an equal amount. Be very careful not to scratch or otherwise damage the bore in the retainer or oil leaks could develop.

4    Make sure the retainer is clean, then apply a thin coat of engine oil to the outer edge of the new seal. The seal must be pressed squarely into the bore, so hammering it into place isn't recommended. If you don't have access to a press, sandwich the housing and seal between two smooth pieces of wood and press the seal into place with the jaws of a large vise. The pieces of wood must be thick enough to distribute the

force evenly around the entire circumference of the seal. Work slowly and make sure the seal enters the bore squarely.

5    As a last resort, the seal can be tapped into the retainer with a hammer. Use a block of wood to distribute the force evenly and make sure the seal is driven in squarely **(see illustration)**.

6    The seal lips must be lubricated with clean engine oil or moly-based grease before the seal/retainer is slipped over the crankshaft and bolted to the block, using a new gasket.

7    Tighten the bolts a little at a time to the torque listed in Chapter 2A Specifications.

## 25   Pistons/connecting rods - installation and rod bearing oil clearance check

*Refer to illustrations 25.3, 25.5, 25.9, 25.11, 25.13, 25.15 and 25.17*

1    Before installing the piston/connecting rod assemblies, the cylinder walls must be perfectly clean, the top edge of each cylinder must be chamfered, and the crankshaft must be in place.

2    Remove the cap from the end of the number one connecting rod (refer to the marks made during removal). Remove the

original bearing inserts and wipe the bearing surfaces of the connecting rod and cap with a clean, lint-free cloth. They must be kept spotlessly clean.

### Connecting rod bearing oil clearance check

3    Clean the back side of the new upper bearing insert, then lay it in place in the connecting rod. Make sure the tab on the bearing fits into the recess in the rod so the oil holes line up **(see illustration)**. Don't hammer the bearing insert into place and be very careful not to nick or gouge the bearing face. Don't lubricate the bearing at this time.

4    Clean the back side of the other bearing insert and install it in the rod cap. Again, make sure the tab on the bearing fits into the recess in the cap, and don't apply any lubricant. It's critically important that the mating surfaces of the bearing and connecting rod are perfectly clean and oil free when they're assembled.

5    Position the piston ring gaps at staggered intervals around the piston **(see illustration)**.

6    Slip a section of plastic or rubber hose over each connecting rod cap bolt.

7    Lubricate the piston and rings with clean engine oil and attach a piston ring compressor to the piston. Leave the skirt protruding about 1/4-inch to guide the piston into the cylinder. The rings must be compressed until they're flush with the piston.

8    Rotate the crankshaft until the number one connecting rod journal is at BDC (bottom

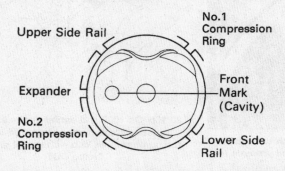

**25.5  Stagger the ring end gaps around the piston, as shown, before installing the pistons**

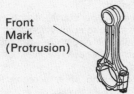
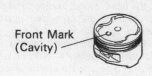

25.9 Check to be sure both the mark on the piston and the mark on the connecting rod are aligned as shown

25.11 Align the mark (arrow) on the top of the piston with the timing belt or timing chain end of the engine and drive the piston into the cylinder bore with a wooden or plastic hammer handle

25.13 Lay the Plastigage strips on each rod bearing journal, parallel to the crankshaft centerline

dead center) and apply a coat of engine oil to the cylinder wall.

9    With the dimple on top of the piston (see illustration) facing the front of the engine, gently insert the piston/connecting rod assembly into the number one cylinder bore and rest the bottom edge of the ring compressor on the engine block.

10    Tap the top edge of the ring compressor to make sure it's contacting the block around its entire circumference.

11    Gently tap on the top of the piston with the end of a wooden hammer handle (see illustration) while guiding the end of the connecting rod into place on the crankshaft journal. The piston rings may try to pop out of the ring compressor just before entering the cylinder bore, so keep some downward pressure on the ring compressor. Work slowly, and if any resistance is felt as the piston enters the cylinder, stop immediately. Find out what's hanging up and fix it before proceeding. Caution: Do not, for any reason, force the piston into the cylinder - you might break a ring and/or the piston.

12    Once the piston/connecting rod assembly is installed, the connecting rod bearing oil

clearance must be checked before the rod cap is permanently bolted in place.

13    Cut a piece of the appropriate size Plastigage slightly shorter than the width of the connecting rod bearing and lay it in place on the number one connecting rod journal, parallel with the journal axis (see illustration).

14    Clean the connecting rod cap bearing face, remove the protective hoses from the connecting rod bolts and install the rod cap. Make sure the mating mark on the cap is on the same side as the mark on the connecting rod. Check the cap to make sure the front mark is facing the timing belt end of the engine.

15    Apply a light coat of oil to the undersides of the nuts, then install and tighten them to the Step 1 torque listed in this Chapter's Specifications, working up to it in three steps. Use a thin-wall socket to avoid erroneous torque readings that can result if the socket is wedged between the rod cap and nut. If the socket tends to wedge itself

between the nut and the cap, lift up on it slightly until it no longer contacts the cap. Do not rotate the crankshaft at any time during this operation. Note: After reaching the specified torque, tighten each nut an additional 90-degrees (1/4-turn) (see illustration).

16    Remove the nuts and detach the rod cap, being very careful not to disturb the Plastigage.

17    Compare the width of the crushed Plastigage to the scale printed on the Plastigage envelope to obtain the oil clearance (see illustration). Compare it to this Chapter's Specifications to make sure the clearance is correct.

18    If the clearance is not as specified, the bearing inserts may be the wrong size (which means different ones will be required). Before deciding that different inserts are needed, make sure that no dirt or oil was between the bearing inserts and the connecting rod or cap when the clearance was measured. Also, recheck the journal diameter. If the Plastigage was wider at one end than the other, the journal may be tapered (refer to Section 19).

**2C**

25.15 Install the connecting rod caps with the front mark (arrow) facing the timing belt or timing chain end of the engine, and torque to Specifications - during the final torque stages, use an angle gauge as shown, or paint reference marks on the rod and the socket

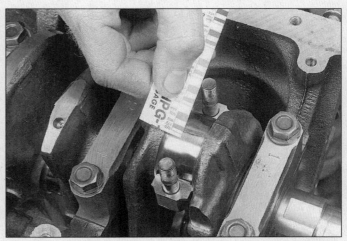

25.17 Measure the width of the crushed Plastigage to determine the rod bearing oil clearance (be sure to use the correct scale - standard and metric scales are included)

## Final connecting rod installation

19   Carefully scrape all traces of the Plastigage material off the rod journal and/or bearing face. Be very careful not to scratch the bearing, use your fingernail or the edge of a credit card to remove the Plastigage.

20   Make sure the bearing faces are perfectly clean, then apply a uniform layer of clean moly-base grease or engine assembly lube to both of them. You'll have to push the piston higher into the cylinder to expose the face of the bearing insert in the connecting rod, be sure to slip the protective hoses over the rod bolts first.

21   Slide the connecting rod back into place on the journal, remove the protective hoses from the rod cap bolts, install the rod cap and tighten the nuts to the torque listed in this Chapter's Specifications. Again, work up to the torque in three steps.

22   Repeat the entire procedure for the remaining pistons/connecting rods.

23   The important points to remember are:

a) *Keep the back sides of the bearing inserts and the insides of the connecting rods and caps perfectly clean when assembling them.*

b) *Make sure you have the correct piston/rod assembly for each cylinder.*

c) *The dimple on the piston must face the front of the engine.*

d) *Lubricate the cylinder walls with clean oil.*

e) *Lubricate the bearing faces when installing the rod caps after the oil clearance has been checked.*

24   After all the piston/connecting rod assemblies have been properly installed, rotate the crankshaft a number of times by hand to check for any obvious binding.

25   As a final step, the connecting rod endplay must be checked. Refer to Section 13 for this procedure.

26   Compare the measured endplay to this Chapter's Specifications to make sure it's correct. If it was correct before disassembly and the original crankshaft and rods were reinstalled, it should still be right. If new rods or a new crankshaft were installed, the endplay may be inadequate. If so, the rods will have to be removed and taken to an automotive machine shop for resizing.

## 26   Initial start-up and break-in after overhaul

**Warning:** *Have a fire extinguisher handy when starting the engine for the first time.*

1   Once the engine has been installed in the vehicle, double-check the engine oil and coolant levels.

2   With the spark plugs out of the engine and the ignition system and fuel pump disabled (see Section 4), crank the engine until oil pressure registers on the gauge or the light goes out.

3   Install the spark plugs, hook up the plug wires and restore the ignition system and fuel pump functions.

4   Start the engine. It may take a few moments for the fuel system to build up pressure, but the engine should start without a great deal of effort.

5   After the engine starts, it should be allowed to warm up to normal operating temperature. While the engine is warming up, make a thorough check for fuel, oil and coolant leaks.

6   Shut the engine off and recheck the engine oil and coolant levels.

7   Drive the vehicle to an area with minimum traffic, accelerate from 30 to 50 mph, then allow the vehicle to slow to 30 mph with the throttle closed. Repeat the procedure 10 or 12 times. This will load the piston rings and cause them to seat properly against the cylinder walls. Check again for oil and coolant leaks.

8   Drive the vehicle gently for the first 500 miles (no sustained high speeds) and keep a constant check on the oil level. It is not unusual for an engine to use oil during the break-in period.

9   At approximately 500 to 600 miles, change the oil and filter.

10   For the next few hundred miles, drive the vehicle normally. Do not pamper it or abuse it.

11   After 2000 miles, change the oil and filter again and consider the engine broken in.

# Chapter 3
# Cooling, heating and air conditioning systems

## Contents

## Specifications

### General

| | |
|---|---|
| Radiator cap pressure rating | 10.7 to 14.9 psi |
| Thermostat rating | |
| 1997 and earlier | 183 to 203 degrees F |
| 1998 and later | 173 to 194 degrees F |
| Refrigerant type | |
| 1993 | R-12 |
| 1994 and later | R-134a |
| Refrigerant capacity | 23 to 26.5 ounces |

### Torque specifications

| | |
|---|---|
| Thermostat housing bolts | 80 to 82 in-lbs |
| Water pump-to-block bolts | |
| 1997 and earlier | 132 in-lbs |
| 1998 and later | 96 in-lbs |

## 1   General information

### Engine cooling system

*Refer to illustrations 1.1 and 1.6*

All vehicles covered by this manual employ a pressurized engine cooling system with thermostatically-controlled coolant circulation **(see illustration)**. An impeller type water pump mounted on the front of the block pumps coolant through the engine. The coolant flows around each cylinder and toward the rear of the engine. Cast-in coolant passages direct coolant around the intake and exhaust ports, near the spark plug areas and in proximity to the exhaust valve guides.

A wax-pellet type thermostat is located in the thermostat housing at the transaxle end of the engine. During warm up, the closed thermostat prevents coolant from circulating through the radiator. When the engine reaches normal operating temperature, the thermostat opens and allows hot

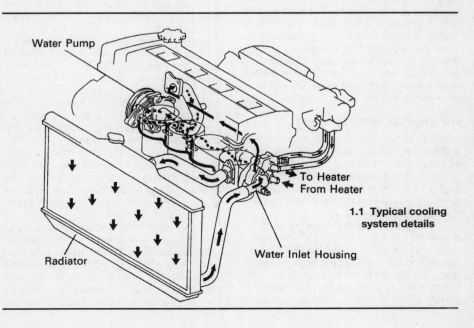

1.1 Typical cooling system details

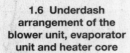

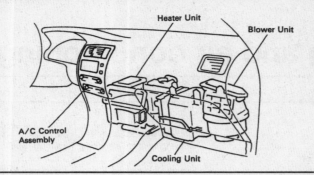

**1.6 Underdash arrangement of the blower unit, evaporator unit and heater core**

**2.4 An inexpensive hydrometer can be used to test the condition of your coolant**

coolant to travel through the radiator, where it is cooled before returning to the engine.

The cooling system is sealed by a pressure-type radiator cap. This raises the boiling point of the coolant, and the higher boiling point of the coolant increases the cooling efficiency of the radiator. If the system pressure exceeds the cap pressure-relief value, the excess pressure in the system forces the spring-loaded valve inside the cap off its seat and allows the coolant to escape through the overflow tube into a coolant reservoir. When the system cools, the excess coolant is automatically drawn from the reservoir back into the radiator.

The coolant reservoir serves as both the point at which fresh coolant is added to the cooling system to maintain the proper fluid level and as a holding tank for overheated coolant.

This type of cooling system is known as a closed design because coolant that escapes past the pressure cap is saved and reused.

### Heating system

The heating system consists of a blower fan and heater core located within the heater box under the right end of the dashboard, the inlet and outlet hoses connecting the heater core to the engine cooling system and the heater/air conditioning control head on the dashboard **(see illustration)**. Hot engine coolant is circulated through the heater core. When the heater mode is activated, a flap door opens to expose the heater box to the passenger compartment. A fan switch on the control head activates the blower motor, which forces air through the core, heating the air.

### Air conditioning system

The air conditioning system consists of a condenser mounted in front of the radiator, an evaporator mounted adjacent to the heater core, a compressor mounted on the engine, a filter-drier which contains a high pressure relief valve and the plumbing connecting all of the above.

A blower fan forces the warmer air of the passenger compartment through the evaporator core (sort of a radiator-in-reverse), transferring the heat from the air to the refrigerant. The liquid refrigerant boils off into low pressure vapor, taking the heat with it when it leaves the evaporator. The compressor

keeps refrigerant circulating through the system, pumping the warmed coolant through the condenser where it is cooled and then circulated back to the evaporator.

## 2   Antifreeze - general information

*Refer to illustration 2.4*

**Warning:** *Do not allow antifreeze to come in contact with your skin or painted surfaces of the vehicle. Rinse off spills immediately with plenty of water. Antifreeze is highly toxic if ingested. Never leave antifreeze lying around in an open container or in puddles on the floor; children and pets are attracted by it's sweet smell and may drink it. Check with local authorities about disposing of used antifreeze. Many communities have collection centers which will see that antifreeze is disposed of safely. Never dump used antifreeze on the ground or into drains.*
**Note:** *Non-toxic antifreeze is now manufactured and available at local auto parts stores, but even these types should be disposed of properly.*

The cooling system should be filled with a water/ethylene-glycol based antifreeze solution, which will prevent freezing down to at least -20 degrees F, or lower if local climate requires it. It also provides protection against corrosion and increases the coolant boiling point.

The cooling system should be drained, flushed and refilled every 30,000 miles or every two years (see Chapter 1). The use of antifreeze solutions for periods of longer than two years is likely to cause damage and encourage the formation of rust and scale in the system. If your tap water is "hard", i.e. contains a lot of dissolved minerals, use distilled water with the antifreeze.

Before adding antifreeze to the system, check all hose connections, because antifreeze tends to leak through very minute openings. Engines do not normally consume coolant. Therefore, if the level goes down, find the cause and correct it.

The exact mixture of antifreeze-to-water you should use depends on the relative weather conditions. The mixture should contain at least 50 percent antifreeze, but should never contain more than 70 percent antifreeze. Consult the mixture ratio chart on the antifreeze container before adding

coolant. Hydrometers are available at most auto parts stores to test the ratio of antifreeze to water **(see illustration)**. Use antifreeze which meets the vehicle manufacturer's specifications.

## 3   Thermostat - check and replacement

**Warning:** *Do not attempt to remove the radiator cap, coolant or thermostat until the engine has cooled completely.*

### Check

1    Before assuming the thermostat is responsible for a cooling system problem, check the coolant level (Chapter 1), drivebelt tension (Chapter 1) and temperature gauge (or light) operation.
2    If the engine takes a long time to warm up (as indicated by the temperature gauge or heater operation), the thermostat is probably stuck open. Replace the thermostat with a new one.
3    If the engine runs hot, use your hand to check the temperature of the lower radiator hose. If the hose is not hot, but the engine is, the thermostat is probably stuck in the closed position, preventing the coolant inside the engine from traveling through the radiator. Replace the thermostat. **Caution:** *Do not drive the vehicle without a thermostat. The computer may stay in open loop and emissions and fuel economy will suffer.*
4    If the lower radiator hose is hot, it means that the coolant is flowing and the thermostat is open. Consult the *Troubleshooting* Section at the front of this manual for further diagnosis.

### Replacement

*Refer to illustrations 3.7, 3.8, 3.9 and 3.10*

5    Disconnect the negative cable from the battery. **Caution:** *If the stereo in your vehicle is equipped with an anti-theft system, make sure you have the correct activation code before disconnecting the battery.*
6    Drain the coolant from the radiator (see

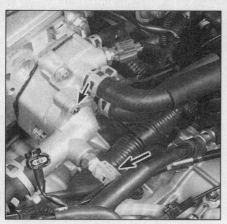

**3.7 Disconnect the ECT connector (arrow) and remove the two nuts (arrow indicates the upper nut) retaining the thermostat cover to the water inlet - 1997 and earlier**

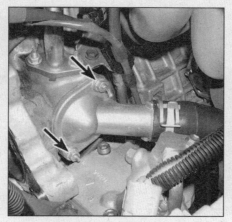

**3.8 Thermostat housing mounting nuts - 1998 and later**

**3.9 The thermostat is installed with the spring end towards the cylinder head**

**3.10 The thermostat gasket, which is actually a grooved sealing ring, fits around the edge of the thermostat**

Chapter 1).

7    On 1997 and earlier models, disconnect the Engine Coolant Temperature (ECT) switch connector from the thermostat housing located just under the distributor at the left end of the cylinder head **(see illustration)**. On 1998 and later models, remove the engine drivebelt and the alternator (see Chapter 5), then disconnect the Engine Coolant Temperature (ECT) switch connector from the thermostat housing (if equipped).

8    Detach the thermostat housing from the engine **(see illustration)**. Be prepared for some coolant to spill as the gasket seal is broken. The radiator hose can be left attached to the housing, unless the housing itself is to be replaced.

9    Remove the thermostat, noting the direction in which it was installed in the housing, and thoroughly clean the sealing surfaces **(see illustration)**.

10    Fit a new gasket onto the thermostat **(see illustration)**. Make sure it is evenly fitted all the way around.

11    Install the thermostat and housing, positioning the jiggle pin at the highest point.

12    Tighten the housing fasteners to the torque listed in this Chapter's Specifications and reinstall the remaining components in the reverse order of removal.

13    Refill the cooling system, run the engine and check for leaks and proper operation.

---

**4    Engine cooling fan and relay - check and replacement**

---

**Warning:** *To avoid possible injury, keep clear of the fan blades, as they may start turning at any time!*

## *Check*

*Refer to illustrations 4.2 and 4.4*

### 1999 and earlier models

1    If the radiator fan won't shut off when the engine is cool, check the cooling fan relay

and the ECT switch and related wires between these two components. Then disconnect the wiring connector from the Engine Coolant Temperature (ECT) switch **(see illustration 3.7)**. With the key ON, the fan should now operate. If not, check the fan relay, wiring between the ECT and the relay, and the fan motor itself.

2    The ECT can be tested for continuity with an ohmmeter. Disconnect the wiring connector and attach one lead of the ohmmeter to the prong on the ECT, and the other lead to the body of the ECT **(see illustration)**. When the engine is cold (below 180-degrees F) there should be continuity. When the engine is warm (above 199-degrees F), there should be NO continuity.

### 2000 and later models

3    On these models the engine cooling fan is controlled by the Powertrain Control Module (PCM) through the inputs it receives from the coolant temperature sensor. Refer to Chapter 6 for coolant temperature sensor testing procedures. The following fan motor and relay checks still apply to these models.

### All models

4    To test an inoperative fan motor (one that doesn't come on when the engine gets hot or when the air conditioner is on), first check the fuses and/or fusible links (see Chapter 12). Then disconnect the electrical

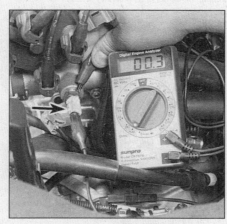

**4.2 The ECT switch (arrow) for the electric fan is mounted in the thermostat housing - test for continuity with one lead on the terminal and one on the ECT body (1999 and earlier models)**

**4.4 Disconnect the fan wiring connector (arrow) and connect jumper wires directly to the positive and negative terminals of the battery - this view is from below on the driver's side (1997 and earlier models shown)**

**3**

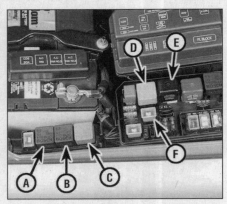

**4.6a  Main relay box details (1997 and earlier models shown)**

A    *Air conditioning relay*
B    *No. 3 cooling fan relay*
C    *No. 2 cooling fan relay*
D    *No. 1 cooling fan relay*
E    *Main engine relay*
F    *Fan fuse*

connector at the motor and use fused jumper wires to connect the fan directly to the battery **(see illustration)**. If the fan still does not work, replace the fan motor. **Warning:** *Do not allow the test clips to contact each other or any metallic part of the vehicle.*

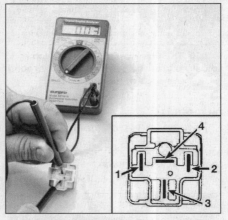

**4.7  Test the cooling fan relay number 1 (Nippondenso) for continuity - with no voltage applied, there should be continuity between 1 and 2, and 3 and 4 - with voltage applied across 1 and 2, there should be NO continuity between 3 and 4 - on Bosch relays, disregard the numbers on the terminals, test it just like this Nippondenso relay**

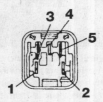

**4.8  Main engine relay terminal guide**

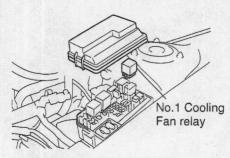

**4.6b  No. 1 cooling fan relay location - 1998 and later models**

5    If the motor tested OK in the previous test but is still inoperative, then the fault lies in the relays, fuse, or wiring. The fan relay and main engine relay can be tested as described below.

**Relay check**

*Refer to illustrations 4.6a, 4.6b, 4.6c, 4.7 and 4.8*

6    Locate the main relay box, in the engine compartment, on the driver's side **(see illustrations)**. Air-conditioned models have extra relays in a separate box next to the battery .

7    Remove the cooling fan relay number 1 and, using an ohmmeter, test for continuity as shown **(see illustration)**. Both Nippondenso and Bosch-made relays are used, and the terminals are numbered differently. Be sure to test either relay as shown in the illustration.

8    If the fan relays check OK, remove the main engine relay and test it **(see illustration)**. There should be continuity between terminals 3 and 5, and between 2 and 4. There should be NO continuity between terminals 1 and 2. With battery voltage applied across terminals 3 and 5, there should be continuity between 1 and 2, and NO continuity between 2 and 4. If both relays test OK, we recommend you take the vehicle to a dealer or other qualified repair facility for further diagnosis, due to the complexity and variety of the circuits involved.

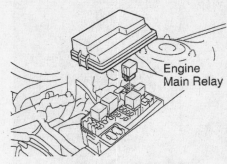

**4.6c  Engine main relay location - 1998 and later models**

## Replacement

*Refer to illustrations 4.12a, 4.12b, 4.12c, 4.12d, 4.13 and 4.14*

9    Disconnect the negative battery cable. **Caution:** *If the stereo in your vehicle is equipped with an anti-theft system, make sure you have the correct activation code before disconnecting the battery.*

10    Disconnect the wiring connector at the fan motor.

11    Drain enough coolant (see Chapter 1) to disconnect the upper radiator hose at the radiator, then disconnect the coolant overflow hose from the top of the radiator.

12    Unbolt the fan shroud from the radiator and lift the fan/shroud assembly from the vehicle. **Note:** *Models equipped with air conditioning have two cooling fans. On 1997 and earlier models, the fan on the front side of the radiator is cooling fan no. 2, which provides additional cooling when the air conditioning is on. This fan and its relay can be tested and removed/replaced with the same procedures as the engine cooling fan no. 1* **(see illustrations)**.

13    Hold the fan blades and remove the fan retaining nut (and spacer, if equipped) **(see illustration)**. **Note:** *On some air-conditioned models, the number 2 (condenser) cooling*

**Electric Cooling Fan Connector**

**Electric Cooling Fan**

**4.12a  Engine cooling fan mounting details - 1997 and earlier**

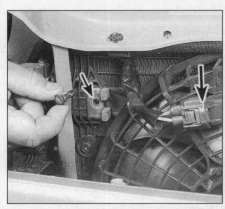

**4.12b  The air conditioning cooling fan on 1997 and earlier vehicles is mounted in front of the radiator. To remove it, simply disconnect the electrical connector (right arrow), detach the plastic retaining pin (left arrow) . . .**

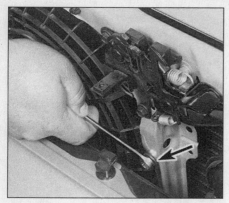

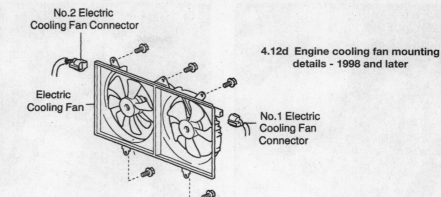

**4.12d  Engine cooling fan mounting details - 1998 and later**

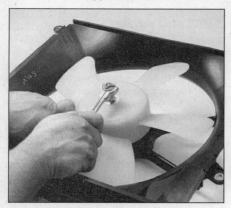

**4.12c  . . . and remove the bolts (arrow) securing the A/C fan to the hood latch support brace**

**4.14  Remove the motor from the shroud**

**5.6  Remove the automatic transaxle cooler lines and the lower radiator hose (arrows)**

**4.13  Remove the fan from the motor**

*fan blade may be retained by a clip or screws, not a nut.*

14  Unbolt the fan motor from the shroud **(see illustration)**.

15  Installation is the reverse of removal.

## 5  Radiator and coolant reservoir - removal and installation

**Warning:** *Do not start this procedure until the engine is completely cool.*

### Radiator

*Refer to illustrations 5.6, 5.7 and 5.8*

1  Disconnect the negative battery cable. **Caution:** *If the stereo in your vehicle is equipped with an anti-theft system, make sure you have the correct activation code before disconnecting the battery.*

2  Drain the coolant into a container (see Chapter 1).

3  Remove both the upper and lower radiator hoses.

4  Disconnect the reservoir hose from the radiator filler neck.

5  Remove the cooling fan (see Section 4).

6  If equipped with an automatic transaxle, disconnect the cooler lines from the radiator **(see illustration)**. Place a drip pan to catch the fluid and cap the fittings.

7  Remove the two upper radiator mounting brackets **(see illustration)**. **Note:** *The bottom of the radiator is retained by grommeted projections that fit into holes in the body.*

8  Lift out the radiator **(see illustration)**. Be aware of dripping fluids and the sharp fins.

9  With the radiator removed, it can be inspected for leaks, damage and internal blockage. If in need of repairs, have a professional radiator shop or dealer service department perform the work as special techniques are required.

10  Bugs and dirt can be cleaned from the radiator with compressed air and a soft brush. Don't bend the cooling fins as this is done. **Warning:** *Wear eye protection when using compressed air.*

11  Installation is the reverse of the removal procedure. Be sure the rubber mounts are in place on the bottom of the radiator.

12  After installation, fill the cooling system with the proper mixture of antifreeze and water. Refer to Chapter 1 if necessary.

13  Start the engine and check for leaks. Allow the engine to reach normal operating

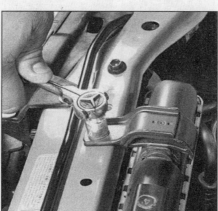

**5.7  Remove the upper hold-down clamps from each end of the radiator**

**5.8  Remove the radiator carefully, and do not lose the rubber mounts - they must be in place on the bottom projections when the radiator is reinstalled**

**3**

**5.15 On most 1997 and earlier vehicles, the coolant reservoir can be removed by detaching the cap (left arrow) and pulling the reservoir bottle out of its bracket (right arrow)**

**7.4 Use a screwdriver to pry the wiring clip out of the dipstick tube bracket, then remove the one bolt retaining the dipstick tube to the engine and pull the tube out**

**7.5 Disconnect the water temperature sender (arrow) from the water neck**

temperature, indicated by both radiator hoses becoming hot. Recheck the coolant level and add more if required.

14   On automatic transmission equipped models, check and add fluid as needed.

### Coolant reservoir

*Refer to illustration 5.15*

15   On early models, the coolant reservoir simply pulls up and out of the bracket next to the battery **(see illustration)**. On later models it will be necessary to remove the battery first, then remove the two reservoir retaining nuts before the reservoir can be removed from the vehicle.

16   Pour the coolant into a container. Wash out and inspect the reservoir for cracks and chafing. Replace it if damaged.

17   Installation is the reverse of removal.

---

**6   Water pump - check**

---

1   A failure in the water pump can cause serious engine damage due to overheating.

2   With the engine running and warmed to normal operating temperature, squeeze the upper radiator hose. If the water pump is working properly, a pressure surge should be felt as the hose is released. **Warning:** *Keep hands away from fan blades!*

3   Water pumps are equipped with weep or vent holes **(see illustration 7.7)**. If a failure occurs in the pump seal, coolant will leak from this hole. In most cases it will be necessary to use a flashlight to find the hole on the water pump by looking through the space behind the pulley just below the water pump shaft.

4   If the water pump shaft bearings fail there may be a howling sound at the front of the engine while it is running. Bearing wear can be felt if the water pump pulley is rocked up and down. Do not mistake drivebelt slippage, which causes a squealing sound, for water pump failure. Spray automotive drive-

belt dressing on the belts to eliminate the belt as a possible cause of the noise.

---

**7   Water pump - removal and installation**

---

**Warning:** *Do not start this procedure until the engine is completely cool.*

1   Disconnect the cable from the negative terminal of the battery. (see Chapter 1). **Caution:** *If the stereo in your vehicle is equipped with an anti-theft system, make sure you have the correct activation code before disconnecting the battery.*

2   With the parking brake applied and the shifter in Park (automatic) or in gear (manual), loosen the lug nuts from the right front wheel, then raise the front of the vehicle and support it securely on jackstands.

3   Remove the right front wheel and the right splash shield from the wheelwell. Referring to Chapter 1, drain the cooling system and remove the engine drivebelt(s) (see Chapter 1).

### 1997 and earlier

*Refer to illustrations 7.4, 7.5, 7.6a, 7.6b and 7.7*

4   Refer to Chapter 2A and remove the upper timing belt covers. Remove the oil dipstick. Remove the bolt retaining the dipstick tube to the engine and pull the dipstick tube from the oil pump housing **(see illustration)**.

5   Behind the water pump, remove the two nuts retaining the right-side water neck to the cylinder head, and unplug the electrical connector from the water temperature gauge sender **(see illustration)**.

6   Remove the three bolts retaining the water pump to the engine block and remove the water pump and water neck assembly. The pulley can remain on the water pump during removal **(see illustrations)**.

7   Disassemble the water pump and replace the gaskets. Thoroughly clean all sealing surfaces, removing the O-ring and cleaning the groove **(see illustration)**. Assemble the water pump.

8   Using a new O-ring, install the pump assembly to the block. Use a new gasket on the water neck. **Note:** *If the pump has been replaced after many miles of usage, it's a*

**7.6a From above, remove the two nuts (smaller arrows) retaining the water neck to the cylinder head, then remove the top bolt (larger arrow) from the water pump**

**7.6b From below, remove the two remaining water pump assembly-to-block bolts - do not remove the bolts retaining the pump halves together until the assembly is removed from the block**

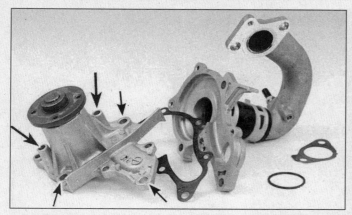

**7.7  Remove the water pump assembly, then disassemble the pump and replace the gaskets - the three small arrows indicate the water pump-to-block bolts, while the two larger arrows indicate the weep holes**

**7.11a  On 1998 and later models, the upper water pump bolts (arrows) are accessible from above . . .**

good idea to also replace the hose connecting the water pump to the water neck.

9    Install the remaining parts in the reverse order of removal. When reinstalling the dipstick tube, use a new O-ring where it pushes into the oil pump.

10    Refill the cooling system (see Chapter 1), run the engine and check for leaks and proper operation.

### 1998 and later

*Refer to illustrations 7.11a, 7.11b and 7.12*

11    Remove the bolts retaining the water pump to the engine block and remove the water pump **(see illustrations)**. Note the length and the position of the bolts as they were originally installed before removing them.

12    Thoroughly clean all sealing surfaces, removing the O-ring from the groove in the timing chain cover **(see illustration)**.

13    Install a new O-ring in the timing chain cover and place the water pump in position. Install the bolts in their original positions and tighten them to the torque listed in this Chapter's Specifications.

14    Install the remaining parts in the reverse order of removal. Refill the cooling system (see Chapter 1), run the engine and check for leaks and proper operation.

### 8    Coolant temperature sending unit - check and replacement

### Check

1    If the coolant temperature gauge is inoperative, check the fuses first (see Chapter 12).

2    If the temperature gauge indicates excessive temperature after running awhile, see the *Troubleshooting* section in the front of the manual.

3    If the temperature gauge indicates Hot as soon as the engine is started cold, disconnect the wire at the coolant temperature sender **(see illustration 7.5)**. If the gauge reading drops, replace the sending unit. If the reading

**7.11b  . . . and the lower bolts (arrows) are accessed through the right wheelwell**

remains high, the wire to the gauge may be shorted to ground or the gauge is faulty.

4    If the coolant temperature gauge fails to show any indication after the engine has been warmed up, (approximately 10 minutes) and the fuses checked out OK, shut off the engine. Disconnect the wire at the sending unit and, using a jumper wire, connect the wire to a clean ground on the engine. Briefly turn on the ignition without starting the engine. If the gauge now indicates Hot, replace the sending unit.

5    If the gauge fails to respond, the circuit may be open or the gauge may be faulty - see Chapter 12 for additional information.

### Replacement

**Warning:** *Do not start this procedure until the engine is completely cool.*

6    Drain the coolant (see Chapter 1).

7    Disconnect the wiring connector from the sending unit on the thermostat housing.

8    Using a deep socket or a box-end wrench, remove the sending unit.

9    Install the new unit and tighten it securely. Do not use thread sealer as it may electrically insulate the sending unit.

**7.12  Always remove the O-ring from the timing chain cover (arrow) and replace it with a new one**

10    Reconnect the wiring connector, refill the cooling system and check for coolant leakage and proper gauge function.

### 9    Blower unit - check, removal and installation

*Refer to illustrations 9.4, 9.6 and 9.9*

**Warning:** *These models are equipped with airbags. The airbag is armed and can deploy (inflate) anytime the battery is connected. To prevent accidental deployment (and possible injury), turn the ignition key to LOCK and disconnect the negative battery cable whenever working near airbag components. After the battery is disconnected, wait at least two minutes before beginning work (the system has a back-up capacitor that must fully discharge). For more information, see Chapter 12.*

1    The blower unit is located in the passenger compartment above the right front footwell. If the blower doesn't work, check the fuse and all connections in the circuit for looseness and corrosion. Make sure the bat-

**3**

**9.4  With the connector disconnected from the blower motor, apply fused power and ground to the motor terminals (arrow) - replace the blower if it doesn't operate**

**9.6  Blower motor resistor location (arrow) - the resistor is retained by two screws**

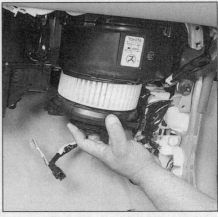

**9.9  Remove the three screws and lower the blower motor from the housing**

tery is fully charged.

2     Remove the glove compartment liner, the glove compartment door and the right lower dash panel (see Chapter 11).

3     If the blower motor does not operate, disconnect the electrical connector at the blower motor, turn the ignition key On (engine not running) and check for battery voltage on the black wire terminal with the blower speed switch ON. If battery voltage is not present, there is a problem in the ignition feed circuit.

4     If battery voltage is present, reconnect the terminal to the blower motor and back-probe the black/white wire with a jumper wire connected to ground. If the motor still does not operate, the motor is probably faulty. Apply fused power and ground connections to the blower motor terminals, if the motor does not operate, replace the motor **(see illustration)**.

5     If the motor is good, but doesn't operate at any speed, the heater/air conditioning control switch is probably faulty. Remove the control assembly (see Section 11) and check for continuity through the switch in each position.

6     If the blower motor operates at High speed, but not at one or more of the lower speeds, check the blower motor resistor, located under the instrument panel on the passenger side **(see illustration)**.

7     Disconnect the electrical connector from the blower motor resistor. Remove the screws and withdraw the resistor from the housing.

8     Using a continuity tester on the blower motor resistor, there should be continuity between all terminals. If not, replace the resistor.

9     If the blower motor must be replaced, remove the bracket retaining the wiring connector to the blower, then remove the three mounting screws and lower the blower assembly from the housing **(see illustration)**. The fan can be removed and reused on the new blower motor.

10     Installation is the reverse of removal. Check for proper operation.

## 10  Heater core - removal and installation

*Refer to illustrations 10.3, 10.7, 10.8, 10.9, 10.10a, 10.10b, 10.10c, 10.11, 10.12a and 10.12b*

**Warning 1:** *These models are equipped with airbags. The airbag is armed and can deploy (inflate) anytime the battery is connected. To prevent accidental deployment (and possible injury), turn the ignition key to LOCK and disconnect the negative battery cable whenever working near airbag components. After the battery is disconnected, wait at least two minutes before beginning work (the system has a back-up capacitor that must fully discharge). For more information see Chapter 12.*

**Warning 2:** *Wait until the engine is completely cool before beginning this procedure.*

**Note 1:** *The following procedure details removing the heater core assembly by itself, but it is much easier to remove the heater core housing if the evaporator (cooling) assembly is removed first, although this necessitates having the refrigerant discharged and recovered before work begins. See Section 16 for evaporator removal, and do not discharge the refrigerant without reading the Warning in Section 12.*

**Note 2:** *The factory recommends removal of the entire instrument panel and dropping the steering column to remove the heater core. This involves disconnecting numerous electrical connectors and there is the potential for breakage of delicate plastic tabs on various components. This is a difficult job for the average home mechanic. It is possible to remove the heater core assembly without removing the dash, and is the procedure outlined below.*

1     Disconnect the negative cable from the battery. **Caution:** *If the stereo in your vehicle is equipped with an anti-theft system, make sure you have the correct activation code before disconnecting the battery.*

2     Drain the cooling system (see Chapter 1).

3     Working in the engine compartment, disconnect the heater hoses at the firewall

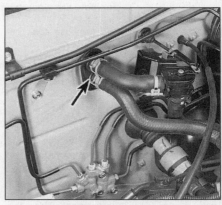

**10.3  Disconnect the heater hoses at the firewall (arrow)**

**(see illustration)**. Push the rubber seal around the hoses toward the inside of the vehicle, releasing it from the sheetmetal.

4     Refer to Chapters 11 and 12 and remove the center console, glove compartment, glove compartment liner, ashtray, radio and center dash bezels.

5     Refer to Section 11 of this Chapter to remove the heater/air conditioning controls.

6     Refer to Chapter 6 and remove the ECM without disconnecting the connectors. Set the ECM aside to allow room under the heater core housing.

7     There are two metal braces running from the dashboard to the floor supporting the center of the instrument panel. The left one should be unbolted from the floor, but it need not be entirely removed from the vehicle. The right brace should be unbolted and removed **(see illustration)**. **Caution:** *There are Phillips screws retaining the brace to the backside of the dash plastic. Use the right size Phillips bit and be careful. Seat your screwdriver squarely into the slots while removing them. They are in tight and the heads could strip.*

8     Unbolt and remove the center ventilation duct **(see illustration)**.

9     Remove the heater duct/control door that is directly above and in front of the heater unit **(see illustration)**.

10.7  Remove the right dashboard-to-floor brace - besides the mounting bolts (arrows), you must also unbolt several ground wires from the brace before removing it

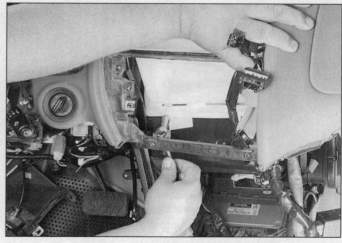

10.8  Remove the center ventilation duct

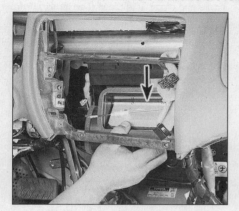

10.9  Pull up and out on this duct/door assembly (arrow) then pull it down and out to allow more working room around the heater unit

10.10a  Remove the nut from the stud near the upper arm of the accelerator pedal (arrow)

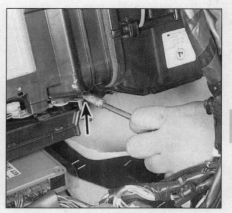

10.10b  Remove the stud/nut located at the lower right corner of the heater core housing (arrow)

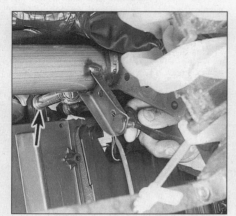

10.10c  Remove the last nut from the stud at the upper right corner (arrow) of the heater core housing

10.11  After the housing tabs clear the firewall studs and the housing is clear of the lip on the evaporator housing, pull the heater unit down and out to the right

10.12a  Remove the two screws and clamps (arrows) . . .

10  Loosen the nuts on the studs that retain the left side of the evaporator housing (see Section 16). There are three studs retaining the heater unit to the firewall. Remove the nuts from these studs (see illustrations).

11  Pull the heater unit out from behind the dash (see illustration). Keep plenty of towels or rags on the carpeting to catch any coolant that may drip. Caution: Work slowly and carefully to avoid breaking any plastic components during removal. The assembly must be pulled out (toward the rear of the vehicle) far enough for the plastic tabs to clear the firewall studs and then moved to the left to clear the lip of the evaporator housing before pulling the unit down and out.

12  Once the heater unit is removed from the vehicle, remove the clamps retaining the

3

**10.12b** . . . then slide the heater core out of the housing

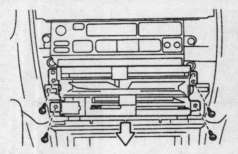

**11.4 Heating/air conditioning control panel mounting screws (1997 and earlier shown, 1998 and later similar)**

**11.5 Twist the plastic flag (arrow) on each cable to release it from the control unit, then lift the cable end eye off the control lever pin**

heater core tubes to the housing. Slide the heater core out of the housing **(see illustrations)**.

13   Installation is the reverse order of removal. **Note:** *The nuts used on the firewall studs for both the heater core housing and the evaporator housing are designed for one-time use on the assembly line. After removing them, purchase new metric nuts and lockwashers to reinstall the housing to the studs.*

14   Refill the cooling system, reconnect the battery and run the engine. Check for leaks and proper system operation.

## 11   Heater and air conditioning control assembly - removal, installation and adjustment

*Refer to illustrations 11.4 and 11.5*
**Warning:** *These models are equipped with airbags. The airbag is armed and can deploy (inflate) anytime the battery is connected. To prevent accidental deployment (and possible injury), turn the ignition key to LOCK and disconnect the negative battery cable whenever working near airbag components. After the battery is disconnected, wait at least two minutes before beginning work (the system has a back-up capacitor that must fully discharge). For more information see Chapter 12.*

### Removal and installation

1   Disconnect the negative cable from the battery. **Caution:** *If the stereo in your vehicle is equipped with an anti-theft system, make sure you have the correct activation code before disconnecting the battery.*

2   Remove the center cluster trim panels (see Chapter 11).

3   On 1997 and earlier models, pull off the control knobs from the control unit.

4   Remove the mounting screws located on the front of the control assembly **(see illustration)**.

5   Pull the control unit out slightly and disconnect the heater control cables from the rear of the control unit **(see illustration)**. Disconnect the electrical connectors from the blower speed switch and the A/C switch (if equipped).

6   Installation is the reverse of the removal procedure.

7   Run the engine and check for proper functioning of the heater (and air conditioning, if equipped).

### Adjustment

*Refer to illustrations 11.9, 11.10 and 11.11*

8   With the cables attached at the heater/air conditioning control unit and the control unit installed in the dash, adjust the cables at their ends. The controls should be set to: RECIRC, COOL, and DEF.

9   To adjust the air inlet (vent) control cable set the damper lever to RECIRC, install the

cable and clamp it in place **(see illustration)**.

10   To adjust the air mix (temperature) control cable, set the air mix damper to COOL, install the cable and lock the clamp while applying slight pressure on the outer cable **(see illustration)**.

11   To adjust the mode control cable, set the mode damper to the DEF mode, hook the cable end on and tighten the clamp **(see illustration)**.

## 12   Air conditioning and heating system - check and maintenance

### Air conditioning system

*Refer to illustration 12.1*
**Warning:** *The air conditioning system is under high pressure. Do not loosen any hose fittings or remove any components until the system has been discharged. Air conditioning refrigerant should be properly discharged into an EPA-approved recovery/recycling unit by a dealer service department or an automotive air conditioning repair facility. Always wear eye protection when disconnecting air conditioning system fittings.*

1   The following maintenance checks should be performed on a regular basis to

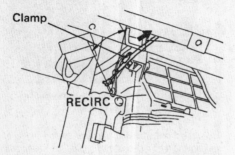

**11.9 To adjust the air inlet (vent) control cable, move the arm away from the firewall, attach the cable end and tighten the clamp**

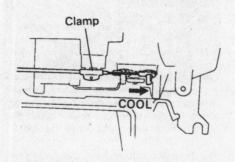

**11.10 To adjust the air mix (temperature) control cable, push the lever away from the cable clamp and tighten the clamp**

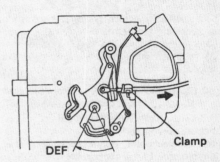

**11.11 Adjust the mode control cable with the lever pulled toward the cable clamp, then tighten the clamp**

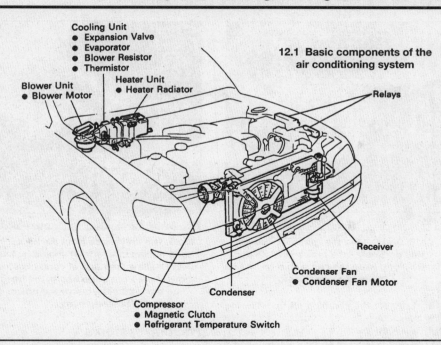

Cooling Unit
- Expansion Valve
- Evaporator
- Blower Resistor
- Thermistor

**12.1 Basic components of the air conditioning system**

Blower Unit
- Blower Motor

Heater Unit
- Heater Radiator

Relays

Receiver

Condenser Fan
- Condenser Fan Motor

Condenser

Compressor
- Magnetic Clutch
- Refrigerant Temperature Switch

**12.9 With the system operating, the evaporator outlet line (large tubing) should feel cold**

**12.10 Check the temperature of the output air in the center register with a thermometer - it should be 35-40 degrees below the ambient air temperature**

**3**

ensure that the air conditioner continues to operate at peak efficiency **(see illustration)**:

a) *Inspect the condition of the compressor drivebelt. If it is worn or deteriorated, replace it* (see Chapter 1).

b) *Check the drivebelt tension and, if necessary, adjust it* (see Chapter 1).

c) *Inspect the system hoses. Look for cracks, bubbles, hardening and deterioration. Inspect the hoses and all fittings for oil bubbles or seepage. If there is any evidence of wear, damage or leakage, replace the hose(s).*

d) *Inspect the condenser fins for leaves, bugs and any other foreign material that may have embedded itself in the fins. Use a "fin comb" or compressed air to remove debris from the condenser.*

e) *Make sure the system has the correct refrigerant charge.*

2    It's a good idea to operate the system for about ten minutes at least once a month. This is particularly important during the winter months because long term non-use can cause hardening, and subsequent failure, of the seals.

3    Leaks in the air conditioning system are best spotted when the system is brought up to operating temperature and pressure, by running the engine with the air conditioning ON for five minutes. Shut the engine off and inspect the air conditioning hoses and connections. Traces of oil usually indicate refrigerant leaks.

4    Because of the complexity of the air conditioning system and the special equipment required to effectively work on it, accurate troubleshooting of the system should be left to a professional technician.

5    If the air conditioning system doesn't operate at all, check the fuse panel and the air conditioning relay, located in the fuse/relay box in the engine compartment. Refer to Sec-

tions 4, 9 and 11 for electrical checks of heating/air conditioning system components.

6    The most common cause of poor cooling is simply a low system refrigerant charge. If a noticeable drop in cool air output occurs, the following quick check will help you determine if the refrigerant level is low. For more complete information on the air conditioning system, refer to the *Haynes Automotive Heating and Air Conditioning Manual.*

## Checking the refrigerant charge

*Refer to illustrations 12.9, 12.10 and 12.11*

7    Warm the engine up to normal operating temperature.

8    Place the air conditioning temperature selector at the coldest setting and put the blower at the highest setting. Open the doors (to make sure the air conditioning system doesn't cycle off as soon as it cools the passenger compartment).

9    With the compressor engaged - the clutch will make an audible click and the center of the clutch will rotate. After the system reaches operating temperature, feel the large pipe exiting from the evaporator at the firewall **(see illustration)**.

10    The large evaporator outlet pipe should feel cold. If the evaporator outlet is warm or moderately warm, the system needs a charge. Insert a thermometer in the center air distribution duct while operating the air conditioning system **(see illustration)** - the temperature of the output air should be 35 to 40 degrees F below the ambient air temperature (down to approximately 40 degrees F). If the ambient (outside) air temperature is very high, say 110 degrees F, the duct air temperature may be as high as 60 degrees F, but generally the air conditioning is 35 to 40 degrees F cooler than the ambient air. If the air isn't as cold as it used to be, the system probably needs a charge. Further inspection or testing

of the system is beyond the scope of the home mechanic and should be left to a professional.

11    Inspect the sight glass, if equipped. If the refrigerant looks foamy when running, it's low **(see illustration)**. When ambient temperatures are very hot, bubbles may show in the sight glass even with the proper amount of refrigerant. With the proper amount of refrigerant, when the air conditioning is turned off,

**12.11 On all Corolla models and 1997 and earlier Prizm models, a sight glass is located on top of the receiver/drier (arrow)**

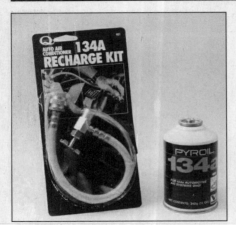

**12.12  A basic charging kit is available at most auto parts stores - it must say R-134a and so must the cans of refrigerant you buy**

**12.15  Add R-134a refrigerant to the low-side port only - the procedure will go faster if you wrap the can with a warm wet towel to prevent icing**

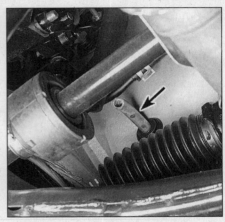

**12.25  The drain hose from the heater/air conditioning unit (arrow) should be kept clear to allow drainage of condensation - shown here from underneath, the hose is on the right side, below the power steering pump**

the sight glass should show refrigerant that foams, then clears.

### Adding refrigerant (models with R134a systems only)

*Refer to illustrations 12.12 and 12.15*
**Caution:** *Refrigerant has changed from the use of R-12, used in some 1993 models, to the "environmentally friendly" R-134a used in 1994 and later models. The two refrigerants are NOT compatible. Even after purging and evacuating an R-12 system, there is enough residual oil and refrigerant in the hoses and components that simply filling the system with R-134a cannot be done. Special fittings and manifold gauge sets are used on the different refrigerant types so that an accidental hookup of the two systems cannot be made. When replacing entire components, additional refrigerant oil should be added equal to the amount that is removed with the component being replaced. Refrigerant oils, just like refrigerant R-12 vs. R-134a, are not compatible. Be sure to read the can before adding any oil to the system, to make sure it is compatible with the type of system being repaired.*
**Note:** *Because of Federal regulations by the Environmental Protection Agency, R-12 refrigerant is not available for home-mechanic use, however, cans of R-134 refrigerant are commonly available in auto parts stores. Models with R-12 systems will have to be serviced at a dealership or air conditioning shop.*
12   Buy an automotive charging kit at an auto parts store. A charging kit includes a 14-ounce can of R-134a refrigerant, a tap valve and a short section of hose that can be attached between the tap valve and the system low side service valve **(see illustration)**. Because one can of refrigerant may not be sufficient to bring the system charge up to the proper level, it's a good idea to buy a couple of additional cans. Try to find at least one can that contains red refrigerant dye. If the system is leaking, the red dye will leak out with the refrigerant and help you pinpoint the location of the leak.

13   Connect the charging kit by following the manufacturer's instructions.
14   Back off the valve handle on the charging kit and screw the kit onto the refrigerant can, making sure first that the O-ring or rubber seal inside the threaded portion of the kit is in place. **Warning:** *Wear protective eyewear when dealing with pressurized refrigerant cans.*
15   Remove the dust cap from the low-side charging port and attach the quick-connect fitting on the kit hose **(see illustration)**. **Warning:** *DO NOT hook the charging kit hose to the system high side! The fittings on the charging kit are designed to fit only on the low side of the system.*
16   Warm the engine to normal operating temperature and turn on the air conditioner. Keep the charging kit hose away from the fan and other moving parts.
17   Turn the valve handle on the kit until the stem pierces the can, then back the handle out to release the refrigerant. You should be able to hear the rush of gas. Add refrigerant to the low side of the system until both the outlet and the evaporator inlet pipe feel about the same temperature. Allow stabilization time between each addition. **Warning:** *Never add more than two cans of refrigerant to the system. The can may tend to frost up, slowing the procedure. Wrap a shop towel wet with hot water around the bottom of the can to keep it from frosting.*
18   If you have an accurate thermometer, you can place it in the center air conditioning duct inside the vehicle to monitor the air temperature. A charged system that is working properly, should output air down to approximately 40 degrees F.
19   When the can is empty, turn the valve handle to the closed position and release the connection from the low-side port. Replace the dust cap.
20   Remove the charging kit from the can and store the kit for future use with the piercing valve in the UP position, to prevent inadvertently piercing the can on the next use.

### Heating systems

*Refer to illustration 12.25*
21   If the air coming out of the heater vents isn't hot, the problem could stem from any of the following causes:
   a) *The thermostat is stuck open, preventing the engine coolant from warming up enough to carry heat to the heater core. Replace the thermostat (see Section 3).*
   b) *A heater hose is blocked, preventing the flow of coolant through the heater core. Feel both heater hoses at the firewall. They should be hot. If one of them is cold, there is an obstruction in one of the hoses or in the heater core, or the heater control valve is shut. Detach the hoses and back flush the heater core with a water hose. If the heater core is clear but circulation is impeded, remove the two hoses and flush them out with a water hose.*
   c) *If flushing fails to remove the blockage from the heater core, the core must be replaced. (see Section 10).*
22   If the blower motor speed does not correspond to the setting selected on the blower switch, the problem could be a bad fuse, circuit, switch, blower motor resistor or motor (see Sections 9 and 11).
23   If there isn't any air coming out of the vents:
   a) *Turn the ignition ON and activate the fan control. Place your ear at the heating/air conditioning register (vent) and listen. Most motors are audible. Can you hear the motor running?*
   b) *If you can't (and have already verified that the blower switch and the blower motor resistor are good), the blower motor itself is probably bad (see Section 9).*
24   If the carpet under the heater core is damp, or if antifreeze vapor or steam is coming through the vents, the heater core is leaking. Remove it (see Section 10) and install a

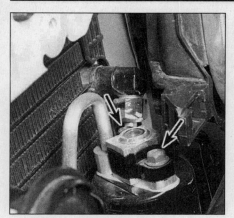

**13.3  After the system has been discharged, unbolt the two refrigerant lines from the top of the receiver-drier and cap them - there is one bolt and one screw to remove (arrows)**

**14.4  Disconnect the wiring harness connector (large arrow) at the compressor, then unbolt the flanges (two smaller arrows) and detach the refrigerant lines from the compressor**

**14.5  Remove the compressor mounting bolts (arrows indicate the two bottom bolts, two more are near the top of the compressor) and remove the compressor**

new unit (most radiator shops will not repair a leaking heater core).

25  Inspect the drain hose from the heater/air conditioning assembly at the right side of the firewall, make sure it is not clogged **(see illustration)**.

## 13  Air conditioning receiver-drier/accumulator - removal and installation

*Refer to illustration 13.3*

**Warning:** *The air conditioning system is under high pressure. Do not loosen any hose fittings or remove any components until the system has been discharged. Air conditioning refrigerant should be properly discharged into an EPA-approved recovery/recycling unit by a dealer service department or an automotive air conditioning repair facility. Always wear eye protection when disconnecting air conditioning system fittings.*

**Note:** *All Corolla models and 1997 and earlier Prizm models are equipped with a receiver-drier in the engine compartment and an expansion valve mounted next the evaporator core in the passenger compartment. All 1998 and later Prizm models are equipped with an accumulator and a expansion (orifice) tube in the engine compartment. The expansion (orifice) tube is located in the refrigerant line between the condensor and the evaporator core.*

1  Have the refrigerant discharged and recovered by an air conditioning technician.

### Receiver-drier

2  On 1997 and earlier models, refer to Chapter 11 and remove the front grille. **Note:** *On 1997 and earlier models, the receiver-drier is mounted on the driver's side of the vehicle in front of the radiator support. On 1998 and later Corolla models, the receiver-drier is mounted on the passenger's side of the vehicle just behind the radiator support.*

3  Disconnect the refrigerant lines **(see illustration)** from the top of the receiver-drier and cap the open fittings to prevent entry of moisture.

4  Loosen the pinch bolt and slip the receiver-drier out of the bracket.

### Accumulator (1998 and later Prizm models)

5  Detach the cruise control actuator from the passenger side inner fenderwell and position it aside.

6  Disconnect the refrigerant (suction) line from the accumulator and cap the open fittings to prevent entry of moisture.

7  Disconnect the accumulator (discharge) line from the evaporator core at the firewall. Cap the open fittings to prevent entry of moisture.

8  Loosen the pinch bolt and slip the accumulator out of the bracket.

### All models

9  Installation is the reverse of removal.

10  Have the system evacuated, charged and leak tested by the shop that discharged it. If the receiver-drier or accumulator was replaced, have them add new refrigeration oil to the compressor, about 1.5 to 2.0 fluid ounces. Use only the refrigerant oil compatible with the refrigerant of your system (R-12 vs. R-134a).

## 14  Air conditioning compressor - removal and installation

*Refer to illustrations 14.4 and 14.5*

**Warning:** *The air conditioning system is under high pressure. Do not loosen any hose fittings or remove any components until the system has been discharged. Air conditioning refrigerant should be properly discharged into an EPA-approved recovery/recycling unit by a dealer service department or an automotive air conditioning repair facility. Always wear*

eye protection when disconnecting air conditioning system fittings.

**Caution:** *The receiver-drier/accumulator should be replaced whenever a new compressor is installed. On 1998 and later Prizm models, the expansion (orifice) tube should also be replaced.*

1  Have the refrigerant discharged by an automotive air conditioning technician.

2  Disconnect the negative cable from the battery. **Caution:** *If the stereo in your vehicle is equipped with an anti-theft system, make sure you have the correct activation code before disconnecting the battery.*

3  Raise the front of the vehicle and support it securely on jackstands. Remove the engine under cover (if equipped). Remove the drivebelt from the compressor (see Chapter 1).

4  Detach the wiring connector and disconnect the refrigerant lines **(see illustration)**.

5  Unbolt the compressor and lift it from the vehicle **(see illustration)**.

6  If a new or rebuilt compressor is being installed, follow the directions supplied with the compressor regarding the proper level of oil prior to installation.

7  Installation is the reverse of removal. Replace any O-rings with new ones specifically made for the type of refrigerant in your system and lubricate them with refrigerant oil, also designed specifically for your system (R-12 vs. R-134a).

8  Have the system evacuated, recharged and leak tested by the shop that discharged it.

## 15  Air conditioning condenser - removal and installation

*Refer to illustrations 15.4a and 15.4b*

**Warning 1:** *The air conditioning system is under high pressure. Do not loosen any hose fittings or remove any components until the system has been discharged. Air conditioning refrigerant should be properly discharged into* *an EPA-approved recovery/recycling unit by a dealer service department or an automotive air conditioning repair facility. Always wear eye protection when disconnecting air conditioning system fittings.*

**Warning 2:** *These models are equipped with airbags. The airbag is armed and can deploy (inflate) anytime the battery is connected. To prevent accidental deployment (and possible injury), turn the ignition key to LOCK and disconnect the negative battery cable whenever working near airbag components. After the battery is disconnected, wait at least two minutes before beginning work (the system has a back-up capacitor that must fully discharge). For more information see Chapter 12.*

1    Have the refrigerant discharged by an air conditioning technician.

2    Remove the radiator as described in Section 5.

3    On 1997 and earlier models, remove the grille and right headlight assembly for access (see Chapter 11) to the condenser mounting bolts and the refrigerant lines.

4    Disconnect the condenser inlet and outlet fittings **(see illustration)**. Cap the open fittings immediately to keep moisture and contamination out of the system. Remove the condenser mounting bolts, pull the condenser back and lift it out **(see illustration)**. **Note:** *Some models have two condenser mounting bolts, while other models use four mounting bolts to retain the condenser.*

5    Install the condenser, brackets and

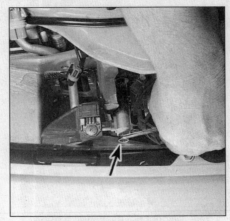

**15.4a  Disconnect the refrigerant line on the driver's side by unbolting this flange (arrow)**

**15.4b  Remove the condenser mounting bolts - there is one bolt at each end (Corolla models shown)**

bolts, making sure the rubber cushions fit on the mounting points properly.

6    Reconnect the refrigerant lines, using new O-rings where needed. If a new condenser has been installed, add approximately 1.5 to 2.0 fluid ounces of new refrigerant oil of the correct type (R-12 vs. R-134a).

7    Reinstall the remaining parts in the reverse order of removal.

8    Have the system evacuated, charged and leak tested by the shop that discharged it.

## 16   Air conditioning evaporator core - removal and installation

*Refer to illustrations 16.1, 16.2, 16.4 and 16.6*

**Warning 1:** *The air conditioning system is under high pressure. Do not loosen any hose fittings or remove any components until the system has been discharged. Air conditioning refrigerant should be properly discharged into an EPA-approved recovery/recycling unit by a dealer service department or an automotive air conditioning repair facility. Always wear eye protection when disconnecting air conditioning system fittings.*

**Warning 2:** *These models are equipped with airbags. The airbag is armed and can deploy (inflate) anytime the battery is connected. To prevent accidental deployment (and possible injury), turn the ignition key to LOCK and disconnect the negative battery cable whenever working near airbag components. After the battery is disconnected, wait at least two minutes before beginning work (the system has a back-up capacitor that must fully discharge). For more information see Chapter 12.*

**Note:** *All Corolla models and 1997 and earlier Prizm models are equipped with a receiver-drier in the engine compartment and an expansion valve mounted next the evaporator core in the passenger compartment. All 1998 and later Prizm models are equipped with an accumulator and an expansion (orifice) tube in the engine compartment. The expansion (orifice) tube is located in the refrigerant line between the condensor and the evaporator core. Refer to Section 17 for the orifice tube replacement procedure on 1998 and later Prizm models.*

1    Remove the glove compartment assembly **(see illustration)**.

2    Disconnect the air conditioning lines at the firewall, use a back-up wrench so as not to damage the fittings **(see illustration)**. Cap the open fittings after disassembly to prevent the entry of air or dirt.

3    Remove two nuts and three screws retaining the evaporator housing to the fire-

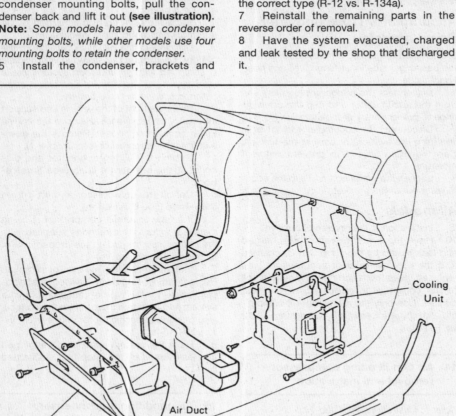

Cooling Unit

Air Duct

Glove Box Assembly

Front Door Scuff Plate

**16.1  Evaporator housing installation details**

**16.2  Use a back-up wrench when disconnecting the air conditioning lines at the firewall - later models use quick connect fittings at the firewall which require special spring lock coupling tools to remove them**

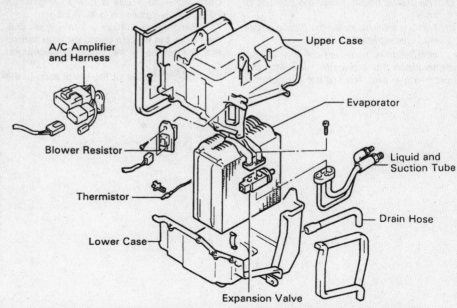

**16.4  Separate the two halves of the evaporator housing and remove the core**

A/C Amplifier and Harness

Upper Case

Evaporator

Blower Resistor

Liquid and Suction Tube

Thermistor

Drain Hose

Lower Case

Expansion Valve

**16.6  The fins of the core can be cleaned with a fin comb - use the side of the tool with the correct fins-per-inch spacing for your core**

wall and pull the unit out of the vehicle **(see illustration 16.1)**.

4    With the evaporator housing on the bench, remove the screws and clips and separate the top and bottom halves of the case and pull out the evaporator **(see illustration)**.

5    Pull the thermistor sensor probe from the evaporator core and unbolt the expansion valve and the two short refrigerant lines **(see illustration 16.4)**.

6    The evaporator core can be cleaned with a "fin comb" and blown off with compressed air **(see illustration)**. **Warning:** *Be sure to wear eye protection when using compressed air.*

7    If the evaporator core is replaced with a new unit, add 1.5 to 2.0 fluid ounces of new refrigerant oil of the correct type (R-12 vs. R-134a) to the system.

8    The remainder of the installation is the reverse of the removal process. Be sure to use new O-rings, and new gaskets on the expansion valve.

9    Have the system evacuated, charged and leak tested by the shop that discharged it.

## 17   Air conditioning expansion (orifice) tube - removal and installation

*Refer to illustration 17.5*
**Warning:** *The air conditioning system is*

under high pressure. DO NOT loosen any fittings or remove any components until after the system has been discharged. Air conditioning refrigerant should be properly discharged into an EPA-approved container at a dealership service department or an automotive air conditioning repair facility. Always wear eye protection when disconnecting air conditioning system fittings.
**Note:** *This procedure applies to 1998 and later Prizm models only.*

1    Have the air conditioning system discharged and the refrigerant recovered (see **Warning** above). Disconnect the negative battery cable. **Caution:** *If the stereo in your vehicle is equipped with an anti-theft system, make sure you have the correct activation code before disconnecting the battery.*

2    Raise the front of the vehicle and support it securely on jackstands.

3    Hold the stationary fitting with one wrench, then loosen the other fitting with another wrench. **Note:** *The orifice tube fitting is located on the small diameter refrigerant line leading from the condenser approximately 10 to 15 inches from the firewall adjacent to the valve cover.*

4    The expansion tube is a tube with a fixed-diameter orifice and a mesh filter at each end. When you separate the pipe at the fitting you will see one end of the orifice tube inside the pipe leading to the evaporator. Use needle-nose pliers to remove the orifice tube.

5    The orifice tube acts to meter the refrigerant, changing it from high-pressure liquid to low-pressure gas. It is possible to reuse the orifice tube if **(see illustration)**:

a) *The screens aren't plugged with grit or foreign material*
b) *Neither screen is torn*

3

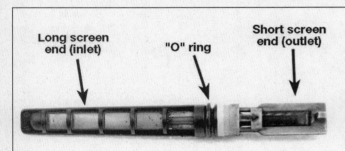

Long screen end (inlet)

"O" ring

Short screen end (outlet)

**17.5  The expansion orifice tube is equipped with a tapered mesh screen that must be cleaned and not have any holes or damage**

c) *The plastic housing over the screens is intact*

d) *The brass orifice inside the plastic housing is unrestricted*

6     Installation is the reverse of removal. Be sure to insert the expansion tube with the shorter end in first, toward the evaporator.

**Caution:** *Always use a new O-ring when installing the expansion (orifice) tube.*

7     Retighten the fitting and refrigerant line, then have the system evacuated, recharged and leak-tested by the shop that discharged it.

8     The remainder of the installation is the reverse of the removal process. Be sure to use new O-rings, and new gaskets on the expansion valve.

9     Have the system evacuated, charged and leak tested by the shop that discharged it.

# Chapter 4
# Fuel and exhaust systems

## Contents

## Specifications

### Fuel system

| | |
|---|---|
| Fuel pressure (key On, engine Off) | |
| 1997 and earlier | 38 to 44 psi |
| 1998 and later | 44 to 50 psi |
| Fuel injector resistance | 13.4 to 14.2 ohms (at 68-degrees F) |
| Fuel level sending unit resistance (approximate) | |
| Empty | 107 ohms |
| 1/2 | 53 ohms |
| Full | 4 ohms |

### Torque specification

| | |
|---|---|
| Air intake plenum mounting bolts | 168 in-lbs |
| Fuel rail mounting bolts | |
| 1997 and earlier | 132 in-lbs |
| 1998 and later | 156 in-lbs |
| Throttle body mounting bolts | 168 in-lbs |

## 1  General information

The fuel system consists of a fuel tank, an electric fuel pump (located in the fuel tank), an EFI/fuel pump relay, fuel injectors, a fuel pressure regulator, an air filter assembly and a throttle body unit. All models covered by this manual are equipped with a multi port fuel injection system.

### Multi port fuel injection system

Multi port fuel injection uses timed impulses to sequentially inject the fuel directly into the intake port of each cylinder.

The injectors are controlled by the Electronic Control Module (ECM). The ECM monitors various engine parameters and delivers the exact amount of fuel, in the correct sequence, into the intake ports. The throttle body serves only to control the amount of air passing into the system. Because each cylinder is equipped with an injector mounted immediately adjacent to the intake valve, much better control of the fuel/air mixture ratio is possible.

### Fuel pump and lines

Fuel is delivered from the fuel tank to the fuel injection system through a metal line running along the underside of the vehicle by an electric fuel pump inside the fuel tank. On 1997 and earlier models, excessive pressure is bled off by the fuel pressure regulator located on the fuel rail and fuel is returned to the fuel tank through a separate return line. On 1998 and later models, the fuel pressure regulator is located in the fuel tank with the fuel pump, and excessive fuel pressure is bled-off directly into the fuel tank.

The fuel pump will operate as long as the engine is cranking or running and the ECM is receiving ignition reference pulses from the electronic ignition system (see Chapter 5). If there are no reference pulses, the fuel pump will shut off after 2 or 3 seconds.

## Exhaust system

The exhaust system includes an exhaust manifold fitted with an exhaust oxygen sensor, a catalytic converter, an exhaust pipe, and a muffler.

The catalytic converter is an emission control device added to the exhaust system to reduce pollutants. A three-way (reduction) catalyst is used to reduce hydrocarbons (HC), carbon monoxide (CO) and oxides of nitrogen (NOx). Refer to Chapter 6 for more information regarding the catalytic converter.

## 2    Fuel pressure relief procedure

*Refer to illustrations 2.3a and 2.3b*
**Warning:** *Gasoline is extremely flammable, so take extra precautions when you work on any part of the fuel system. Don't smoke or allow open flames or bare light bulbs near the work area, and don't work in a garage where a gas-type appliance (such as a water heater or a clothes dryer) is present. Since gasoline is carcinogenic, wear latex gloves when there's a possibility of being exposed to fuel, and, if you spill any fuel on your skin, rinse it off immediately with soap and water. Mop up any spills immediately and do not store fuel-soaked rags where they could ignite. The fuel system is under constant pressure, so, if any fuel lines are to be disconnected, the fuel pressure in the system must be relieved first. When you perform any kind of work on the fuel system, wear safety glasses and have a Class B type fire extinguisher on hand.*

1    Before servicing any fuel system component, you must relieve the fuel pressure to minimize the risk of fire or personal injury.

2    Remove the fuel filler cap - this will relieve any pressure built up in the tank.

3    On 1997 and earlier models, remove the radio (see Chapter 12) and cup holder for access to the circuit opening relay. On 1998 and later models, locate the circuit opening relay under the left side of the instrument panel, above the kick panel. Unplug the electrical connector from the relay (1997 and earlier) or remove the relay from the panel (1998

**3.3  Check for battery voltage directly at the fuel pump electrical connector on the top of the fuel tank**

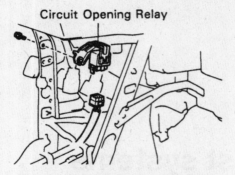

Circuit Opening Relay

**2.3a  To depressurize the fuel system, unplug the electrical connector from the circuit opening relay, start the engine and allow it to stall - 1997 and earlier model shown**

and later) **(see illustrations)**.

4    Attempt to start the engine, the engine will immediately stall. Crank the engine for 3 or 4 seconds, then turn the ignition key to Off.

5    The fuel system is now depressurized.
**Note:** *Place a rag around the fuel line before removing any hose clamp or fitting to prevent any residual fuel from spilling onto the engine.*

6    Before working on the fuel system, disconnect the cable from the negative terminal of the battery. **Caution:** *If the stereo in your vehicle is equipped with an anti-theft system, make sure you have the correct activation code before disconnecting the battery.*

## 3    Fuel pump/fuel pressure - check

**Warning:** *Gasoline is extremely flammable, so take extra precautions when you work on any part of the fuel system. See the* **Warning** *in Section 2.*

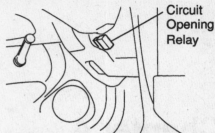

Circuit Opening Relay

**2.3b  Circuit opening relay location - 1998 and later models**

## Preliminary check

*Refer to illustration 3.3*
**Note:** *When performing the following checks on the wiring harness and connectors, refer to the wiring diagrams at the end of Chapter 12 to determine the correct wire for testing if necessary.*

1    If you suspect insufficient fuel delivery, first inspect all fuel lines to ensure that the problem is not simply a leak in a line. Check that there is adequate fuel in the fuel tank.

2    Set the parking brake and have an assistant turn the ignition switch to the ON position while you listen to the fuel pump (inside the fuel tank). You should hear a "whirring" sound, lasting for a couple of seconds indicating the fuel pump is operating. Cycle the ignition switch On and Off several times to verify operation. If the fuel pump is operating, proceed to the pressure check.

3    If there is no sound coming from the fuel pump, remove the fuel pump access cover (see Section 4). Disconnect the fuel pump electrical connector and connect a test light or voltmeter to the fuel pump voltage supply terminal of the harness connector (blue wire with a black stripe on most models) and a

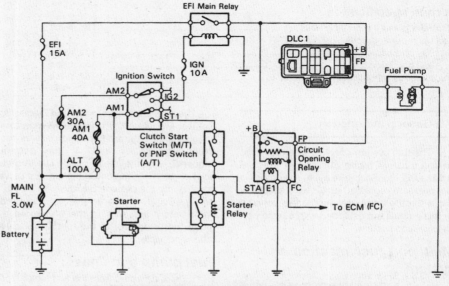

**3.5a  Typical electrical schematic of the fuel pump circuit**

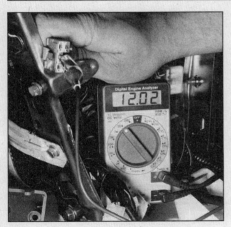

3.5b  Remove the circuit opening relay and check for battery voltage

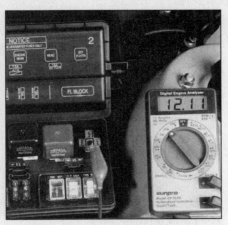

3.5c  Locate and remove the EFI relay and check for battery voltage to the relay with the ignition key ON

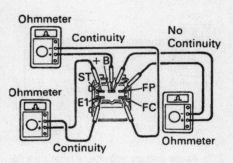

3.7a  Circuit opening relay checks - 1997 and earlier models

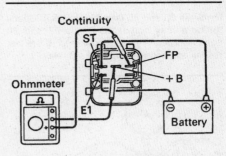

3.7b  Apply battery voltage to ST and E1 and check for continuity across terminals FP and +B

good chassis ground **(see illustration)**. Turn the ignition key On, battery voltage should be indicated for approximately two seconds. Cycle the ignition switch On and Off several times to verify operation. Check the fuel pump ground wire for continuity to a good chassis ground (white wire with a black stripe on most models).

4    If the power and ground circuits are good and the fuel pump does not operate when connected, replace the fuel pump (see Section 5). If no voltage is present at the fuel pump connector, check the fuel pump circuit, referring to Chapter 12 and the wiring diagrams if necessary. Check the related fuses and the related wiring to ensure power is reaching the fuel pump connector. Check the circuit opening relay and the EFI main relay as follows. If the relays and circuits are good and the ECM is not controlling the circuit opening relay, have the ECM diagnosed by a dealer service department or other qualified repair shop.

### EFI main relay and circuit opening relay checks

*Refer to illustrations 3.5a, 3.5b and 3.5c*

**Note:** *Accessing the circuit opening relay on 1997 and earlier models will require removal of the radio (see Chapter 12).*

5    There are two relays involved in the fuel pump circuit, the circuit opening relay and the EFI main relay **(see illustration)**. If voltage was not present at the fuel pump connector in Step 3, remove the circuit opening relay and check for battery voltage at the relay connector or relay panel socket with the ignition key On **(see illustrations 2.3a and 2.3b for relay locations)**. If battery voltage is not present at the circuit opening relay, remove the EFI main relay from the underhood fuse/relay panel and check for battery voltage at the EFI main relay socket with the ignition key On **(see illustrations)**.

6    If battery voltage is present at the relay connector or relay panel socket, check the relays as follows.

### 1997 and earlier models

#### Circuit opening relay

*Refer to illustrations 3.7a and 3.7b*

7    Using an ohmmeter, check for continuity across terminals ST and E1 **(see illustration)**. Check for continuity across terminals +B and FC. Check that there is no continuity across terminals +B and FP. Apply battery voltage across terminals ST and E1 **(see illustration)**. Check for continuity across terminals +B and FP again. Continuity should now exist. If the test results are incorrect, replace the relay.

#### EFI main relay (Nippondenso type)

*Refer to illustrations 3.8a and 3.8b*

8    Using an ohmmeter, check for continuity across terminals 1 and 2 **(see illustration)**. Check that there is no continuity across terminals 3 and 5. Apply battery voltage across terminals 1 and 2 **(see illustration)**. Check for continuity across terminals 3 and 5 again. Continuity should now exist. If the test results are incorrect, replace the relay.

#### EFI main relay (Bosch type)

*Refer to illustrations 3.9a and 3.9b*

9    Using an ohmmeter, check for continuity across terminals 86 and 85 **(see illustration)**. Check that there is no continuity across terminals 87 and 30. Apply battery voltage across

**4**

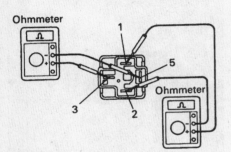

3.8a  EFI main relay checks on Nippondenso style relay - 1997 and earlier models

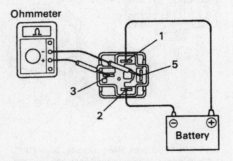

3.8b  Apply battery voltage to terminals 1 and 2 and check for continuity between terminals 3 and 5

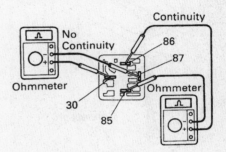

3.9a  EFI main relay checks on Bosch style relay - 1997 and earlier models

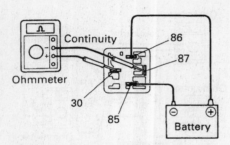

**3.9b Apply battery voltage to terminals 85 and 86 and confirm that continuity exists across terminals 87 and 30**

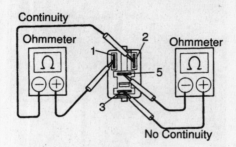

**3.10a EFI main relay checks - 1998 and later models**

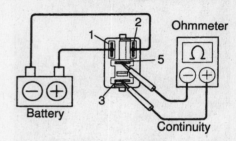

**3.10b Apply battery voltage to terminals 1 and 2 and check for continuity between terminals 3 and 5**

terminals 86 and 85 **(see illustration)**. Check for continuity across terminals 87 and 30 again. Continuity should now exist. If the test results are incorrect, replace the relay.

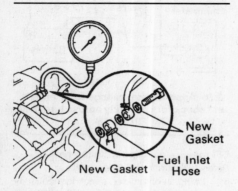

**3.12 Fuel gauge installation details - 1997 and later models**

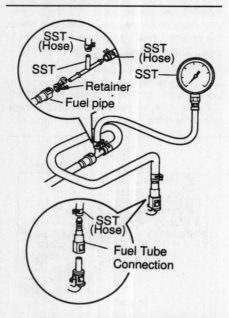

**3.13 Fuel gauge installation details - 1998 and later models**

### 1998 and later models

*Refer to illustrations 3.10a and 3.10b*
**Note:** *On 1998 and later models the tests for both relays are identical.*
10    Using an ohmmeter, check for continuity across terminals 1 and 2 **(see illustration)**. Check that there is no continuity across terminals 3 and 5. Apply battery voltage across terminals 1 and 2 **(see illustration)**. Check for continuity across terminals 3 and 5 again. Continuity should now exist. If the test results are incorrect, replace the relay.

## Pressure check

*Refer to illustrations 3.12, 3.13, 3.14 and 3.16*
**Note:** *In order to perform the fuel pressure test, you will need to obtain a fuel pressure gauge capable of measuring high fuel pressure and the necessary fittings to connect the fuel gauge to the fuel line.*
11    Relieve the fuel system pressure (see Section 2).
12    On 1997 and earlier models, disconnect the fuel supply line union bolt fitting from the fuel rail and install a fuel pressure gauge with the appropriate adapters **(see illustration)**.
13    On 1998 and later models, disconnect the fuel line quick-connect fitting (see Section 6) and install a fuel pressure gauge with the appropriate adapters **(see illustration)**.
14    Turn the ignition switch On. The fuel pump should run for approximately two sec-

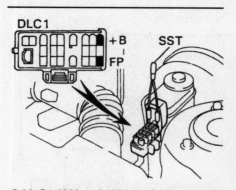

**3.14 On 1993 and 1994 models and 1995 1.6L models, bridge terminals +B and FP at the test connector with a jumper wire to operate the fuel pump**

onds - pressure should register on the gauge and hold steady. Compare the pressure reading with the value listed in this Chapter's Specifications. Cycle the ignition key On and Off several times to obtain the highest reading, if necessary. **Note:** *On 1993 and 1994 models and 1995 1.6L models, it is possible to power the fuel pump by bridging terminals +B and FP at the test connector* **(see illustration)**.
15    Start the engine and allow it to idle. On 1997 and earlier models, the fuel pressure should be 7 to 10 psi lower with the engine idling. Disconnect and plug the vacuum hose from the fuel pressure regulator (see Section 14) - the pressure should increase to the value recorded in Step 14. On 1998 and later models, the fuel pressure should remain within the key On, engine Off specifications with the engine idling. If the pressure readings are correct, the system is operating properly.
16    If the fuel pressure is not within specifications on a 1997 or earlier model, check the following **(see illustration)**:
a)    *If the pressure is higher than specified, check for vacuum to the fuel pressure regulator. Vacuum must fluctuate with the increase or decrease in the engine rpm. If vacuum is present, check for a*

**3.16 Connect a vacuum gauge to the vacuum line leading to the fuel pressure regulator and check the vacuum source**

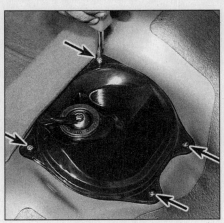

**4.4  Remove the fuel pump/fuel level sending unit access cover screws (arrows) and lift the cover off**

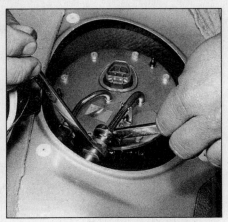

**4.5  Use a flare-nut wrench and a back-up wrench to remove the fuel feed line; remove the clamp and hose from the fuel return line fitting**

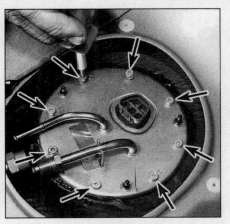

**4.6  Remove the fuel pump/sending unit assembly mounting bolts (arrows)**

pinched or clogged fuel return hose or pipe. If the return line is OK, replace the fuel pressure regulator (see Section 14).

b) *If the pressure is lower than specified, check for a restriction in the fuel filter or fuel line. If the fuel filter and lines are OK, start the engine (if possible) and slowly pinch the return hose shut. If the pressure rises above 44 psi, replace the fuel pressure regulator (see Section 14).*

**Warning:** *Don't allow the fuel pressure to exceed 60 psi. If the fuel pressure is still low, replace the fuel pump (see Section 5).*

17   If the fuel pressure is not within specifications on a 1998 or later model, check the following:

a) *If the pressure is higher than specified, replace the fuel pressure regulator (see Section 14).*

b) *If the pressure is lower than specified, check for a restriction in the fuel line. If the fuel lines are OK, replace the fuel pump, fuel filter and fuel pressure regulator (see Section 4).*

18   After completing the testing, relieve the

fuel pressure (see Section 2) and remove the fuel pressure gauge.

---

## 4   Fuel pump - removal and installation

**Warning:** *Gasoline is extremely flammable, so take extra precautions when you work on any part of the fuel system. See the* **Warning** *in Section 2.*

1   Remove the fuel tank cap.

2   Disconnect the cable from the negative terminal of the battery. **Caution:** *If the stereo in your vehicle is equipped with an anti-theft system, make sure you have the correct activation code before disconnecting the battery.*

3   Remove the rear seat from inside the passenger compartment (see Chapter 11).

### *1997 and earlier models*

*Refer to illustrations 4.4, 4.5, 4.6, 4.7, 4.8, 4.13 and 4.14*

4   Remove the fuel pump/sending unit

access cover (**see illustration**).

5   Disconnect the electrical connector. Disconnect the fuel lines (**see illustration**).

6   Remove the fuel pump/sending unit retaining bolts (**see illustration**).

7   Carefully withdraw the fuel pump/fuel level sending unit assembly from the fuel tank (**see illustration**).

8   Pry the lower end of the fuel pump loose from the bracket (**see illustration**).

9   Remove the rubber cushion from the lower end of the fuel pump.

10   Remove the clip securing the inlet screen to the pump.

11   Remove the screen and inspect it for contamination. If it is dirty, replace it.

12   If you are only replacing the fuel pump inlet screen, install the new screen, the clip and the rubber cushion, push the lower end of the pump back into the bracket and install the pump/sending unit assembly in the fuel tank.

13   If you are replacing the fuel pump, remove the hose clamp at the upper end of the pump and disconnect the pump from the hose (**see illustration**).

**4**

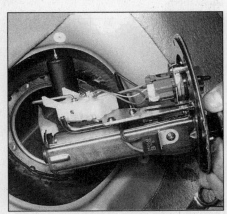

**4.7  Lift the fuel pump/sending unit assembly from the fuel tank at an angle so as not to damage the inlet screen or float arm**

**4.8  Separate the fuel pump from the frame by gently prying outward**

**4.13  Remove the clamp from the fuel pump outlet hose**

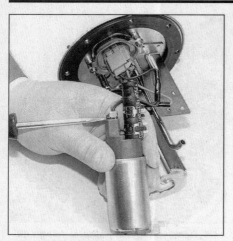

**4.14   Remove the electrical connector from the fuel pump**

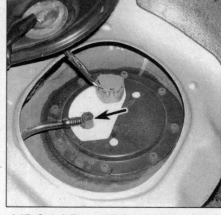

**4.17   Carefully pry off the clip (arrow) and pull the fuel line fitting from the top of the fuel pump cover**

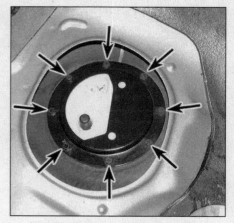

**4.18   Remove the fuel pump flange retaining bolts (arrows) and remove the flange**

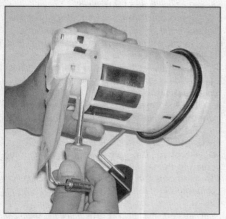

**4.20   Carefully pry the lower support from the fuel pump unit and remove the rubber cushion**

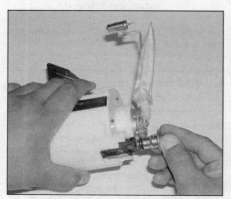

**4.21   Pull the fuel pressure regulator straight off the fuel filter**

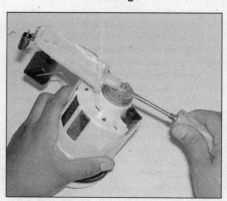

**4.22   Carefully pry off the retaining clip and remove the fuel inlet screen**

14   Disconnect the wires from the pump terminals and remove the pump **(see illustration)**.
15   Installation is the reverse of removal.

### 1998 and later models

*Refer to illustrations 4.17, 4.18, 4.20, 4.21, 4.22, 4.23, 4.24 and 4.25*
16   Remove the fuel pump/sending unit access cover.
17   Disconnect the electrical connector. Disconnect the fuel line **(see illustration)**.
18   Remove the fuel pump/sending unit

retaining bolts and remove the mounting flange **(see illustration)**.
19   Carefully withdraw the fuel pump/fuel level sending unit assembly from the fuel tank.
20   Carefully pry the lower support from the fuel pump unit and remove the rubber cushion **(see illustration)**.
21   Remove the fuel pressure regulator **(see illustration)**.

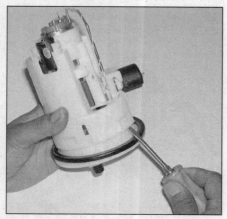

**4.23   Pry the upper support off the retaining tabs and remove the upper support from the unit**

**4.24   Pull the fuel pump and fuel filter assembly off the top cover and disconnect the electrical connector from the fuel pump**

**4.25   Separate the fuel pump from the fuel filter**

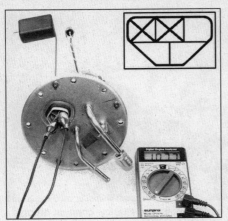

**5.2 Connect an ohmmeter to the fuel level sending unit terminals and measure the resistance with the float down (empty) and then up (full)**

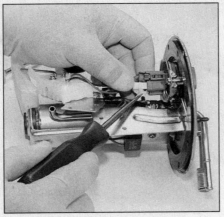

**5.6 Remove the electrical connector from the fuel level sending unit**

**5.7 Remove the screws that retain the fuel level sending unit**

22   Pry off the retaining clip and remove the fuel inlet screen from the unit **(see illustration)**.
23   Carefully pry off the upper support **(see illustration)**.
24   Remove the fuel pump and fuel filter from the top plate and disconnect the electrical connector from the fuel pump **(see illustration)**.
25   Separate the fuel pump from the fuel filter **(see illustration)**.
26   Installation is the reverse of removal.

## 5   Fuel level sending unit - check and replacement

**Warning:** *Gasoline is extremely flammable, so take extra precautions when you work on any part of the fuel system. See the* **Warning** *in Section 2.*

### Check

*Refer to illustration 5.2*

1   Remove the fuel pump/fuel level sending unit assembly (see Sections 4).
2   Connect the probes of an ohmmeter to the two indicated terminals of the fuel sending unit electrical connector **(see illustration)**.
3   Position the float in the down (empty) position. Measure the resistance and compare it to the value listed in this Chapter's Specifications.
4   Move the float up to the full position. Measure the resistance and compare it to the value listed in this Chapter's Specifications.
5   If the fuel level sending unit resistance is incorrect or if the resistance does not change smoothly as the float travels from empty to full, the fuel level sending unit assembly is defective.

### Replacement

#### 1997 and earlier models

*Refer to illustrations 5.6 and 5.7*

6   Disconnect the electrical connector

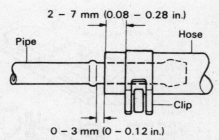

**6.6 When attaching a section of rubber hose to a metal fuel line, be sure to overlap the hose as shown, secure it to the line with a new hose clamp of the proper type**

from the sending unit **(see illustration)**.
7   Remove the two screws and remove the sending unit from the fuel pump unit **(see illustration)**.
8   Installation is the reverse of removal.

#### 1998 and later models

9   Refer to Section 4 and remove the fuel pump and fuel filter components from the sending unit assembly. Transfer the fuel pump and fuel filter components to the new unit.
10   Install the assembly into the fuel tank with a new O-ring (see Section 4).
11   The remainder of installation is the reverse of removal.

## 6   Fuel lines and fittings - replacement

*Refer to illustrations 6.6, 6.7a, 6.7b, 6.7c and 6.7d*
**Warning:** *Gasoline is extremely flammable, so take extra precautions when you work on any part of the fuel system. See the* **Warning** *in Section 2.*
1   At periodic intervals, inspect fuel lines and all fittings and connections for damage or deterioration.
2   Check all hoses and pipes for cracks,

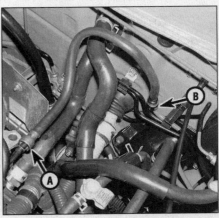

**6.7a Two types of quick-connect fitting are used on later models - the metal collar type (A) and the plastic type (B)**

kinks, deformation or obstructions.
3   Make sure all hoses and pipe clips attach their associated hoses or pipes securely to the underside of the vehicle.
4   Verify all hose clamps attaching rubber hoses to metal fuel lines or pipes are snug enough to assure a tight fit between the hoses and pipes.
5   If you must replace any damaged sections, use original equipment replacement hoses or pipes constructed from exactly the same material as the section you are replacing. Do not install substitutes constructed from inferior or inappropriate material or you could cause a fuel leak or a fire.
6   Always, before detaching or disassembling any part of the fuel line system, note the routing of all hoses and pipes and the orientation of all clamps and clips to assure that replacement sections are installed in exactly the same manner. When attaching hoses to metal lines, overlap them as shown **(see illustration)**.
7   Later models are equipped with quick-connect fittings at many of the fuel line junctions. The fittings are an integral part of the hose and are not serviced separately. Two types of fittings are used, a metal collar type

**4**

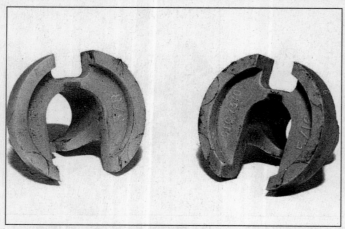

**6.7b  A special tool (available at most auto parts stores) is required to disconnect the metal collar type quick-connect fitting**

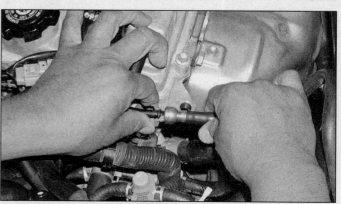

**6.7c  To disconnect a metal collar type quick-connect fitting, remove the safety tether and place the tool squarely over the fuel line; push the tool into the fitting and pull the lines apart (the tool is not required to connect the fitting)**

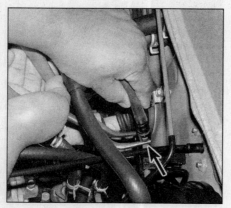

**6.7d  To disconnect a plastic type fitting, squeeze the retaining tabs in and pull the lines apart**

and a plastic type. A special tool is required to disconnect the metal collar fitting **(see illustrations)**. Clean dirt and debris from around the fitting before attempting to disconnect the fitting. Before reconnecting a quick-connect fitting, clean the fitting and metal pipe and apply a thin film of clean engine oil to the metal pipe. Inspect the plastic retainer on the metal pipe for damage and replace it if necessary.

8    No tools are necessary to connect a quick-connect fitting. Push the fitting onto the fuel pipe until the fitting clicks into place. Check the integrity of the connection by attempting to pull the lines apart.

9    Before detaching any part of the fuel system, be sure to relieve the fuel line and tank pressure (see Section 2) and disconnect the battery. Cover the fitting being disconnected with a rag to absorb any fuel that may leak out.

## 7   Fuel tank - removal and installation

*Refer to illustrations 7.6, 7.9a, 7.9b and 7.10*
**Warning:** *Gasoline is extremely flammable, so take extra precautions when you work on*

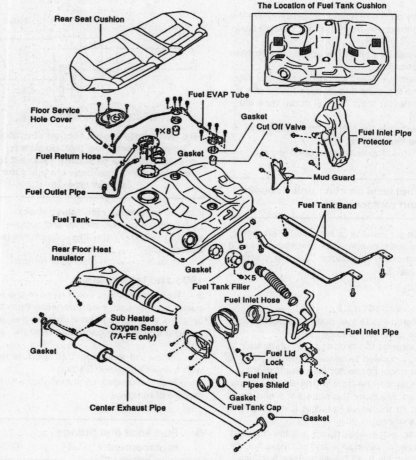

**7.6  Typical fuel tank and related components**

*any part of the fuel system. See the **Warning** in Section 2.*

1    Remove the fuel filler cap to relieve fuel tank pressure. Relieve the fuel system pressure (see Section 2)

2    Detach the cable from the negative terminal of the battery. **Caution:** *If the stereo in your vehicle is equipped with an anti-theft*

*system, make sure you have the correct activation code before disconnecting the battery.*

3    Siphon or pump the fuel into an approved container using a siphoning kit or hand operated pump (available at most auto parts stores). **Warning:** *Do not start the siphoning action by mouth!*

4    Remove the fuel pump access cover

7.9a  Remove the clamp that retains the fuel filler hose

7.9b  Loosen the clamp and detach the fuel vapor line from the fuel tank

7.10  Remove the tank strap bolts (arrows) from the body

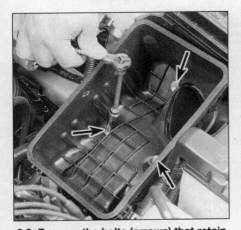

9.3  Remove the bolts (arrows) that retain the air filter housing to the engine compartment

10.2  Loosen the locknuts on the accelerator cable

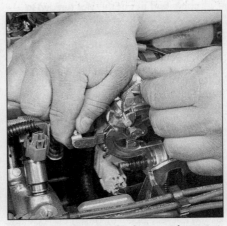

10.3  Rotate the throttle lever and remove the cable end from the slot

and disconnect the fuel pump electrical connector and fuel lines from the fuel pump unit (see Section 4).

5    Raise the vehicle and place it securely on jackstands.

6    Familiarize yourself with the layout of the fuel tank assembly before proceeding (see illustration).

7    Remove the center exhaust pipe and the heat insulator from the vehicle (see Section 14).

8    Detach the parking brake cables from the underbody and position them aside.

9    Disconnect the fuel lines, the vapor return line and the fuel filler hose (see illustrations). Note: Be sure to plug the hoses to prevent leakage and contamination of the fuel system.

10   Support the fuel tank with a floor jack. Place a sturdy plank between the jack head and the fuel tank to protect the tank. Remove the bolts from the fuel tank retaining straps (see illustration).

11   Lower the tank enough to disconnect the electrical connector and ground strap from the fuel pump/fuel gauge sending unit, if you have not already done so.

12   Remove the tank from the vehicle.

13   Installation is the reverse of removal.

## 8    Fuel tank cleaning and repair - general information

1    Any repairs to the fuel tank or filler neck should be carried out by a professional who has experience in this critical and potentially dangerous work. Even after cleaning and flushing of the fuel system, explosive fumes can remain and ignite during repair of the tank.

2    If the fuel tank is removed from the vehicle, it should not be placed in an area where sparks or open flames could ignite the fumes coming out of the tank. Be especially careful inside garages where a gas-type appliance is located, because the pilot light or burner could cause an explosion.

## 9    Air filter housing - removal and installation

Refer to illustration 9.3

1    Disconnect the air intake hose and breather hose from the housing cover. Disconnect the electrical connector from the

intake air temperature sensor.

2    Detach the clips and remove the air filter housing cover and the air filter element (see Chapter 1).

3    Remove the three bolts and remove the air filter housing from the engine compartment (see illustration).

4    Installation is the reverse of removal.

## 10   Accelerator cable - removal, installation and adjustment

Refer to illustrations 10.2, 10.3 and 10.4

### Removal

1    Detach the cable from the negative terminal of the battery. Caution: If the stereo in your vehicle is equipped with an anti-theft system, make sure you have the correct activation code before disconnecting the battery.

2    Loosen the locknut on the threaded portion of the throttle cable at the throttle body (see illustration).

3    Rotate the throttle lever and slip the throttle cable end out of the slot in the lever (see illustration).

**10.4 Separate the cable from the accelerator pedal assembly and slide the cable end out of the slot (arrow) in the housing**

4    Detach the throttle cable from the accelerator pedal **(see illustration)**. Remove the two bolts securing the cable retainer to the firewall.
5    From inside the vehicle, pull the cable through the firewall.

## Installation and adjustment

6    Installation is the reverse of removal. Make sure the cable casing grommet seats properly in the firewall.
7    To adjust the cable, fully depress the accelerator pedal and check that the throttle is fully opened.
8    If not fully opened, loosen the locknuts, depress accelerator pedal and adjust the cable until the throttle is fully open.
9    Tighten the locknuts and recheck the adjustment. Make sure the throttle closes fully when the pedal is released.

## 11  Fuel injection system - general information

*Refer to illustration 11.1*

1    All models are equipped with a multi port fuel injection system. The fuel injection system is composed of three basic subsystems: fuel system, air induction system and electronic engine control system **(see illustration)**.

## Fuel system

2    An electric fuel pump located inside the fuel tank supplies fuel under constant pressure to the fuel rail, which distributes fuel evenly to all injectors. From the fuel rail, fuel is injected into the intake ports, just above the intake valves, by fuel injectors. The amount of fuel supplied by the injectors is precisely controlled by an electronic Engine Control Module (ECM). On 1997 and earlier models, a pressure regulator controls system pressure in relation to intake manifold vacuum. A fuel filter between the fuel pump and

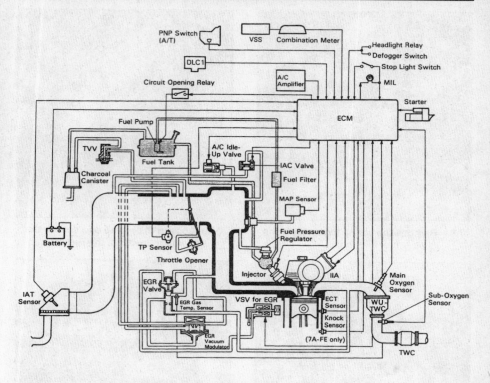

**11.1 Typical Fuel Injection (EFI) system schematic**

the fuel rail filters fuel to protect the components of the system. On 1998 and later models, the fuel pressure regulator maintains a constant pressure to the fuel rail regardless of engine load. The fuel pressure regulator and fuel filter are located in the fuel tank along with the fuel pump.

## Air induction system

3    The air induction system consists of an air filter housing, the throttle body and the duct connecting the two. The throttle plate inside the throttle body is controlled by the driver. As the throttle plate opens, additional air is drawn into the cylinders. The information provided by the various sensors allows the ECM determine the exact amount of fuel to be injected by the injectors during the various operating conditions.

## Electronic engine control system

4    The electronic engine control system controls the fuel injection, ignition and emissions systems by means of an Engine Control Module (ECM), which employs a microprocessor. The ECM receives signals from a number of information sensors which monitor such variables as intake air temperature, throttle angle, coolant temperature, engine rpm, engine load, vehicle speed and exhaust oxygen content. These signals help the ECM determine the injection duration necessary for the optimum air/fuel ratio. The sensors and their corresponding ECM-controlled

relays, are located throughout the engine compartment. For further information regarding the ECM and its relationship to the fuel injection and ignition systems, see Chapter 6.

## 12  Fuel injection system - check

*Refer to illustrations 12.6, 12.7, 12.8 and 12.9*
**Note:** *The following procedure is based on the assumption that the fuel pump is working and the fuel pressure is adequate (see Section 3).*
1    Check the ground wire connections for tightness. Check all wiring and electrical connectors that are related to the system. Loose electrical connectors and poor grounds can cause many problems that resemble more serious malfunctions.
2    Check to see that the battery is fully charged, as the control unit and sensors depend on an accurate supply voltage in order to properly meter the fuel.
3    Check the air filter element - a dirty or partially blocked filter will severely impede performance and economy (see Chapter 1).
4    Check the fuses. If a blown fuse is found, replace it and see if it blows again. If it does, search for a grounded wire in the fuel injection system wiring harness (see Chapter 12 and the wiring diagrams).
5    Check the air intake duct from the air cleaner housing to the intake manifold for leaks, which will result in an excessively lean mixture. Also check the condition of the vac-

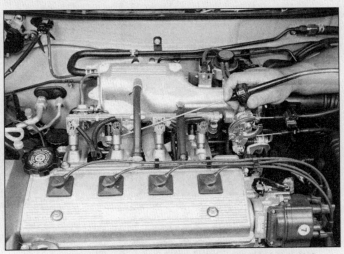

12.6  With the engine off, use aerosol carburetor cleaner (make sure it is safe for use with the catalytic converter and oxygen sensor), a toothbrush and a rag to clean the throttle body - open the throttle plate so you can clean behind it

12.7  Use a stethoscope or a screwdriver to determine if the injectors are working properly - they should make a steady clicking sound that rises and falls with engine speed changes

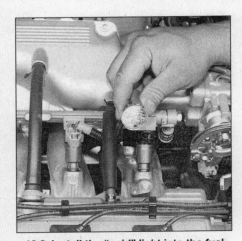

12.8  Install the "noid" light into the fuel injector electrical connector and check to see that it flashes with the engine running

12.9  Using an ohmmeter, measure the resistance across the terminals of the injector

13.8  A typical throttle body is retained by three or four bolts/nuts (arrows)

uum hoses connected to the intake manifold.

6    Remove the air intake duct from the throttle body and check for carbon and residue build-up. If it's dirty, clean it with aerosol carburetor cleaner (make sure the can says it's safe for use on a system with an oxygen sensor and catalytic converter) and a toothbrush (see illustration).

7    With the engine running, place a stethoscope against each injector, one at a time, and listen for a clicking sound, indicating operation (see illustration). If you don't have an automotive stethoscope you can use a long screwdriver; just place the tip of the screwdriver against the injector body and press your ear against the handle.

8    If there is a problem with an injector, purchase a special injector test light (noid light) and install it into the injector electrical connector (see illustration). Start the engine and make sure that each injector connector flashes the noid light. This will test for the proper voltage signal to the injector.

9    With the engine OFF and the fuel injector electrical connectors disconnected, measure the resistance of each injector (see illustration). Compare your measurement with the value listed in this Chapter's Specifications. If the injector resistance is excessive or an open or short circuit is indicated, replace the injector (see Section 15).

10    Additional fuel injection control system component checks can be found in Chapter 6.

## 13    Throttle body - removal and installation

Refer to illustrations 13.8 and 13.9
**Warning:** Wait until the engine is completely cool before beginning this procedure.

1    Detach the cable from the negative terminal of the battery **Caution:** If the stereo in your vehicle is equipped with an anti-theft

system, make sure you have the correct activation code before disconnecting the battery.

2    Loosen the hose clamps and remove the air intake duct.

3    Detach the accelerator cable from the throttle lever (see Section 10).

4    Detach the throttle cable bracket and set it aside (it's not necessary to detach the throttle cable from the bracket).

5    If equipped, detach the throttle valve cable from the throttle linkage (see Chapter 7B), detach the throttle valve cable brackets from the engine and set the cable and brackets aside.

6    Clearly label, then detach, all vacuum and coolant hoses from the throttle body. Plug the coolant hoses to prevent coolant loss.

7    Disconnect the electrical connectors from the throttle position sensor, idle air control valve, manifold absolute pressure sensor and/or mass airflow sensor, as required.

8    Remove the throttle body mounting bolts (see illustration).

4

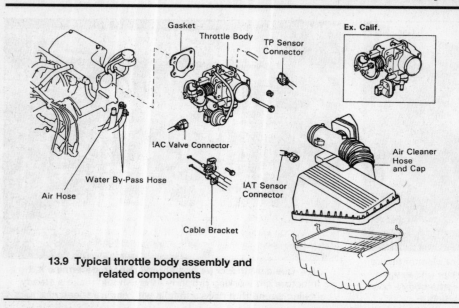

**13.9 Typical throttle body assembly and related components**

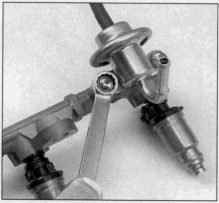

**14.5 To remove the fuel pressure regulator from the fuel rail on a 1997 and earlier model, detach the fuel return hose, remove the two regulator bolts and separate the regulator from the fuel rail (fuel rail removed for clarity)**

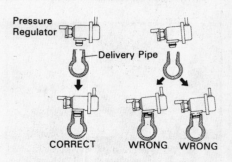

**14.6 If the fuel pressure regulator is cocked during installation, it will not seal properly**

9   Detach the throttle body and gasket from the intake manifold (see illustration).
10   Using a soft brush and carburetor cleaner, thoroughly clean the throttle body casting, then blow out all passages with compressed air. **Caution:** *Do not clean the throttle position sensor with anything. Just wipe it off carefully with a clean, soft cloth.*
11   Installation of the throttle body is the reverse of removal.
12   Tighten the throttle body mounting bolts to the torque listed in this Chapter's Specifications.
13   Check the coolant level, adding as necessary (see Chapter 1).

## 14   Fuel pressure regulator - removal and installation

**Warning:** *Gasoline is extremely flammable, so take extra precautions when you work on any part of the fuel system. See the **Warning** in Section 2.*

### 1997 and earlier models

Refer to illustrations 14.5 and 14.6
1   Relieve the fuel pressure (see Section 2) and detach the cable from the negative terminal of the battery. **Caution:** *If the stereo in your vehicle is equipped with an anti-theft system, make sure you have the correct activation code before disconnecting the battery.*
2   Detach the vacuum hose from the regulator.
3   Place a metal container or shop towel under the fuel return hose.
4   Slide the clamp down the hose and remove the fuel return hose from the regulator.
5   Remove the pressure regulator mounting bolts **(see illustration)** and detach the pressure regulator from the fuel rail.
6   Use a new O-ring and make sure that the pressure regulator is installed properly on the fuel rail **(see illustration)**.

7   The remainder of installation is the reverse of removal.

### 1998 and later models

8   The fuel pressure regulator on 1998 and later models is located on the fuel pump unit. Refer to Section 4 for fuel pressure regulator removal and installation procedures.

## 15   Fuel rail and injectors - removal and installation

**Warning:** *Gasoline is extremely flammable, so take extra precautions when you work on any part of the fuel system. See the **Warning** in Section 2.*
1   Relieve the fuel pressure (see Section 2) and detach the cable from the negative terminal of the battery. **Caution:** *If the stereo in your vehicle is equipped with an anti-theft system, make sure you have the correct activation code before disconnecting the battery.*

### 1997 and earlier models

Refer to illustrations 15.7, 15.9, 15.11, 15.12, 15.14a, 15.14b and 15.14c
1   Detach the accelerator cable (see Section 10) and throttle valve cable (see Chapter 7B) from the throttle body assembly.
2   Disconnect the electrical connectors

from the idle air control valve, throttle position sensor and EGR/EVAP canister control solenoids.
3   Clearly label, then detach, the vacuum lines from air intake plenum, the EGR valve and the fuel pressure regulator.
4   Disconnect the PCV system hoses from the fittings on the air intake plenum.
5   Remove the plenum bracket and bolt (if equipped).
6   Detach the coolant hoses from the throttle body and plug them.
7   Remove the air intake plenum retaining nuts/bolts **(see illustration)**.

**15.7 Remove the bolts (arrows) that retain the air intake plenum to the intake manifold**

15.9 Number each injector electrical connector before disconnecting them from the fuel rail

15.11 Remove the bolts (arrow) that retain the fuel rail to the intake manifold

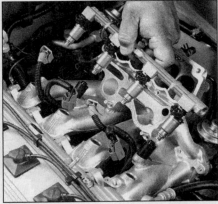

15.12 Lift the fuel rail assembly from the engine. Beware of any fuel that may spill out of the fuel pressure regulator or fuel rail while you are lifting it out

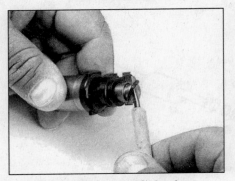

15.14a Remove the O-ring from the injector

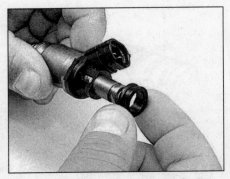

15.14b Remove the grommet from the injector

15.14c Remove the insulator from each bore in the intake manifold

8    Remove the air intake plenum and throttle body as an assembly from the lower intake manifold.

9    Carefully mark each injector connector with a felt pen or paint (see illustration). Disconnect the fuel injector electrical connectors and set the injector wire harness aside.

10    Disconnect the fuel lines from the fuel pressure regulator and the fuel rail.

11    Remove the fuel rail mounting bolts (see illustration).

12    Remove the fuel rail with the fuel injectors attached (see illustration).

13    Remove the fuel injector(s) from the fuel rail and set them aside in a clearly labeled storage container.

14    Replace the grommets and O-rings on the fuel injectors and the insulator in each intake manifold bore (see illustrations).

15    Lubricate the O-rings with clean engine oil and install the fuel injectors onto the fuel rail. Make sure the spacers are in place and install the fuel rail/injector assembly onto the intake manifold. Install the fuel rail mounting bolts hand-tight and make sure the injectors rotate smoothly in the bores; if an injector is binding, check the O-ring installation. Position the electrical connectors pointing up.

16    Tighten the fuel rail mounting bolts to the torque listed in this Chapter's Specifications.

17    Be sure to clean and inspect the mount-ing surface of the lower intake manifold and the air intake plenum before positioning the new gasket onto the lower intake mounting face. Install the air intake plenum and throttle body assembly onto the intake manifold. Ensure the gasket remains in place. Install the air intake plenum retaining bolts and tighten the bolts to the torque listed in this Chapter's Specifications following a criss-cross pattern.

18    The remainder of installation is the reverse of removal.

## 1998 and later models

*Refer to illustrations 15.24 and 15.26*

19    On 1998 and 1999 models, detach the PCV hose from the valve cover and remove the fuel injector wiring harness cover. On 2000 and later models, remove the engine cover.

20    On 1998 and 1999 models, detach the spark plug wires from the ignition coils and position the wires aside. **Caution:** *If the spark plugs wires are not numbered, apply numbered tags to the spark plug wires corresponding to the cylinder numbers.*

21    Detach the accelerator cable and throttle valve cable brackets from the cylinder head and position the cables aside.

22    Remove the fuel line fitting cover and disconnect the fuel line quick-connect fitting (see Section 6). Cap the open fuel lines to prevent the entry of dirt or moisture.

23    Disconnect the electrical connectors from the fuel injectors. Remove the wiring harness protector retaining screws and position the wiring harness aside.

24    Remove the fuel rail mounting bolts and carefully withdraw the fuel rail and injectors from the cylinder head (see illustration).

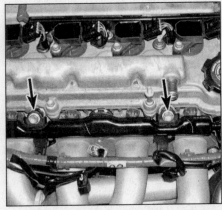

15.24 Fuel rail mounting bolts - 1998 and later models

4

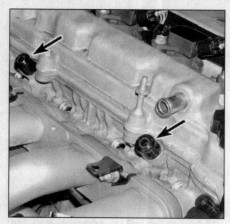

**15.26 Make sure the spacers are in place before installing the fuel rail**

25   Replace the O-ring and grommet on each injector (see step 14). **Note:** *The O-ring and grommet arrangement is slightly different on later models. Later models do not use insulators in the intake manifold.*

26   Lubricate the O-rings with clean engine oil and install the fuel injectors onto the fuel rail. Make sure the spacers are in place and install the fuel rail/injector assembly onto the intake manifold **(see illustration)**. Install the fuel rail mounting bolts and tighten the bolts to the torque listed in this Chapter's Specifications.

27   The remainder of installation is the reverse of removal.

## 16   Exhaust system servicing - general information

*Refer to illustration 16.1*

**Warning:** *Inspection and repair of exhaust system components should be done only after the system components have cooled completely.*

1    The exhaust system consists of the exhaust manifold, catalytic converter, the muffler, the tailpipe and all connecting pipes, brackets, hangers and clamps **(see illustration)**. The exhaust system is attached to the body with mounting brackets and rubber hangers. If any of these parts are damaged or deteriorated, excessive noise and vibration will be transmitted to the body.

2    Conducting regular inspections of the exhaust system will keep it safe and quiet. Look for any damaged or bent parts, open seams, holes, loose connections, excessive corrosion or other defects which could allow exhaust fumes to enter the vehicle. Deteriorated exhaust system components should not be repaired - they should be replaced with new parts.

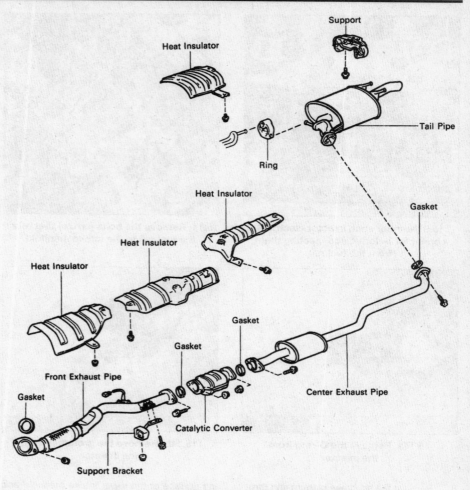

**16.1  Typical exhaust system and related components**

3    If the exhaust system components are extremely corroded or rusted together, they will probably have to be cut from the exhaust system. The convenient way to accomplish this is to have a muffler repair shop remove the corroded sections with a cutting torch. If, however, you want to save money by doing it yourself and you don't have an oxy/acetylene welding outfit with a cutting torch, simply cut off the old components with a hack-saw. If you have compressed air, special pneumatic cutting chisels can also be used. If you do decide to tackle the job at home, be sure to wear eye protection to protect your eyes from metal chips and work gloves to protect your hands.

4    Here are some simple guidelines to apply when repairing the exhaust system:

a)  *Work from the back to the front when removing exhaust system components.*

b)  *Apply penetrating oil to the exhaust sys-*

tem component fasteners to make them easier to remove.

c)  *Use new gaskets, hangers and clamps when installing exhaust system components.*

d)  *Apply anti-seize compound to the threads of all exhaust system fasteners during reassembly.*

e)  *Be sure to allow sufficient clearance between newly installed parts and all points on the underbody to avoid overheating the floor pan and possibly damaging the interior carpet and insulation. Pay particularly close attention to the catalytic converter and its heat shield.* **Warning:** *The catalytic converter operates at very high temperatures and takes a long time to cool. Wait until it's completely cool before attempting to remove the converter. Failure to do so could result in serious burns.*

# Chapter 5
# Engine electrical systems

## Contents

## Specifications

### Ignition timing (1997 and earlier)
With test terminals TE1 and E1 grounded ............................ 10-degrees BTDC

### Ignition coil resistance (cold)
1997 and earlier
  Distributors with internal igniter
    Primary resistance ............................ 1.11 to 1.75 ohms
    Secondary resistance ............................ 9.0 to 15.7K ohms
  Distributors with external igniter
    Primary resistance ............................ 0.36 to 0.55 ohms
    Secondary resistance ............................ 9.0 to 15.4K ohms
1998 and 1999
  Primary resistance ............................ Not applicable
  Secondary resistance ............................ 9.7 to 16.7K ohms
2000 and later ............................ Not applicable

### Distributor
Air gap ............................ 0.008 to 0.016 inch
Pick-up coil resistance (Toyota)
  Distributors with internal igniter
    Cold (below 103-degrees F)
      Terminals G+ and G- ............................ 185 to 275 ohms
      Terminals NE+ and NE- ............................ 370 to 550 ohms
    Hot (above 104-degrees F)
      Terminals G+ and G- ............................ 240 to 325 ohms
      Terminals NE+ and NE- ............................ 475 to 650 ohms
  Distributors with external igniter
    Cold (below 103-degrees F)
      Terminals G+ and G- ............................ 185 to 275 ohms
      Terminals NE+ and NE- ............................ 1,630 to 2,740 ohms
    Hot (above 104-degrees F)
      Terminals G+ and G- ............................ 240 to 325 ohms
      Terminals NE+ and NE- ............................ 2,065 to 3,225 ohms

5

## Distributor

Pick-up coil resistance (Geo)
    1993 and 1994
        Cold (below 103-degrees F)

| | |
|---|---|
|             Terminals 3 and 6 | 185 to 275 ohms |
|             Terminals 2 and 5 | 370 to 550 ohms |

        Hot (above 104-degrees F)

| | |
|---|---|
|             Terminals 3 and 6 | 240 to 325 ohms |
|             Terminals 2 and 5 | 475 to 650 ohms |

    1995 and later
        Cold (below 103-degrees F)

| | |
|---|---|
|             Terminals 3 and 6 (VIN 6) or 1 and 2 (VIN 8) | 185 to 275 ohms |
|             Terminals 2 and 5 | 370 to 550 ohms |

        Hot (above 104-degrees F)

| | |
|---|---|
|             Terminals 3 and 6 (VIN 6) or 1 and 2 (VIN 8) | 240 to 325 ohms |
|             Terminals 2 and 5 | 475 to 650 ohms |

## Charging system

| | |
|---|---|
| Charging voltage | 13.5 to 15.1 volts |
| Standard amperage | |
|     All lights and accessories turned off | Less than 10 amps |
|     Headlights (hi-beam) and heater blower motor turned on | 30 amps or more |
| Alternator brush length | |
|     Standard | 13/32-inch |
|     Minimum | 1/16-inch |

## 1   General information

The engine electrical systems include all ignition, charging and starting components. Because of their engine related functions, these components are discussed separately from chassis electrical devices such as the lights, the instruments, etc. (which are included in Chapter 12).

Always observe the following precautions when working on the electrical systems:

a) *Be extremely careful when servicing engine electrical components. They are easily damaged if checked, connected or handled improperly.*

b) *Never leave the ignition switch on for long periods of time (10 minutes maximum) with the engine off.*

c) *Don't disconnect the battery cables while the engine is running.*

d) *Maintain correct polarity when connecting a battery cable from another vehicle during jump starting.*

e) *Always disconnect the negative cable first and hook it up last or the battery may be shorted by the tool being used to loosen the cable clamps.*

It's also a good idea to review the safety-related information regarding the engine electrical systems located in the *Safety first* section near the front of this manual before beginning any operation included in this Chapter.

## 2   Battery - emergency jump starting

Refer to the *Booster battery (jump) starting* procedure at the front of this manual.

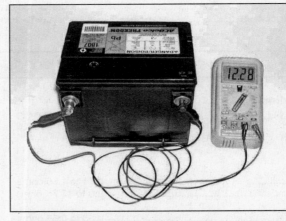

**3.2  To test the open circuit voltage of the battery, connect a voltmeter to the battery - a fully charged battery should measure at least 12.4 volts (depending on outside air temperature)**

## 3   Battery - check and replacement

**Warning:** *Hydrogen gas is produced by the battery, so keep open flames and lighted cigarettes away from it at all times. Always wear eye protection when working around a battery. Rinse off spilled electrolyte immediately with large amounts of water.*

### Check

*Refer to illustrations 3.2 and 3.3*

1    The battery's surface charge must be removed before accurate voltage measurements can be made. Turn On the high beams for ten seconds, then turn them Off, let the vehicle stand for two minutes. Remove the battery from the vehicle (see Steps 4 through 10).

2    Check the battery state of charge. Visually inspect the indicator eye on the top of the battery, if the indicator eye is clear, charge the battery as described in Chapter 1. Next per-form an open voltage circuit test using a digital voltmeter **(see illustration)**. With the engine and all accessories Off, connect the negative probe of the voltmeter to the negative terminal of the battery and the positive probe to the positive terminal of the battery. The battery voltage should be 12.4 volts or more. If the battery is less than the specified voltage, charge the battery before proceeding to the next test. Do not proceed with the battery load test unless the battery charge is correct.

3    Perform a battery load test. An accurate check of the battery condition can only be performed with a load tester (available at most auto parts stores). This test evaluates the ability of the battery to operate the starter and other accessories during periods of heavy amperage draw (load). Install a special battery load testing tool onto the terminals **(see illustration)**. Load test the battery according to the tool manufacturer's instructions. This tool utilizes a carbon pile to increase the load demand (amperage draw) on the battery. Maintain the load on the bat-

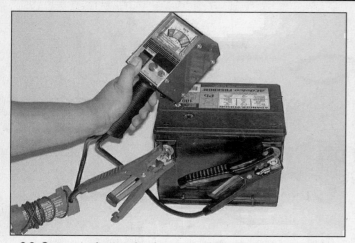

**3.3 Connect a battery load tester to the battery and check the battery condition under load following the tool manufacturer's instructions**

**3.6 To remove the battery, detach the negative battery cable first, then the positive cable, remove the hold-down clamp nut and bolt (arrows) and remove the hold-down clamp**

tery for 15 seconds or less and observe that the battery voltage does not drop below 9.6 volts. If the battery condition is weak or defective, the tool will indicate this condition immediately. **Note:** *Cold temperatures will cause the minimum voltage requirements to drop slightly. Follow the chart given in the tool manufacturer's instructions to compensate for cold climates. Minimum load voltage for freezing temperatures (32 degrees F) should be approximately 9.1 volts.*

### Replacement

*Refer to illustration 3.6*

4    Disconnect the cable from the negative battery terminal. **Caution:** *If the stereo in your vehicle is equipped with an anti-theft system, make sure you have the correct activation code before disconnecting the battery.*

5    Disconnect the positive battery cable.

6    Remove the battery hold-down clamp **(see illustration)**.

7    Remove the battery and place it on a workbench. **Note:** *Battery handling tools are available at most auto parts stores for a reasonable price. They make it easier to remove and carry the battery.*

8    While the battery is removed, inspect the tray, retainer brackets and related fasteners for corrosion or damage.

9    If corrosion is evident, remove the battery tray and use a baking soda/water solution to clean the corroded area to prevent further oxidation. Repaint the area as necessary using rust resistant paint.

10    Clean and service the battery and cables (see Chapter 1).

11    If you are replacing the battery, make sure you purchase one that is identical to yours, with the same dimensions, amperage rating, cold cranking amps rating, etc. Make sure it is fully charged prior to installation in the vehicle.

12    Installation is the reverse of removal. Connect the positive cable first and the negative cable last.

13    After connecting the cables to the battery, apply a light coating of battery terminal corrosion inhibitor to the connections to help prevent corrosion.

---

### 4    Battery cables - replacement

**Caution:** *If the stereo in your vehicle is equipped with an anti-theft system, make sure you have the correct activation code before disconnecting the battery.*

1    Periodically inspect the entire length of each battery cable for damage, cracked or burned insulation and corrosion. Poor battery cable connections can cause starting problems and decreased engine performance.

2    Check the cable-to-terminal connections at the ends of the cables for cracks, loose wire strands and corrosion. The presence of white, fluffy deposits under the insulation at the cable terminal connection is a sign that the cable is corroded and should be replaced. Check the terminals for distortion, missing mounting bolts and corrosion.

3    When removing the cables, always disconnect the negative cable first and hook it up last or the battery may be shorted by the tool used to loosen the cable clamps. Even if only the positive cable is being replaced, be sure to disconnect the negative cable from the battery first (see Chapter 1 for further information regarding battery cable removal).

4    Disconnect the old cables from the battery, then trace each of them to their opposite ends and detach them from the starter solenoid and ground terminals. Note the routing of each cable to ensure correct installation.

5    If you are replacing either or both of the old cables, take them with you when buying new cables. It is vitally important that you replace the cables with identical parts. Cables have characteristics that make them easy to identify: positive cables are usually red, larger in cross-section and have a larger

diameter battery post clamp; ground cables are usually black, smaller in cross-section and have a slightly smaller diameter clamp for the negative post.

6    Clean the threads of the solenoid or ground connection with a wire brush to remove rust and corrosion. Apply a light coat of battery terminal corrosion inhibitor, or petroleum jelly, to the threads to prevent future corrosion.

7    Attach the cable to the solenoid or ground connection and tighten the mounting nut/bolt securely.

8    Before connecting a new cable to the battery, make sure that it reaches the battery post without having to be stretched.

9    Connect the positive cable first, followed by the negative cable.

---

### 5    Ignition system - general information and precautions

1    There are two types of ignition systems used on the models covered by this manual - 1997 and earlier models are equipped with a conventional distributor ignition system and 1998 and later models are equipped with a distributorless ignition system.

### *1997 and earlier models*

*Refer to illustrations 5.3a and 5.3b*

2    The distributor ignition system includes the ignition switch, the battery, the igniter, the pick-up coil, the ignition coil, the primary (low voltage) and secondary (high voltage) wiring circuits, the distributor and the spark plugs. The ignition system is controlled by the Engine Control Module (ECM). Using data provided by information sensors which monitor various engine functions (such as rpm, intake air volume, engine temperature, etc.), the ECM ensures a perfectly timed spark under all conditions.

3    The distributor ignition system is divided into two groups; distributor with external

**5**

igniter and distributor with internal igniter **(see illustrations)**. On the latter ignition system, the igniter is located within the distributor housing. On both systems the ignition coil is mounted inside the distributor. When diagnosing these units, be sure to make all the necessary ignition system checks before replacing any components, as they are expensive and usually non-returnable.

## 1998 and later models

4    1998 and later models are equipped with a distributorless ignition system. The ignition system consists of the battery, ignition coils, spark plugs, camshaft position sensor, crankshaft position sensor and the Engine Control Module (ECM). The ECM controls the ignition timing and spark advance characteristics for the engine. The ignition timing is not adjustable.

5    The crankshaft position sensor and camshaft position sensor generate pulses that are input to the Engine Control Module. The ECM determines piston position and engine speed from these two sensors. The ECM calculates injector sequence and ignition timing from the piston position. Refer to Chapter 6 for testing and replacement procedures for the crankshaft position sensor and camshaft position sensor.

6    1998 and 1999 models utilize a coil pack consisting of two ignition coils with an igniter incorporated into each unit. This type of ignition system uses a "waste spark" method of spark distribution. Each cylinder is paired with its opposing cylinder in the firing order (1-4, 2-3) so one cylinder under compression fires simultaneously with its opposing cylinder, where the piston is on the exhaust stroke. Since the cylinder on the exhaust stroke requires very little of the available voltage to fire its plug, most of the voltage is used to fire the plug of the cylinder on the compression stroke. In a conventional ignition system, one end of the ignition coil secondary winding is connected to engine ground. In a waste spark system, neither end of the secondary winding is grounded - instead, one end of the coil secondary winding is directly attached to the spark plug and the other end is attached to the spark plug of the companion cylinder.

7    2000 and later models utilize an individual ignition coil/igniter unit for each cylinder. The unit is positioned directly over each spark plug. The ECM fires each coil sequentially in the firing order sequence.

8    The ECM controls the ignition system by opening and closing the primary ignition coil control circuit. The computerized ignition system provides complete control of the ignition timing by determining the optimum timing in response to engine speed, coolant temperature, throttle position and engine load. These parameters are relayed to the ECM by the camshaft position sensor, crankshaft position sensor, throttle position sensor, coolant temperature sensor and manifold absolute pressure sensor or mass airflow sensor. Refer to Chapter 6 for addi-

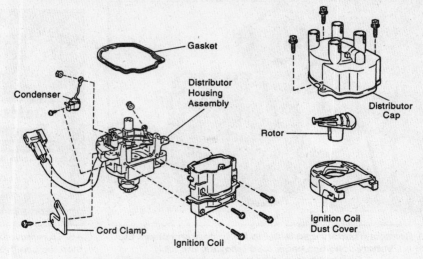

**5.3a  Distributor components with an external igniter**

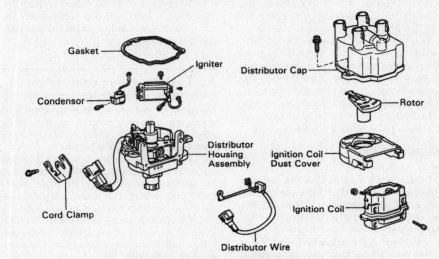

**5.3b  Distributor components with an internal igniter**

tional information on the various sensors.

9    The ignition system is also integrated with a knock sensor system. The system uses a knock sensor in conjunction with the ECM to control spark timing. The knock sensor system allows the engine to use maximum spark advance without spark knock, which improves driveability and fuel economy.

10   When working on the ignition system, take the following precautions:

a)  *Do not keep the ignition switch on for more than 10 seconds if the engine will not start.*

b)  *Always connect a tachometer in accordance with the manufacturer's instructions. Some tachometers may be incompatible with this ignition system. Consult the tachometer manufacturer's consultant before buying a tachometer for use with this vehicle.*

c)  *Never allow the ignition coil terminals to touch ground. Grounding the coil could*

result in damage to the igniter and/or the ignition coil.

d)  *Do not disconnect the battery when the engine is running.*

e)  *Make sure the igniter is properly grounded.*

## 6    Ignition system - check

**Warning:** *Because of the high voltage generated by the ignition system, extreme care should be taken whenever an operation is performed involving ignition components.*

1    If a malfunction occurs and the vehicle won't start, do not immediately assume that the ignition system is causing the problem. First, check the following items:

a)  *Make sure the battery cable clamps, where they connect to the battery, are clean and tight.*

6.2  To use a calibrated ignition tester (available at most auto parts stores), remove a spark plug wire from a spark plug, connect the spark plug boot to the tester and clip the tester to a good ground, then crank the engine over - if there is enough voltage to fire the plug, bright blue spark will be clearly visible between the electrode tip and the tester body

6.7  Check for battery voltage to the positive side (+) of the ignition coil

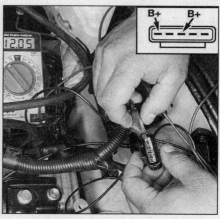

6.8  Disconnect the igniter electrical connector and with the ignition key ON (engine not running), check for battery voltage to the igniter (external igniter shown - on models with an internal igniter, check for voltage on the brown wire to the igniter)

*b) Test the condition of the battery (see Section 3). If it does not pass all the tests, replace it with a new battery.*
*c) Check the ignition system wiring and connections.*
*d) Check the related fuses inside the fuse box (see Chapter 12). If they're burned, determine the cause and repair the circuit.*

### 1997 and earlier models

*Refer to illustrations 6.2, 6.7 and 6.8*

2    If the engine turns over but won't start, disconnect the spark plug wire from any spark plug and attach it to a calibrated tester (available at most auto parts stores) **(see illustration)**. Connect the clip on the tester to a bolt or metal bracket on the engine.

3    Crank the engine and watch the end of the tester or spark plug wire to see if a bright blue, well-defined spark occurs (weak spark or intermittent spark is the same as no spark).

4    If spark occurs, sufficient voltage is reaching the plug to fire it (repeat the check at the remaining plug wires to verify that the distributor cap and rotor are OK).

5    If the ignition system is operating properly the problem lies elsewhere; i.e. a mechanical or fuel system problem. However, the plugs themselves may be fouled, so remove and check them as described in Chapter 1.

6    If no sparks or intermittent sparks occur, remove the distributor cap and check the cap, rotor and spark plug wires as described in Chapter 1. If moisture is present, dry out the cap and rotor, then reinstall the cap and repeat the spark test. If sparks now occur, the distributor cap, rotor or plug wire(s) may be defective.

7    If no sparks occur, check the primary wire connections at the coil to make sure

they're clean and tight. Check for voltage to the coil on the primary circuit from the ignition switch **(see illustration)**. Check the ignition coil (see Section 7) and the distributor pick-up coil (see Section 11). Make any necessary repairs, then repeat the check again.

8    Check for battery voltage to the igniter and ignition coil with the ignition key On (engine not running) **(see illustration)**. If voltage is available and there is still no spark, the igniter or the ECM may be defective. Have the system diagnosed by a dealer service department or other qualified repair shop.

### 1998 and later models

*Refer to illustration 6.12*

9    If the engine turns over but won't start, make sure there is sufficient secondary ignition voltage to fire the spark plug. On 1998 and 1999 models, disconnect a spark plug wire from one of the spark plugs and attach a calibrated ignition system tester (available at most auto parts stores) to the spark plug boot. Connect the clip on the tester to a bolt or metal bracket on the engine **(see illustration 6.2)**. On 2000 and later models, remove an ignition coil (see Section 7) and attach the calibrated ignition system tester to the spark plug boot. Reconnect the electrical connector to the coil and clip the tester to a good ground. Crank the engine and watch the end of the tester to see if a bright blue, well-defined spark occurs (weak spark or intermittent spark is the same as no spark).

10    If spark occurs, sufficient voltage is reaching the plug to fire it (repeat the check at the remaining spark plug wires or ignition coils to verify that the spark plug wires, connectors and ignition coils are good). If the ignition system is operating properly the problem lies elsewhere; i.e. a mechanical or fuel system problem. However, the plugs themselves may be fouled, so remove and check them as described in Chapter 1.

11    If no spark occurs, remove the spark plug wire or boot from the suspected ignition coil and check the terminals for damage. Using an ohmmeter, check the wire or boot for an open or high resistance (on 1998 and

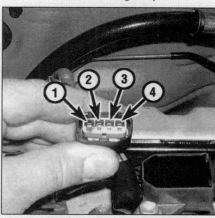

6.12  Disconnect the electrical connector from the ignition coil and check for battery voltage at terminal no. 1 with the ignition key On - also check for continuity to a good engine ground point at terminal no. 4

1999 models, the spark plug wire resistance should not be greater than 25,000 ohms). On 1998 and 1999 models, disconnect the spark plug wires from all the ignition coil towers and check the ignition coil secondary resistance across each pair of coil towers (see Section 7). Compare your measurement with the values listed in this Chapter's Specifications. Replace the ignition coil assembly if either coil is not within specifications.

12    Check for battery voltage to the ignition coil with the ignition key On (engine not running). Disconnect the coil electrical connector and check for power at the indicated terminal of the coil connector **(see illustration)**.

13    Battery voltage should be available with the ignition key On. If there is no battery voltage present, check the wiring and/or circuit between the fuse box and ignition coil (don't forget to check the fuses). Also check the ground circuit for continuity. **Note:** *Refer to*

**5**

**7.2  Carefully pry the plastic shield at the locking tab to release it from the ignition coil**

**7.4a  To check the primary resistance of the coil, measure the resistance between the positive and the negative terminals**

the wiring diagrams at the end of Chapter 12 for wire color identification for testing and additional information on the circuits.

14   If battery voltage is available to the ignition coil, check the ignition coil control circuits as follows **(see illustration 6.12): Caution:** *Use only an LED test light to avoid damaging the ECM.*

  a) *Remove the EFI fuse from the fuse box (this disables the fuel injectors so the engine will not start or flood).*

  b) *On 1998 and 1999 models, connect an LED test light to terminal no. 2 of one of the coil harness connectors. Perform the remainder of the test, then connect the LED test light to terminal no. 2 of the other coil harness connector and perform the test.*

  c) *On 2000 and later models, connect an LED test light to terminal no. 3 of one of the coil harness connectors. Perform the remainder of the test at each coil connector in turn.*

  d) *Crank the engine with the starter and confirm that the LED test light flashes as the engine rotates. A flashing LED test light indicates the ECM, camshaft position sensor and crankshaft position sensor are functioning properly.*

15   If a control signal is not present at the ignition coil, refer to Chapter 6 and check the camshaft position sensor and the crankshaft position sensor. If the sensors are good and there is no control signal, have the ECM checked by a dealer service department or other qualified repair shop.

16   If battery voltage and a control signal exist at the ignition coil and there is no spark, replace the ignition coil.

## 7   Ignition coil - check and replacement

1   Detach the cable from the negative terminal of the battery. **Caution:** *If the stereo in your vehicle is equipped with an anti-theft system, make sure you have the correct activation code before disconnecting the battery.*

**7.4b  To check the secondary resistance of the coil, measure the resistance between the positive terminal and the high tension terminal**

## *1997 and earlier models*
### Check

*Refer to illustrations 7.2, 7.4a and 7.4b*

2   Remove the plastic shield from the coil **(see illustration)**.

3   Remove the nuts from the coil electrical terminals and pull the wires back off the terminals.

4   Using an ohmmeter, measure the primary and secondary resistance **(see illustrations)**:

  a) *Measure the resistance between the positive and negative terminals of the coil. Compare your reading with the specified coil primary resistance listed in this Chapter's Specifications.*

  b) *Measure the resistance between the positive terminal and the high tension terminal. Compare your reading with the specified coil secondary resistance listed in this Chapter's Specifications.*

5   If either of the above tests yield resistance values outside the specified amount, replace the coil.

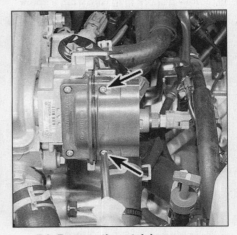

**7.9  Remove the retaining screws (arrows) and lift the coil from the distributor housing**

### Replacement

*Refer to illustration 7.9*

6   Detach the cable from the negative terminal of the battery (see the **Caution** above).

7   Remove the heat shield from the coil.

8   Label and disconnect the wires from the coil terminals.

9   Remove the coil mounting screws **(see illustration)**.

10   Installation is the reverse of removal.

## *1998 and 1999 models*
### Check

*Refer to illustration 7.12*

11   If the spark plug wires are not marked, label them corresponding to the cylinder numbers. Detach the spark plug wires from the ignition coils.

12   Using an ohmmeter, measure the resistance across each pair of spark plug wire terminals **(see illustration)**. Compare your measurement with the secondary resistance value listed in this Chapter's specifications.

13   If the test yields a resistance value outside the specified amount, replace the coil.

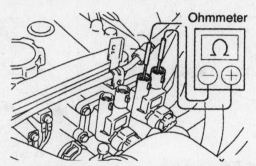

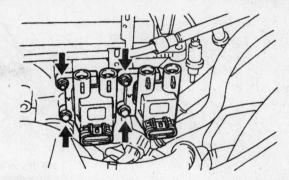

**7.12  To check the ignition coil secondary resistance on a 1998 or 1999 model, detach the spark plug wires and measure the resistance across the spark plug wire terminals of each coil**

**7.16  Ignition coil mounting bolts (arrows) - 1998 and 1999 models**

**7.21  Ignition coil mounting bolts (arrows) - 2000 and later models**

check the ignition coil.

18  Remove the engine cover.

19  Disconnect the electrical connectors from the coils.

20  Remove the nuts and position the wire harness aside if necessary.

21  Remove the mounting bolts and carefully withdraw the coil from the cylinder head **(see illustration)**.

22  Installation is the reverse of removal.

## Replacement

*Refer to illustration 7.16*

14  If the spark plug wires are not marked, label them corresponding to the cylinder numbers. Detach the spark plug wires from the ignition coils.

**8.5  Paint or scribe a mark (arrow) on the edge of the distributor housing immediately below the rotor tip to ensure that the rotor is pointing in the same direction when the distributor is reinstalled - also, paint or scribe another mark across the cylinder head and the distributor body (arrow) to ensure that the distributor is aligned correctly when it is reinstalled**

15  Disconnect the electrical connectors from the ignition coils.

16  Remove the mounting bolts/nuts and remove the coils **(see illustration)**. If reinstalling the original coils, label the coils so they may be installed in their original locations.

17  Installation is the reverse of removal.

## *2000 and later models*

*Refer to illustration 7.21*

**Note:** *Coil resistance cannot be measured on 2000 and later models; refer to Section 6 to*

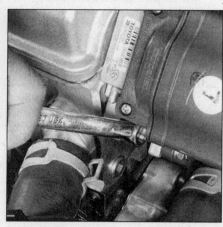

**8.7  Remove the hold-down bolt from the distributor body and pull the distributor straight out**

## 8  Distributor (1997 and earlier models) - removal and installation

## *Removal*

*Refer to illustrations 8.5 and 8.7*

1  Detach the cable from the negative battery terminal. **Caution:** *If the stereo in your vehicle is equipped with an anti-theft system, make sure you have the correct activation code before disconnecting the battery.*

2  Disconnect the electrical connectors from the distributor.

3  Look for a raised "1" on the distributor cap. This marks the location for the number one cylinder spark plug wire terminal. If the cap does not have a mark for the number one terminal, locate the number one spark plug and trace the wire back to the terminal on the cap.

4  Remove the distributor cap (see Chapter 1) and turn the engine over until the rotor is pointing toward the number one spark plug terminal (see locating TDC procedure in Chapter 2A).

5  Make a mark on the edge of the distributor base directly below the rotor tip and in line with it. Also, mark the distributor base and the engine block to ensure that the distributor is installed correctly **(see illustration)**.

6  If equipped with collar bolts, loosen but do not remove the two bolts in the distributor collar. This will give the distributor shaft clearance.

7  Remove the distributor hold-down bolt **(see illustration)**, then pull the distributor straight out to remove it. **Caution:** *DO NOT*

**5**

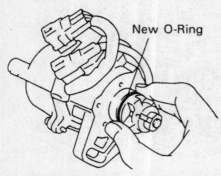

8.8  Install a new O-ring in the groove in the distributor housing

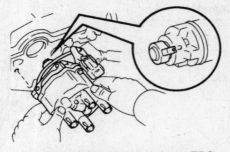

8.9  If you have set the engine at TDC compression for number one cylinder, align the cut-out portion of the coupling with the groove in the distributor housing

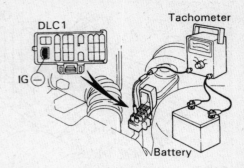

9.1  Connect the tachometer lead to the IG terminal of the test connector

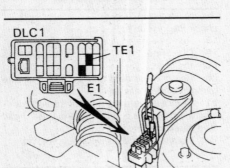

9.2  Attach a jumper wire between terminals E1 and TE1 of the test connector

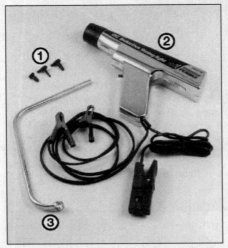

9.3  Tools needed to check and adjust the ignition timing

1    *Vacuum plugs* - *Vacuum hoses will, in most cases, have to be disconnected and plugged. Molded plugs in various shapes and sizes are available for this*

2    *Inductive pick-up timing light -  Flashes a bright, concentrated beam of light when the number one spark plug fires. Connect the leads according to the instructions supplied with the light*

3    *Distributor wrench - On some models, the hold-down bolt for the distributor is difficult to reach and turn with conventional wrenches or sockets. A special wrench like this must be used*

9.5  Point the timing light at the timing marks with the engine at idle

*turn the crankshaft while the distributor is out of the engine, or the alignment marks will be useless.*

## Installation

*Refer to illustrations 8.8 and 8.9*

**Note:** *If the crankshaft has been moved while the distributor is out, locate Top Dead Center (TDC) for the number one piston (see Chapter 2A) and position the distributor and the rotor accordingly.*

8    Install a new O-ring onto the distributor housing **(see illustration)**.

9    Align the cut-out portion of the coupling with the groove in the housing **(see illustration)**.

10    Insert the distributor into the engine in exactly the same relationship to the block that it was in when removed.

11    If the distributor does not seat completely, recheck the alignment marks between the distributor base and the block to verify that the distributor is in the same position it was in before removal. Also check the rotor to see if it's aligned with the mark you made on the edge of the distributor base.

12    Loosely install the distributor hold-down bolt(s).

13    Installation is the reverse of removal.

14    Check the ignition timing (see Section 9) and tighten the distributor hold-down bolt securely.

## 9    Ignition timing (1997 and earlier models) - check and adjustment

*Refer to illustrations 9.1, 9.2, 9.3 and 9.5*

**Note:** *Ignition timing is not adjustable on 1998 and later models. If the procedure specified on the VECI label of your vehicle differs from this one, use the procedure found on the VECI label.*

1    Connect a tachometer according to the manufacturer's instructions **(see illustration)**.

2    Locate the diagnostic test connector and insert a jumper wire between terminals

E1 and TE1 **(see illustration)**.

3    With the ignition switch off, connect a timing light according to the manufacturer's instructions **(see illustration)**. Most timing lights are powered by the battery. Also, an inductive pick-up is installed onto the number one cylinder spark plug wire.

4    Locate the timing marks on the front cover and the crankshaft pulley.

5    Start the engine and allow it to warm up to normal operating temperature (upper radiator hose hot). Verify that the engine idle is correct (750 rpm with an automatic transaxle and 700 rpm with a manual transaxle. Aim the timing light at the timing scale on the front cover **(see illustration)**. The mark on the crankshaft pulley should line up with the 10-degree mark on the scale. If necessary, loosen the distributor hold-down bolt and slowly rotate the distributor until the timing marks align. Tighten the hold-down bolt and recheck the timing.

6    Remove the jumper wire from the test connector.

7    Turn the engine off and remove the tachometer and the timing light.

## 10    Igniter (1997 and earlier models) - replacement

1    Detach the cable from the negative terminal of the battery. **Caution:** *If the stereo in your vehicle is equipped with an anti-theft*

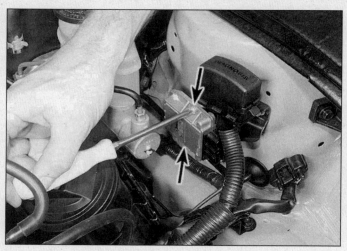

10.3  Remove the igniter screws (arrows) and remove the igniter from the engine compartment

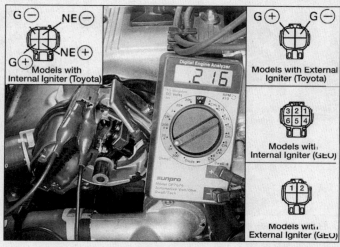

11.1  Check the resistance between the terminals on the pick-up coil indicated in this Chapter's Specifications

system, make sure you have the correct activation code before disconnecting the battery.

### Externally mounted igniter

*Refer to illustration 10.3*

2    Disconnect the electrical connector from the igniter.

3    Remove the screws from the bracket assembly and pull the igniter/bracket assembly out of the engine compartment **(see illustration)**.

4    Installation is the reverse of removal.

### Internally mounted igniter

5    Remove the ignition coil from the distributor (see Section 7).

6    Disconnect the electrical connector from the igniter.

7    Remove the screws from the igniter assembly and pull the igniter/bracket assembly out of the engine compartment **(see illustration 5.3b)**.

8    Installation is the reverse of removal.

---

### 11  Pick-up coil (1997 and earlier models) - check and replacement

### Pick-up coil check

*Refer to illustration 11.1*

1    Disconnect the electrical connector at the distributor and using an ohmmeter, measure the resistance between the pick-up coil terminals **(see illustration)**.

2    Compare the measurements to the values listed in this Chapter's Specifications. If the resistance is not as specified, replace the pick-up coil.

### Air gap check

*Refer to illustration 11.5*

3    Detach the cable from the negative terminal of the battery. **Caution:** *If the stereo in your vehicle is equipped with an anti-theft*

system, make sure you have the correct activation code before disconnecting the battery.

4    Remove the distributor from the engine (see Section 8).

5    Using a brass feeler gauge, measure the gap between the signal rotor and the pick-up coil projection **(see illustration)**. Compare your measurement to the air gap listed in this Chapter's Specifications. If the air gap is not as specified, replace the distributor, as the air gap is not adjustable. Excessive air gap is usually a sign of wear in the distributor shaft bushing.

### Replacement

*Refer to illustrations 11.9, 11.10 and 11.11*

6    Detach the cable from the negative terminal of the battery. **Caution:** *If the stereo in your vehicle is equipped with an anti-theft system, make sure you have the correct activation code before disconnecting the battery.*

7    Remove the distributor from the engine (see Section 8).

8    Remove the coil from the distributor (see Section 7).

9    Remove the set screws that retain the

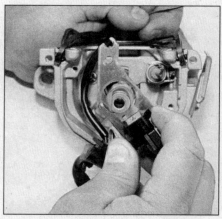

11.5  Measure the air gap between the signal rotor and the pick-up coil projection - if the gap is not within specification, replace the distributor

pick-up coil to the distributor **(see illustration)**.

10    Remove the pick-up coil from the distributor **(see illustration)**.

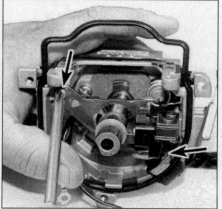

11.9  Remove the pick-up coil set screws (arrows)

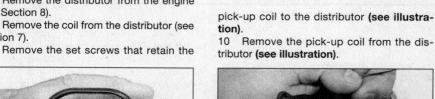

11.10  Remove the pick-up coil from the distributor

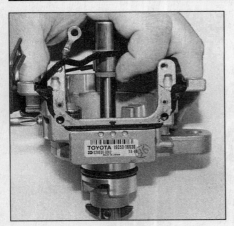

**11.11  Remove the ignition condenser from the distributor**

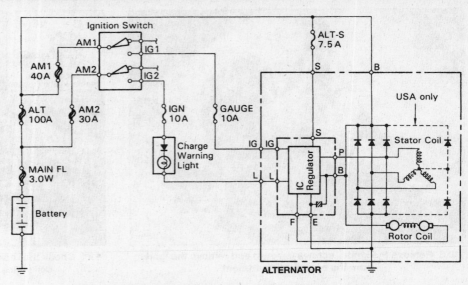

**12.1  Typical alternator charging system schematic (Nippondenso)**

11   Remove the condenser from the distributor **(see illustration)** and replace it.
12   Installation is the reverse of removal.

## 12   Charging system - general information and precautions

*Refer to illustrations 12.1 and 12.4*

The charging system includes the alternator, an internal voltage regulator, a charge indicator, the battery, a fusible link and the wiring between all the components **(see illustration)**. The charging system supplies electrical power for the ignition system, the lights, the radio, etc. The alternator is driven by a drivebelt at the front of the engine.

These models are equipped with either a Nippondenso or a Delco alternator.

The purpose of the voltage regulator is to limit the alternator's voltage to a preset value. This prevents power surges, circuit overloads, etc., during peak voltage output.

The fusible link is a short length of insulated wire integral with the engine compartment wiring harness **(see illustration)**. The link is several wire gauges smaller in diameter than the circuit it protects. See Chapter 12 for additional information regarding fusible links.

The charging system doesn't ordinarily require periodic maintenance. However, the drivebelt, battery and wires and connections should be inspected at the intervals outlined in Chapter 1.

The dashboard warning light should come on when the ignition key is turned to Start, then should go off immediately. If it remains on, there is a malfunction in the charging system. Some vehicles are also equipped with a voltage gauge. If the voltage gauge indicates abnormally high or low voltage, check the charging system (see Section 13).

Be very careful when making electrical circuit connections to a vehicle equipped with an alternator and note the following:

a) *When reconnecting wires to the alternator from the battery, be sure to note the polarity.*
b) *Before using arc welding equipment to repair any part of the vehicle, disconnect the wires from the alternator and the battery terminals.*
c) *Never start the engine with a battery charger connected.*
d) *Always disconnect both battery leads before using a battery charger.*
e) *The alternator is driven by an engine*

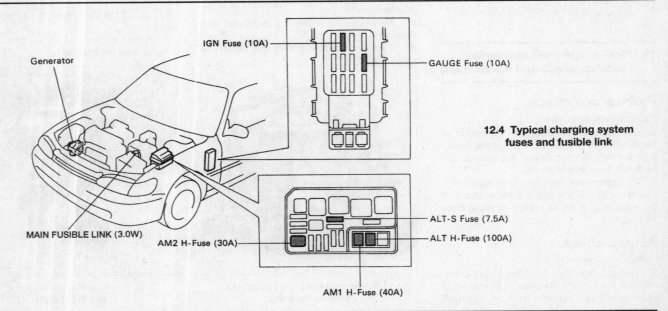

**12.4  Typical charging system fuses and fusible link**

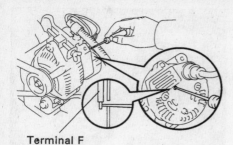

Terminal F

**13.7 If the alternator is putting out less than standard voltage, ground terminal F, start the engine and check the voltage at the battery - if the reading is greater than standard voltage, replace the regulator; if the reading is less than standard, check the alternator or have it checked by a dealer service department or other qualified repair shop (Nippondenso alternators only)**

drivebelt which could cause serious injury if your hand, hair or clothes become entangled in it with the engine running.

f) Because the alternator is connected directly to the battery, it could arc or cause a fire if overloaded or shorted out.

g) Wrap a plastic bag over the alternator and secure it with rubber bands before steam cleaning the engine.

## 13 Charging system - check

*Refer to illustrations 13.7 and 13.8*

1 If a malfunction occurs in the charging circuit, don't automatically assume that the alternator is causing the problem. First check the following items:

a) Check the drivebelt tension and its condition. Replace it if worn or deteriorated.

b) Make sure the alternator mounting and adjustment bolts are tight.

c) Inspect the alternator wiring harness and the electrical connectors at the alterna-

tor and voltage regulator. They must be in good condition and tight.

d) Check the fusible link located at the positive battery cable or the large main fuses in the engine compartment. If it's burned, determine the cause, repair the circuit and replace the link or fuse (the vehicle won't start and/or the accessories won't work if the fusible link or fuse blows).

e) Check all the fuses that are in series with the charging system circuit (see illustration 12.4). The location of these fuses and fusible links may vary from year and model but the designations are the same; main fusible link 3.0W, Alt H fuse (100 amp), AM1 (40 amp), AM2 (30 amp), IG (10 amp), Gauge 10A, and ALT-S (7.5 amp).

f) Start the engine and check the alternator for abnormal noises (a shrieking or squealing sound indicates a bad bushing).

g) Check the specific gravity of the battery electrolyte. If it's low, charge the battery (doesn't apply to maintenance free batteries).

h) Make sure that the battery is fully charged (one bad cell in a battery can cause overcharging by the alternator).

i) Disconnect the battery cables (negative first, then positive). **Caution:** If the stereo in your vehicle is equipped with an anti-theft system, make sure you have the correct activation code before disconnecting the battery. Inspect the battery posts and the cable clamps for corrosion. Clean them thoroughly if necessary (see Section 4 and Chapter 1). Reconnect the positive cable, then the negative cable.

2 Using a voltmeter, check the battery voltage with the engine off. It should be approximately 12 volts.

3 Start the engine and check the battery voltage again. It should now be approximately 13.5 to 15.1 volts.

4 Turn on the headlights. The voltage should drop and then come back up, if the charging system is working properly.

5 If the voltage reading is greater than the specified charging voltage, replace the voltage regulator (see Section 15).

6 If the voltmeter reading is less than standard voltage, check the regulator and alternator on Nippondenso models as follows.

7 Remove the rear cover from the alternator. Ground terminal F **(see illustration)**, start the engine, check the voltage at the battery and compare your reading to the standard voltage.

a) If the voltmeter reading is greater than standard voltage, replace the regulator.

b) If the voltmeter reading is less than standard voltage, check the alternator (or have it checked by a dealer service department if you do not have an ammeter).

8 If you have an ammeter, hook it up to the charging system as shown **(see illustration)**. If you don't have an ammeter, you can also use an inductive-type current indicator. This device is inexpensive, readily available at auto parts stores and accurate enough to perform simple amperage checks like the following test.

9 With the engine running at 2,000 rpm, check the reading on the ammeter with all accessories and lights off, then again with the high-beam headlights on and the heater blower switch turned to the HI position. Compare your readings to the standard amperage listed in this Chapter's Specifications.

10 If the ammeter reading is less than standard amperage, repair or replace the alternator.

## 14 Alternator - removal and installation

*Refer to illustrations 14.2, 14.3 and 14.4*

1 Detach the cable from the negative terminal of the battery. **Caution:** If the stereo in your vehicle is equipped with an anti-theft system, make sure you have the correct activation code before disconnecting the battery.

2 Disconnect the electrical connectors from the alternator **(see illustration)**.

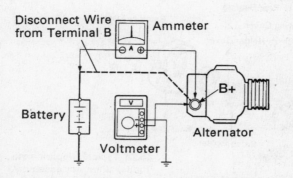

**13.8 Hook up an ammeter as shown to check alternator output**

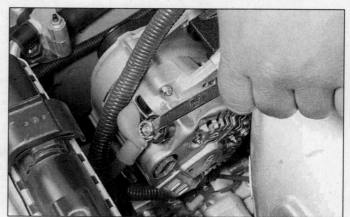

**14.2 Disconnect the electrical connections from the alternator**

5

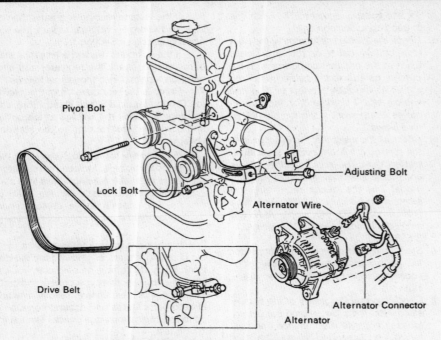

14.3  Alternator installation details - 1997 and earlier models

14.4  Alternator mounting bolts (arrows) -
1998 and later models

3    On 1997 and earlier models, loosen the alternator pivot and lock bolts. Loosen the adjustment bolt and detach the drivebelt. Remove the pivot bolt and lock bolt from the alternator adjustment bracket (see illustration).
4    On 1998 and later models, rotate the drivebelt tensioner and remove the drivebelt (see Chapter 1). Remove the alternator mounting bolts (see illustration).
5    Remove the alternator from the engine.
6    If you are replacing the alternator, take the old alternator with you when purchasing a replacement unit. Make sure that the

new/rebuilt unit is identical to the old alternator. Look at the terminals - they should be the same in number, size and locations as the terminals on the old alternator. Finally, look at the identification markings - they will be stamped in the housing or printed on a tag or plaque affixed to the housing. Make sure that these numbers are the same on both alternators.
7    Many new/rebuilt alternators do not have a pulley installed, so you may have to switch the pulley from the old unit to the new/rebuilt one. When buying an alternator, find out the shop's policy regarding installation of pulleys - some shops will perform this

service free of charge.
8    Installation is the reverse of removal.
9    After the alternator is installed, install the drivebelt and on 1997 and earlier models, adjust the drivebelt tension (see Chapter 1).
10   Check the charging voltage to verify proper operation of the alternator (see Section 13).

## 15   Alternator components - check and replacement

**Note:** *This procedure is for Nippondenso alternators only. If your vehicle is equipped with a Delco alternator, exchange the alternator for a new or rebuilt unit in the event of failure.*

### *Disassembly*

*Refer to illustrations 15.2a, 15.2b, 15.2c, 15.3, 15.4a, 15.4b, 15.5 and 15.7*
1    Remove the alternator (see Section 14) and place it on a clean workbench.

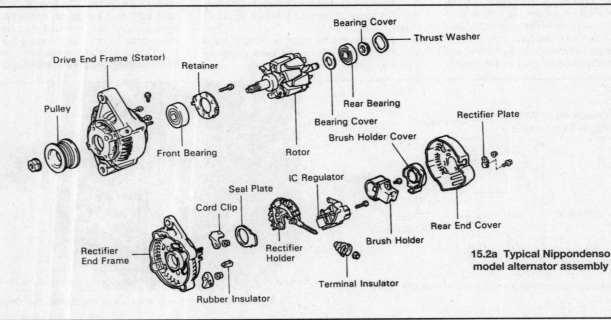

15.2a  Typical Nippondenso
model alternator assembly

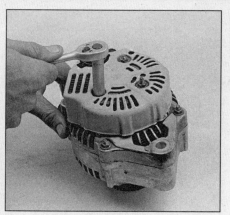

**15.2b Remove the three nuts from the rear cover**

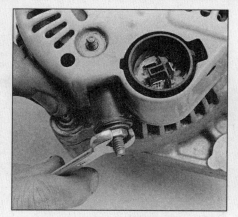

**15.2c Remove the nut, washer and insulator from terminal B and remove the alternator end cover**

**15.3 Remove the five screws (arrows) that retain the voltage regulator and the brush holder**

**15.4a Remove the brush holder**

**15.4b Remove the regulator**

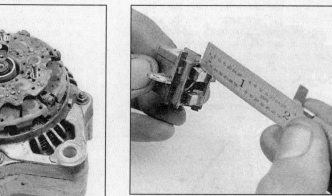

**15.5 Measure the exposed length of the brushes and compare your measurements to the specified minimum length to determine if they should be replaced**

**5**

2    Remove the rear cover nuts, the nut and terminal insulator and the rear cover **(see illustrations)**.

3    Remove the five voltage regulator and brush holder mounting screws **(see illustration)**.

4    Remove the brush holder and the regulator from the rear end frame **(see illustrations)**. If you are only replacing the regulator, proceed to Step 8, install the new unit, reassemble the alternator and install it on the engine (see Section 14). If you are going to replace the brushes, proceed with the next Step.

5    Measure the exposed length of each brush **(see illustration)** and compare it to the minimum length listed in this Chapter's Specifications. If the length of either brush is less than the specified minimum, replace the brushes and brush holder assembly. **Note:** *On some models, it may be necessary to solder the new brushes in place.*

6    Make sure that each brush moves smoothly in the brush holder.

7    Remove the rectifier assembly **(see illustration)**. Remove the four rubber insulators and the seal plate.

8    Scribe or paint marks on the front and rear end frame housings of the alternator to facilitate reassembly.

9    Remove the nut retaining the pulley to the rotor shaft and remove the pulley.

10    Remove the four nuts retaining the front and rear end frames together, then separate the rear end frame assembly from the front end frame **(see illustration 15.2a)**.

11    Remove the thrust washer and remove the rotor from the front end frame.

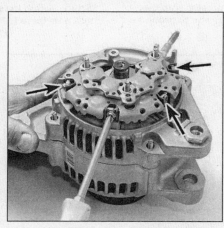

**15.7 Remove the screws (arrows) retaining the rectifier assembly to the stator windings**

## Component checks

*Refer to illustrations 15.12a, 15.12b, 15.13, 15.14a, 15.14b, 15.14c and 15.14d*

12    Check for an open between the two slip rings **(see illustration)**. There should be 2 to 4 ohms resistance between the slip rings.

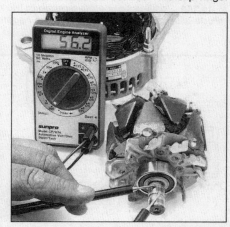

**15.12a Continuity should exist between the two rotor slip rings**

Check for grounds between each slip ring and the rotor **(see illustration)**. There should be no continuity (infinite resistance) between the rotor and either slip ring. If the rotor fails either test, or if the slip rings are excessively worn, the rotor is defective.

13 Check for opens between each end terminal of the stator windings **(see illustration)**. If either reading is high (infinite resistance), the stator is defective. Check for a grounded stator winding between each stator terminal and the frame. If there's continuity between any stator winding and the frame, the stator is defective.

14 Check the positive and negative rectifiers as follows:

a) Check the positive diode assembly by touching the positive probe of the ohmmeter onto the diode terminal and the negative probe onto one of the other designated diode terminals **(see illustrations)**. Then reverse the probes and check again. The diode should have continuity with the ohmmeter one way and no continuity when the probes are reversed. Check each of the terminals in this manner. If any of the diodes fail the test, the diode assembly is defective.

b) Check the negative diode assembly by touching the negative probe of the ohm-

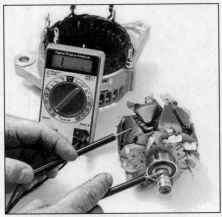

**15.12b Check the continuity between the rotor frame and the slip rings - there should be NO continuity**

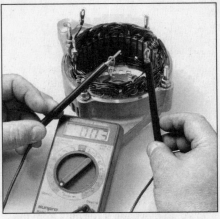

**15.13 Check for continuity between the stator winding terminals - continuity should exist between all the stator winding terminals**

meter onto the NEGATIVE TERMINALS and the other probe onto each rectifier terminal **(see illustration)**. Reverse the polarity (reverse probes) and check to make sure there is no continuity in one position and continuity in the other position. Check each of the terminals in this manner. If any of the diodes fail the test, the diode assembly is defective.

### Reassembly

*Refer to illustration 15.16*

15 Install the components in the reverse order of removal, noting the following:

16 Install the brush holder by depressing each brush with a small screwdriver to clear the shaft **(see illustration)**.

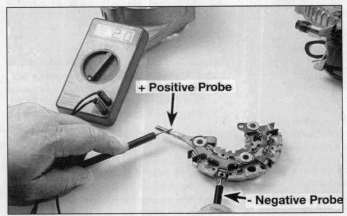

**15.14a Position the positive probe of the ohmmeter onto the diode assembly positive post and the negative probe to ground - continuity should exist**

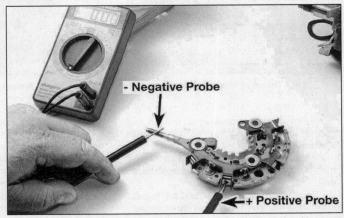

**15.14b Switch the polarity of the ohmmeter probes and confirm that now there is NO continuity within the diodes - check each diode (four total) individually**

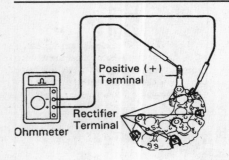

**15.14c Positive rectifier diode check**

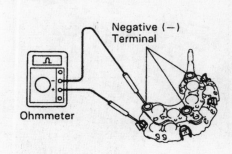

**15.14d Negative rectifier diode check**

**15.16 To facilitate installation of the brush holder, depress each brush with a small screwdriver to clear the shaft**

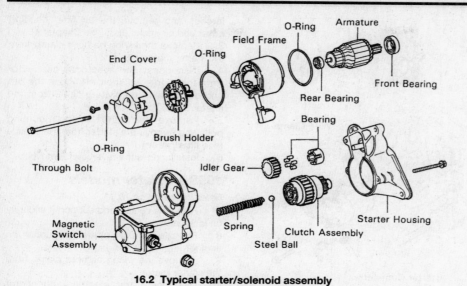

**16.2  Typical starter/solenoid assembly**

tions when working on the starting system:

a) *Excessive cranking of the starter motor can overheat it and cause serious damage. Never operate the starter motor for more than 15 seconds at a time without pausing to allow it to cool for at least two minutes.*

b) *The starter is connected directly to the battery and could arc or cause a fire if mishandled, overloaded or short circuited.*

c) *Always detach the cable from the negative terminal of the battery before working on the starting system.* **Caution:** *If the stereo in your vehicle is equipped with an anti-theft system, make sure you have the correct activation code before disconnecting the battery.*

17    Install the voltage regulator and brush holder screws into the rear frame.
18    Install the rear cover and tighten the three nuts securely.
19    Install the terminal insulator and tighten it with the nut.
20    Install the alternator (see Section 14).

## 16    Starting system - general information and precautions

*Refer to illustration 16.2*

The sole function of the starting system is to turn over the engine quickly enough to allow it to start.

The starting system consists of the battery, the starter motor, the starter solenoid and the electrical circuit connecting the components. The solenoid is mounted directly on the starter motor **(see illustration)**.

On 1997 and earlier models, the starter motor assembly is installed on the rear (firewall) side of the engine on the transaxle bellhousing above the driveaxle. On 1998 and later models, the starter motor assembly is installed on the front (radiator) side of the engine on the lower section of transaxle bellhousing.

When the ignition key is turned to the START position, the starter solenoid is actuated through the starter control circuit. The starter solenoid then connects the battery to the starter. The battery supplies the electrical energy to the starter motor, which does the actual work of cranking the engine.

The starter motor on a vehicle equipped with a manual transaxle can be operated only when the clutch pedal is depressed; the starter on a vehicle equipped with an automatic transaxle can be operated only when the transaxle selector lever is in Park or Neutral.

Always observe the following precau-

## 17    Starter motor and circuit - check

*Refer to illustration 17.5*
**Note:** *Before diagnosing starter problems, make sure the battery is fully charged.*
1    If the starter motor does not turn at all when the switch is operated, make sure that the shift lever is in Neutral or Park (automatic transaxle) or that the clutch pedal is depressed (manual transaxle).
2    Make sure that the battery is charged and that all cables, both at the battery and starter solenoid terminals, are clean and secure.
3    If the starter motor spins but the engine is not cranking, the overrunning clutch in the starter motor is slipping and the starter motor must be replaced.
4    If, when the switch is actuated, the starter motor does not operate at all but the solenoid clicks, then the problem lies with either the battery, the main solenoid contacts or the starter motor itself (or the engine is seized).
5    If the solenoid plunger cannot be heard when the switch is actuated, the battery is

**5**

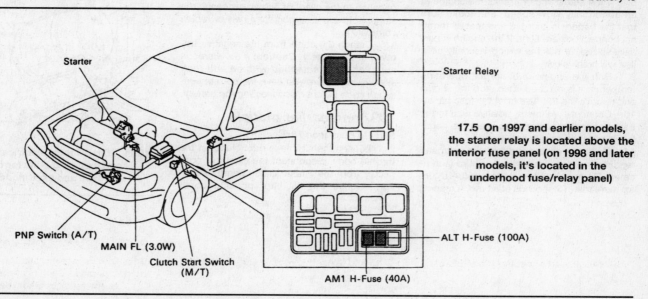

**17.5  On 1997 and earlier models, the starter relay is located above the interior fuse panel (on 1998 and later models, it's located in the underhood fuse/relay panel)**

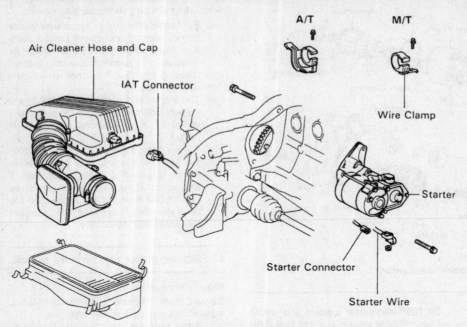

**18.5  Starter motor installation details - 1997 and earlier models**

bad, the fusible link is burned (the circuit is open), the starter relay is defective or the starter solenoid itself is defective **(see illustration)**. **Note:** *Follow the relay testing procedure in Chapter 12 to diagnose a defective relay.*

6  To check the solenoid, connect a jumper lead between the battery (+) and the ignition switch terminal (the small terminal) on the solenoid. If the starter motor now operates, the solenoid is OK and the problem is in the ignition switch, Neutral start switch or in the wiring.

7  If the starter motor still does not operate, remove the starter/solenoid assembly for disassembly, testing and repair.

8  If the starter motor cranks the engine at an abnormally slow speed, first make sure that the battery is charged and that all terminal connections are tight. If the engine is partially seized, or has the wrong viscosity oil in it, it will crank slowly.

9  Run the engine until normal operating temperature is reached, then stop the engine and remove the EFI fuse from the fuse panel.

10  Connect a voltmeter positive lead to the battery positive post and connect the negative lead to the negative post.

11  Crank the engine and take the voltmeter readings as soon as a steady figure is indicated. Do not allow the starter motor to turn for more than 15 seconds at a time. A reading of nine volts or more, with the starter motor turning at normal cranking speed, is normal. If the reading is nine volts or more but the cranking speed is slow, the motor is faulty. If the reading is less than nine volts and the cranking speed is slow, the solenoid contacts are probably burned, the starter motor is bad, the battery is discharged or there is a bad connection.

## 18  Starter motor - removal and installation

**Note:** *The starter motor and solenoid assembly cannot be repaired using separate components. In the event of failure, exchange the starter/solenoid assembly for a new or rebuilt complete unit.*

1  Detach the cable from the negative terminal of the battery. **Caution:** *If the stereo in your vehicle is equipped with an anti-theft system, make sure you have the correct activation code before disconnecting the battery.*

### 1997 and earlier models

*Refer to illustration 18.5*

2  Loosen the air intake duct clamp at the throttle body, disconnect the electrical connector from the intake air temperature sensor, release the air filter housing cover latches and remove the air filter housing cover and air intake duct (see Chapter 4).

3  Remove the wiring harness clamp from the starter motor.

4  Disconnect the electrical connector from the starter solenoid. Remove the nut and disconnect the battery cable from the starter motor.

5  Remove the starter motor mounting bolts and remove the starter from the vehicle **(see illustration)**.

6  Installation is the reverse of removal.

### 1998 and later models

*Refer to illustration 18.11*

7  Raise the vehicle and support it securely on jackstands.

8  Remove the slash shield from under the engine.

9  Remove the wiring harness clamp from the starter motor.

10  Disconnect the electrical connector from the starter solenoid. Remove the nut and disconnect the battery cable from the starter motor.

11  Remove the starter motor mounting bolts and remove the starter from the vehicle **(see illustration)**.

12  Installation is the reverse of removal.

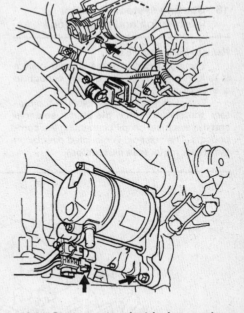

**18.11  Starter motor electrical connector and mounting bolts (arrows) - 1998 and later models**

# Chapter 6
# Emissions and engine control systems

## Contents

## Specifications

### Camshaft position sensor resistance
| | |
|---|---|
| Cold | 835 to 1400 ohms |
| Hot | 1060 to 1645 ohms |

### Crankshaft position sensor resistance
| | |
|---|---|
| Cold | 1630 to 2740 ohms |
| Hot | 2065 to 3225 ohms |

### EGR temperature sensor resistance
| | |
|---|---|
| 112-degrees F | 69 to 89 K-ohms |
| 212-degrees F | 11 to 15 K-ohms |
| 302-degrees F | 2000 to 4000 ohms |

### Idle air control valve resistance (1997 and earlier models)
| | |
|---|---|
| Cold | 17.0 to 24.5 ohms |
| Hot | 21.5 to 28.5 ohms |

### Idle speed
| | |
|---|---|
| Automatic transaxle | 750 rpm |
| Manual transaxle | 700 rpm |

**1.1a Emissions and engine control components - 1997 and earlier models**

| | | | | | |
|---|---|---|---|---|---|
| *1* | *Crankshaft position sensor* | *5* | *EGR valve* | *8* | *Air intake resonator* |
| *2* | *MAP sensor* | *6* | *Throttle Position Sensor* | *9* | *IAC valve* |
| *3* | *EGR vacuum switching valve* | *7* | *Test connector* | *10* | *Oxygen sensor* |
| *4* | *EGR vacuum modulator* | | | *11* | *PCV valve* |

**1.1b Emissions and engine control components - 2000 and later models**

| | | | | | |
|---|---|---|---|---|---|
| *1* | *Positive Crankcase Ventilation valve* | *4* | *Engine coolant temperature sensor* | *6* | *Camshaft position sensor* |
| *2* | *EVAP purge and vent valves* | *5* | *Throttle body with Throttle Position* | *7* | *Knock sensor* |
| *3* | *Mass Airflow sensor and intake air temperature sensor* | | *Sensor and IAC valve* | *8* | *Crankshaft position sensor* |

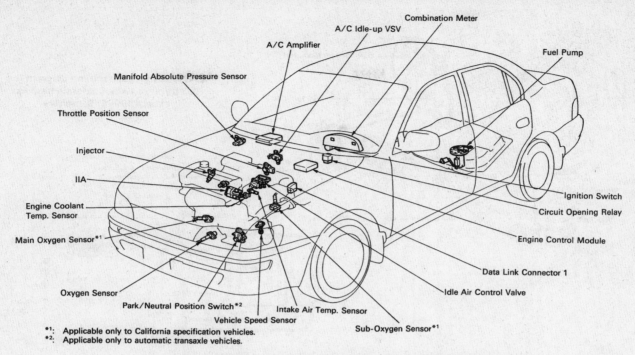

*1: Applicable only to California specification vehicles.
*2: Applicable only to automatic transaxle vehicles.

**1.1c Typical locations of emissions and engine control components**

## 1    General information

*Refer to illustrations 1.1a, 1.1b, 1.1c, 1.4a, 1.4b, 1.4c, 1.6a and 1.6b*

To minimize pollution of the atmosphere from incompletely burned and evaporating gases and to maintain good driveability and fuel economy, a number of emission control systems are used on these vehicles **(see illustrations)**. They include the:

*Positive Crankcase Ventilation (PCV) system*
*Evaporative Emission Control (EVAP) system*
*Exhaust Gas Recirculation (EGR) system*
*Three-way catalytic converter (TWC) system*
*Electronic Fuel Injection (EFI) system*

The Sections in this Chapter include general descriptions, checking procedures within the scope of the home mechanic and component replacement procedures (when possible) for each of the systems listed above.

Before assuming an emissions control system is malfunctioning, check the fuel and ignition systems carefully (see Chapters 4 and 5). The diagnosis of some emission control devices requires specialized tools, equipment and training. If checking and servicing become too difficult or if a procedure is beyond the scope of your skills, consult your dealer service department or other repair shop.

This doesn't mean, however, that emission control systems are particularly difficult to maintain and repair. You can quickly and easily perform many checks and do most of the regular maintenance at home, provided you have the necessary tools and equipment. **Note:** *The most frequent cause of emissions problems is simply a loose or broken electrical connector or vacuum hose, so always check the electrical connectors and vacuum hoses first* **(see illustrations)**.

Pay close attention to any special precautions outlined in this Chapter. It should be noted that the illustrations of the various systems may not exactly match the system installed on your vehicle because of changes made by the manufacturer during production or from year-to-year.

The Vehicle Emissions Control Informa-

**6**

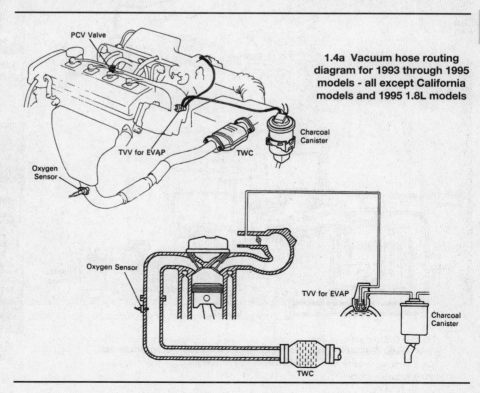

**1.4a Vacuum hose routing diagram for 1993 through 1995 models - all except California models and 1995 1.8L models**

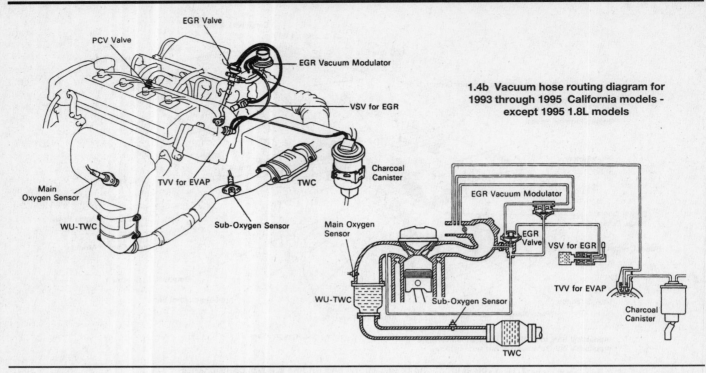

**1.4b  Vacuum hose routing diagram for 1993 through 1995 California models - except 1995 1.8L models**

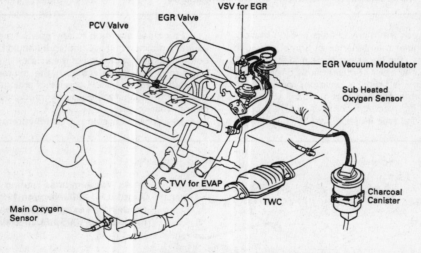

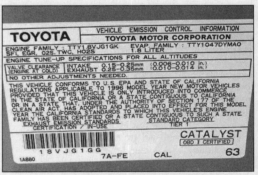

**1.6a  The Vehicle Emission Control Information (VECI) label is located in the engine compartment and contains information on the emission devices on your vehicle**

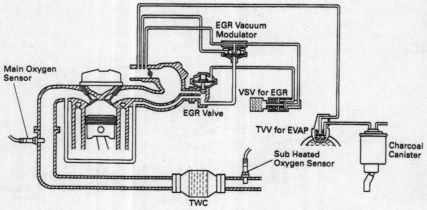

**1.4c  Vacuum hose routing diagram for 1995 1.8L models and all 1996 and 1997 models**

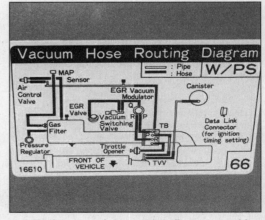

**1.6b  A vacuum hose routing diagram provides the specific vacuum hose routing information for the vehicle**

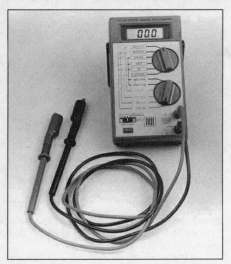

2.1 Digital multimeters can be used for testing all types of circuits; because of their high impedance, they are much more accurate than analog meters for measuring millivolts in low-voltage computer circuits

2.2 Scanners like the Actron Scantool and the AutoXray XP240 are powerful diagnostic aids - programmed with comprehensive diagnostic information, they can tell you just about anything you want to know about your engine management system

tion (VECI) label and a vacuum hose diagram are located on the hood (see illustrations). These contain important emissions specifications and setting procedures, and a vacuum hose schematic with emissions components identified. When servicing the engine or emissions systems, the VECI label in your particular vehicle should always be checked for up-to-date information.

## 2  On Board Diagnosis (OBD) system and trouble codes

### Diagnostic tool information

*Refer to illustrations 2.1, 2.2 and 2.4*

1   A digital multimeter is necessary for checking fuel injection and emission related components (see illustration). A digital volt-ohmmeter is preferred over the older style analog multimeter for several reasons. The analog multimeter cannot display the volts-ohms or amps measurement in hundredths and thousandths increments. When working with electronic circuits which are often very low voltage, this accurate reading is most important. Another good reason for the digital multimeter is the high impedance circuit. The digital multimeter is equipped with a high resistance internal circuitry (10 million ohms). Because a voltmeter is hooked up in parallel with the circuit when testing, it is vital that none of the voltage being measured should be allowed to travel the parallel path set up by the meter itself. This dilemma does not show itself when measuring larger amounts of voltage (9 to 12 volt circuits) but if you are measuring a low voltage circuit such as the oxygen sensor signal voltage, a fraction of a volt may be a significant amount when diagnosing a problem.

2   Hand-held scanners are the most powerful and versatile tools for analyzing engine management systems used on later model vehicles (see illustration). Each brand scan tool must be examined carefully to match the year, make and model of the vehicle you are working on. Often interchangeable cartridges are available to access the particular manufacturer; Ford, GM, Chrysler, etc.). Some manufacturers will specify by continent; Asia, Europe, USA, etc. Seek the advice of your

local auto parts retailer.

3   With the arrival of the federally mandated emission control system (OBD II), a specially designed scanner has been developed. Several tool manufacturers have released OBD II scan tools for the home mechanic. Ask the parts salesperson at a local auto parts store for additional information concerning availability and cost.

4   Another type of code reader may be available at parts stores (see illustration). These tools simplify the procedure for extracting codes from the engine management computer by simply "plugging in" to the diagnostic connector on the vehicle wiring harness and are much less expensive (however, for use on 1996 and later models, they must be designed for an OBD-II system).

### On Board Diagnostic system general information

5   The fuel injection system is controlled by means of a microcomputer known as the Engine Control Module (ECM).

6   The ECM receives signals from various sensors which monitor changing engine operating conditions such as intake air volume, intake air temperature, coolant temperature, engine rpm, acceleration/deceleration, exhaust oxygen content, etc. These signals are utilized by the ECM to determine the correct fuel injection duration.

7   The system is analogous to the central nervous system in the human body: The sensors (nerve endings) constantly relay signals to the ECM (brain), which processes the data and, if necessary, sends out a command to change the operating parameters of the engine (body).

8   Here's a specific example of how one portion of this system operates: An oxygen sensor, located in the exhaust manifold, constantly monitors the oxygen content of the exhaust gas. If the percentage of oxygen in the exhaust gas is incorrect, an electrical signal is sent to the ECM. The ECM takes this information, processes it and then sends a command to the fuel injection system telling it to change the air/fuel mixture. This happens in a fraction of a second and it goes on continuously when the engine is running. The end result is an air/fuel mixture ratio which is constantly maintained at a predetermined ratio, regardless of driving conditions.

9   In the event of a sensor malfunction, a backup circuit will take over to provide driveability until the problem is identified and fixed.

### Precautions

10   When working on the system, follow these steps:

a) *Always disconnect the power by either turning off the ignition switch or disconnecting the battery terminals before unplugging any electrical connectors.* **Warning:** *These models are equipped with airbags. The airbag is armed and can deploy (inflate) anytime the battery is*

2.4 Trouble code tools simplify the task of extracting the trouble codes

6

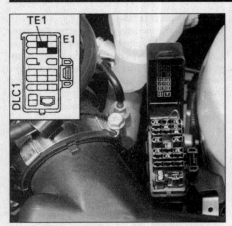

2.15 To access the self diagnosis system on 1995 and earlier models, locate the test connector in the engine compartment and using a jumper wire, bridge terminals TE1 and E1

2.19 On 1996 and later models, the diagnostic connector is typically located under the instrument panel

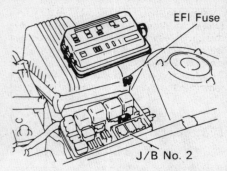

2.20 Typical location of the 15A EFI fuse

*connected. To prevent accidental deployment (and possible injury), turn the ignition key to LOCK and disconnect the negative battery cable whenever working near airbag components. After the battery is disconnected, wait at least two minutes before beginning work. This system has a back-up capacitor that must fully discharge. For more information, see Chapter 12.* **Caution:** *If the stereo in your vehicle is equipped with an anti-theft system, make sure you have the correct activation code before disconnecting the battery.*

b) *When installing a battery, be particularly careful to avoid reversing the positive and negative battery cables.*

c) *Do not subject EFI components, emissions-related components or the ECM to severe impact during removal or installation.*

d) *Do not be careless during troubleshooting. Even slight terminal contact can invalidate a testing procedure and damage one of the numerous transistor circuits.*

e) *Never attempt to work on the ECM or open the ECM cover. While the vehicle is under warranty, the ECM is protected by a government-mandated warranty that will be nullified if you tamper with or damage the ECM.*

f) *If you are inspecting electronic control system components during rainy weather, make sure that water does not enter any part. When washing the engine compartment, do not spray these parts or their electrical connectors with water.*

11   The ECM contains a built-in self-diagnosis system which detects and identifies malfunctions occurring in the network. When the ECM detects a problem, three things happen: the CHECK ENGINE light comes on, the trouble is identified and a diagnostic code is recorded and stored. The ECM stores the

failure code assigned to the specific problem area until the diagnosis system is canceled by removing the EFI fuse with the ignition switch off.

12   The CHECK ENGINE warning light, which is located on the instrument panel, comes on when the ignition switch is turned to ON and the engine is not running. When the engine is started, the warning light should go out. If the light remains on, the self-diagnosis system has detected a malfunction.

## Obtaining diagnostic system trouble codes

**Note:** *1995 and earlier models are equipped with the OBD I self-diagnosis system, while 1996 and later models are equipped with the second generation OBD II self-diagnosis system. 1996 and later models require the use of a scan tool to access trouble codes.*

### 1995 and earlier models
*Refer to illustration 2.15*

13   To obtain an output of diagnostic codes, verify first that the battery voltage is above 11 volts, the throttle is fully closed, the transaxle is in Neutral, the accessory switches are off and the engine is at normal operating temperature.

14   Turn the ignition switch to ON (engine not running). Do not start the engine.

15   Use a jumper wire to bridge terminals TE1 and E1 of the test connector **(see illustration)**. **Note:** *The self-diagnosis system can be accessed by using either test terminal number 1 (engine compartment) or test terminal number 2 (under driver's dash).*

16   Read the diagnosis code as indicated by the number of flashes of the CHECK ENGINE light on the dash. Normal system operation is indicated by Code No. 1 (no malfunctions) for all models. The CHECK ENGINE light displays a Code No. 1 by blinking once every 0.25 seconds. Each code will be displayed by first blinking the first digit of the code, then pause, and blink the second digit of the code. For example; Code 24 (IAT sensor) will flash two times, pause, and then flash four times. Each flash will be the exact same length but the distinction will be the

pause that separates the digits of the code. Only code 1 (normal operation) will flash continuously without a pause.

17   If there are any malfunctions in the system, their corresponding trouble codes are stored in computer memory and the light will blink the requisite number of times for the indicated trouble codes. If there's more than one trouble code in the memory, they'll be displayed in numerical order (from lowest to highest) with a pause between each one. After the code with the largest number of flashes has been displayed, there will be another pause and then the sequence will begin all over again.

18   To ensure correct interpretation of the flashing CHECK ENGINE light, watch carefully for the interval between the end of one code and the beginning of the next; otherwise, you will become confused by the apparent number of flashes and misinterpret the display (the length of this interval varies with the model year).

### 1996 and later models
*Refer to illustration 2.19*

19   1996 and later (OBD-II equipped) vehicles require a special scan tool to obtain the diagnostic trouble codes. The scan tool is programmed to interface with the OBD-II system by plugging into the Data Link Connector (DLC) **(see illustration)**. When used, the scan tool has the ability to diagnose in-depth driveability problems and allows freeze frame data to be retrieved from the PCM stored memory. Freeze frame data is an OBD-II feature that records all related sensor and actuator activity on the PCM data stream whenever an engine control or emissions fault is detected and a DTC is set. This ability to look at the circuit conditions and values when the malfunction occurs provides a valuable tool when trying to diagnose intermittent driveability problems. If the tool is not available and the CHECK ENGINE light is illuminated, have the vehicle checked at a dealer service department or other qualified repair shop.

## Clearing codes
*Refer to illustration 2.20*

20   After the malfunctioning component has been repaired/replaced, the trouble codes stored in ECM memory must be canceled. To

accomplish this, remove the 15A EFI fuse **(see illustration)** for at least 10 seconds with the ignition switch off. **Note:** *On 1996 and later models, the scan tool may be used to clear the diagnostic trouble codes.*
21   A stored code can also be canceled by removing the cable from the negative battery terminal, but other memory systems (such as the clock and radio presets) will also be canceled. **Caution:** *If the stereo in your vehicle is equipped with an anti-theft system, make sure you have the correct activation code before disconnecting the battery.*
22   If the diagnosis code is not canceled, it will be stored by the ECM and appear with any new codes in the event of future trouble.
23   Should it become necessary to work on engine components requiring removal of the battery terminal, always check to see if a diagnostic code has been recorded before disconnecting the battery.

## Diagnostic Trouble Codes - 1995 and earlier models

| Code | Circuit or system | Diagnosis | Trouble area |
|---|---|---|---|
| Code 1 | Normal | The CHECK ENGINE light flashes on and off rapidly when no codes are identified | |
| Code 12 | RPM signal | No rpm signal to the ECM within several seconds after the engine is cranked | Distributor or circuit<br>Crankshaft position sensor or circuit<br>ECM or circuit |
| Code 13 | RPM signal | No rpm signal to the ECM with engine speed above 1,500 rpm | Distributor or circuit<br>Crankshaft position sensor or circuit<br>ECM or circuit |
| Code 14 | Ignition signal | No ignition signal to the ECM | Igniter or circuit<br>Ignition coil<br>Ignition switch or circuit<br>ECM |
| Code 16 | ECM control signal | Problem between the engine controls and transaxle controls inside the ECM | ECM |
| Code 21 | Main oxygen sensor | Problem in the main oxygen sensor circuit | Main oxygen sensor or circuit<br>ECM |
| Code 22 | Coolant temperature | Open or short in the coolant temperature sensor circuit | Coolant temperature sensor or circuit<br>ECM |
| Code 24 | Intake air temperature sensor | Open or short in the intake air temperature sensor circuit | Intake air temperature sensor or circuit<br>ECM |
| Code 25 | Oxygen sensor or circuit | An excessively lean air/fuel ratio has been indicated by the oxygen sensor circuit | Injector or circuit<br>Oxygen sensor or circuit<br>ECM<br>Fuel pressure regulator<br>Coolant temperature sensor or circuit<br>Intake air temperature sensor or circuit<br>Vacuum or exhaust leak<br>Contaminated fuel<br>Ignition system |
| Code 26 | Oxygen sensor or circuit | An overly rich air/fuel ratio has been indicated by the oxygen sensor circuit | Injector or injector circuit<br>Coolant temperature sensor or circuit<br>Oxygen sensor or circuit<br>Intake air temperature sensor or circuit<br>Fuel pressure regulator<br>EVAP system<br>EGR system<br>MAP sensor or circuit<br>ECM<br>Air intake system |
| Code 27 | Post-converter oxygen sensor | Open or shorted circuit in the post-converter oxygen sensor circuit | Post-converter oxygen sensor or circuit<br>ECM |

**6**

## Diagnostic Trouble Codes - 1995 and earlier models

| Code | Circuit or system | Diagnosis | Trouble area |
|------|-------------------|-----------|--------------|
| Code 31 | MAP sensor | Open or short in MAP sensor circuit | MAP sensor or circuit<br>ECM |
| Code 41 | Throttle position sensor | Open or short in the throttle position sensor circuit | Throttle position sensor or circuit<br>ECM |
| Code 42 | Vehicle speed sensor | No speed signal for 8 seconds when the engine speed is between 3,000 and 5,000 rpm and the transaxle is in gear | Vehicle speed sensor or circuit<br>ECM<br>Speedometer<br>Instrument panel printed circuit |
| Code 43 | Starter signal | No starter signal to the ECM until engine speed reaches 800 rpm with the vehicle not moving | Starter signal circuit<br>Ignition switch<br>ECM |
| Code 51 | Switch condition signal | No throttle position signal, gear selector signal or air conditioning signal to the ECM | Air conditioning switch or circuit<br>Air conditioning amplifier<br>Neutral Start switch<br>Throttle Position sensor<br>ECM |
| Code 52 | Knock sensor signal | Open or short circuit in knock sensor circuit | Knock sensor or circuit<br>ECM |
| Code 71 | EGR system | EGR temperature signal is too low | EGR system (EGR valve, hoses, etc.)<br>EGR temperature sensor or circuit<br>EGR vacuum switching valve<br>ECM |

## Diagnostic Trouble Codes - 1996 and later models

| Trouble code | Code identification |
|--------------|---------------------|
| P0100 | Mass airflow sensor or circuit fault |
| P0101 | Mass airflow sensor range or performance problem |
| P0105 | Manifold absolute pressure sensor or circuit fault |
| P0106 | Manifold absolute pressure range or performance problem |
| P0110 | Intake air temperature sensor or circuit fault |
| P0115 | Engine coolant temperature sensor or circuit fault |
| P0116 | Engine coolant temperature sensor range or performance problem |
| P0120 | Throttle position sensor or circuit fault |
| P0121 | Throttle Position sensor range or performance problem |
| P0125 | Insufficient coolant temperature for closed loop; oxygen sensor heater malfunction |
| P0128 | Thermostat malfunction |
| P0130 | Pre-converter oxygen sensor or circuit fault |
| P0133 | Pre-converter oxygen sensor circuit slow response fault |

| Trouble code | Code identification |
|---|---|
| P0135 | Pre-converter oxygen sensor heater fault |
| P0136 | Post-converter oxygen sensor or circuit fault |
| P0141 | Post-converter oxygen sensor heater or circuit fault |
| P0171 | Fuel injection system lean |
| P0172 | Fuel injection system rich |
| P0300 | Multiple cylinder misfire detected |
| P0301 | Cylinder no. 1 misfire detected |
| P0302 | Cylinder no. 2 misfire detected |
| P0303 | Cylinder no. 3 misfire detected |
| P0304 | Cylinder no. 4 misfire detected |
| P0325 | Knock sensor or circuit fault |
| P0335 | Crankshaft position sensor or circuit fault |
| P0340 | Camshaft position sensor or circuit fault |
| P0401 | EGR insufficient flow detected |
| P0402 | EGR excessive flow detected |
| P0420 | Catalyst system fault |
| P0440 | EVAP system malfunction |
| P0441 | EVAP system incorrect purge flow detected |
| P0442 | EVAP system leak detected |
| P0446 | EVAP canister vent control valve circuit fault |
| P0450 | EVAP system pressure sensor or circuit fault |
| P0451 | EVAP system pressure sensor range or performance problem |
| P0500 | Vehicle speed sensor or circuit fault |
| P0505 | Idle air control valve or circuit fault |
| P0751 | Automatic transaxle shift solenoid no.1 stuck open or closed |
| P0753 | Automatic transaxle shift solenoid no.1 circuit malfunction |
| P0755 | Automatic transaxle shift solenoid no.2 stuck open or closed |
| P0758 | Automatic transaxle shift solenoid no.2 circuit malfunction |
| P0770 | Automatic transaxle shift solenoid valve SL stuck open or closed |
| P0773 | Automatic transaxle shift solenoid valve SL circuit malfunction |

| Trouble code | Code identification |
|---|---|
| P1300 (1998 and 1999) | Ignition system malfunction (igniter circuit fault) |
| P1300 (2000 and 2001) | Igniter no. 1 circuit fault |
| P1305 | Igniter no. 2 circuit fault |
| P1310 | Igniter no. 3 circuit fault |
| P1315 | Igniter no. 4 circuit fault |
| P1335 | Crankshaft position sensor or circuit fault |
| P1346 | VVT sensor circuit fault |
| P1349 | VVT system malfunction |
| P1500 | Starter signal circuit malfunction |
| P1520 | Brake light signal malfunction |
| P1600 | ECM battery supply malfunction |
| P1656 | OCV circuit malfunction |
| P1780 | Park/Neutral position switch or circuit fault |

## 3   Engine Control Module (ECM) - removal and installation

*Refer to illustration 3.4*
**Warning:** *On models equipped with a Supplemental Restraint System (SRS) (more commonly known as airbags), always disable the airbag system before working in the vicinity of the airbag system components to avoid the possibility of accidental deployment of the airbag, which could cause personal injury (see Chapter 12).*
**Caution:** *Static electricity can damage the ECM. Be sure to ground yourself to the vehicle body before touching the ECM or the electrical connector. Store the ECM on an anti-static pad once it is removed.*
**Note:** *Anytime the battery is disconnected, stored operating parameters may be lost from the ECM causing the engine to run rough for sometime while the ECM relearns the information.*
1    Disconnect the negative cable from the battery (see Chapter 5). **Caution:** *If the stereo in your vehicle is equipped with an anti-theft system, make sure you have the correct activation code before disconnecting the battery.*
2    Remove the lower finish panel on the passenger side under the glove compartment (see Chapter 11).
3    Remove the center console from the passenger compartment (see Chapter 11).
4    Disconnect the electrical connectors from the ECM **(see illustration)**. Each connector has a locking tab which must be dis-

**3.4  Carefully disconnect the electrical connectors (arrows) from the ECM**

engaged before the connector is unplugged.
5    Remove the bolts from the ECM brackets.
6    Remove the ECM from the instrument panel frame.
7    Installation is the reverse of removal.

## 4   Information sensors

## *Coolant temperature sensor*
### General description
*Refer to illustration 4.1*
1    The coolant temperature sensor is a

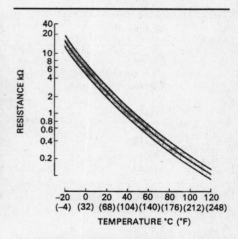

**4.1  Compare the indicated resistance values specified on this graph - note that as the temperature increases (as the engine warms up) the resistance decreases**

thermistor (a resistor which varies the value of its voltage output in accordance with temperature changes). As the sensor temperature DECREASES, the resistance values will INCREASE. As the sensor temperature INCREASES, the resistance values will DECREASE **(see illustration)**.

### Check
*Refer to illustrations 4.2a, 4.2b and 4.3*
2    To check the sensor, disconnect the

**4.2a  To check the coolant temperature sensor, use an ohmmeter to measure the resistance across the two sensor terminals (1997 and later models)**

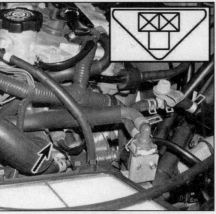

**4.2b  On 1998 and later models, disconnect the electrical connector from the engine coolant temperature sensor (arrow) and measure the resistance across the indicated terminals of the sensor**

**4.3  Use a voltmeter and probe the coolant temperature sensor connector for reference voltage with the ignition key ON (engine not running) - it should be approximately 5.0 volts**

electrical connector and measure the resistance across the sensor terminals **(see illustrations)**. With the engine completely cold (68-degrees F) the resistance should be 2,000 to 3,000 ohms. Next, start the engine and warm it up until it reaches operating temperature (180-degrees F) - the resistance should be 200 to 400 ohms. **Note:** *If necessary, remove the sensor and perform the tests in a pan of heated water to simulate the conditions. Compare the resistance values with the accompanying graph* **(see illustration 4.1)**.

3    If the resistance values of the coolant temperature sensor are correct, check the circuit for the proper signal voltage. Turn the ignition key ON (engine not running) and check for reference voltage **(see illustration)**. It should be approximately 5 volts.

### Replacement

4    Partially drain the cooling system (see Chapter 1).

5    Disconnect the electrical connector, then carefully unscrew the sensor.

6    Before installing the new sensor, wrap the threads with Teflon sealing tape to prevent leakage and thread corrosion. Installation is the reverse of removal.

## *Oxygen sensor*

### General description

*Refer to illustrations 4.7a and 4.7b*

7    The models covered by this manual are equipped with either a single oxygen sensor system or a dual-stage oxygen sensor system. Prior to 1996, most models were equipped with a single sensor and only California emissions models were equipped with the dual sensor system. On 1996 and later models, all are equipped with two heated oxygen sensors **(see illustrations)**. On dual-stage systems, the main (pre-converter) oxygen sensor is mounted ahead of the front catalytic converter and monitors the exhaust gases exiting the engine. The post-converter oxygen sensor monitors the exhaust gases after they have passed through the catalytic

converter. The oxygen sensor monitors the oxygen content of the exhaust gas stream. The oxygen content in the exhaust reacts with the oxygen sensor to produce a voltage output which varies from 0.1-volt (high oxygen, lean mixture) to 0.9-volts (low oxygen, rich mixture). The ECM constantly monitors this variable voltage output to determine the ratio of oxygen to fuel in the mixture. The ECM alters the air/fuel mixture ratio by controlling the pulse width (open time) of the fuel injectors. A mixture ratio of 14.7 parts air to 1 part fuel is the ideal mixture ratio for minimizing exhaust emissions, thus allowing the catalytic converter to operate at maximum efficiency. It is this ratio of 14.7 to 1 which the ECM and the oxygen sensor attempt to maintain at all times.

8    The post-converter oxygen sensor in the exhaust system has no effect on ECM control of the air/fuel ratio. This sensor is identical to the pre-converter sensor and operates in the same way. The ECM uses the post-converter signal, however, for the catalyst monitor system. A post-converter oxygen sensor will produce a slower fluctuating voltage signal that reflects the lower oxygen content in the post-catalyst exhaust.

9    The oxygen sensor produces no voltage when it is below its normal operating temperature of about 600-degrees F. During this initial period before warm-up, the ECM operates in open loop mode. If the engine reaches normal operating temperature and/or has been running for two or more minutes, and if the main oxygen sensor is producing a steady signal voltage below 0.70-volts at 1,500 or more rpm, the ECM will set a Diagnostic Trouble Code.

10    When there is a problem with the oxygen sensor or its circuit, the ECM operates in the open loop mode - that is, it controls fuel delivery in accordance with a programmed default value instead of feedback information from the oxygen sensor.

11    The proper operation of the oxygen sen-

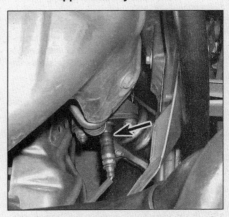

**4.7a  The pre-converter oxygen sensor is located in the exhaust pipe near the exhaust manifold**

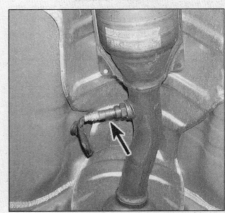

**4.7b  The post-converter oxygen sensor is located in the exhaust pipe after the catalytic converter**

sor depends on four conditions:

a)  *Electrical - The low voltages generated by the sensor depend upon good, clean connections which should be checked whenever a malfunction of the sensor is suspected or indicated.*

**6**

**4.13 Insert a pin into the backside of the oxygen sensor connector on the correct terminal and check for a millivolt output signal generated by the sensor**

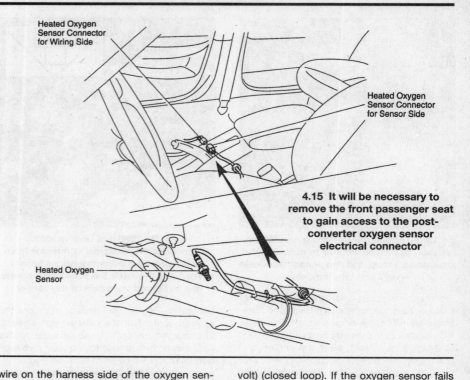

Heated Oxygen Sensor Connector for Wiring Side

Heated Oxygen Sensor Connector for Sensor Side

Heated Oxygen Sensor

**4.15 It will be necessary to remove the front passenger seat to gain access to the post-converter oxygen sensor electrical connector**

b) *Outside air supply - The sensor is designed to allow air circulation to the internal portion of the sensor. Whenever the sensor is removed and installed or replaced, make sure the air passages are not restricted.*

c) *Proper operating temperature - The ECM will not react to the sensor signal until the sensor reaches approximately 600-degrees F. This factor must be taken into consideration when evaluating the performance of the sensor.*

d) *Unleaded fuel - The use of unleaded fuel is essential for proper operation of the sensor. Make sure the fuel you are using is of this type.*

12    In addition to observing the above conditions, special care must be taken whenever the sensor is serviced.

a) *The oxygen sensor has a permanently attached pigtail and electrical connector which should not be removed from the sensor. Damage to or removal of the pigtail or electrical connector can adversely affect operation of the sensor.*

b) *Grease, dirt and other contaminants should be kept away from the electrical connector and the louvered end of the sensor.*

c) *Do not use cleaning solvents of any kind on the oxygen sensor.*

d) *Do not drop or roughly handle the sensor.*

e) *The silicone boot must be installed in the correct position to prevent the boot from being melted and to allow the sensor to operate properly.*

## Check

*Refer to illustrations 4.13, 4.15, 4.16, 4.17, 4.18 and 4.19*

13    To check the oxygen sensor use a digital voltmeter to monitor the millivolt signal from the oxygen sensor during actual operating conditions. Locate the oxygen sensor electrical connector and backprobe the black

wire on the harness side of the oxygen sensor connector **(see illustration)**. To properly backprobe the connector insert a long straight pin (a T-pin is preferred) alongside the wire until the pin contacts the metal wire terminal inside the connector. Connect the positive probe of a voltmeter onto the pin and the negative probe to ground.

14    Start the engine and monitor the voltage signal of the main (pre-converter) oxygen sensor as the engine warms-up. The oxygen sensor will produce a steady voltage signal at first (open loop) of approximately 100 to 200 millivolts (0.1 to 0.2 volt) with the engine cold. After a period of approximately two minutes, the engine will reach operating temperature and the oxygen sensor voltage will fluctuate between 100 and 900 millivolts (0.1 to 0.9

volt) (closed loop). If the oxygen sensor fails to operate as described, replace it.

15    To check the post-converter oxygen sensor, locate the electrical connector **(see illustration)** and check it in the same manner as the main oxygen sensor. The voltage signal from a post-converter oxygen sensor should also read between 100 to 900 millivolts but it should not switch actively. The post-converter sensor voltage may stay toward the center of its range (approximately 400 millivolts) or remain at the upper or lower limits for a relatively longer period of time.

16    Also check the oxygen sensor heater (if equipped) as follows: Disconnect the oxygen sensor electrical connector and connect an ohmmeter between the +B and HT terminals on the oxygen sensor side of the connector

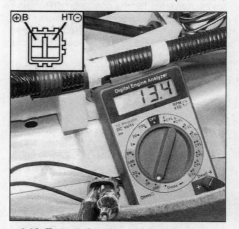

**4.16 To test the oxygen sensor heater, disconnect the harness connector and check the resistance across terminals HT and +B of the oxygen sensor connector - resistance should be 11 to 17 ohms**

**4.17 Check for the voltage to the oxygen sensor heater on the black/red or black wire (+) and the blue/black or pink wire (-) - it should be approximately 12 volts (battery voltage)**

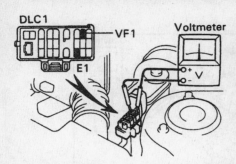

4.18 Connect the probes of the voltmeter to terminals VF1 (+) and E1 (-), raise the engine speed to 2,500 rpm and jump terminals TE1 and E1 with a jumper wire or paper clip

4.19 With terminals TE1 and E1 of the test connector jumpered, observe the number of needle sweeps in a 10-second time period

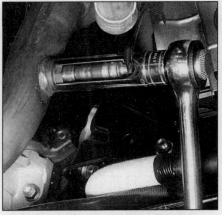

4.26 Slotted sockets are available for easing oxygen sensor removal

(see illustration). It should measure approximately 11.0 to 17.0 ohms. **Note:** *Not all models are equipped with a heated oxygen sensor. Models with heated oxygen sensors will be equipped with a four-wire electrical connector.*

17 Check for proper supply voltage to the oxygen sensor heater. With the ignition key ON (engine not running), check for battery voltage at the black/red (+) (1997 and earlier) or black (+) (1998 and later) and the blue/black wire (-) (1997 and earlier) or pink (-) (1998 and later) wire on the harness side of the connector **(see illustration).**

### 1993 and 1994 models

18 On 1993 and 1994 models, it's also possible to check the oxygen sensor in another manner. With the engine completely warmed up and the oxygen sensor connected, connect a voltmeter to the test connector VF1 (positive probe +) and E1 (negative probe -) **(see illustration). Note:** *Use only an analog type voltmeter because it will be necessary to watch the needle fluctuations.*

19 Run the engine at 2,500 rpm for approximately two minutes and then jump terminals TE1 and E1 of the test connector **(see illustration).** Check the number of times the needle fluctuates in 10 seconds. It should fluctuate eight times or more. If it does not, warm the engine up again and repeat the test.

20 If the voltmeter still does not fluctuate eight times or more, remove the jumper wire from terminals TE1 and E1 of the test connector. Maintain engine speed at 2,500 rpm and measure the voltage between terminals VF1 and E1. If the voltage reading is more than 0 volts, replace the oxygen sensor with a new part. If the voltage reading is 0 volts, access the self diagnostic codes (see Section 2) and check for any malfunctions.

21 If codes 21, 25 or 26 are obtained, then remove the PCV hose from the valve cover (see Section 8) and measure the voltage between VF1 and E1. If the voltage is 0 volts, replace the oxygen sensor. If the voltage reading is more than 0 volts, repair the over-rich running condition.

22 If codes other than 21, 25 or 26 are

obtained, repair the particular sensor or circuit.

### Replacement

*Refer to illustration 4.26*

**Caution:** *The oxygen sensor may be very difficult to loosen when the engine is hot and excessive force may damage the threads. Allow the engine to cool completely before performing this procedure.*

23 Disconnect the cable from the negative terminal of the battery. **Caution:** *If the stereo in your vehicle is equipped with an anti-theft system, make sure you have the correct activation code before disconnecting the battery.*

24 Raise the vehicle and place it securely on jackstands.

25 Carefully disconnect the electrical connector from the sensor pigtail lead. If removing the post-converter oxygen sensor, remove the passenger seat (1997 and earlier) or center console (1998 and later) to access the electrical connector.

26 Remove the oxygen sensor from the exhaust system **(see illustration). Note:** *Some oxygen sensors are threaded directly into the exhaust manifold while others are mounted in the exhaust manifold or pipe with two bolts.*

27 Anti-seize compound must be used on the threads of the sensor to facilitate future

removal. The threads of new sensors will already be coated with this compound, but if an old sensor is removed and reinstalled, recoat the threads. Install the sensor and tighten it securely. Reconnect the electrical connector of the pigtail lead to the main engine wiring harness.

28 Lower the vehicle and reconnect the cable to the negative terminal of the battery.

## Throttle Position Sensor (TPS)

### General description

29 The Throttle Position Sensor (TPS) is located on the end of the throttle shaft on the throttle body (see Chapter 4). By monitoring the output voltage from the TPS, the ECM determines fuel delivery based on throttle valve angle (driver demand). A broken or loose TPS can cause intermittent bursts of fuel from the injector and an unstable idle because the ECM thinks the throttle is moving.

### Check

*Refer to illustrations 4.31a, 4.31b and 4.34*

### 1997 and earlier models

30 Disconnect the electrical connector from the TPS. On California models, apply vacuum to the throttle positioner.

31 Insert a feeler gauge of the specified thickness between the throttle stop screw and the stop lever. Using an ohmmeter, measure the resistance between the indicated

**6**

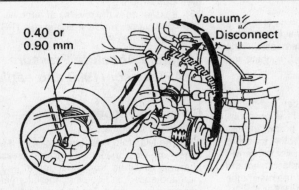

4.31a To check the TPS on 1997 and earlier models, insert a feeler gauge of the specified thickness between the throttle stop screw and the stop lever, then measure the resistance between the proper terminals

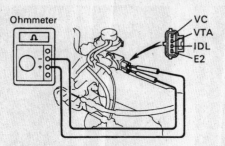

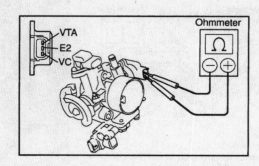

| Clearance between lever and stop screw | Between terminals | Resistance |
|---|---|---|
| 0 mm (0 in.) | VTA − E2 | 0.2 − 6.0 kΩ |
| 0.40 mm (0.016 in.) | IDL − E2 | 2.3 kΩ or less |
| 0.90 mm (0.035 in.) | IDL − E2 | Infinity |
| Throttle valve fully open | VTA − E2 | 3.3 − 10.0 kΩ |
| − | VC − E2 | 4.0 − 8.5 kΩ |

**4.31b Throttle Position Sensor terminal guide and continuity table - 1997 and earlier models**

| Clearance between lever and stop screw | Between terminals | Resistance |
|---|---|---|
| 0 mm (0 in.) | VTA − E2 | 0.2 − 5.7 kΩ |
| Throttle valve fully open | VTA − E2 | 2.0 − 10.2 kΩ |
| − | VC − E2 | 2.5 − 5.9 kΩ |

**4.34 Throttle Position Sensor terminal guide and continuity table - 1998 and later models**

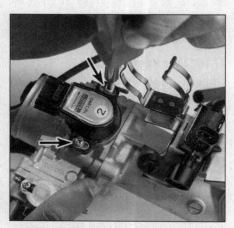

**4.35 Remove the two screws (arrows) and remove the TPS from the throttle body**

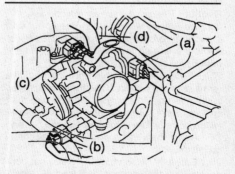

**4.37 Throttle body components - 1998 and 1999 models**

a  *Throttle Position Sensor*
b  *Idle Air Control valve*
c  *Manifold Absolute Pressure sensor*
d  *PCV hose*

**4.38 Check for reference voltage on the MAP sensor connector yellow wire terminal - it should be approximately 5.0 volts**

terminals on the TPS **(see illustrations)**.

32   If the resistance is not as specified, insert a 0.90 mm (0.035 in.) feeler gauge between the throttle stop screw and lever, and connect the ohmmeter to terminals IDL and E2. Loosen the TPS mounting screws and slowly rotate the sensor clockwise until the ohmmeter reads infinity.

33   Tighten the mounting screws, and using the proper feeler gauge, recheck the resistance between the specified terminals. If the resistance is not as specified, replace the TPS.

### 1998 and later models

34   Disconnect the electrical connector from the TPS. Make sure there is no clearance between the throttle top screw and the throttle lever. Using an ohmmeter, measure the resistance between the indicated terminals on the TPS with the throttle closed and fully open **(see illustration)**. If the resistance is not as specified, replace the TPS.

### Replacement

*Refer to illustration 4.35*

35   Disconnect the electrical connector from the TPS. Remove the mounting screws and remove the TPS from the throttle body **(see illustration)**.

36   Installation is the reverse of removal. On 1997 and earlier models, adjust the TPS as described in Step 32.

## Manifold Absolute Pressure (MAP) sensor (1999 and earlier models)

### General Information

*Refer to illustration 4.37*

37   The Manifold Absolute Pressure (MAP) sensor monitors the intake manifold pressure changes resulting from changes in engine load and speed and converts the information into a voltage output. The ECM uses the MAP

sensor to control fuel delivery and ignition timing. The ECM will receive information as a voltage signal. This signal can be detected using a voltmeter. Under ideal conditions, the voltage will vary from 4.0 volts with the engine off (no vacuum) to 0.5 volts with the engine idling (25 in-Hg vacuum). On 1997 and earlier models the MAP sensor is mounted on the firewall **(see illustration 1.1a)**. On 1998 and 1999 models, the MAP sensor is located on the throttle body **(see illustration)**.

### Check

*Refer to illustrations 4.38, 4.39 4.40a and 4.40b*

38   Disconnect the electrical connector from the MAP sensor. Turn the ignition On (do not start the engine) and check the reference voltage from the MAP sensor on the yellow wire terminal. It should be approximately

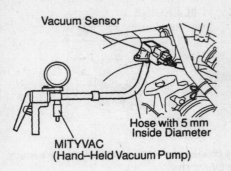

4.39  To test the MAP sensor on 1998 and 1999 models, remove the sensor from the throttle body and connect a vacuum pump to the fitting on the sensor

4.40a  Install a vacuum pump to the MAP sensor and check for signal voltage on the light green/red (+) wire and brown wire (-) without vacuum applied. It should be approximately 3.0 to 4.0 volts

4.40b  Now apply vacuum to the MAP sensor an observe that the voltage decreases to approximately 0.5 to 1.5 volts

5.0 volts (see illustration).

39  On 1997 and earlier models, disconnect the vacuum hose from the MAP sensor. On 1998 and 1999 models, remove the sensor from the throttle body (see illustration). Install a hand-held vacuum pump to the fitting on the sensor. Reconnect the electrical connector to the MAP sensor.

40  Using a voltmeter and the appropriate probes, backprobe the MAP sensor harness connector and connect the voltmeter to the light green/red wire (+) and brown wire (-)

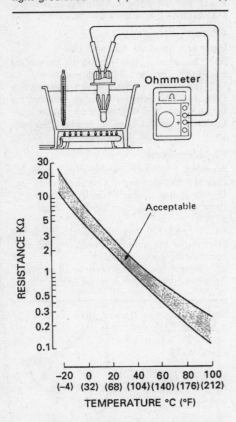

4.42  The air intake temperature sensor resistance will DECREASE when the temperature of the air INCREASES

(see illustration). Note: *Refer to Chapter 12 for additional information on how to back-probe a connector.* Turn the ignition key On. Without vacuum, the sensor voltage should be approximately 3.0 to 4.0 volts. Using the hand-held vacuum pump, apply 15 to 25 in-Hg of vacuum to the MAP sensor and observe the voltage readings. The voltage should decrease to approximately 0.5 to 1.5 volts (see illustration). If the test results are incorrect, replace the MAP sensor.

### Replacement

41  Disconnect the electrical connector from the MAP sensor. On 1997 and earlier models, disconnect the vacuum hose from the sensor. Remove the mounting screws and the sensor. Installation is the reverse of removal.

## Intake Air Temperature (IAT) sensor (1999 and earlier models)

### General description

*Refer to illustration 4.42*

Note: *On 2000 and later models, the IAT sensor is incorporated into the MAF sensor.*

42  The intake air temperature sensor is

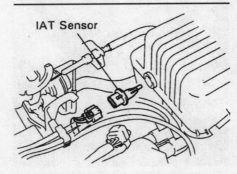

4.43  Location of the IAT sensor

located on the air filter housing. This sensor is a resistor which changes value according to the temperature of the air entering the engine. Low temperatures produce a high resistance value (for example, at 68-degrees F the resistance is 2,000 to 3,000 ohms) while high temperatures produce low resistance values (at 176-degrees F the resistance is 200 to 400 ohms (see illustration). The ECM supplies approximately 5-volts (reference voltage) to the air temperature sensor. The IAT sensor alters the voltage according to the temperature of the incoming air. The signal voltage sent back to the ECM will be high when the air temperature is cold and low when the air temperature is warm.

### Check

*Refer to illustration 4.43, 4.44 and 4.45*

43  Disconnect the electrical connector from the air temperature sensor (see illustration). Turn the ignition key ON, but do not start the engine.

44  Measure the voltage (reference voltage) across the two terminals in the harness connector. The voltage should read approximately 5-volts (see illustration). If the refer-

4.44  Check the IAT sensor reference voltage on the yellow/black (+) wire

**4.45  To check the IAT sensor, use an ohmmeter to measure the resistance across the two sensor terminals**

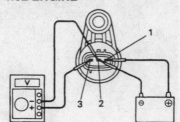

**1.6L ENGINE**

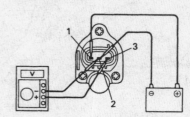

**1.8L ENGINE**

**4.50  Check the VSS output with a voltmeter by connecting a 12-volt battery to the sensor and rotating the pinion shaft**

ence voltage is not correct, have the ECM diagnosed by a dealer service department or other repair shop.

45   Measure the resistance across the air temperature sensor terminals **(see illustration)**. The resistance should be HIGH when the air temperature is LOW. Next, start the engine and let it idle. Wait awhile and let the engine reach operating temperature. Turn the ignition OFF, disconnect the air temperature sensor and measure the resistance across the terminals. The resistance should be LOW when the air temperature is HIGH. Compare your measurements with the resistance chart **(see illustration 4.42)** If the sensor resistance is incorrect, replace the sensor.

### Replacement

46   Disconnect the electrical connector from the sensor. Remove the air filter cover and remove the sensor from the cover. Installation is the reverse of removal.

## *Vehicle Speed Sensor (VSS)*

### General description

47   The Vehicle Speed Sensor (VSS) is located on the output section of the

transaxle. The sensor sends a pulsing voltage signal to the speedometer, which is converted into miles per hour. The ECM monitors the VSS signal for various engine control operations.

### Check

*Refer to illustration 4.50*

48   Disconnect the electrical connector from the VSS. Connect a voltmeter to the two outside terminals of the harness connector. Turn the ignition key On, battery voltage should be indicated on the meter. If it isn't, check the gauge fuse, the related circuits and the instrument cluster ground.

49   Remove the VSS from the transaxle. Inspect the pinion gear for damage and replace it if necessary.

50   Using jumper wires, connect a 12-volt battery to terminals 1 and 2 of the VSS **(see illustration)**. Connect a voltmeter to terminals 2 and 3. Rotate the pinion shaft while watching the voltmeter; the voltage should fluctuate from zero to 11 volts approximately four times per revolution of the pinion shaft. If the sensor fails to operate as described, replace the sensor.

### Replacement

*Refer to illustration 4.51*

51   Disconnect the electrical connector from the sensor **(see illustration)**. Remove the mounting bolts and withdraw the sensor from the transaxle. Replace the O-ring and

lubricate it with transmission fluid. Installation is the reverse of removal.

## *Knock sensor*

### General Description

52   The knock sensor detects abnormal vibration (spark knock or pinging) in the engine. The knock control system is designed to reduce spark knock during periods of heavy detonation. This allows the engine to use maximum spark advance to improve driveability. The knock sensor produces an AC output voltage which increases with the severity of the knock. The signal is fed into the PCM and the timing is retarded to compensate for the severe detonation. The knock sensor is located on the side of the engine block below the intake manifold.

### Check

*Refer to illustrations 4.53 and 4.54*

53   Disconnect the electrical connector from the knock sensor **(see illustration)**.

54   Using an ohmmeter, check that there is no continuity between the terminal on the knock sensor and the sensor body **(see illustration)**. If continuity exists, replace the sensor.

### Replacement

**Warning:** *Wait until the engine is completely cool before beginning this procedure.*

55   Drain the cooling system (see Chapter 1). Remove the intake manifold (see Chapter 2A or 2B).

56   Disconnect the electrical connector from the knock sensor and remove the sensor from the engine block.

57   Installation is the reverse of removal. Refill the cooling system (see Chapter 1).

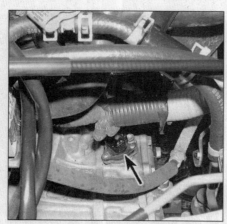

**4.51  The VSS (arrow) is located on the transaxle**

**4.53  The knock sensor (arrow) is located under the intake manifold**

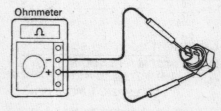

**4.54  Check that NO continuity exists between the sensor terminal and the body of the sensor**

4.58a  Location of the crankshaft position sensor (arrow) - 1997 and earlier models

4.58b  Location of the crankshaft position sensor (arrow) - 1998 and later models

4.60a  Check the resistance of the crankshaft sensor on the electrical connector (1997 and earlier model shown)

## Crankshaft Position Sensor (1995 1.8L models and all 1996 and later models)

### General Description

*Refer to illustrations 4.58a and 4.58b*

58   The crankshaft position sensor sends a signal to the ECM to indicate the exact position (angle) of the crankshaft. The ECM uses this signal to determine engine speed (rpm), control ignition timing and fuel injection synchronization. The crankshaft position sensor is located at the front of the engine near the crankshaft pulley **(see illustrations)**.

### Check

*Refer to illustrations 4.60a and 4.60b*

59   Disconnect the electrical connector from the sensor.
60   Using an ohmmeter, measure the resistance of the crankshaft position sensor **(see illustrations)**. It should be between 1,630 to 3,225 ohms depending on the temperature; the warmer the temperature of the sensor, the higher the resistance value. If the resistance is not within the specified range, replace the sensor.

### Replacement

61   On 1997 and earlier models, remove the crankshaft pulley (see Chapter 2A).
62   On 1998 and later models, remove the intake manifold (see Chapter 2B).
63   Disconnect the electrical connector from the sensor, remove the mounting bolt and withdraw the crankshaft position sensor from the engine front cover.
64   Installation is the reverse of removal.

## Camshaft position sensor (1998 and later models)

### General description

*Refer to illustration 4.65*

65   The camshaft position sensor sends a signal to the ECM to indicate the exact position of the camshaft. The ECM uses this signal to fine-tune ignition timing and fuel injec-

4.60b  Crankshaft position sensor electrical connector location (arrow) - 1998 and later models

tion synchronization. The camshaft position sensor is located at the rear of the cylinder head near the number four fuel injector **(see illustration)**.

### Check

66   Disconnect the electrical connector from the camshaft position sensor. Using an ohmmeter, measure the resistance across the two terminals of the sensor. Resistance should be 835 to 1,645 ohms depending on the temperature; the warmer the temperature of the sensor, the higher the resistance value. If the resistance is not within the specified range, replace the sensor.

### Replacement

67   Disconnect the electrical connector from the sensor. Remove the mounting bolt and withdraw the sensor from the cylinder head. Installation is the reverse of removal.

## Mass Airflow (MAF) sensor (2000 and later models)

### General description

68   The Mass Airflow (MAF) sensor is installed in the air intake duct. This sensor

4.65  Camshaft position sensor location (arrow) - 1998 and later models

uses a hot-wire sensing element to measure the molecular mass (or weight) of air entering the engine. The air passing over the hot wire causes it to cool, and the sensor converts this temperature change into a voltage signal to the ECM. The ECM in turn calculates the required fuel injector pulse width to obtain the necessary air/fuel ratio. A defective MAF sensor can cause surging, stalling, rough idle and other driveability problems.

### Check

*Refer to illustrations 4.70 and 4.71*

69   Before checking the MAF sensor operation, check the power and ground circuits to the MAF sensor. Disconnect the MAF sensor electrical connector and connect the positive lead of a voltmeter to the black wire terminal and the negative lead to the brown wire terminal of the harness connector. Turn the ignition On but do not start the engine. The meter should indicate approximately battery voltage. If battery voltage is not present, check the EFI relay and related circuits (see Chapter 12 and the wiring diagrams). Check for continuity to ground on the harness connector brown wire terminal. If continuity is not indicated, check the ground circuit.

6

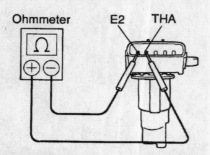

**4.70 On 2000 and later models, measure the resistance across the indicated terminals on the MAF sensor**

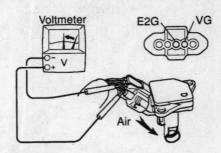

**4.71 Backprobe the indicated terminals with a voltmeter to check the MAF sensor operation**

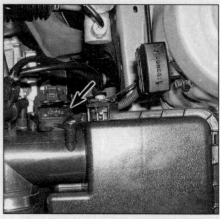

**4.74 Mass Airflow (MAF) sensor location (arrow) - 2000 and later models**

70  Disconnect the electrical connector from the MAF sensor. Using an ohmmeter, measure the resistance across the indicated terminals of the MAF sensor **(see illustration)**. Compare your measurement with the intake air temperature sensor resistance chart **(see illustration 4.42)**. The resistance should vary depending on temperature. If the resistance is not as specified, replace the MAF sensor.

71  Reconnect the electrical connector to the MAF sensor. Using suitable probes, backprobe the MAF sensor connector and connect a voltmeter to the indicated terminals **(see Illustration)**. **Note:** *See Step 13 for additional information on how to backprobe a connector.* Turn the ignition key On.

72  Start the engine and note the voltage at idle. Increase the engine rpm. The MAF signal voltage should fluctuate. It is impossible to simulate driving conditions in the driveway, but it is necessary to watch the voltmeter for a fluctuation in signal voltage as the engine speed is raised and lowered. The engine is not under load, but signal voltage should vary slightly. If the voltage does not fluctuate, replace the MAF sensor.

73  If the voltage readings fluctuate, the MAF sensor is operating properly. Refer to the wiring diagrams and check the wiring harness for open circuits or a damaged harness. If the circuits are good, have the ECM diagnosed by a dealer service department or other qualified repair shop. **Note:** *If MAF related driveability problems continue but these general tests don't indicate a MAF fault, have the sensor tested by a dealer service department or other qualified repair shop. A MAF sensor can develop voltage signal problems that can't be seen on a voltmeter. The ECM can see such signal faults and a driveability problem will result.*

### Replacement

*Refer to illustration 4.74*

74  Disconnect the electrical connector from the MAF sensor, remove the two mounting screws and withdraw the sensor from the air filter housing **(see illustration)**.

75  Replace the O-ring and install the MAF sensor.

## 5  Idle control system

### General description

1  Engine idle speed is controlled by the Idle Air Control (IAC) valve and the ECM. The Idle Air Control (IAC) valve controls the amount of air that bypasses the throttle valve; i.e. more air, higher idle speed. The IAC valve is mounted on the throttle body and is controlled by the ECM. The ECM adjusts the idle speed depending upon the running conditions of the engine (cold, hot, load, no-load, etc.).

2  The minimum idle speed is pre-set at the factory and should not require adjustment under normal operating conditions; however if the throttle body has been replaced or you suspect the minimum idle speed has been tampered with (for example, if the idle speed screw was removed from the throttle body) have the vehicle checked by a dealer service department or a qualified automotive repair shop.

3  Some 1997 and earlier models may be equipped with a vacuum operated throttle opener and/or an air conditioning idle-up valve. The throttle opener opens the throttle plate slightly when the engine is started (no vacuum), when the engine starts and vacuum is applied to the throttle opener the throttle plate return to its normal setting. The air conditioning idle-up valve is energized by the

ECM when the air conditioning is switched on. When the valve opens, additional air is allowed to enter the intake manifold raising the idle speed slightly to allow for the additional load of the air conditioning compressor.

### Idle Air Control (IAC) valve
#### Check

*Refer to illustrations 5.4, 5.5, 5.8 and 5.9*

4  Apply the parking brake, shift the transaxle to Neutral (manual) or Park (automatic) and block the drive wheels. Connect a tachometer according to the tool manufacturer's instructions. Install the lead of the tachometer to the IG terminal on the test connector **(see illustration)**. Start the engine and allow it to reach normal operating temperature. Check the idle speed and compare it to the idle speed listed in this Chapter's Specifications.

5  Using a jumper wire, bridge terminals TE1 and E1 of the test connector **(see illustration)**.

6  The engine speed should increase to approximately 1,000 to 1,200 rpm for five seconds then return to normal idle speed.

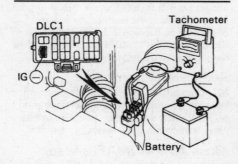

**5.4 Install the lead from the tachometer into the IG terminal of the test connector located in the corner of the engine compartment**

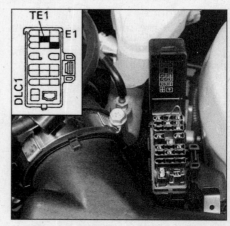

**5.5 To test the IAC motor, locate the test connector and using a jumper wire or paper clip, bridge terminals TE1 and E1**

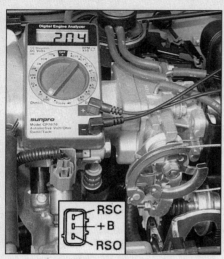

5.8  On 1997 and earlier models, use an ohmmeter to measure the resistance between +B and RSC or RSO

a) If the engine speed changes as described, the IAC valve is okay.
b) If the engine speed does not change as described, continue checking the IAC valve.

7    Remove the jumper wire.
8    On 1997 and earlier models, disconnect the IAC valve electrical connector. Measure the resistance between the middle terminal and each of the other two outer terminals **(see illustration)**. Compare your results to the IAC valve resistance in this Chapter's Specifications. If the resistance is not as specified, replace the IAC valve.
9    On 1998 and later models, remove the IAC valve and inspect the valve position through the air passage **(see illustration)**.

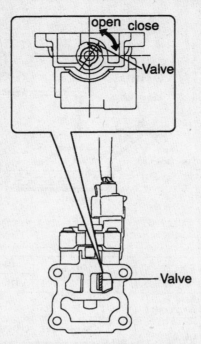

5.9  On 1998 and later models, remove the IAC valve and inspect the valve position through the air passage

The valve should be opened half-way. Connect the electrical connector to the IAC valve and turn the ignition key On. The valve should open. Repeatedly disconnect and connect the connector several times, if necessary, to verify the IAC valve is operating. If the IAC valve does not operate as described, replace the valve.

## Replacement
Refer to illustration 5.11
10   Remove the throttle body (see Chapter 4).
11   Remove the mounting screws and detach the IAC valve and gasket **(see illustration)**.
12   If the IAC assembly was replaced, be sure to install the Thermal Vacuum Valve (TVV) from the original assembly into the new unit (if equipped).
13   Installation of the IAC valve is the

5.11  Remove the mounting screws that retain the IAC valve to the throttle body

reverse of removal. Be sure to use a new gasket when installing the IAC valve.

## Air conditioning idle-up valve
### Check
Refer to illustrations 5.15, 5.16 and 5.17
14   Connect a tachometer according to the tool manufacturer's instructions (see Step 4). Start the engine and allow it to reach normal operating temperature and idle speed, then turn the air conditioning system ON. The engine idle speed should increase slightly.

a) If the engine speed increases as described, the air conditioning idle-up system is operating properly.
c) If the engine speed does not change or if it decreases, measure the air conditioning idle-up valve resistance.

15   Disconnect the air conditioning idle-up valve electrical connector **(see illustration)**.
16   Measure the resistance across the two terminals on the valve **(see illustration)**. It should be between 30 and 34 ohms. If the resistance is not as specified, replace the valve.
17   Apply battery voltage across the terminals and check that air flows from port E to port F **(see illustration)**. Remove battery voltage and observe that air does not flow from port E through port F.
18   If the test results are incorrect, replace

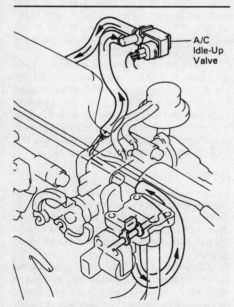

5.15  Schematic of the air conditioning idle-up system

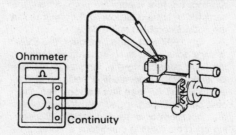

5.16  Measure the resistance across the two terminals on the air conditioning idle-up valve

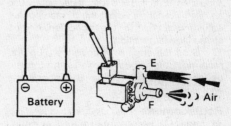

5.17  With battery voltage applied to the two terminals on the valve, air should flow through port E and out port F

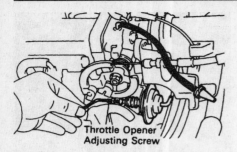

**5.26  Use an Allen wrench to adjust the throttle opener adjusting screw to the specified engine speed**

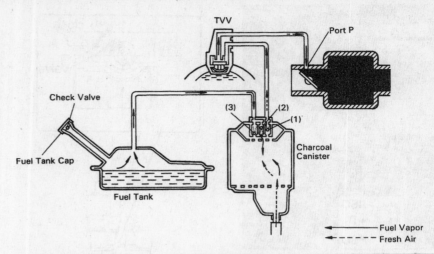

**6.2  Typical EVAP system and operation chart - 1997 and earlier models**

| ECT | TVV | Throttle Position | Canister Check Valve | | | Check Valve in Cap | Evaporated Fuel (HC) |
|-----|-----|-------------------|------|------|------|---------------------|----------------------|
| | | | (1) | (2) | (3) | | |
| Below 35°C (95°F) | CLOSED | — | — | — | — | — | HC from tank is absorbed into the canister. |
| Above 54°C (129°F) | OPEN | Below port P | CLOSED | — | — | — | HC from tank is absorbed into the canister. |
| | | Above port P | OPEN | — | — | — | HC from canister is led into air intake chamber. |
| High pressure in tank | — | — | OPEN | CLOSED | CLOSED | | HC from tank is absorbed into the canister. |
| High vacuum in tank | — | — | CLOSED | OPEN | OPEN | | Air is led into the fuel tank. |

the air conditioning idle-up valve. If the idle-up valve tests good, have the ECM and air conditioning control system diagnosed by a dealer service department or other qualified repair shop.

### Replacement

19   Disconnect the electrical connector and vacuum hoses from the idle-up valve.
20   Remove the mounting screw and detach the idle-up valve from the firewall.
21   Installation is the reverse of removal.

## Throttle opener

### Check

22   Disconnect the vacuum hose from the throttle opener. Connect a hand-held vacuum pump to the fitting on the throttle opener. Apply vacuum to the throttle opener; the plunger should extend against the throttle lever and open the throttle slightly. Release the vacuum and the plunger should retract to its normal position.
23   If the throttle opener does not operate as described, replace it.

### Adjustment

*Refer to illustration 5.26*
24   Connect a tachometer according to the tool manufacturer's instructions (see Step 4). Start the engine and allow it to reach normal operating temperature and idle speed.
25   Stop the engine, disconnect and plug the vacuum hose from the throttle opener, then start the engine.
26   Raise the engine speed to approximately 2500 rpm, then release the throttle. Engine idle speed with the throttle opener set should be 1300 to 1500 rpm. If necessary use an Allen wrench to adjust the throttle opener adjusting screw to the specified engine speed **(see illustration)**.
27   Stop the engine, remove the plug and reconnect the vacuum hose to the throttle opener. Start the engine and check the engine idle speed.

### Replacement

28   Remove the throttle body (see Chapter 4).
29   Remove the throttle opener from the bracket.
30   Installation is the reverse of removal.

### 6   Evaporative Emission Control (EVAP) system

## General description

*Refer to illustrations 6.2 and 6.3*
1   The Evaporative Emission Control (EVAP) system is designed to trap and store fuel that evaporates from the fuel tank, throttle body and intake manifold that would normally enter the atmosphere in the form of hydrocarbon (HC) emissions. Fuel vapors are transferred from the fuel tank and throttle body to a canister where they're stored when the engine isn't running. When the engine is running, the fuel vapors are purged from the canister by intake airflow and consumed in the normal combustion process.
2   On 1997 and earlier models, the EVAP system consists of a charcoal-filled canister, the lines connecting the canister to the fuel tank and intake manifold and the Thermal Vacuum Valve (TVV) **(see illustration)**. The charcoal canister is equipped with a check valve that incorporates three check balls. Depending upon the running conditions and the pressure in the fuel tank, the check balls open and close the passageways to the TVV and fuel tank. When the engine is at operating temperature the TVV opens and manifold vacuum purges vapors from the charcoal canister.

3   On 1998 and later models, the EVAP system consists of a charcoal-filled canister, the lines connecting the canister to the fuel tank and intake manifold, the canister purge valve, the fuel tank pressure sensor and the pressure sensor switching valve **(see illustration)**. 2000 and later models are equipped with a canister vent valve. The system is controlled electronically by the ECM.
4   1998 and later systems perform a self-diagnostic leak check under certain operating conditions. The ECM monitors the voltage signal from the fuel tank pressure sensor to determine if a leak has developed in the system. If a leak is detected the ECM will store a Diagnostic Trouble Code and illuminate the CHECK ENGINE light (see Section 2).

## Check

5   Poor idle, stalling and poor driveability can be caused by an inoperative valve, a damaged canister, split or cracked hoses or hoses connected to the wrong fittings. Check the fuel filler cap for a damaged or deformed gasket.
6   Evidence of fuel loss or fuel odor can be caused by liquid fuel leaking from fuel lines, a cracked or damaged canister, an inoperative valve, disconnected, misrouted, kinked, deteriorated or damaged vapor or control hoses.

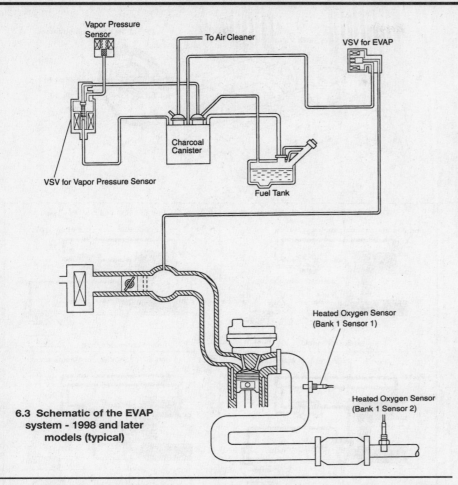

**6.3  Schematic of the EVAP system - 1998 and later models (typical)**

**6.10  Apply air pressure into the charcoal canister purge control valve A (inlet) and confirm that the valve allows the air to pass into the charcoal canister**

**6.11  Apply air pressure to port A on the TVV (top port) and confirm that air does not pass through the valve when the temperature is below 95-degrees F**

**6**

7    Inspect each hose attached to the canister for kinks, leaks and cracks along its entire length. Repair or replace as necessary.

8    Look for fuel leaking from the bottom of the canister. If fuel is leaking, replace the canister and check the hoses and hose routing.

9    Inspect the canister. If it's cracked or damaged, replace it.

### 1997 and earlier models

*Refer to illustrations 6.10 and 6.11*

10    Check for a clogged filter or a stuck check valve. Using low pressure compressed air, blow into the canister tank pipe **(see illustration)**. Air should flow freely from the other pipes. If a problem is found, replace the canister.

11    Check the operation of the TVV. With the engine completely cold, use a hand-held pump and direct air into port A **(see illustration)**. Air should not pass through the TVV. Now warm the engine to operating temperature (above 129-degrees F) and observe that air passes through the TVV. Replace the valve if the test results are incorrect.

### 1998 and later models

*Refer to illustrations 6.13a and 6.13b*

12    The canister purge valve, pressure switching valve and canister vent valve are all vacuum switching valves and can be

checked in the same manner. The fuel tank pressure sensor is similar in operation to a Manifold Absolute Pressure sensor and can be checked in a similar manner.

13    To check the canister purge valve, pressure switching valve or canister vent valve, remove the valve from the vehicle **(see illustrations)**. It will be necessary to remove the charcoal canister to access the pressure switching valve.

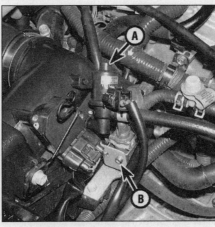

**6.13a  EVAP canister vent valve (A) and purge valve (B) locations - 1998 and later models (typical)**

**6.13b  On 1998 and later models, the fuel tank pressure sensor (A) and pressure switching valve (B) are located on the charcoal canister**

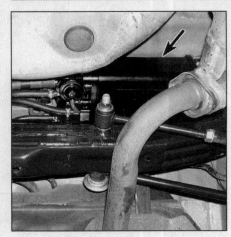

**6.31 Location of the charcoal canister - 1998 and later models**

14   Measure the resistance across the two terminals of the valve solenoid. the resistance should be as follows:
 a) *Purge valve - 27 to 33 ohms at 68-degrees F*
 b) *Vent valve - 25 to 30 ohms at 68-degrees F*
 c) *Pressure switching valve - 37 to 44 ohms at 68-degrees F*

15   If the resistance is not within specifications, replace the valve.

16   Check for continuity between each terminal and the valve body or bracket. If continuity exists, replace the valve.

17   Attempt to blow air through the valve ports. No air should pass through the ports on the purge valve and the pressure switching valve. On the vent valve, air should pass through the ports.

18   Using jumper wires, apply battery voltage to the two terminals. Air should pass through the ports on the purge valve and the pressure switching valve with the solenoid energized. On the vent valve, no air should pass through the ports.

19   If the valve does not operate as described, replace the valve.

20   To check the fuel tank pressure sensor, disconnect the electrical connector from the sensor. Connect a voltmeter to the two outside terminals of the harness connector, turn the ignition key On, approximately 5.0 volts should be indicated on the meter. If the correct voltage is not available at the harness connector, check the circuits from the ECM to the pressure sensor. If the circuits are good have the ECM diagnosed by a dealer service department or other qualified repair shop.

21   Turn the ignition key off and reconnect the electrical connector to the pressure sensor. Disconnect the vacuum hose from the sensor. Using a suitable probe, backprobe the center wire terminal and connect the positive lead of a voltmeter to the probe. Connect the negative lead to a good chassis ground point.

22   Turn the ignition key On, 3.0 to 3.6 volts

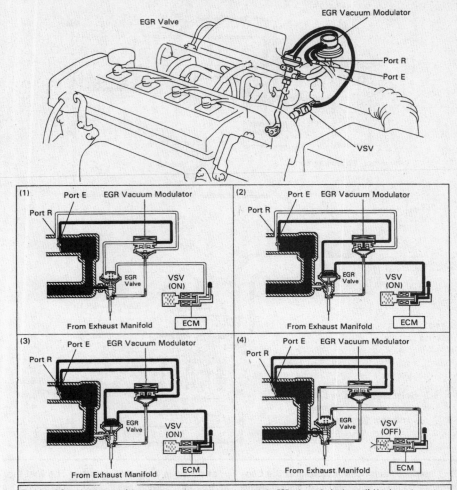

To reduce NOx emissions, part of the exhaust gases are recirculated through the EGR valve to the intake manifold to lower the maximum combustion temperature.

| ECT | RPM | VSV | Throttle Position | Pressure in the EGR Valve Pressure Chamber | | EGR Vacuum Modulator | EGR Valve | Exhaust Gas |
|---|---|---|---|---|---|---|---|---|
| Below 47°C (117°F) | – | **** OFF | – | – | | – | CLOSED | Not recirculated |
| | | | – | – | | – | CLOSED | Not recirculated |
| | | | Below port E | – | | – | CLOSED | Not recirculated |
| Above 53°C (127°F) | Below 4,000 rpm | *** ON | Between port E and port R | (1) LOW | * Pressure constantly alternating between low and high | OPENS passage to atmosphere | CLOSED | Not recirculated |
| | | | | (2) HIGH | | CLOSES passage to atmosphere | OPEN | Recirculated |
| | | | Above port R | (3) HIGH | ** | CLOSES passage to atmosphere | OPEN | Recirculated (increase) |
| | Above 4,400 rpm | (4) OFF | – | – | | – | CLOSED | Not recirculated |

Remarks:  * Pressure increase ──── Modulator closes ──── EGR valve opens ──── Pressure drops
                              EGR valve closes ──── Modulator opens
           ** When the throttle valve is positioned above port R, the EGR vacuum modulator will close the atmosphere passage and open the EGR valve to increase the exhaust gas, even if the exhaust pressure is insufficiently low.
          *** VSV switched ON when product of engine speed multiplied by vacuum sensor valve exceeds a specified valve.
         **** If terminals TE1 and E1 of data link connector 1 are connected, the VSV switches ON.

**7.1 Typical EGR system and operation chart**

should be indicated on the meter. Apply a small amount of vacuum (less than 1 in-Hg) to the port on the sensor, 1.3 to 2.1 volts should be indicated on the meter. Apply a small amount of pressure (less than 1 psi) to

the port on the sensor, 4.2 to 4.8 volts should be indicated on the meter.

23   If the sensor does not operate as described, replace the sensor.

7.4a  To remove the EGR vacuum modulator filters for cleaning, remove the cap . . .

7.4b  . . . then pull out the two filters and blow them out with compressed air - be sure the coarse side of the outer filter faces the atmosphere (out) when reinstalling the filters

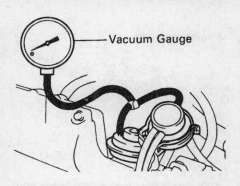

7.6  Install a vacuum gauge between the vacuum modulator and the EGR valve using a three way adapter

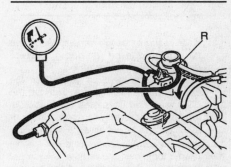

7.8  Disconnect the vacuum hose from port R of the vacuum modulator and plug the hose - using a hand-held vacuum pump, apply vacuum to port R and raise the engine speed to 2,500 rpm; high vacuum should be indicated on the gauge

## Charcoal canister replacement

### 1997 and earlier models

24   Remove the air filter cover and intake duct (see Chapter 4).
25   Clearly label, then detach the hoses from the canister.
26   Remove the mounting clamp bolts, lower the canister with the bracket, disconnect the hoses from the check valve and remove it from the vehicle.
27   Installation is the reverse of removal.

### 1998 and later models

*Refer to illustration 6.31*
28   Raise the vehicle and support it securely on jackstands.
29   Remove the muffler (see Chapter 4).
30   Remove the fuel tank vent hose from the canister by pinching the tabs on the hose together.
31   Clearly label, then detach the hoses from the canister **(see illustration)**.
32   Disconnect the electrical connectors from the pressure sensor and switching valve.
33   Remove the mounting bracket bolts and remove the canister from the vehicle.
34   Installation is the reverse of removal.

## 7   Exhaust Gas Recirculation (EGR) system (1997 and earlier models)

## General description

*Refer to illustration 7.1*
1   To reduce oxides of nitrogen emissions, the Exhaust Gas Recirculation (EGR) system recirculates some of the exhaust gases through the EGR valve to the intake manifold to lower combustion temperatures **(see illustration)**. 1.6L engines with California emissions equipment and all 1.8L engines are equipped with an EGR system. The two systems differ slightly.
2   The EGR system consists of the EGR valve, the EGR modulator, vacuum switching valve (VSV), the Engine Control Module (ECM) and on some models, an EGR gas temperature sensor.
3   If equipped, the EGR temperature sensor is mounted on the EGR valve. This sensor detects the temperature of the exhaust as it moves through the EGR valve. The information is sent to the ECM and in turn the EGR control is regulated precisely and more efficiently.

## Check

### EGR system

*Refer to illustrations 7.4a, 7.4b, 7.6, 7.8 and 7.10*
4   Remove the cover from the vacuum modulator and check the filters **(see illustrations)**.
5   Clean or replace the filters, if necessary, and reinstall the cover.
6   Disconnect the hose from the EGR valve and install a three-way union and vacuum gauge between the EGR valve and the vacuum modulator **(see illustration)**.
7   Start the engine and connect terminals TE1 and E1 on the test connector **(see illustration 2.15)**.

a)  *With the engine cold (coolant temperature below 117-degrees F), verify that the vacuum gauge indicates zero (no vacuum) at 2,500 rpm.*
b)  *With the engine warm (coolant temperature above 127-degrees F), raise the rpm to 2,500 and confirm the vacuum gauge indicates low vacuum.*

8   Disconnect the vacuum hose from port R of the vacuum modulator and plug the hose. Using a hand-held vacuum pump, apply vacuum to port R and raise the engine speed to 2,500 rpm; high vacuum should be indicated on the gauge **(see illustration)**. **Note:** *The engine will run rough as the EGR valve opens.*
9   Stop the engine, remove the vacuum gauge and connect all the hoses to their proper locations.

7.10  Apply vacuum to the EGR valve and confirm that the valve opens and allows exhaust gases to circulate - once it is activated, the EGR valve should hold steady (no loss in vacuum)

10   Detach and plug the vacuum hose from the EGR valve and attach a hand-held vacuum pump to the EGR valve fitting **(see illustration)**. Start the engine and apply vacuum to the EGR valve. Vacuum should remain steady and the engine should run poorly or stall.

6

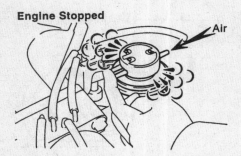

**Engine Stopped**

Air

**7.12 Disconnect the vacuum hoses from ports P, Q and R at the top section of the vacuum modulator, cover ports P and R with your finger while blowing air into port Q - air should flow through the filters**

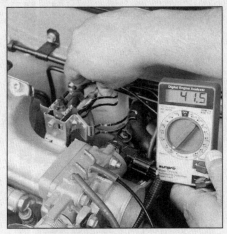

**7.15 Check the resistance on the EGR VSV using an ohmmeter - it should be between 37 and 44 ohms**

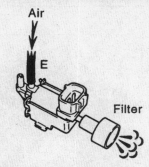

Air

E

Filter

**7.17 When the solenoid is not energized, air should flow through the valve to the filter**

a) *If the vacuum doesn't remain steady and the engine doesn't run poorly, replace the EGR valve and recheck it.*

b) *If the vacuum remains steady but the engine doesn't run poorly, remove the EGR valve and check the valve and the intake manifold for blockage. Clean or replace parts as necessary and recheck.*

11    Stop the engine, connect the hoses to their proper locations and remove the jumper wire from the test connector. If the system does not operate as described, check the individual components.

### EGR vacuum modulator

*Refer to illustration 7.12*

12    Disconnect the vacuum hoses from ports P, Q and R at the top section of the vacuum modulator. Cover ports P and R with your finger while blowing air into port Q. Air should flow through the filters **(see illustration)**.

13    Start the engine and raise the engine speed to 2,500 rpm and repeat the test. No air should flow through the filters. Replace the vacuum modulator if it does not operate as described.

### Vacuum Switching Valve (VSV)

*Refer to illustrations 7.15 and 7.17*

14    Disconnect the electrical connector from the VSV.

15    Measure the resistance between the two terminals of the VSV. Resistance should be between 37 and 44 ohms **(see illustration)**. If the resistance is incorrect, replace the VSV.

16    Check for continuity between each terminal and the valve body or bracket. If continuity exists, replace the VSV.

17    Attempt to blow air through the valve port E **(see illustration)**. Air should flow through the valve to the filter.

18    Using jumper wires, apply battery voltage to the two terminals. Air should not pass through the valve to the filter with the solenoid energized. On 1.8L models, air should flow from port E to port F.

19    Replace the VSV if it does not operate as described.

### EGR temperature sensor

*Refer to illustration 7.20*

20    Disconnect the harness connector for

the EGR temperature sensor **(see illustration)** and measure the resistance of the sensor at various temperatures. Refer to resistance values listed in this Chapter's Specifications. Replace the temperature sensor if the resistance value is not as specified.

## *Component replacement*

### EGR valve

*Refer to illustration 7.25*

21    Disconnect the cable from the negative terminal on the battery.

22    Remove the air filter housing cover and the air intake duct (see Chapter 4).

23    On 1.6L models, perform the following:

a) *Remove the throttle body (see Chapter 4).*

b) *Remove the EGR vacuum modulator.*

c) *Remove the intake manifold support bracket and engine lift bracket.*

d) *Loosen the EGR pipe-to-EGR valve fitting.*

24    Disconnect the vacuum hose from the EGR valve. If equipped, disconnect the EGR temperature sensor electrical connector.

25    Remove the EGR mounting bolts and remove the EGR valve **(see illustration)**.

26    Installation is the reverse of removal. Be sure to install a new gasket.

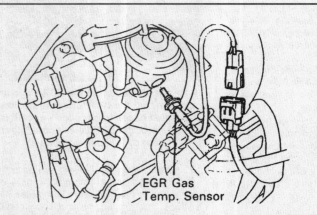

EGR Gas
Temp. Sensor

**7.20 Disconnect the EGR gas temperature sensor connector and check the resistance of the sensor**

**7.25 Remove the nuts that retain the EGR valve assembly to the engine**

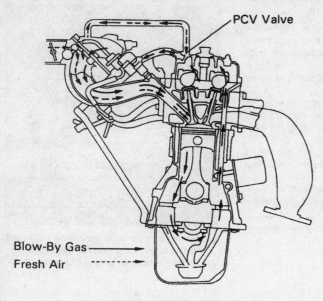

PCV Valve

Blow-By Gas ⟶

Fresh Air ⇢

**8.1 Diagram of a typical PCV system**

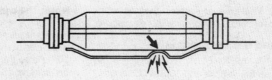

**9.4 Periodically inspect the shield for dents and other damage - if a dent is deep enough to touch the surface of the converter, replace the shield**

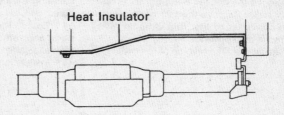

Heat Insulator

**9.5 Periodically inspect the heat insulator to make sure there's adequate clearance between it and the converter**

## EGR vacuum modulator

27 Label and disconnect the vacuum hoses and remove the EGR vacuum modulator from its bracket.
28 Installation is the reverse of removal.

## Vacuum Switching Valve (VSV)

29 Disconnect the electrical connector from the VSV. Label and disconnect the vacuum hoses from the VSV.
30 Remove the mounting screw and remove the VSV and bracket.
31 Installation is the reverse of removal.

## EGR temperature sensor

32 Disconnect the harness connector for the EGR temperature sensor and using an open-end wrench, remove the sensor from the EGR valve (see illustration 7.20).
33 Installation is the reverse of removal.

## 8 Positive Crankcase Ventilation (PCV) system

### General information

*Refer to illustration 8.1*

1 The Positive Crankcase Ventilation (PCV) system reduces hydrocarbon emissions by scavenging crankcase vapors. It does this by circulating fresh air from the air cleaner through the crankcase, where it mixes with blow-by gases and is then rerouted through the PCV valve to the intake manifold (see illustration).
2 The main components of the PCV system are the PCV valve, a fresh air intake and the hoses connecting these components to the engine.
3 To maintain idle quality, the PCV valve restricts the flow when the intake manifold

vacuum is high. If abnormal operating conditions (such as piston ring problems) arise, the system is designed to allow excessive amounts of blow-by gases to flow back through the crankcase vent tube into the air cleaner to be consumed by normal combustion.
4 This system directs the blow-by into the throttle body which, over time, can cause an oily residue build up in the area near the throttle plate. Consequently, it's a good idea to periodically clean this residue from the throttle body with an approved throttle body cleaning fluid and a soft towel.

### Check

5 To check the valve, remove the PCV valve from the valve cover and shake the valve. It should rattle, indicating that it's not clogged with deposits. If the valve does not rattle, replace it with a new one.
6 Start the engine and allow it to idle, then place your finger over the valve opening. If vacuum is felt, the PCV valve is working properly. If no vacuum is felt, the PCV valve may be bad or the hose may be plugged. Also, check for vacuum leaks at the valve, filler cap and all the hoses.

### Replacement

**Note:** *Refer to Chapter 1 for PCV valve locations.*
7 On 1999 and earlier models, carefully pull the PCV valve out of the grommet. Check the rubber grommet for cracks and distortion. If it's damaged, replace it. On 2000 and later models, remove the hose and unthread the PCV valve from the valve cover fitting.
8 If the valve is clogged, the hose is also probably plugged. Remove the hose and clean it with solvent.

9 After cleaning the hose, inspect it for damage, wear and deterioration. Make sure it fits snugly on the fittings.
10 If necessary, install a new PCV valve.
11 Install the clean PCV hose. Make sure that the PCV valve and hose are secure.

## 9 Catalytic converter

**Note:** *Because of a federally mandated warranty which covers emissions-related components such as the catalytic converter, check with a dealer service department before replacing the converter at your own expense.*

### General description

1 To reduce hydrocarbon, carbon monoxide and oxides of nitrogen emissions, all vehicles are equipped with a three-way catalyst system which oxidizes and reduces these chemicals, converting them into harmless nitrogen, carbon dioxide and water.
2 The catalytic converter is mounted in the exhaust system much like a muffler.

### Check

*Refer to illustrations 9.4 and 9.5*
3 Periodically inspect the catalytic converter-to-exhaust pipe mating flanges and bolts. Make sure that there are no loose bolts and no leaks between the flanges.
4 Look for dents in or damage to the catalytic converter protector (see illustration). If any part of the protector is damaged or dented enough to touch the converter, repair or replace it.
5 Inspect the heat insulator for damage. Make sure that there is adequate clearance between the heat insulator and the catalytic converter (see illustration).

**6**

## Replacement

**Caution:** *The exhaust system must be completely cool before performing this procedure.*

### 1997 and earlier models

*Refer to illustration 9.7*

6     Raise the vehicle and support it securely on jackstands.

7     Remove the two bolts securing the converter to the center exhaust pipe and separate the center exhaust pipe from the converter **(see illustration).**

8     Remove the converter support bracket-to-body bolts. Remove the two bolts securing the converter to the front exhaust pipe and remove the converter.

9     Installation is the reverse of removal.

### 1998 and later models

**Note:** *The converter is welded to the center exhaust pipe. Cutting and welding equipment will be required to remove the converter from the center exhaust pipe.*

10     Raise the vehicle and support it securely on jackstands.

11     Disconnect the electrical connector from the post-converter oxygen sensor.

12     Remove the two bolts securing the center pipe to the muffler/tail pipe.

13     Remove the bolt and loosen the clamp securing the converter to the front exhaust pipe.

14     Detach the exhaust pipe from the supports and remove the converter and center pipe assembly.

15     Installation is the reverse of removal.

**9.7  Be sure to spray penetrating lubricant onto the catalytic converter flange bolts before attempting to remove them**

# Chapter 7  Part A
# Manual transaxle

## Contents

## Specifications

### Torque specifications

**Ft-lbs** (unless otherwise indicated)

| | |
|---|---|
| Back-up light switch | 30 |
| Transaxle-to-engine bolts - Corolla | |
| 1995 and earlier models **(see illustration 6.32a)** | |
| Bolts A | 47 |
| Bolt B | 34 |
| Stiffener plate bolts | 17 |
| 1996 and 1997 models | |
| Upper mounting bolts | 47 |
| Lower mounting bolts | |
| 1.6L engine | 34 |
| 1.8L engine **(see illustration 6.32b)** | |
| Bolts A | 17 |
| Bolt B | 34 |
| Stiffener plate bolts | 17 |
| 1998 and later models | |
| Upper mounting bolts | 47 |
| Lower mounting bolts **(see illustration 6.32c)** | |
| Bolts A | 17 |
| Bolts B | 35 |
| Transaxle-to-engine bolts - Prizm | |
| 1997 and earlier models | 34 |
| 1998 and later models | |
| Upper mounting bolts | 47 |
| Lower mounting bolts **(see illustration 6.32c)** | |
| Bolts A | 132 in-lbs |
| Bolts B | 35 |
| Shift rod through-bolt nuts (1998 and later Prizm only) | 15 |
| Shift extension rod stud nut (1998 and later Prizm only) | 29 |
| Shifter mounting bolts | |
| Corolla | 108 in-lbs |
| Prizm | |
| 1996 and earlier | 15 |
| 1997 through 1999 | 106 in-lbs |
| 1998 and later | 89 in-lbs |

7A

**2.1  To disconnect the shift cables from the transaxle linkage, remove these two clips and washers (arrows)**

**2.2  To detach the shift cable assembly from the cable bracket, remove these two retainers (arrows) with a pair of pliers**

**2.5  To disconnect the shift cables from the shift lever, remove these clips and washers (arrow)**

## 1   General information

The vehicles covered by this manual are equipped with a 5-speed manual transaxle or a 3- or 4-speed automatic transaxle. Information on the manual transaxle is included in this Part of Chapter 7. Service procedures for the automatic transaxle are contained in Chapter 7, Part B.

The manual transaxle is a compact, two-piece, lightweight aluminum alloy housing containing both the transmission and differential assemblies.

Because of the complexity, unavailability of replacement parts and special tools necessary, internal repair procedures for the manual transaxle are beyond the scope of this manual. For readers who wish to tackle a transaxle rebuild, exploded views and a brief *Manual transaxle overhaul - general information* Section are provided. The bulk of information in this Chapter is devoted to removal and installation procedures.

## 2   Shift cables - removal and installation

*Refer to illustrations 2.1, 2.2 and 2.5*
**Note 1:** *This procedure applies to all except 1998 and later Prizm models, which have a shift rod.*
**Note 2:** *The shift cables are not adjustable. If they become stretched, causing shifting problems, they must be replaced.*
1    In the engine compartment, remove the retaining clips and washers and disconnect the shift cables from the selecting bellcrank **(see illustration)**.
2    Remove the cable retainers from the cable bracket **(see illustration)**.
3    Inside the vehicle, remove the floor console (see Chapter 11).
4    Remove the cable retainers from the shift lever base **(see illustration 3.3a or 3.3b)**.
5    Remove the retaining clips and washers

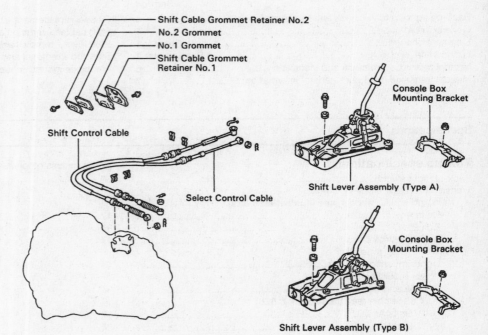

**4.4a  Shift control cables and shift lever assembly details (1993 through 1995 models)**

from the cable ends **(see illustration)** and disconnect the cables from the shift lever assembly.
6    Trace the cable assembly to the firewall and remove the weatherproofing grommet. Pull the cable assembly through the firewall.
7    Installation is the reverse of removal.

## 3   Shift rod - removal and installation

**Note:** *This procedure applies to 1998 and later Prizm models only.*
1    Apply the parking brake. Raise the front of the vehicle and support it securely on

jackstands.
2    Remove the nut and through-bolt connecting the shift rod to the shift lever.
3    Remove the nut and through-bolt connecting the shift rod to the transaxle shift lever. Remove the shift rod.
4    Installation is the reverse of removal. Be sure to tighten the through-bolt nuts to the torque listed in this Chapter's Specifications.

## 4   Shift lever - removal and installation

*Refer to illustrations 4.4a and 4.4b*
1    Remove the center console (see Chap-

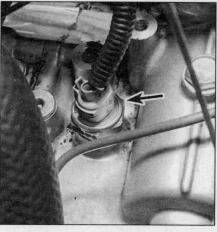

5.8 Unplug the electrical connector from the back-up light switch (arrow)

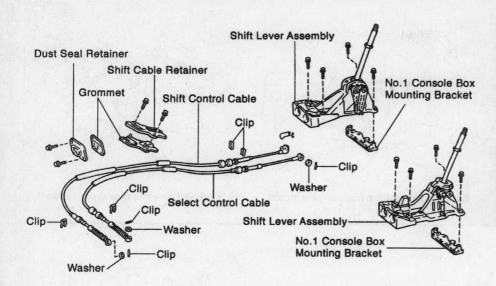

**4.4b  Shift control cables and shift lever assembly details (1996 and later models, except 1998 and later Prizms)**

ter 11).

2    On all models except 1998 and later Prizms, remove the shift cable retainers and disconnect both cables from the shift lever (see Section 2). If you're working on a 1998 or later Prizm, remove the through-bolt connecting the shift rod to the shift lever.

3    If you're working on a 1998 or later Prizm, remove the nut and detach the shifter extension rod from the transaxle.

4    Remove the retaining bolts from the shift lever base **(see illustrations)** and detach the shift lever from the vehicle. **Note:** On 1998 and later Prizm models the retaining bolts are accessed from underneath the vehicle.

5    Installation is the reverse of removal.

---

## 5    Back-up light switch - check and replacement

### Check

1    The back-up light switch is located on top of the transaxle.

2    Turn the ignition key to the On position and move the shift lever to the Reverse position. The switch should close the back-up light circuit and turn on the back-up lights.

3    If it doesn't, check the back-up light fuse (see Chapter 12).

4    If the fuse is okay, verify that there's voltage available on one side of the switch (with the ignition turned to On).

5    If there's no voltage to the switch, check the wire between the fuse and the switch; if there is voltage, put the shift lever in reverse and see if there's voltage on the other side of the switch.

6    If there's no voltage on the other side of the switch, replace the switch (see below); if

there is voltage, note whether one or both back-up lights are out.

7    If only one bulb is out, replace it; if they're both out, the bulbs could be the problem, but it's more likely that the wire between the switch and the bulbs has an open somewhere.

### Replacement

*Refer to illustration 5.8*

8    Disconnect the electrical connector from the back-up light switch **(see illustration)**.

9    Unscrew and remove the switch.

10    To test the new switch before installation, simply check continuity across the switch terminals: with the plunger depressed, there should be continuity; with the plunger free, there should be no continuity.

11    Screw in the new switch and tighten it securely.

12    Connect the electrical connector.

13    Check the operation of the back-up lights.

---

## 6    Manual transaxle - removal and installation

### Removal

*Refer to illustrations 6.11a, 6.11b, 6.18, 6.19, 6.20, 6.22, 6.23 and 6.24*

1    Disconnect the negative cable from the battery. **Caution:** *If the stereo in your vehicle is equipped with an anti-theft system, make sure you have the correct activation code before disconnecting the battery.*

2    Remove the air filter housing (see Chapter 4). Remove the coolant reservoir (see

Chapter 3).

3    Loosen the driveaxle/hub nuts (see Chapter 8).

4    Remove the clutch release cylinder and the clutch hydraulic line (see Chapter 8).

5    Unplug the electrical connectors from the back-up light switch (see Section 4) and the speed sensor (see Chapter 6).

6    Locate the ground cable on top of the transaxle. Remove the cable retaining bolt and detach the ground cable from the transaxle.

7    Disconnect the shift cables from the transaxle (see Section 2).

8    Detach any wire harness clamps from the engine and/or transaxle and set the harnesses aside.

9    Remove the upper starter mounting bolt (see Chapter 5).

10    Remove the two upper and the single front transaxle-to-engine mounting bolts. Note the location of any ground connectors or brackets, so that they may be installed in their original location.

11    Remove the left engine mount stay and the left engine mount retaining bolt **(see illustrations)**.

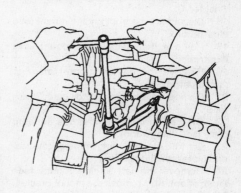

**6.11a  Remove the left engine mount stay**

**7A**

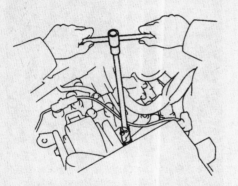

**6.11b  Remove the left engine mount retaining bolt**

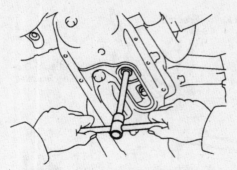

**6.18  Disconnect the front engine mount**

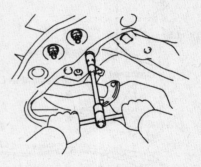

**6.19  Remove the rear engine mount**

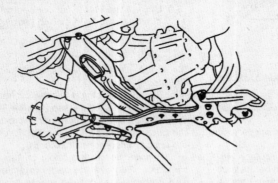

**6.20  Remove the center support member**

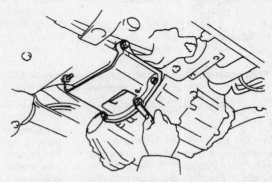

**6.22  On 1.6L models, remove the stiffener plate**

12   Loosen the wheel lug nuts. Raise the vehicle and support it securely on jackstands. Remove the wheels.

13   Support the engine. This can be done from above by using an engine support fixture or an engine hoist, or by placing a jack (with a wood block as an insulator) under the engine oil pan. The engine must be supported at all times while the transaxle is out of the vehicle.

14   Remove the under-vehicle splash shields (see Chapter 2).

15   Drain the transaxle fluid (see Chapter 1).

16   Remove the driveaxles (see Chapter 8).

17   Remove the front exhaust pipe (see Chapter 4).

18   Disconnect the front engine mount **(see illustration)**.

19   Remove the rear engine mount **(see illustration)**.

20   Remove the center support member **(see illustration)**.

21   Remove the starter motor (see Chapter 5).

22   On 1.6L models, remove the stiffener plate **(see illustration)**.

23   On 1.8L models, remove the engine rear end plate **(see illustration)**.

24   Remove the left engine mount **(see illustration)**. Also remove the mount bracket.

25   Support the transaxle with a jack (preferably a special jack made for this purpose; these jacks are available at most

equipment rental yards). Safety chains will help steady the transaxle on the jack.

26   Remove the rest of the bolts securing the transaxle to the engine.

27   Make a final check that all wires and hoses have been disconnected from the transaxle.

28   Lower the left (driver's) end of the engine, then roll the transaxle and jack toward the side of the vehicle. Once the input shaft is clear of the splines in the clutch hub, lower the transaxle and remove it from under the vehicle. Try to keep the transaxle as level as possible. **Caution:** *Do not depress the clutch pedal while the transaxle is removed from the vehicle.*

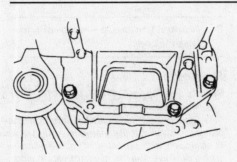

**6.23  On 1.8L models, remove the engine rear end plate**

29   The clutch components can now be inspected (see Chapter 8). In most cases, new clutch components should be routinely installed whenever the transaxle is removed.

### Installation

*Refer to illustrations 6.32a, 6.32b and 6.32c*

30   If removed, install the clutch components (see Chapter 8).

31   With the transaxle secured to the jack as on removal, raise it into position and then carefully slide it forward, engaging the input shaft with the splines in the clutch hub. Do not use excessive force to install the transaxle - if the input shaft does not slide into place, readjust the angle of the transaxle

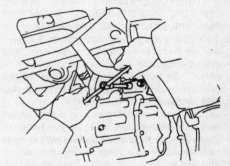

**6.24  Remove the left engine mount**

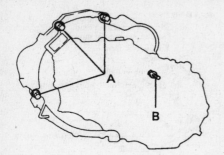

**6.32a  Tighten bolts A and B to the torque listed in this Chapter's Specifications (1995 and earlier models)**

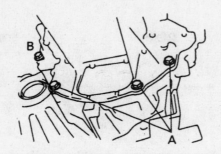

**6.32b  Tighten bolts A and B to the torque listed in this Chapter's Specifications (1996 and 1997 models with a 1.8L engine)**

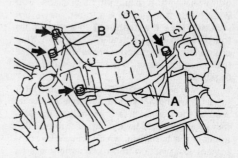

**6.32c  Tighten bolts A and B to the torque listed in this Chapter's Specifications (1998 and later models)**

so it is level and/or turn the input shaft so the splines engage properly with the clutch.

32   Install the transaxle-to-engine bolts **(see illustrations)**. Tighten the bolts to the torque listed in this Chapter's Specifications.

33   Install the transaxle mount nuts and bolts. Tighten all nuts and bolts securely.

34   Install any suspension components which were detached or removed. Tighten all nuts and bolts to the torque listed in the Chapter 10 Specifications.

35   Remove the jacks supporting the transaxle and the engine.

36   Install the various items removed previously. Refer to Chapter 4 for information regarding the exhaust pipe, Chapter 5 for the starter motor and Chapter 8 for the driveaxles.

37   Make sure that the wiring harness connectors for the back-up light switch and the speed sensor, and any other electrical devices, are plugged in. And make sure that all harness clamps are reattached to the engine and/or transaxle.

38   If the transaxle was drained, fill it with the specified lubricant to the proper level (see Chapter 1).

39   Lower the vehicle.

40   Connect the shift cables (see Section 2).

41   Connect the negative battery cable. Road test the vehicle to check for proper transaxle operation and check for leakage.

---

## 7   Manual transaxle overhaul - general information

*Refer to illustrations 7.4a, 7.4b and 7.4c*

1   Overhauling a manual transaxle is a difficult job for the do-it-yourselfer. It involves the disassembly and reassembly of many small parts. Numerous clearances must be precisely measured and, if necessary, changed with select-fit spacers and snap-rings. As a result, if transaxle problems arise, it can be removed and installed by a competent do-it-yourselfer, but overhaul should be left to a transmission repair shop. Rebuilt transaxles may be available - check with your dealer parts department and auto parts stores. At any rate, the time and money involved in an

overhaul is almost sure to exceed the cost of a rebuilt unit.

2   Nevertheless, it's not impossible for an inexperienced mechanic to rebuild a transaxle if the special tools are available and the job is done in a deliberate step-by-step manner so nothing is overlooked.

3   The tools necessary for an overhaul include internal and external snap-ring pliers, a bearing puller, a slide hammer, a set of pin punches, a dial indicator and possibly a hydraulic press. In addition, a large, sturdy workbench and a vise or transaxle stand will be required.

4   During disassembly of the transaxle, make careful notes of how each piece comes

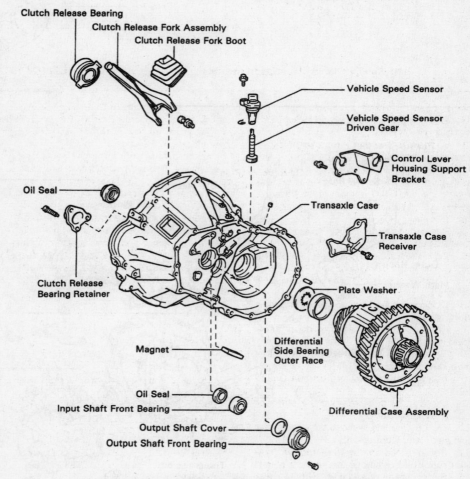

**7.4a  An exploded view of the manual transaxle assembly (1 of 3)**

7A

off, where it fits in relation to other pieces and what holds it in place. Exploded views are included **(see illustrations)** to show where the parts go - but actually noting how they are installed when you remove the parts will make it much easier to get the transaxle back together.

5    Before taking the transaxle apart for repair, it will help if you have some idea what area of the transaxle is malfunctioning. Certain problems can be closely tied to specific areas in the transaxle, which can make component examination and replacement easier. Refer to the *Troubleshooting* section at the front of this manual for information regarding possible sources of trouble.

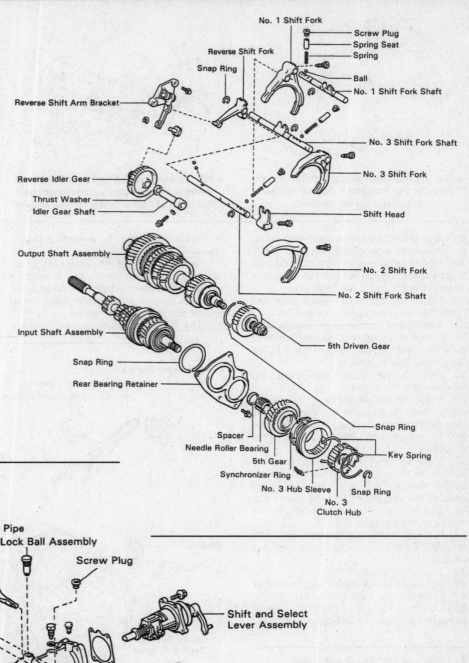

**7.4b An exploded view of the manual transaxle assembly (2 of 3)**

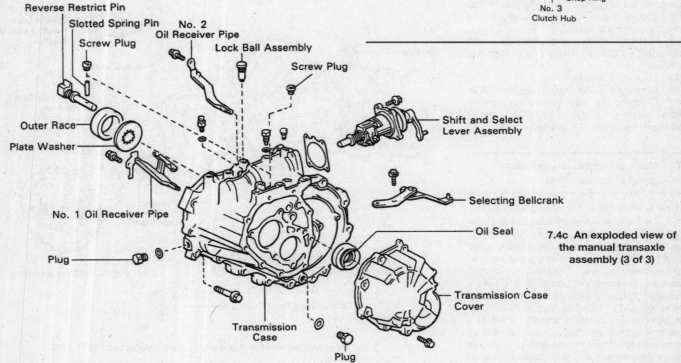

**7.4c An exploded view of the manual transaxle assembly (3 of 3)**

# Chapter 7 Part B
# Automatic transaxle

## Contents

## Specifications

### Shift lock system

Shift lock solenoid resistance
  1993 and 1994
    A131L .......................................................................... 30 to 35 ohms
    A245E .......................................................................... 21 to 27 ohms
  1995 and later ................................................................ 21 to 27 ohms
Key interlock solenoid resistance ........................................ 12.5 to 16.5 ohms

### Torque specifications

**Ft-lbs** (unless otherwise indicated)

Park/Neutral Position switch .............................................. 48 in-lbs
Torque converter-to-driveplate bolts
  A131L
    1997 and earlier ......................................................... 156 in-lbs
    1998 and later ............................................................ 18
  A245E .......................................................................... 18

## Torque specifications

| | Ft-lbs (unless otherwise indicated) |
|---|---|
| Transaxle-to-engine bolts - Corolla models | |
| 1993 and 1994 models | |
| A131L | 47 |
| A245E (see illustration 9.27a) | |
| Bolt A | 47 |
| Bolt B | 18 |
| Bolt C | 17 |
| Bolt D | 47 |
| 1995 through 1997 models | |
| A131L | |
| Upper and lower bolts | 47 |
| Stiffener plate bolts | 17 |
| A245E (see illustration 9.27b) | |
| Bolt A | 17 |
| Bolt B | 18 |
| Bolt C | 34 |
| Bolt D | 47 |
| 1998 and later models (A131L and A245E) (see illustration 9.27c) | |
| Upper bolts | 47 |
| Lower bolts | |
| Bolts A | 34 |
| Bolts B | 17 |
| Transaxle-to-engine bolts - Prizm models | 47 |
| Valve body bolts | 84 in-lbs |
| Manual valve retaining bolts (A131L) | 84 in-lbs |
| Wire retainer bolt (A245E) | 56 in-lbs |
| Detent spring bolt(s) | 84 in-lbs |

## 1    General information

All vehicles covered in this manual are equipped with either a 5-speed manual transaxle or a 3- or 4-speed automatic transaxle. All information on the automatic transaxle is included in this Part of Chapter 7. Information for the manual transaxle can be found in Part A of this Chapter.

The A131L is a three-speed automatic transaxle. Its shift points are controlled by the governor and the throttle valve. The A245E is an electronically controlled four-speed automatic transaxle with fourth gear being an overdrive gear. Shifts are attained by the use of shift solenoids, which are controlled by the Engine Control Module (ECM). Both transaxles utilize a lock-up torque converter.

Because of the complexity of the automatic transaxles and the specialized equipment necessary to perform most service operations, this Chapter contains only those procedures related to general diagnosis, routine maintenance, adjustment and removal and installation.

If the transaxle requires major repair work, it should be left to a dealer service department or an automotive or transmission repair shop. You can, however, remove and install the transaxle yourself and save the expense, even if the repair work is done by a transmission shop.

## 2    General diagnosis and trouble codes

**Note:** *Automatic transaxle malfunctions may be caused by five general conditions: poor*
*engine performance, improper adjustments, hydraulic malfunctions, mechanical malfunctions or malfunctions in the computer or its signal network. Diagnosis of these problems should always begin with a check of the easily repaired items: fluid level and condition (see Chapter 1), shift linkage adjustment and throttle linkage adjustment. Next, perform a road test to determine if the problem has been corrected or if more diagnosis is necessary. If the problem persists after the preliminary tests and corrections are completed, additional diagnosis should be done by a dealer service department or transmission repair shop. Refer to the Troubleshooting section at the front of this manual for information on symptoms of transaxle problems.*

### Preliminary checks

1    Drive the vehicle to warm the transaxle to normal operating temperature.

2    Check the fluid level as described in Chapter 1:

a) *If the fluid level is unusually low, add enough fluid to bring the level within the designated area of the dipstick, then check for external leaks (see below).*

b) *If the fluid level is abnormally high, drain off the excess, then check the drained fluid for contamination by coolant. The presence of engine coolant in the automatic transmission fluid indicates that a failure has occurred in the internal radiator walls that separate the coolant from the transmission fluid (see Chapter 3).*

c) *If the fluid is foaming, drain it and refill the transaxle, then check for coolant in the fluid, or a high fluid level.*

3    Check the engine idle speed. **Note:** *If the engine is malfunctioning, do not proceed*
*with the preliminary checks until it has been repaired and runs normally.*

4    Check the throttle valve cable for freedom of movement. Adjust it if necessary (see Section 4). **Note:** *The throttle valve cable may function properly when the engine is shut off and cold, but it may malfunction once the engine is hot. Check it cold and at normal engine operating temperature.*

5    Inspect the shift cable (see Section 5). Make sure that it's properly adjusted and that the cable operates smoothly.

### Fluid leak diagnosis

6    Most fluid leaks are easy to locate visually. Repair usually consists of replacing a seal or gasket. If a leak is difficult to find, the following procedure may help.

7    Identify the fluid. Make sure it's transmission fluid and not engine oil or brake fluid (automatic transmission fluid is a deep red color).

8    Try to pinpoint the source of the leak. Drive the vehicle several miles, then park it over a large sheet of cardboard. After a minute or two, you should be able to locate the leak by determining the source of the fluid dripping onto the cardboard.

9    Make a careful visual inspection of the suspected component and the area immediately around it. Pay particular attention to gasket mating surfaces. A mirror is often helpful for finding leaks in areas that are hard to see.

10    If the leak still cannot be found, clean the suspected area thoroughly with a degreaser or solvent, then dry it.

11    Drive the vehicle for several miles at normal operating temperature and varying speeds. After driving the vehicle, visually inspect the suspected component again.

12  Once the leak has been located, the cause must be determined before it can be properly repaired. If a gasket is replaced but the sealing flange is bent, the new gasket will not stop the leak. The bent flange must be straightened.

13  Before attempting to repair a leak, check to make sure that the following conditions are corrected or they may cause another leak. **Note:** *Some of the following conditions cannot be fixed without highly specialized tools and expertise. Such problems must be referred to a transmission shop or a dealer service department.*

### Gasket leaks

14  Check the pan periodically. Make sure the bolts are tight, no bolts are missing, the gasket is in good condition and the pan is flat (dents in the pan may indicate damage to the valve body inside).

15  If the pan gasket is leaking, the fluid level may be too high, the vent may be plugged, the pan bolts may be too tight, the pan sealing flange may be warped, the sealing surface of the transaxle housing may be damaged, the gasket may be damaged or the transaxle casting may be cracked or porous. If sealant instead of gasket material has been used to form a seal between the pan and the transaxle housing, it may be the wrong sealant.

### Seal leaks

16  If a transaxle seal is leaking, the fluid level or pressure may be too high, the vent may be plugged, the seal bore may be damaged, the seal itself may be damaged or improperly installed, the surface of the shaft protruding through the seal may be damaged or a loose bearing may be causing excessive shaft movement.

17  Make sure the dipstick tube seal is in good condition and the tube is properly seated. Periodically check the area around the speedometer gear or sensor for leakage. If transmission fluid is evident, check the O-ring for damage.

### Case leaks

18  If the case itself appears to be leaking, the casting is porous and will have to be repaired or replaced.

19  Make sure the oil cooler hose fittings are tight and in good condition.

### Fluid comes out vent pipe or fill tube

20  If this condition occurs, the transaxle is overfilled, there is coolant in the fluid, the case is porous, the dipstick is incorrect, the vent is plugged or the drain-back holes are plugged.

## Self-diagnosis system and trouble codes

### 1995 and earlier models

*Refer to illustration 2.22*

21  1995 and earlier Corollas and Prizms with the A245E (four-speed transaxle) have a self-diagnosis feature that can be accessed without the use of special tools. If certain problems develop with the transaxle or its control system, the O/D OFF indicator light on the instrument panel will blink, and a trouble code will be stored in the ECM's memory.

22  To obtain any stored trouble codes, turn the ignition switch to the On position (but don't start the engine). Push the O/D main

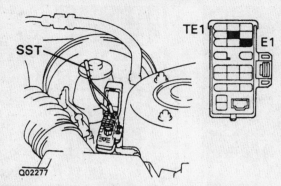

**2.22  To extract any stored transaxle trouble codes, connect terminals TE1 and E1 of the data link connector with a jumper wire (with the O/D switch on) - 1995 and earlier models with the A245E transaxle only**

switch (on the shift lever) to the On position, then connect a jumper wire to terminals TE1 and E1 of the data link connector, located in the engine compartment on the left strut tower **(see illustration)**.

23  The diagnostic code is the number of flashes indicated on the O/D OFF light. If any malfunction has been detected, the light will blink the first digit(s) of the code, pause 1.5 seconds, then blink the second digit of the code. For example, a code 42 (vehicle speed sensor) will first blink four flashes, pause 1.5 seconds, then blink two flashes. If there is more than one code stored in the ECM, the ECM will pause 2.5 seconds before flashing the next code. If the system is operating normally (no malfunctions), the warning light will blink two times per second.

24  The accompanying tables explain the code that will be flashed for each of the malfunctions. The accompanying table indicates the diagnostic code - in blinks - along with the system, diagnosis and specific areas. Check the indicated system or component or take the vehicle to a dealer service department to have the malfunction repaired.

25  After the diagnosis check, clear the trouble codes. To do this, turn the ignition switch Off and remove the EFI fuse from the engine compartment fuse box for at least 10 seconds. Reinstall the fuse and recheck for trouble codes.

**7B**

## A245E trouble codes (1995 and earlier models only)

| Code number | Trouble area | Action to take |
| --- | --- | --- |
| **Code 42**<br>(4 flashes, pause, 2 flashes) | Vehicle Speed Sensor (VSS) or circuit | Check VSS (see Chapter 6), wiring between VSS and ECM, combination meter |
| **Code 62**<br>(6 flashes, pause, 2 flashes) | No. 1 Solenoid valve or circuit | Check wiring between transaxle and ECM. If OK, take the vehicle to a transmission specialist |
| **Code 63**<br>(6 flashes, pause, 3 flashes) | No. 2 Solenoid valve or circuit | Check wiring between transaxle and ECM. If OK, take the vehicle to a transmission specialist |
| **Code 64**<br>(6 flashes, pause, 4 flashes) | SL Solenoid valve or circuit | Check wiring between transaxle and ECM. If OK, take the vehicle to a transmission specialist |

**3.4  Carefully pry out the driveaxle oil seal with a seal removal tool or a large screwdriver; make sure you don't damage the seal bore or the new seal may leak**

### 1996 and later models

26   If certain problems occur on a 1996 and later model with a A245E transaxle, the CHECK ENGINE light on the instrument panel will come on. To obtain any stored trouble codes on these models, a special scan tool is required. Refer to Chapter 6 for more information on scan tools, how to connect the scan tool and how to extract trouble codes. Also refer to the trouble code chart in that Chapter.

## 3   Oil seal replacement

1   Oil leaks frequently occur due to wear of the driveaxle oil seals and/or the speedometer drive gear oil seal and O-rings. Replacement of these seals is relatively easy, since the repairs can usually be performed without removing the transaxle from the vehicle.

### Driveaxle oil seals

*Refer to illustrations 3.4 and 3.6*

2   The driveaxle oil seals are located on the sides of the transaxle, where the inner ends of the driveaxles are splined into the differential side gears. If you suspect that a driveaxle

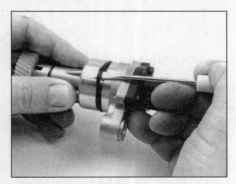

**3.10  Remove the oil seal O-ring from the speed sensor with a small screwdriver; make sure you don't scratch the surface of the sensor or gouge the O-ring groove**

**3.6  Use a seal installer, a large socket or a piece of pipe to install the new seal**

oil seal is leaking, raise the vehicle and support it securely on jackstands. If the seal is leaking, you'll see lubricant on the side of the transaxle, below the seal.

3   Remove the driveaxle (see Chapter 8).
4   Using a screwdriver or prybar, carefully pry the oil seal out of the transaxle bore **(see illustration)**.
5   If the oil seal cannot be removed with a screwdriver or pry bar, a special oil seal removal tool (available at auto parts stores) will be required.
6   Using a seal installer, a large section of pipe or a large deep socket as a drift, install the new oil seal. Drive it into the bore squarely and make sure it's completely seated **(see illustration)**. On A131L transaxles, a fully-seated seal should be flush with the surface of the transaxle housing; on A245E units, a fully-seated left seal should be recessed about 13/64-inch, a right seal about 1/8-inch.
7   Lubricate the lip of the new seal with multi-purpose grease, then install the driveaxle (see Chapter 8). Be careful not to damage the lip of the new seal.

### Speed sensor O-ring

*Refer to illustrations 3.9 and 3.10*

8   The speed sensor is located on the transaxle housing. Look for lubricant around

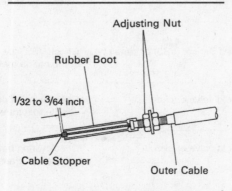

**4.3a  Throttle cable adjustment details (A131L)**

**3.9  Unplug the speed sensor electrical connector (arrow), remove the bolt (arrow) and pull the speed sensor unit straight out of the transaxle**

the sensor housing to determine if the O-ring is leaking.
9   Unplug the electrical connector and unbolt the vehicle speed sensor from the transaxle **(see illustration)**.
10   Using a scribe or a small screwdriver, remove the O-ring from the sensor **(see illustration)** and install a new O-ring. Lubricate the new O-ring with automatic transmission fluid to protect it during installation of the sensor.
11   Installation is the reverse of removal.

## 4   Throttle Valve (TV) cable - check, adjustment and replacement

### Check

*Refer to illustrations 4.3a and 4.3b*

1   Remove the duct between the air cleaner and the throttle body (see Chapter 5).
2   Have an assistant press the accelerator pedal all the way to the floor and hold it while you measure the distance between the end of the boot and the stopper on the cable.
3   If the measurement taken is as shown **(see illustrations)**, the cable is properly adjusted. If it's out-of-range, adjust it as follows.

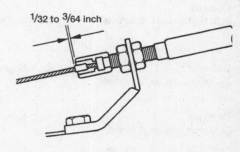

**4.3b  Throttle cable adjustment details (A245E)**

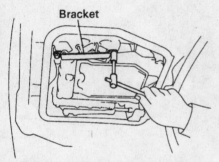

**4.11a** On A131L models, remove the two bolts that attach the apply tube bracket, remove the bracket . . .

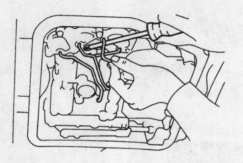

**4.11b** . . . and carefully pry out both ends of all four oil tubes - make sure you don't bend or kink the tubes

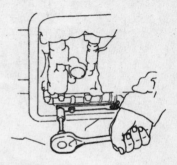

**4.13** Remove the manual detent spring

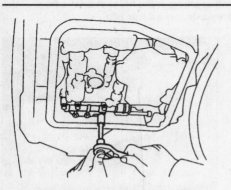

**4.14a** On A131L models, remove the manual valve . . .

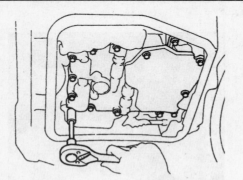

**4.14b** . . . remove all 14 valve body retaining bolts and remove the valve body

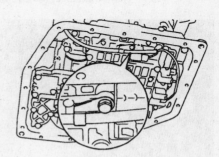

**4.14c** On A245E models, remove the wire retainer . . .

## Adjustment

4    Have your assistant continue to hold the pedal down while you loosen the adjusting nuts and adjust the cable housing so that the distance between the end of the boot and the stopper on the cable is within the range shown.

5    Tighten the adjusting nuts securely, recheck the clearance and make sure the throttle valve opens all the way when the throttle is depressed.

## Replacement

*Refer to illustrations 4.11a, 4.11b, 4.13, 4.14a, 4.14b, 4.14c, 4.14d and 4.15*

6    Loosen the cable locknut and detach the cable from the bracket at the throttle body.

7    Disconnect the cable from the throttle linkage.

8    Detach the cable from the bracket on the transaxle.

9    Follow the cable down to the front of the transaxle, where it enters the transaxle housing right behind the Park/Neutral Position switch. Remove the cable hold-down bolt.

10    Remove the pan, drain the transaxle fluid and remove the filter (see Chapter 1).

11    On A131L models, remove the oil tubes **(see illustrations)**.

12    On A245E models, unplug the electrical connectors for the solenoids.

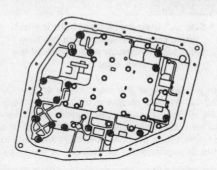

**4.14d** . . . remove all 17 valve body retaining bolts and remove the valve body

13    Remove the manual detent spring **(see illustration)**.

14    Remove the manual valve (A131L) and valve body **(see illustrations)**.

15    Disconnect the throttle valve cable from the cam on top of the valve body **(see illustration)**.

16    Installation is the reverse of removal. Be sure to tighten the valve body, manual valve, wire retainer and detent spring bolts to the torque listed in this Chapter's Specifications. Refer to Chapter 1 for the torque specifications for the pan bolts and the type and quantity of transaxle fluid required to refill the transaxle.

**4.15** Disengage the plug on the lower end of the throttle valve cable from the cam on top of the valve body

## 5    Shift cable - removal, installation and adjustment

**Warning:** *These models are equipped with airbags. Always disable the airbag system before working in the vicinity of any airbag system component to avoid the possibility of accidental deployment of the airbag(s), which could cause personal injury (see Chapter 12).*

## Removal and installation

*Refer to illustrations 5.1a, 5.1b, 5.3a, 5.3b, 5.3c, 5.3d, 5.4a, 5.4b, 5.5a, 5.5b and 5.6*

1    Disconnect the shift cable from the manual shift lever at the transaxle and detach

**7B**

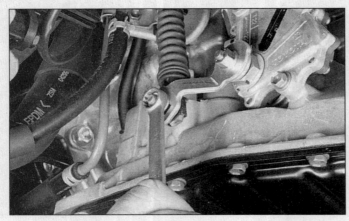

5.1a  To disconnect the shift cable from the manual shift lever at the transaxle, remove this nut

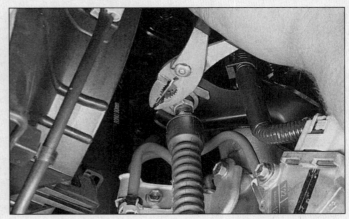

5.1b  To detach the shift cable from the bracket on the front of the transaxle, pry off this clip

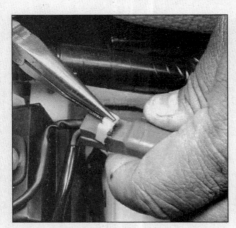

5.3a  Before attempting to remove the shift lever handle, remove this lock . . .

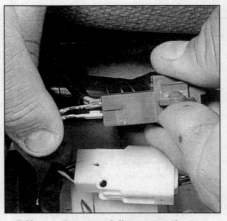

5.3b  . . . then carefully extract the two white overdrive leads from the blue electrical connector at the back of the shift lever base

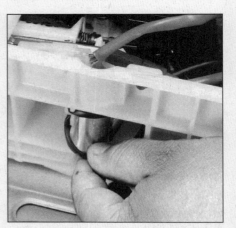

5.3c  Remove the four shift lever base bolts, tip the base on its side and remove the overdrive leads; note how they're routed forward, then back - they must be routed exactly the same way when installing the shift lever handle and overdrive switch

it from the bracket on the front of the transaxle **(see illustrations)**.

2    Remove the center console (see Chapter 11).

3    **Caution:** *Do not attempt to remove the shift lever handle until the overdrive switch leads have been disconnected from the blue connector at the rear of the shift lever base,*

*or you will break the leads to the switch.* On A245E models, removing the shift lever handle is tricky because the leads for the overdrive switch are routed down through the handle, under the shift lever base and back to the blue connector at the rear of the base.

And there is no connector between the switch in the handle and the blue connector. The two white leads are an integral part of the overdrive switch. If you try to pull them from the switch, you will have to buy a new switch. Fortunately, it's not absolutely necessary to remove the shift lever handle, *unless* you need to replace the handle itself, the gear-position indicator panel or the shift lever base. If necessary, unbolt the shift lever base (four bolts), follow the two white leads back to the blue connector at the rear of the shift lever base, carefully extract both leads from the connector, remove the shift lever handle retaining screws and remove the handle **(see illustrations)**. Note how the overdrive switch leads are routed forward, then back, underneath the shift lever base, and make sure you route them the same way during installation.

4    Remove the gear-position indicator panel retaining screws, lift up the gear-position indicator panel and disconnect the light bulb **(see illustrations)**. If you removed the shift lever handle, remove the gear-position

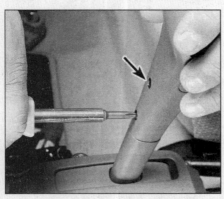

5.3d  To remove the shift lever handle, remove these two screws and pull the handle straight up, being extremely careful not to damage the two white leads for the overdrive switch

5.4a  To detach the gear position indicator panel from the shift lever base, remove these four screws (arrows) . . .

5.4b . . . then detach the shift indicator panel light and remove the panel

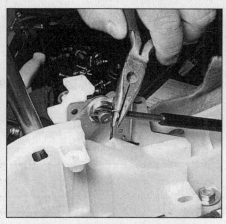

5.5a To disconnect the shift cable from the shift lever, pull out the retaining clip . . .

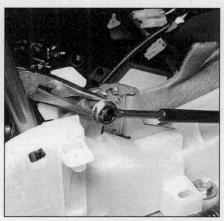

5.5b . . . and pry the cable from the pin on the shift lever

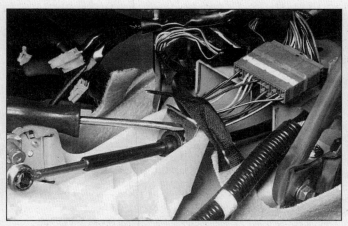

5.6 To detach the shift cable from the shift lever base, pry off this clip

6.2a Unplug the electrical connector from the Park/Neutral Position switch and check the continuity of the switch with an ohmmeter (use the proper terminal guide and continuity chart)

indicator panel; if you didn't remove the shift lever handle, tie the gear-position indicator panel to the shift lever handle so that you have room to work.

5    Remove the retaining clip and disconnect the shift cable from the shift lever (see illustrations).

6    Remove the retaining clip from the front edge of the shift lever base (see illustration).

7    Pull the cable through the grommet in the firewall.

8    Installation is the reverse of removal.

9    When you're done, adjust the shift cable.

## Adjustment

10    Loosen the nut on the manual shift lever at the transaxle (see illustration 5.1).

11    Push the lever toward the right side of the vehicle until it stops, then return it two notches to the Neutral position.

12    Move the shift lever inside the vehicle to the Neutral position.

13    While holding the lever with a slight pressure toward the Reverse position, tighten the nut securely.

14    Check the operation of the transaxle in each shift lever position (try to start the engine in each gear - the starter should operate in the Park and Neutral positions only).

**7B**

## 6    Park/Neutral Position (PNP) switch - check, adjustment and replacement

## Check

*Refer to illustrations 6.2a, 6.2b, 6.2c and 6.2d*

1    Apply the parking brake and block the rear wheels, raise the front of the vehicle and support it securely on jackstands. Remove the engine under covers.

2    Unplug the electrical connector from the

O——O Continuity

| Terminal<br>Shift Position | 3 | 2 | 9 | 1 | 4 | 6 | 5 | 7 | 8 |
|---|---|---|---|---|---|---|---|---|---|
| P | O—|—O | O—|—O | | | | | |
| R | | | O | | O | | | | |
| N | O—|—O | O | | | O | | | |
| D | | | O | | | | O | | |
| 2 | | | O | | | | | O | |
| L | | | O | | | | | | O |

6.2b PNP switch terminal guide and continuity chart - 1993 and 1994 models

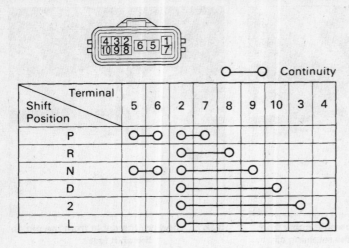

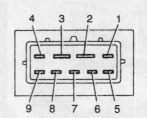

| Shift Position | Terminal 5 | 6 | 2 | 7 | 8 | 9 | 10 | 3 | 4 |
|---|---|---|---|---|---|---|---|---|---|
| P | O—O | | O—O | | | | | | |
| R | | | O——O | | O | | | | |
| N | O—O | | O—O | | | O | | | |
| D | | | O——O | | | | O | | |
| 2 | | | O——O | | | | | O | |
| L | | | O——O | | | | | | O |

O——O  Continuity

**6.2c  PNP switch terminal guide and continuity chart – 1995 through 1997 models**

| Shift position | Terminal No. to continuity | |
|---|---|---|
| P | 1 – 6 | 2 – 3 |
| R | 5 – 6 | – |
| N | 2 – 3 | 6 – 7 |
| D | 6 – 8 | – |
| 2 | 6 – 9 | – |
| L | 4 – 6 | – |

**6.2d  PNP switch terminal guide and continuity chart - 1998 and later models**

PNP switch. Using an ohmmeter, check the continuity of the switch across the indicated terminals in each gear position **(see illustrations)**.

3    If the switch fails any of the checks, try adjusting it as described in the following procedure, then recheck it. If it still fails, replace it.

### Adjustment

4    If the engine will start with the shift lever in any position other than Park or Neutral, adjust the Park/Neutral Position switch.

5    Apply the parking brake and block the rear wheels. Raise the front of the vehicle and place it securely on jackstands. Shift the transaxle into Neutral.

6    Unplug the electrical connector from the switch and loosen the switch retaining bolts.

7    Touch the ohmmeter leads to the Neutral terminals of the switch (terminals 2 and 3 on 1993 and 1994 models, terminals 5 and 6 on 1995 through 1997 models, and terminals 2 and 3 on 1998 and later models) and rotate the switch until there is continuity between the terminals, indicating that it's now in the

Neutral position **(see illustration)**. Tighten the bolts securely.

### Replacement

*Refer to illustration 6.14*

8    Disconnect the negative cable from the battery. **Caution:** *If the stereo in your vehicle is equipped with an anti-theft system, make sure you have the correct activation code before disconnecting the battery.*

9    Shift the transaxle into Neutral.

10    Remove the nut and lift off the shift lever.

11    Unplug the electrical connector.

12    Remove the retaining bolts and lift the switch off the shift shaft.

13    To install, line up the flats on the shift shaft with the flats in the switch and push the switch onto the shaft.

14    Rotate the switch until the neutral basic line aligns with the groove **(see illustration)**. Tighten the bolts securely and connect the electrical connector.

15    Install the shift lever, connect the negative battery cable and verify the engine will not start with the shift lever in any position other than Park or Neutral, if necessary follow the adjustment procedure (Steps 5 through 7).

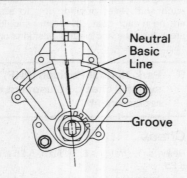

**6.14  Rotate the switch until the neutral basic line aligns with the groove, then tighten the bolts**

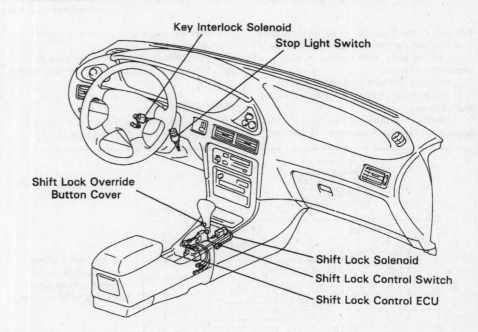

**7.1a  Shift lock system components - 1997 and earlier models**

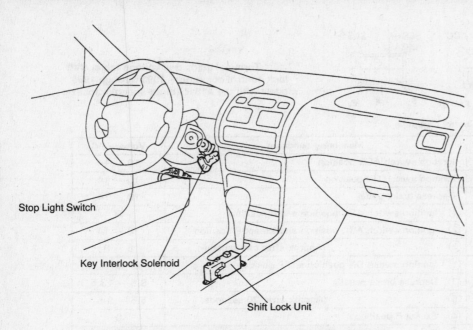

**7.1b Shift lock system components - 1998 and later models**

## 7  Shift lock system - description, check and component replacement

### Description

*Refer to illustrations 7.1a and 7.1b*

1   The shift lock system (see illustrations) prevents the shift lever from being shifted out of Park or Neutral until the brake pedal is applied.

### Key interlock solenoid

*Refer to illustration 7.4*

2   Remove the steering column covers (see Chapter 11).
3   Unplug the electrical connector for the key interlock solenoid.
4   Using an ohmmeter, measure the resistance between the indicated terminals on the solenoid side of the connector (see illustration), and compare your measurements with the resistance listed in this Chapter's Specifications.
5   Apply battery voltage between the same two terminals. Verify that the solenoid makes an audible "click" when energized. **Caution:** *Don't apply battery voltage any longer than necessary for this check, as the solenoid*

could be damaged.
6   If the key interlock solenoid fails either of these tests, replace it.
7   Install the steering column covers.

### Shift lock solenoid (1997 and earlier models)

8   Remove the floor console (see Chapter 11).
9   Unplug the shift lock solenoid connector.
10   Using an ohmmeter, measure the resistance between the two terminals of the connector (on the solenoid side of the connector), and compare your measurement with the resistance listed in this Chapter's Specifications.
11   Apply battery voltage to the same two terminals and verify that there's an audible "click" from the solenoid. **Caution:** *Don't apply battery voltage any longer than necessary for this check, as the solenoid could be damaged.*
12   If the shift lock solenoid fails either of these tests, replace it.

### Shift lock control switch

*Refer to illustration 7.13*

13   Unplug the electrical connector and verify that there's continuity between the indicated terminals in each shift lever position (see illustration).
14   If the shift lock control switch fails any of these tests, replace it.

### Shift lock control computer

*Refer to illustrations 7.15a, 7.15b, 7.15c and 7.15d*

15   Turn the ignition key to On and, using a

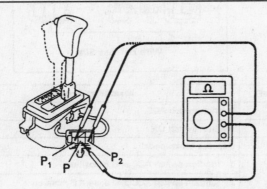

**7.13 Terminal guide and continuity table, shift lock control switch**

**7B**

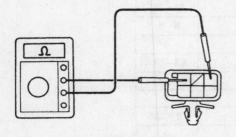

**7.4 Terminal guide, key interlock solenoid (1994 and earlier models shown; later models have a two-terminal connector)**

○—○ : Continuity

| Shift Position \ Terminal | P | P₁ | P₂ |
|---|---|---|---|
| P position (Release button not is pushed) | ○ | ○ | |
| P position (Release button is pushed) | ○ | ○ | |
| | ○ | | ○ |
| R,N,D,2,L position | ○ | | ○ |

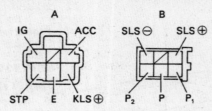

Control Computer

7.15a  Terminal guide and voltage table, shift lock control computer - 1993 through 1995 models with the A131L (three-speed) transaxle

| Connector | Terminal | Measuring condition | | Voltage (V) |
|---|---|---|---|---|
| A | ACC — E | Ignition switch ACC position | | 10 — 14 |
| | IG — E | ignition switch ON position | | 10 — 14 |
| | STP — E | Depress brake pedal | | 10 — 14 |
| | KLS — E | ① | Ignition switch ACC position and P position | 0 |
| | | ② | Ignition switch ACC position and except P position | 8 — 14 |
| | | ③ | (Approx-after one second) | 6 — 9 |
| B | SLS ⊕ — SLS ⊖ | ① | Ignition switch ON position and P position | 0 |
| | | ② | Depress brake pedal | 8.5 — 13.5 |
| | | ③ | (Approx-after 20 seconds) | 5.5 — 9.5 |
| | | ④ | Except P position | 0 |
| | P₁ — P | ① | Ignition switch ON, P position and depress brake pedal | 0 |
| | | ② | Except P position | 9 — 13.5 |
| | P₂ — P₁ | ③ | Ignition switch ACC position and P position | 9 — 13.5 |
| | | ④ | Except P position | 0 |

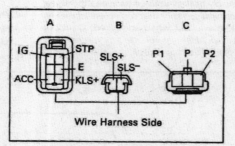

Wire Harness Side

7.15b  Terminal guide and voltage table, shift lock control computer - 1996 and 1997 models with the A131L (three-speed) transaxle

| Connector | Terminal | Measuring condition | | Voltage (V) |
|---|---|---|---|---|
| A | ACC – E | Ignition switch ACC | | 10 – 14 |
| | IG – E | Ignition switch ON | | 10 – 14 |
| | STP – E | Depress brake pedal | | 10 – 14 |
| | KLS⁺ – E | 1 | Ignition switch ACC and P position | 0 |
| | | 2 | Ignition switch ACC and except P position | 10 – 14 |
| | | 3 | (After approx 1 second) | 6 – 9 |
| B | SLS⁺ – SLS⁻ | 1 | Ignition switch ON and P position | 0 |
| | | 2 | Depress brake pedal | 8.5 – 13.5 |
| | | 3 | (After approx 20 seconds) | 5.5 – 9.5 |
| | | 4 | Except P position | 0 |
| C | P1 – P | 1 | Ignition switch ON, P position and depress brake pedal | 9 – 13.5 |
| | | 2 | Shift except P position under conditions above | 0 |
| | P2 – P | 1 | Ignition switch ACC and P position | 0 |
| | | 2 | Shift except P position under condition above | 9 – 13.5 |

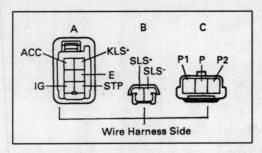

**7.15c Terminal guide and voltage table, shift lock control computer - 1997 and earlier models with the A245E (four-speed) transaxle**

| Connector | Terminal | Measuring condition | Voltage (V) |
|---|---|---|---|
| A | ACC – E | Ignition switch ACC position | 10 – 14 |
| | IG – E | Ignition switch ON position | 10 – 14 |
| | STP – E | Depress brake pedal | 10 – 14 |
| | KLS$^+$ – E | 1  Ignition switch ACC position and P position | 0 |
| | | 2  Ignition switch ACC position and except P position | 10 – 14 |
| | | 3      (Approx-after one second) | 6 – 9 |
| B | SLS$^+$ – SLS$^-$ | 1  Ignition switch ON position and P position | 0 |
| | | 2  Depress brake pedal | 8.5 – 13.5 |
| | | 3      (Approx-after 20 seconds) | 5.5 – 9.5 |
| | | 4  Except P position | 0 |
| C | P1 – P | 1  Ignition switch ON, P position and depress brake pedal | 0 |
| | | 2  Shift except P position under conditions above | 9 – 13.5 |
| | P2 – P | 1  Ignition switch ACC position and P position | 9 – 13.5 |
| | | 2  Shift except P position under condition above | 0 |

voltmeter, measure the voltage at the indicated terminals **(see illustrations)** and compare your measurements with the voltage values specified in the **accompanying illustrations. Note:** *Do NOT unplug the computer connector; instead, "backprobe" the connec-* tor *(probe through the back of the wiring harness connector with a small T-head needle).*
16   If the shift lock control computer fails any of the tests, replace it.
17   Install the floor console (see Chapter 11).

## 8   Transaxle mount - check and replacement

*Refer to illustration 8.1*
1   Insert a large screwdriver or prybar

**7B**

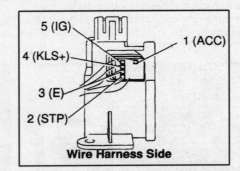

**Wire Harness Side**

**7.15d Terminal guide and voltage table, shift lock control computer - 1998 and later models**

| Terminal | Measuring Condition | Voltage (V) |
|---|---|---|
| 1 – 3 (ACC – E) | Ignition switch ACC | 10 – 14 |
| 5 – 3 (IG – E) | Ignition switch ON | 10 – 14 |
| 2 – 3 (STP – E) | Depressing brake pedal | 10 – 14 |
| 4 – 3 (KLS$^+$ – E) | (1) Ignition switch ACC and P position | 0 |
| | (2) Ignition switch ACC and except P position | 7.5 – 11 |
| | (2) Ignition switch ACC and except P position (After approx. 1 second) | 6 – 9.5 |

**8.1  To check any of the three transaxle mounts, insert a large screwdriver or prybar between the mount and the body as shown, and try to lever the mount from side to side or up and down - there should be a little movement, but it should be firm; if the mount moves too easily or looks cracked and torn, replace it (upper left transaxle mount shown)**

between the transaxle mount and the body and try to pry it away from the body **(see illustration)**.

2    The transaxle mount should not move excessively. If it does, replace the mount.

3    To replace a mount, support the transaxle with a jack, remove the nuts and bolts and remove the mount. It may be necessary to raise the transaxle slightly to provide enough clearance to remove the mount.

4    Installation is the reverse of removal.

## 9    Automatic transaxle - removal and installation

### *Removal*

*Refer to illustrations 9.4, 9.5, 9.6, 9.14a, 9.14b, 9.14c, 9.16, 9.19a, 9.19b, 9.20 and 9.21*

1    Detach the cable from the negative bat-

**9.4  Remove these bolts (arrows) from the upper side of the left transaxle mount (there are two more bolts, but they're underneath, on the front side)**

tery terminal. **Caution:** *If the stereo in your vehicle is equipped with an anti-theft system, make sure you have the correct activation code before disconnecting the battery.*

2    If you're planning to reuse the same transaxle, simply remove the cable retaining clip from the bracket and disconnect the throttle valve cable from the linkage on the throttle body; if you're planning to replace the transaxle, disconnect the throttle valve cable from the valve body (in either case, see Section 4).

3    Remove the air cleaner housing (see Chapter 4).

4    Remove the upper bolts from the left transaxle mount **(see illustration)**. (There are two more bolts, but you won't be able to reach them until the vehicle is raised.)

5    Detach the ground cable **(see illustration)** from the transaxle. Unplug the electrical connectors for the speed sensor (see Section 3), Park/Neutral Position switch (see Section 6) and solenoid (A245E only), next to the Park/Neutral Position switch on the front left corner of the transaxle. Detach any wiring harness clamps from the transaxle and set the wiring harnesses aside.

**9.5  Remove this bolt (arrow) and detach the ground cable from the top of the transaxle**

6    Remove the upper transaxle-to-engine bolts **(see illustration)**. Remove the upper transaxle-to-starter motor bolt (see Chapter 5).

7    Raise the vehicle and support it securely on jackstands.

8    Remove the engine under covers.

9    Disconnect the shift cable from the transaxle (see Section 5).

10   Remove the exhaust pipe section between the exhaust manifold and the catalytic converter (see Chapter 4).

11   Drain the transaxle fluid and on three-speed models, the differential fluid (see Chapter 1).

12   Remove the driveaxles (see Chapter 8). **Note:** *It's not absolutely necessary to completely remove the driveaxles; you can detach the inner CV joints and suspend them out of the way. However, you'll have more room to work if you remove the driveaxles. And this is a good time to inspect the CV joint boots for tears and deterioration and, if necessary, repack them with new CV joint grease (see Chapter 8).*

13   Support the engine using an engine support fixture or a hoist from above, or a jack and a wood block under the oil pan to spread the load.

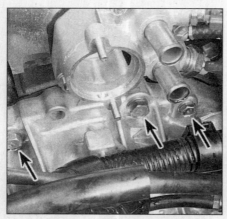

**9.6  Remove the two upper transaxle-to-engine bolts (arrows); the bolt on the right (arrow) is the upper starter bolt**

**9.14a  Remove the four bolts at the front of the support brace (arrows) . . .**

**9.14b  . . . the three nuts at the rear of the support brace (arrows) . . .**

9.14c . . . and the five bolts (arrows) from each side of the
suspension crossmember

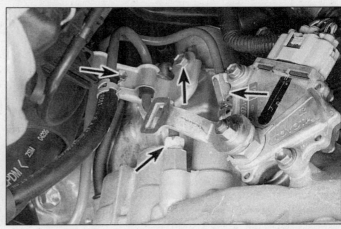

9.16 Unbolt the oil cooler line clamps (arrows) from the transaxle,
unscrew the line fittings (arrows) and disconnect the lines from
the transaxle

14   Support the suspension crossmember
with a floor jack, then remove the bolts and
nuts that attach the center support brace
(engine crossmember) and suspension crossmember to the vehicle (see illustrations).
15   Detach the starter motor leads, remove
the lower starter-to-transaxle bolt and
remove the starter (see Chapter 5).
16   Detach the oil cooler line retaining
clamps from the transaxle, then disconnect
the two line fittings from the transaxle (see
illustration).
17   To detach the dipstick tube from the
transaxle, remove the single bracket bolt
from the front of the bellhousing, then pull
straight up on the tube.
18   Support the transaxle with a jack -
preferably a special jack made for this purpose. Safety chains will help steady the
transaxle on the jack.
19   Remove the remaining two bolts from
the underside of the transaxle mount (see
illustration). Remove the bolts from the front
and rear mounts (see illustration) so that the
engine can be tilted slightly to the left to allow
easier removal and installation of the
transaxle.
20   Remove the torque converter inspection
cover. Mark the relationship of the torque

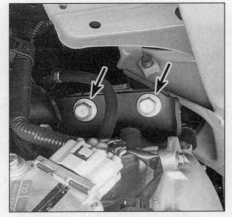

9.19a Remove these two bolts (arrows)
from the underside of the left
transaxle mount

converter to the driveplate so they can be
installed in the same position (see illustration). Remove the six torque converter
mounting bolts. Turn the crankshaft for
access to each one in turn.
21   Remove the lower engine-to-transaxle
bolts: the three lowest bolts are shown in
illustration 9.20, and there's another bolt on

9.19b Remove the nut and through bolt
(arrows) from the front and rear mounts
(front mount shown, rear mount similar)

the backside of the engine, right below the
lower starter bolt. Remove the lower
transaxle-to-engine bolt, on the front side of
the transaxle (see illustration).

9.20 Mark the
relationship of the
torque converter to the
driveplate to ensure
proper dynamic
balance when it's
reattached, then
remove all six torque
converter bolts (bolt
showing in access
window) by rotating
the crankshaft to bring
each bolt to the
bottom, where you can
get at it

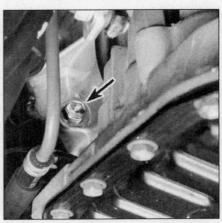

9.21 Remove this front lower transaxle-
to-engine bolt (arrow)

7B

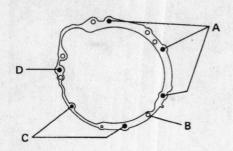

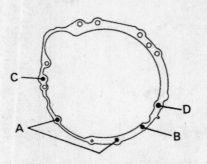

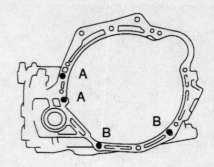

9.27a  1993 and 1994 A245E transaxle-to-engine bolt torque guide - refer to the torque listed in this Chapter's Specifications for each bolt (Corolla models only)

9.27b  1995 through 1997 A245E transaxle-to-engine bolt torque guide - refer to the torque listed in this Chapter's Specifications for each bolt (Corolla models only)

9.27c  1998 and later transaxle-to-engine bolt torque guide (A131L and A245E) - refer to the torque listed in this Chapter's Specifications for each bolt (Corolla models only)

22   Move the transaxle to the side to disengage it from the engine block dowel pins and make sure the torque converter is detached from the driveplate. Secure the torque converter to the transaxle so that it will not fall out during removal. Lower the transaxle from the vehicle.

## Installation

*Refer to illustrations 9.27a, 9.27b and 9.27c*

23   Make sure that the torque converter is securely engaged in the transaxle prior to installation.
24   With the transaxle secured to the jack, raise it into position. Be sure to keep it level so the torque converter does not slide forward. Connect the cooler lines.
25   Move the transaxle carefully into place until the dowel pins are engaged and the torque converter is engaged.
26   Rotate the torque converter to align the bolt holes with the holes in the driveplate. The match marks on the torque converter and driveplate, made during step 20, must align.

27   Install the lower transaxle-to-engine bolts and tighten them to the torque listed in this Chapter's Specifications. If you're working on a Corolla model, refer to the accompanying bolt torque guides **(see illustrations)**.
28   Install the torque converter-to-driveplate bolts. **Note:** *Install all of the bolts before tightening any of them.* Tighten them to the torque listed in this Chapter's Specifications. Install the torque converter cover.
29   Install the support brace and suspension member. Tighten the bolts and nuts securely.
30   Remove the jacks supporting the transaxle and the engine.
31   Install the starter motor (see Chapter 5). (It's easier to install the upper bolt after the vehicle has been lowered.)
32   Install the dipstick tube into the transaxle and attach the dipstick bracket to the transaxle.
33   Connect the oil cooler line fittings and the line retaining clamps to the transaxle.
34   Install and/or connect the driveaxles to the transaxle (see Chapter 8).

35   Install and adjust the Park/Neutral Position switch (see Section 6).
36   Connect and adjust the shift cable (see Section 5) and the throttle valve cable (see Section 4).
37   Plug in the electrical connectors for the solenoid (A245E), Park/Neutral Position switch and vehicle speed sensor. Make sure that the wiring harnesses are routed properly and clamped to the transaxle housing.
38   Install the exhaust pipe between the exhaust manifold and the catalytic converter (see Chapter 5).
39   Remove the jackstands and lower the vehicle.
40   Install the upper transaxle bolts **(see illustration 9.27)** and tighten them to the torque listed in this Chapter's Specifications. Install the upper starter motor bolt, if you haven't already done so, and tighten it to the torque listed in the Chapter 5 Specifications.
41   Fill the transaxle with the proper type and amount of fluid (see Chapter 1). Run the vehicle and check for fluid leaks.

# Chapter 8
# Clutch and driveaxles

## Contents

## Specifications

### Clutch

| | |
|---|---|
| Fluid type | See Chapter 1 |
| Pedal freeplay | See Chapter 1 |
| Driveaxle standard length | |
| Saginaw type | |
|     Left driveaxle | 21-17/64 inches |
|     Right driveaxle | 33-3/4 inches |
| Toyota type | |
|     Left driveaxle | 20-51/64 inches |
|     Right driveaxle | 33-11/64 inches |

### Torque specifications

**Ft-lbs** (unless otherwise indicated)

| | |
|---|---|
| Clutch master cylinder mounting nuts | 108 in-lbs |
| Clutch pressure plate-to-flywheel bolts | |
|   Corolla | 168 in-lbs |
|   Prizm | 17 |
| Clutch release cylinder mounting bolts | 108 in-lbs |
| Clutch release cylinder banjo bolt (1998 and later models only) | 18 |
| Driveaxle/hub nut | |
|   1997 and earlier | 159 |
|   1998 and later | 166 |
| Wheel lug nuts | See Chapter 1 |

8

## 1   General information

The information in this Chapter deals with the components from the rear of the engine to the front wheels, except for the transaxle, which is dealt with in Chapters 7A and 7B. For the purposes of this Chapter, these components are grouped into two categories: clutch and driveaxles. Separate Sections within this Chapter offer general descriptions and checking procedures for both groups.

Since nearly all the procedures covered in this Chapter involve working under the vehicle, make sure it's securely supported on sturdy jackstands or a hoist where the vehicle can be easily raised and lowered.

## 2   Clutch - description and check

*Refer to illustration 2.1*

1    All vehicles with a manual transaxle use a single dry plate, diaphragm spring type clutch **(see illustration)**. The clutch disc has a splined hub which allows it to slide along the splines of the transaxle input shaft. The clutch and pressure plate are held in contact by spring pressure exerted by the diaphragm in the pressure plate.

2    The clutch release system is operated by hydraulic pressure. The hydraulic release system consists of the clutch pedal, a master cylinder and fluid reservoir, the hydraulic line, a slave cylinder which actuates the clutch release lever and the clutch release (or throw-out) bearing.

3    When pressure is applied to the clutch pedal to release the clutch, hydraulic pressure is exerted against the outer end of the clutch release lever. As the lever pivots, the shaft fingers push against the release bearing. The bearing pushes against the fingers of the diaphragm spring of the pressure plate assembly, which in turn releases the clutch plate.

4    Terminology can be a problem regarding the clutch components because common names have in some cases changed from that used by the manufacturer. For example, the driven plate is also called the clutch plate or disc, the pressure plate assembly is sometimes referred to as the clutch cover, the clutch release bearing is sometimes called a throw-out bearing, and the release cylinder is sometimes called the operating or slave cylinder.

5    Other than replacing components that have obvious damage, some preliminary checks should be performed to diagnose a clutch system failure.

a)  *The first check should be of the fluid level in the clutch master cylinder (see Chapter 1). If the fluid level is low, add fluid as necessary and inspect the hydraulic clutch system for leaks. If the master cylinder reservoir has run dry, bleed the system (see Section 7) and retest the clutch operation.*

b)  *To check "clutch spin down time," run the engine at normal idle speed with the transaxle in Neutral (clutch pedal up - engaged). Disengage the clutch (pedal down), wait several seconds and shift the transaxle into Reverse. No grinding noise should be heard. A grinding noise would most likely indicate a problem in the pressure plate or the clutch disc.*

c)  *To check for complete clutch release, run the engine (with the parking brake applied to prevent movement) and hold the clutch pedal approximately 1/2-inch from the floor. Shift the transaxle between 1st gear and Reverse several times. If the shift is not smooth, component failure is indicated. Check the release cylinder pushrod travel. With the clutch pedal depressed completely the release cylinder pushrod should extend substantially. If it doesn't, check the fluid level in the clutch master cylinder.*

d)  *Visually inspect the clutch pedal bushing at the top of the clutch pedal to make sure there is no sticking or excessive wear.*

e)  *Under the vehicle, check that the clutch release lever is solidly mounted on the ball stud.*

## 3   Clutch components - removal, inspection and installation

**Warning:** *Dust produced by clutch wear and deposited on clutch components is hazardous to your health. DO NOT blow it out with compressed air and DO NOT inhale it. DO NOT use gasoline or petroleum based solvents to remove the dust. Brake system cleaner should be used to flush the dust into a drain pan. After the clutch components are wiped clean with a rag, dispose of the contaminated rags and cleaner in a labeled, covered container.*

### Removal

*Refer to illustration 3.6*

1    Access to the clutch components is normally accomplished by removing the transaxle, leaving the engine in the vehicle. If, of course, the engine is being removed for major overhaul, then the opportunity should always be taken to check the clutch for wear and replace worn components as necessary. However, the relatively low cost of the clutch components compared to the time and labor involved in gaining access to them warrants their replacement any time the engine or transaxle is removed, unless they are new or in near-perfect condition. The following procedures assume that the engine will stay in place.

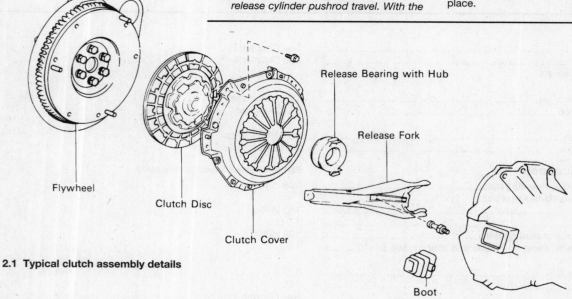

**2.1  Typical clutch assembly details**

Flywheel

Clutch Disc

Clutch Cover

Release Bearing with Hub

Release Fork

Boot

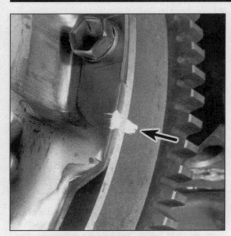

**3.6  Mark the relationship of the pressure plate to the flywheel (in case you're going to reuse the same pressure plate)**

2    Remove the release cylinder (see Section 6). Hang it out of the way with a piece of wire - it isn't necessary to disconnect the hose.

3    Remove the transaxle from the vehicle (see Chapter 7A). Support the engine while the transaxle is out. Preferably, an engine hoist should be used to support it from above. However, if a jack is used underneath the engine, make sure a piece of wood is used between the jack and oil pan to spread the load. **Caution:** *The pick-up for the oil pump is very close to the bottom of the oil pan. If the pan is bent or distorted in any way, engine oil starvation could occur.*

4    The release fork and release bearing can

**3.10  Examine the clutch disc for evidence of excessive wear, such as smeared friction material, loose rivets, worn hub splines and distorted damper cushions or springs**

remain attached to the transaxle for the time being.

5    To support the clutch disc during removal, install a clutch alignment tool through the clutch disc hub.

6    Carefully inspect the flywheel and pressure plate for indexing marks. The marks are usually an X, an O or a white letter. If they cannot be found, scribe marks yourself so the pressure plate and the flywheel will be in the same alignment during installation **(see illustration)**.

7    Slowly loosen the pressure plate-to-flywheel bolts. Work in a diagonal pattern and loosen each bolt a little at a time until all

spring pressure is relieved. Then hold the pressure plate securely and completely remove the bolts, followed by the pressure plate and clutch disc.

## Inspection

*Refer to illustrations 3.10, 3.12a and 3.12b*

8    Ordinarily, when a problem occurs in the clutch, it can be attributed to wear of the clutch driven plate assembly (clutch disc). However, all components should be inspected at this time.

9    Inspect the flywheel for cracks, heat checking, score marks and other damage. If the imperfections are slight, a machine shop can resurface it to make it flat and smooth. Refer to Chapter 2 for the flywheel removal procedure.

10    Inspect the lining on the clutch disc. There should be at least 1/32-inch of lining above the rivet heads. Check for loose rivets, distortion, cracks, broken springs and other obvious damage **(see illustration)**. As mentioned above, ordinarily the clutch disc is replaced as a matter of course, so if in doubt about the condition, replace it with a new one.

11    The release bearing should be replaced along with the clutch disc (see Section 4).

12    Check the machined surface and the diaphragm spring fingers of the pressure plate **(see illustrations)**. If the surface is grooved or otherwise damaged, replace the pressure plate assembly. Also check for obvious damage, distortion, cracking, etc. Light glazing can be removed with emery cloth or sandpaper. If a new pressure plate is indicated, new or factory rebuilt units are available.

## Installation

*Refer to illustration 3.14*

13    Before installation, carefully wipe the flywheel and pressure plate machined surfaces clean. It's important that no oil or grease is on these surfaces or the lining of the clutch disc. Handle these parts only with clean hands.

14    Position the clutch disc and pressure

**NORMAL FINGER WEAR**

**BROKEN OR BENT FINGERS**

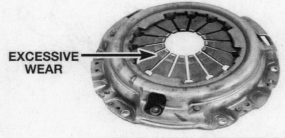

EXCESSIVE WEAR

**EXCESSIVE FINGER WEAR**

**3.12a  Replace the pressure plate if excessive wear or damage are noted**

**3.12b  Examine the pressure plate friction surface for score marks, cracks and evidence of overheating (blue spots)**

**8**

**3.14  Center the clutch disc in the pressure plate with a clutch alignment tool**

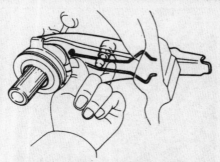

**4.3  Reach behind the release lever and disengage the lever from the ball stud by pulling on the retention spring, then remove the lever and bearing**

**4.4  To check the operation of the bearing, hold it by the outer race and rotate the inner race while applying pressure - the bearing should turn smoothly - if it doesn't, replace it**

plate with the clutch held in place with an alignment tool **(see illustration)**. Make sure it's installed properly (most replacement clutch plates will be marked "flywheel side" or something similar - if not marked, install the clutch disc with the damper springs or cushion toward the transaxle).

15   Tighten the pressure plate-to-flywheel bolts only finger tight, working around the pressure plate.

16   Center the clutch disc by ensuring the alignment tool is through the splined hub and into the recess in the crankshaft. Wiggle the tool up, down or side-to-side as needed to bottom the tool. Tighten the pressure plate-to-flywheel bolts a little at a time, working in a crisscross pattern to prevent distortion of the cover. After all of the bolts are snug, tighten them to the torque listed in this Chapter's Specifications. Remove the alignment tool.

17   Using high-temperature grease, lubricate the inner groove of the release bearing (see Section 4). Also place grease on the release lever contact areas and the transaxle input shaft bearing retainer.

18   Install the clutch release bearing (see Section 4).

19   Install the transaxle, release cylinder and all components removed previously, tightening all fasteners to the proper torque specifications.

## 4   Clutch release bearing and lever - removal, inspection and installation

**Warning:** *Dust produced by clutch wear and deposited on clutch components is hazardous to your health. DO NOT blow it out with compressed air and DO NOT inhale it. DO NOT use gasoline or petroleum-based solvents to remove the dust. Brake system cleaner should be used to flush it into a drain pan. After the clutch components are wiped clean with a rag, dispose of the contaminated rags and cleaner in a labeled, covered container.*

### *Removal*

*Refer to illustration 4.3*

1   Disconnect the negative cable from the battery. **Caution:** *If the stereo in your vehicle is equipped with an anti-theft system, make sure you have the correct activation code before disconnecting the battery.*

2   Remove the transaxle (see Chapter 7).

3   Remove the clutch release lever from the ball stud, then remove the bearing from the lever **(see illustration)**.

### *Inspection*

*Refer to illustration 4.4*

4   Hold the bearing by the outer race and rotate the inner race while applying pressure **(see illustration)**. If the bearing doesn't turn

smoothly or if it's noisy, replace the bearing/hub assembly with a new one. Wipe the bearing with a clean rag and inspect it for damage, wear and cracks. Don't immerse the bearing in solvent - it's sealed for life and to do so would ruin it. Also check the release lever for cracks and bends.

### *Installation*

*Refer to illustrations 4.5 and 4.6*

5   Fill the inner groove of the release bearing with high-temperature grease. Also apply a light coat of the same grease to the transaxle input shaft splines and the front bearing retainer **(see illustration)**.

6   Lubricate the release lever ball socket, lever ends and release cylinder pushrod socket with high-temperature grease **(see illustration)**.

7   Attach the release bearing to the release lever. On models that use a retaining clip, make sure it engages properly with the lever.

8   Slide the release bearing onto the transaxle input shaft front bearing retainer while passing the end of the release lever through the opening in the clutch housing. Push the clutch release lever onto the ball stud until it's firmly seated.

9   Apply a light coat of high-temperature grease to the face of the release bearing

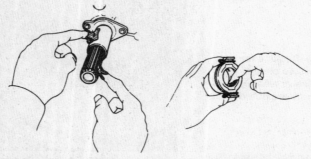

**4.5  Apply a light coat of high-temperature grease to the transaxle bearing retainer and also fill the release bearing groove**

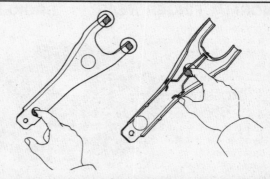

**4.6  Apply high-temperature grease to the release lever in the areas indicated**

**5.2 To release the clutch pushrod from the clutch pedal, remove this clip (arrow) and the clevis pin**

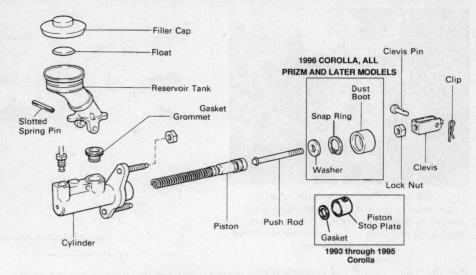

**5.5 Typical clutch master cylinder details**

where it contacts the pressure plate diaphragm fingers.

10   The remainder of installation is the reverse of the removal procedure.

---

**5   Clutch master cylinder - removal, overhaul and installation**

**Note:** *Before beginning this procedure, contact local parts stores and dealer service departments concerning the purchase of a rebuild kit or a new master cylinder. Availability and cost of the necessary parts may dictate whether the cylinder is rebuilt or replaced with a new one. If you decide to rebuild the cylinder, inspect the bore as described in Step 11 before purchasing parts.*

## Removal

*Refer to illustration 5.2*

1   Disconnect the negative cable from the battery. **Caution:** *If the stereo in your vehicle is equipped with an anti-theft system, make sure you have the correct activation code before disconnecting the battery.*

2   Under the dashboard, disconnect the pushrod from the top of the clutch pedal. It's held in place with a clevis pin. To remove the clevis pin, remove the clip **(see illustration)**.

3   Disconnect the hydraulic line at the clutch master cylinder. If available, use a flare-nut wrench on the fitting, to protect the fitting from being rounded off. Have rags handy as some fluid will be lost as the line is removed. **Caution:** *Don't allow brake fluid to come into contact with paint, as it will damage the finish.*

4   From under the dash, remove the nuts which attach the master cylinder to the firewall. Remove the master cylinder, again being careful not to spill any of the fluid. **Note:** *It may be necessary to loosen or partially remove the power brake booster to provide clearance for clutch master cylinder removal (see Chapter 9).*

## Overhaul

*Refer to illustrations 5.5, 5.6 and 5.8*

5   Remove the reservoir cap and drain all fluid from the master cylinder. Drive out the spring pin **(see illustration)** with a hammer and punch, then carefully pry off the reservoir.

6   If you're working on a 1995 or earlier Corolla, use a small screwdriver and bend out the staked part of the piston stop plate until it's flush with the surface of the stop plate **(see illustration)**. Remove the stop plate, gasket and pushrod from the cylinder.

7   If you're working on a Prizm or a 1996 or later Corolla, remove the dust boot, depress the pushrod and remove the snap-ring with a pair of snap-ring pliers. Pull out the pushrod and washer.

8   Tap the master cylinder on a block of wood to eject the piston assembly from inside the bore **(see illustration)**. **Note:** *If the rebuild kit supplies a complete piston assembly, ignore the Steps which don't apply.*

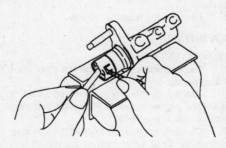

**5.6  If you're working on a 1995 or earlier Corolla, use a small screwdriver to bend the staked part of the piston stop plate out until it's flush with the surface of the stop plate, then remove the stop plate, gasket and pushrod from the cylinder bore**

9   Separate the spring from the piston.

10   Carefully remove the seal from the piston.

11   Inspect the bore of the master cylinder for deep scratches, score marks and ridges. The surface must be smooth to the touch. If the bore isn't perfectly smooth, the master cylinder must be replaced with a new or factory rebuilt unit.

12   If the cylinder will be rebuilt, use the new parts contained in the rebuild kit and follow any specific instructions which may have accompanied the rebuild kit. Wash all parts to be re-used with brake cleaner, denatured alcohol or clean brake fluid. DO NOT use petroleum-based solvents.

13   Attach the seal to the piston. The seal lips must face away from the pushrod end of the piston.

14   Assemble the spring on the other end of the piston.

15   Lubricate the bore of the cylinder and the seals with plenty of fresh brake fluid.

16   Carefully guide the piston assembly into

**5.8  Invert the cylinder and tap it against a block of wood to eject the piston**

**8**

**6.3  Use a flare-nut wrench when disconnecting the hydraulic line fitting (left arrow) to prevent rounding off the corners of the tubing nut, then remove the two mounting bolts (right arrows)**

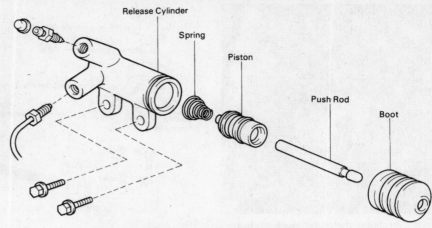

**6.6  Typical clutch release cylinder details**

the bore, being careful not to damage the seals. Make sure the spring end is installed first, with the pushrod end of the piston closest to the opening.

17   If you're working on a 1995 or earlier Corolla, position the pushrod and a new gasket in the bore, compress the spring and install a new stop plate. If you're working on a Prizm or a 1996 or later Corolla, install the pushrod and washer, depress the pushrod and install a new snap-ring, making sure it seats completely in its groove. Install a new dust boot.

18   Install the fluid reservoir with a new grommet. Drive in the spring pin with a small hammer and punch. Make sure the pin protrudes about 1/8-inch, or less, on each side of the reservoir bracket.

## Installation

19   Position the master cylinder on the firewall, installing the mounting nuts finger-tight.

20   Connect the hydraulic line to the master cylinder, moving the cylinder slightly as necessary to thread the fitting properly into the bore. Don't cross-thread the fitting as it's installed.

21   Tighten the mounting nuts and the hydraulic line fitting securely.

22   Connect the pushrod to the clutch pedal.

23   Fill the clutch master cylinder reservoir with brake fluid conforming to DOT 3 specifications and bleed the clutch system (see Section 7).

24   Check the clutch pedal height and freeplay (see Chapter 1).

## 6   Clutch release cylinder - removal, overhaul and installation

**Note:** *Before beginning this procedure, contact local parts stores and dealer service departments concerning the purchase of a rebuild kit or a new release cylinder. Availability and cost of the necessary parts may dictate whether the cylinder is rebuilt or replaced*

*with a new one. If it's decided to rebuild the cylinder, inspect the bore as described in Step 8 before purchasing parts.*

## Removal

*Refer to illustration 6.3*

1   Disconnect the negative cable from the battery. **Caution:** *If the stereo in your vehicle is equipped with an anti-theft system, make sure you have the correct activation code before disconnecting the battery.*

2   Raise the vehicle and support it securely on jackstands.

3   Disconnect the hydraulic line at the release cylinder. If available, use a flare-nut wrench on the fitting, which will prevent the fitting from being rounded off **(see illustration)**. Have a small can and rags handy, as some fluid will be spilled as the line is removed. **Note:** *The hydraulic line on 1998 and later models threads into a banjo fitting on the release cylinder. It's easier to unscrew the banjo bolt on these models, rather than loosen the line fitting (and it doesn't require a flare-nut wrench). The banjo fitting has a sealing washer on each side of it; these washers should be replaced with new ones during installation.*

4   Remove the release cylinder mounting bolts.

5   Remove the release cylinder.

## Overhaul

*Refer to illustration 6.6*

6   Remove the pushrod and the boot **(see illustration)**.

7   Tap the cylinder on a block of wood to eject the piston and seal. Remove the spring from inside the cylinder.

8   Carefully inspect the bore of the cylinder. Check for deep scratches, score marks and ridges. The bore must be smooth to the touch. If any imperfections are found, the release cylinder must be replaced with a new one.

9   Using the new parts in the rebuild kit, assemble the components using plenty of fresh brake fluid for lubrication. Note the installed direction of the spring and the seal.

## Installation

10   Install the release cylinder on the clutch housing. Make sure the pushrod is seated in the release fork pocket.

11   Connect the hydraulic line to the release cylinder. Tighten the connection securely. If you're working on a 1998 and later model and detached line fitting by removing the banjo bolt, install new sealing washers and tighten the banjo bolt to the torque listed in this Chapter's Specifications.

12   Fill the clutch master cylinder with brake fluid (conforming to DOT 3 specifications).

13   Bleed the system (see Section 7).

14   Lower the vehicle and connect the negative battery cable.

## 7   Clutch hydraulic system - bleeding

1   The hydraulic system should be bled of all air whenever any part of the system has been removed or if the fluid level has been allowed to fall so low that air has been drawn into the master cylinder. The procedure is similar to bleeding a brake system.

2   Fill the master cylinder with new brake fluid conforming to DOT 3 specifications. **Caution:** *Do not re-use any of the fluid coming from the system during the bleeding operation or use fluid which has been inside an open container for an extended period of time.*

3   Raise the vehicle and place it securely on jackstands to gain access to the release cylinder, which is located on the left side of the clutch housing.

4   Locate the bleeder valve on the clutch release cylinder (right above the fitting for the hydraulic fluid line). Remove the dust cap which fits over the bleeder valve and push a length of clear tubing over the valve. Place the other end of the hose into a clear container with about two inches of brake fluid in it. The hose end must be submerged in the fluid.

5   Have an assistant depress the clutch pedal and hold it. Open the bleeder valve on the release cylinder, allowing fluid to flow

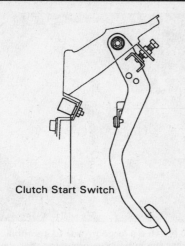

**8.4 The clutch start switch is located under the dash on a bracket in front of the clutch pedal**

Clutch Start Switch

through the hose. Close the bleeder valve when fluid stops flowing from the hose. Once closed, have your assistant release the pedal.
6    Continue this process until all air is evacuated from the system, indicated by a full, solid stream of fluid being ejected from the bleeder valve each time and no air bubbles in the hose or container. Keep a close watch on the fluid level inside the clutch master cylinder reservoir; if the level drops too low, air will be sucked back into the system and the process will have to be started all over again.
7    Install the dust cap and lower the vehicle. Check carefully for proper operation before placing the vehicle in normal service.

## 8   Clutch start switch - check and replacement

### Check

*Refer to illustrations 8.4 and 8.5*
1    Check the clutch pedal height and freeplay and the pushrod freeplay (see Chapter 1).
2    Verify that the engine will not start when the clutch pedal is released. Verify that the engine will start when the clutch pedal is depressed all the way.
3    If the clutch start switch doesn't perform as described, adjust and, if necessary, replace it.
4    Locate the switch **(see illustration)** and unplug the electrical connector.
5    Verify that there is continuity between the clutch start switch terminals when the switch is On (pedal depressed) **(see illustration)**.
6    Verify that no continuity exists between the switch terminals when the switch is Off (pedal released).
7    If the switch fails either of the tests, replace it.

### Replacement

8    Remove the nut nearest the plunger end of the switch and unscrew the switch. Unplug the electrical connector.
9    Installation is the reverse of removal. The switch is self-adjusting, so there's no need for adjustment.
10    Verify again that the engine doesn't start when the clutch pedal is released, and does start when the pedal is depressed.

## 9   Driveaxles - general information and inspection

1    Power is transmitted from the transaxle to the wheels through a pair of driveaxles. The inner end of each driveaxle is splined into the differential side gears. The outer ends of the driveaxles are splined to the axle hubs and locked in place by a large nut.
2    The inner ends of the driveaxles are equipped with sliding constant velocity joints, which are capable of both angular and axial motion. Each inner joint assembly consists of a tripod bearing and a joint housing (outer race) in which the joint is free to slide in and out as the driveaxle moves up and down with the wheel. The joints can be disassembled and cleaned in the event of a boot failure (see Section 11), but if any parts are damaged, the joints must be replaced as a unit. When buying parts for a driveaxle, or a complete replacement driveaxle assembly, make sure you get the right components: some 1993 through 1995 models use Saginaw driveaxles, others use Toyota units; all 1996 and later models use Saginaw units. Both driveaxles are similar in design, but the parts are not necessarily interchangeable. Unless an authorized dealer parts department says otherwise, make sure you obtain Saginaw parts for Saginaw units, and Toyota parts for Toyota units.
3    The outer CV joints are the "Rzeppa" type, which consists of ball bearings running between an inner race and an outer cage, is capable of angular but not axial movement. The outer joints should be cleaned, inspected and repacked, but they cannot be disassembled. If an outer joint is damaged, it must be replaced along with the axleshaft (the outer joint and axleshaft are sold as a single component).
4    The boots should be inspected periodically for damage and leaking lubricant. Torn CV joint boots must be replaced immediately or the joints can be damaged. Boot replacement involves removal of the driveaxle (see Section 10). **Note:** *Some auto parts stores carry "split" type replacement boots, which can be installed without removing the driveaxle from the vehicle. This is a convenient alternative; however, the driveaxle should be removed and the CV joint disassembled and cleaned to ensure the joint is free from contaminants such as moisture and dirt which will accelerate CV joint wear. The most common symptom of worn or damaged CV joints, besides lubricant leaks, is a click-*

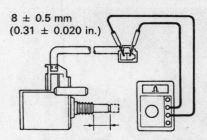

8 ± 0.5 mm
(0.31 ± 0.020 in.)

**8.5 Using an ohmmeter, check the continuity of the clutch start switch - there should be continuity when the switch is On (pedal depressed) and no continuity when it's Off (pedal released)**

ing noise in turns, a clunk when accelerating after coasting and vibration at highway speeds. To check for wear in the CV joints and driveaxle shafts, grasp each axle (one at a time) and rotate it in both directions while holding the CV joint housings, feeling for play indicating worn splines or sloppy CV joints. Also check the driveaxle shafts for cracks, dents and distortion.

## 10   Driveaxle - removal and installation

### Removal

*Refer to illustrations 10.4a, 10.4b, 10.5a, 10.5b, 10.6, 10.9 and 10.10*
1    Disconnect the cable from the negative terminal of the battery. **Caution:** *If the stereo in your vehicle is equipped with an anti-theft system, make sure you have the correct activation code before disconnecting the battery.*
2    Set the parking brake.
3    If the vehicle is equipped with stock steel wheels, remove the wheel cover and skip to Step 4. If the vehicle is equipped with alloy wheels, loosen the front wheel lug nuts, raise the vehicle and support it securely on jackstands. Remove the wheel.
4    Remove the cotter pin and nut lock from the driveaxle/hub nut **(see illustrations)**. If

**10.4a  Remove the cotter pin . . .**

8

10.4b  . . . and the nut lock

10.5a  The driveaxle/hub nut is very tight -
it should be loosened with the wheel on
the ground (otherwise, the vehicle might
topple off the jackstands)

10.5b  Use a large prybar to immobilize
the hub while loosening the
driveaxle/hub nut

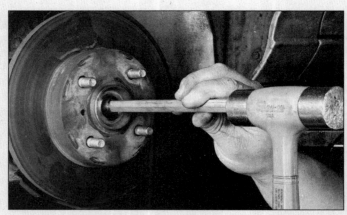

10.6  Using a brass punch, strike the end of the driveaxle sharply
with a hammer; when it breaks free, it will move noticeably

10.9  Pull the steering knuckle out and slide the end of the
driveaxle out of the hub

the vehicle is equipped with alloy wheels, reinstall the wheel and lug nuts, then lower the vehicle.

5    Loosen the driveaxle/hub nut (see illus-

10.10  To separate the inner end of the
driveaxle from the transaxle, pry on the
CV joint housing like this with a large
screwdriver or prybar - you may need to
give the prybar a sharp rap with
a brass hammer

tration), raise the vehicle and support it on jackstands, then remove the wheel. Remove the driveaxle/hub nut and washer. To prevent the hub from turning, wedge a prybar between two of the wheel studs and allow the prybar to rest against the ground or the floor pan of the vehicle (see illustration).

6    To loosen the driveaxle from the hub splines, tap the end of the driveaxle with a soft-faced hammer or a hammer and a brass punch (see illustration). Note: Don't attempt to push the end of the driveaxle through the hub yet. Applying force to the end of the driveaxle, beyond just breaking it loose from the hub, can damage the driveaxle or trans-axle. If the driveaxle is stuck in the hub splines and won't move, it may be necessary to remove the brake disc (see Chapter 9) and push it from the hub with a two-jaw puller after Step 8 is performed.

7    Remove the engine splash shields (see Chapter 1). Place a drain pan underneath the transaxle to catch any lubricant that may spill out when the driveaxle is removed.

8    Remove the nuts and bolt securing the balljoint to the control arm, then pry the con-trol arm down to separate the components (see Chapter 10).

9    Pull out on the steering knuckle and

detach the driveaxle from the hub (see illus-tration). Don't let the driveaxle hang by the inner CV joint after the outer end has been detached from the steering knuckle, as the inner joint could become damaged. Support the outer end of the driveaxle with a piece of wire, if necessary.

10   Carefully pry the inner CV joint out of the transaxle (see illustration).

11   Refer to Chapter 7 for the driveaxle oil seal replacement procedure.

## Installation

12   Installation is the reverse of the removal procedure, but with the following additional points:

a)  Push the driveaxle sharply in to seat the retaining ring on the inner CV joint in the groove in the differential side gear.

b)  With the wheel on the ground, tighten the driveaxle/hub nut to the torque listed in this Chapter's Specifications, then install the nut lock and a new cotter pin.

c)  Tighten the lug nuts to the torque listed in the Chapter 1 Specifications.

d)  Check the transaxle or differential lubri-cant and add, if necessary, to bring it to the proper level (see Chapter 1).

**11.3  Lift the tabs on all the boot clamps with a screwdriver, then open the clamps**

**11.4a  Remove the boot from the inner CV joint and slide the joint housing from the tripod**

**11.4b  On 1995 and earlier Saginaw driveaxles, the outer joint can be removed from the shaft by expanding the snap-ring**

**11.6  Remove the snap-ring with a pair of snap-ring pliers**

**11.7  Drive the tripod joint from the driveaxle with a brass punch and hammer; be careful not to damage the bearing surfaces or the splines on the shaft**

**11.10a  Wrap the splined area of the axleshaft with tape to prevent damage to the boots when removing or installing them**

## 11  Driveaxle boot replacement and CV joint inspection

**Note:** *If the CV joint boots must be replaced, explore all options before beginning the job. Complete rebuilt driveaxles are available on an exchange basis, which eliminates much time and work. Whichever route you choose to take, check on the cost and availability of parts before disassembling the vehicle.*

1    Remove the driveaxle (see Section 10).

### Disassembly

*Refer to illustrations 11.3, 11.4a, 11.4b, 11.6 and 11.7*

2    Mount the driveaxle in a vise with wood lined jaws (to prevent damage to the axleshaft). Check the CV joint for excessive play in the radial direction, which indicates worn parts. Check for smooth operation throughout the full range of motion for each CV joint. If a boot is torn, disassemble the joint, clean the components and inspect for damage due to loss of lubrication and possible contamination by foreign matter.

3    Using a small screwdriver, pry the retaining tabs on the clamps up to loosen them and slide them off **(see illustration)**.

4    Using a screwdriver, carefully pry up on the edge of the outer boot and push it away from the CV joint. Old and worn boots can be cut off. Pull the inner CV joint boot back from the housing and slide the housing from the tripod **(see illustration)**. **Note:** *If you are only replacing the outer boot on a 1995 or earlier Saginaw driveaxle, it isn't necessary to remove the inner joint. Mark the relationship of the outer joint to the shaft, then expand the snap-ring and slide the joint from the shaft* **(see illustration)**. *Use a new snap-ring upon installation.*

5    Mark the tripod and axleshaft to ensure that they are reassembled properly.

6    Remove the tripod joint snap-ring with a pair of snap-ring pliers **(see illustration)**.

7    Use a hammer and a brass punch to drive the tripod joint from the driveaxle **(see illustration)**.

8    If you haven't already cut them off, remove both boots. If you're working on a right-side Toyota-type (or 1996 and later) driveaxle, you'll also have to cut off the clamp for the dynamic damper and slide the damper off. **Note:** *The damper may have to be pressed off with a hydraulic press. Also, before removing the damper, measure its position from the end of the driveaxle - when reassembling, it must be returned to the same spot.*

### Check

9    Thoroughly clean all components, including the outer CV joint assembly, with solvent until the old CV joint grease is completely removed. Inspect the bearing surfaces of the inner tripods and housings for cracks, pitting, scoring and other signs of wear. It's very difficult to inspect the bearing surfaces of the inner and outer races of the outer CV joint, but you can at least check the surfaces of the ball bearings themselves. If they're in good shape, the races probably are too; if they're not, neither are the races. If the inner CV joint is worn, you can buy a new inner CV joint and install it on the old axleshaft; if the outer CV joint on a Toyota driveaxle is worn, you'll have to purchase a new outer CV joint *and* axleshaft (they're sold preassembled). The outer joint on Saginaw driveaxles can be replaced separately.

### Reassembly

*Refer to illustrations 11.10a, 11.10b, 11.10c, 11.10d, 11.11, 11.12a, 11.12b, 11.12c and 11.12d*

10    Wrap the splines on the end the axleshaft with electrical tape to protect the boots from the sharp edges of the splines **(see**

**8**

11.10b  Install the tripod with the recessed portion of the splines facing the axleshaft

11.10c  Place grease at the bottom of the CV joint housing

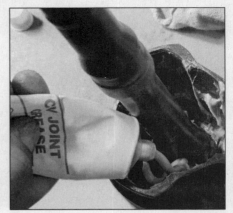

11.10d  Install the boot and clamps onto the axleshaft, then insert the tripod into the housing, followed by the rest of the grease

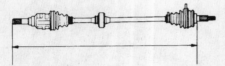

11.11  The driveaxle standard length should be set to the dimension listed in this Chapter's Specifications before the boot clamps are tightened

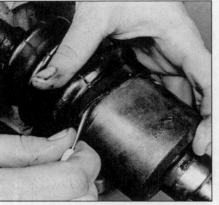

11.12a  Equalize the pressure inside the boot by inserting a small, dull screwdriver between the boot and the outer race

**illustration).** Slide the clamps and boot(s) onto the axleshaft, then place the tripod on the shaft. **Note:** *If you removed the dynamic damper, be sure to install it in its original location and secure it with a new clamp before installing the inner boot.* Pack the CV joint(s) with CV joint grease, and also apply some to the inside of the boot(s). Insert the tripod into the housing and pack the remainder of the grease around the tripod **(see illustrations).** **Caution:** *Use CV joint grease only.*

11     Slide the boot into place, making sure both ends seat in their grooves. Adjust the length of the driveaxle to the dimension listed in this Chapter's Specifications **(see illustration).**

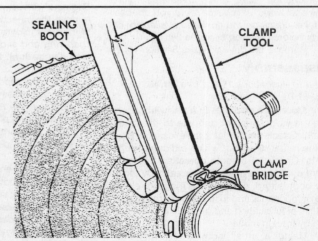

11.12b  To install the new clamps, bend the tang down . . .

12     Equalize the pressure in the boot, then tighten and secure the boot clamps **(see illustrations).**

11.12c  . . . then tap the tabs over to hold it in place

11.12d  If your replacement boot came with crimp-type clamps, a special tool such as this one (available at most auto parts stores) will be required to tighten them properly

# Chapter 9   Brakes

## Contents

## Specifications

### General

| | |
|---|---|
| Brake fluid type | See Chapter 1 |
| Brake pedal height | |
| 1997 and earlier models | 5-21/32 to 6-1/2 inches |
| 1998 and later models | 5-27/32 to 6-1/4 inches |
| Brake pedal freeplay | 3/64 to 1/4 inch |
| Brake pedal reserve distance | |
| 1997 and earlier models | At least 2-3/4 inches |
| 1998 and later models (except 1999 and later models with ABS) | At least 3-11/32 inches |
| 1999 and later models with ABS | At least 3-35/64 inches |
| Brake light switch-to-pedal clearance | 1/64 to 3/32 inch |
| Power brake booster pushrod-to-master cylinder piston clearance | 0.0 inch |

### Disc brakes

| | |
|---|---|
| Minimum brake pad thickness | See Chapter 1 |
| Disc minimum thickness | Cast or stamped into disc |
| Disc runout limit | 0.002 inch |

### Drum brakes

| | |
|---|---|
| Drum maximum diameter | Cast or stamped into drum |

### Parking brake

| | |
|---|---|
| Parking brake lever travel | 4 to 7 clicks |

### Torque specifications

**Ft-lbs (unless otherwise indicated)**

| | |
|---|---|
| Brake hose-to-caliper banjo bolts | 22 |
| Caliper mounting bolts | 25 |
| Caliper torque plate bolts | 65 |
| Master cylinder-to-brake booster nuts | 108 in-lbs |
| Power brake booster mounting nuts | 108 in-lbs |
| Wheel cylinder mounting bolts | 84 in-lbs |
| Wheel speed sensor bolt (front or rear) | 71 in-lbs |
| Wheel lug nuts | See Chapter 1 |

9

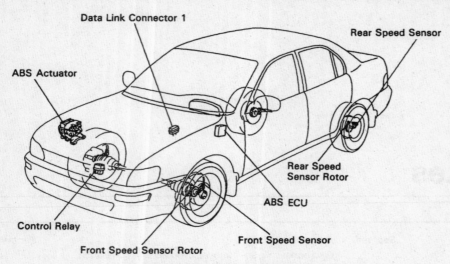

2.1a Anti-Lock Brake System (ABS) component locations -
1997 and earlier Corolla models

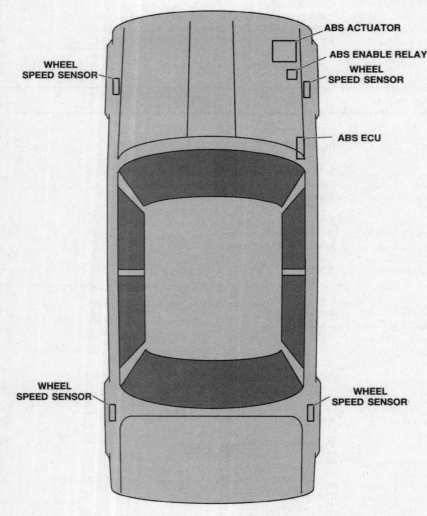

2.1b Anti-Lock Brake System (ABS) component locations -
1997 and earlier Prizm models

# 1  General information

The vehicles covered by this manual are equipped with hydraulically operated front and rear brake systems. The front brakes are disc type and the rear brakes are drum type. Both the front and rear brakes are self adjusting. The disc brakes automatically compensate for pad wear, while the drum brakes incorporate an adjustment mechanism which is activated as the parking brake is applied.

## Hydraulic system

The hydraulic system consists of two separate circuits. The master cylinder has separate reservoirs for the two circuits, and, in the event of a leak or failure in one hydraulic circuit, the other circuit will remain operative. A dual proportioning valve on the firewall provides brake balance between the front and rear brakes.

## Power brake booster

The power brake booster, utilizing engine manifold vacuum and atmospheric pressure to provide assistance to the hydraulically operated brakes, is mounted on the firewall in the engine compartment.

## Parking brake

The parking brake operates the rear brakes only, through cable actuation. It's activated by a lever mounted in the center console.

## Service

After completing any operation involving disassembly of any part of the brake system, always test drive the vehicle to check for proper braking performance before resuming normal driving. When testing the brakes, perform the tests on a clean, dry, flat surface. Conditions other than these can lead to inaccurate test results.

Test the brakes at various speeds with both light and heavy pedal pressure. The vehicle should stop evenly without pulling to one side or the other. Avoid locking the brakes, because this slides the tires and diminishes braking efficiency and control of the vehicle.

Tires, vehicle load and wheel alignment are factors which also affect braking performance.

# 2  Anti-lock Brake System (ABS) - general information and trouble codes

*Refer to illustrations 2.1a, 2.1b and 2.1c*

1  The Anti-lock Brake System (ABS) **(see illustrations)** is designed to maintain vehicle steerability, directional stability and optimum deceleration under severe braking conditions and on most road surfaces. It does so by monitoring the rotational speed of each

wheel and controlling the brake line pressure to each wheel during braking. This prevents the wheel from locking up.

## Components

### Actuator assembly

2    The actuator assembly consists of an electric hydraulic pump and four (Corolla) or two (Prizm) solenoid valves. The electric pump provides hydraulic pressure to charge the reservoirs in the actuator, which supplies pressure to the braking system. The pump and reservoirs are housed in the actuator assembly. The solenoid valves modulate brake line pressure during ABS operation.

### Speed sensors

*Refer to illustrations 2.4 and 2.5*

3    The speed sensors, which are located at each wheel, generate small electrical pulsations when the toothed sensor rotors are turning, sending a variable voltage signal to the ABS electronic control unit (ECU) indicating wheel rotational speed.

4    The front speed sensors **(see illustration)** are mounted on the steering knuckles in close relationship to the toothed sensor rotors, which are integral with the outer constant velocity (CV) joints.

5    The rear wheel sensors are bolted to the rear axle carriers **(see illustration)**. The sensor rotors are integral with the rear brake drum assemblies.

### ABS computer

6    The ABS electronic control unit (ECU), which is mounted under the dashboard or behind the passenger's side kick panel on 1997 and earlier models, or fastened to the ABS actuator on 1998 and later models, is the "brain" of the ABS system. The function of the ECU is to accept and process information received from the wheel speed sensors to control the hydraulic line pressure, avoiding wheel lock up. The ECU also constantly monitors the system, even under normal driving conditions, to find faults within the system.

7    If a problem develops within the system, an "ABS" light will glow on the dashboard. A diagnostic code will also be stored in the ECU, which will indicate the problem area or component. On Corolla models these codes can be retrieved without the use of special tools. On Prizm models, however, a special scan tool is required.

## Diagnosis and repair

8    If a dashboard warning light comes on and stays on while the vehicle is in operation, the ABS system requires attention. If you have a Corolla model you can check for stored trouble codes. Although on Prizm models a special electronic scan tool is necessary to properly diagnose the system, you can perform a few preliminary checks before taking the vehicle to a dealer service department or other repair shop which is equipped with a tester.

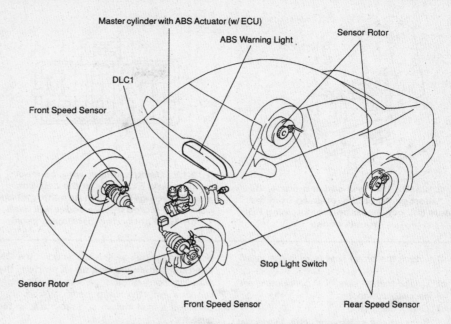

**2.1c  Anti-Lock Brake System (ABS) component locations - 1998 and later Corolla shown, Prizm similar**

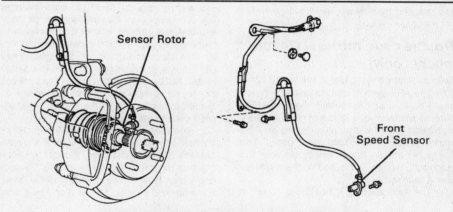

**2.4  ABS front wheel speed sensor and sensor rotor**

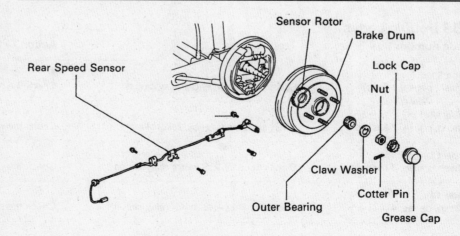

**2.5  ABS rear wheel speed sensor and sensor rotor**

**9**

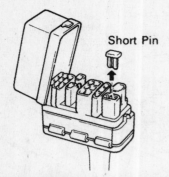

**2.12a  On 1997 and earlier models, the "short pin" must be removed from the data link connector before obtaining ABS trouble codes**

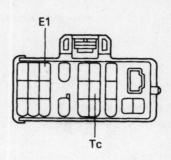

**2.12b  Using a jumper wire, connect terminals E1 and Tc of the data link connector together, then turn the ignition On - any stored trouble codes will flash the ABS light on the instrument panel**

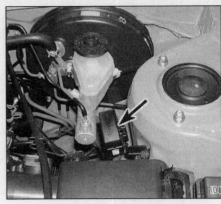

**2.12c  The data link connector is located in the engine compartment, mounted to the left strut tower**

a) *Check the brake fluid level in the reservoir.*
b) *Check that all electrical connectors are securely connected.*
c) *Check the fuses.*

9   If the above preliminary checks do not rectify the problem, or on Corolla models, if any stored trouble codes don't lead you to the problem, the vehicle should be diagnosed and repaired by a dealer service department or other repair shop.

## Trouble code retrieval (Toyota models only)

*Refer to illustrations 2.12a, 2.12b and 2.12c*

10   The ABS system control unit (computer) has a built-in self-diagnosis system which detects malfunctions in the system sensors and alerts the driver by illuminating an ABS warning light in the instrument panel. The computer stores the failure code until the diagnostic system is cleared or the malfunction is repaired.

11   The ABS warning light should come on when the ignition switch is placed in the ON position. When the engine is started, the warning light should go out. If the light remains on, the diagnostic system has detected a malfunction or abnormality in the system.

12   The codes for the ABS can be accessed by turning the ignition key to the OFF position (engine not running). On 1997 and earlier models, remove the short pin from the data link connector **(see illustration)**. On all models, install a jumper wire or paper clip onto terminals E1 and Tc of the data link connector **(see illustration)** and turn the ignition key ON (engine not running). Observe the codes on the ABS warning light. **Note:** *The data link connector is mounted to the strut tower on the left (driver's) side of the engine compartment* **(see illustration)**.

13   The diagnostic code is the number of flashes indicated on the ABS light. If any malfunction has been detected, the light will blink the first digit(s) of the code, pause 1.5 seconds, then blink the second digit of the code. For example, a code 34 (left rear wheel sensor) will first blink three flashes, pause 1.5 seconds, then blink four flashes. If there is more than one code stored in the ECM, the ECM will pause 2.5 seconds before flashing the next code. If the system is operating normally (no malfunctions), the warning light will blink once every 0.5 seconds.

14   The accompanying tables explain the code that will be flashed for each of the malfunctions. The accompanying table indicates the diagnostic code - in blinks - along with the system, diagnosis and specific areas. Check the indicated system or component or take the vehicle to a dealer service department to have the malfunction repaired.

15   After the diagnosis check, clear the trouble codes. First, jump terminals E1 and Tc on the data link connector. Turn the ignition key ON (engine not running) and clear the codes by depressing the brake pedal eight or more times within five seconds. Remove the jumper wire and, on 1997 and earlier models, reinstall the short pin. Close the cap on the data link connector.

## ABS trouble codes

| Code number | Trouble area | Action to take |
|---|---|---|
| **Code 11**<br>(1 flash, pause, 1 flash) | Open circuit in solenoid relay circuit | Check the solenoid relay and the relay circuit |
| **Code 12**<br>(1 flash, pause, 2 flashes) | Short circuit in solenoid relay circuit | Check the solenoid relay and the relay circuit |
| **Code 13**<br>(1 flash, pause, 3 flashes) | Open circuit in ABS motor relay circuit | Check the pump motor relay and circuit |
| **Code 14**<br>(1 flash, pause, 4 flashes) | Short circuit in ABS motor relay circuit | Check the solenoid relay and the relay circuit |
| **Code 21**<br>(2 flashes, pause, 1 flash) | Problem in right front wheel solenoid circuit | Check the actuator solenoid and circuit |

| Code number | Trouble area | Action to take |
|---|---|---|
| **Code 22**<br>(2 flashes, pause, 2 flashes) | Problem in left front wheel solenoid circuit | Check the actuator solenoid and circuit |
| **Code 23**<br>(2 flashes, pause, 3 flashes) | Problem in right rear wheel solenoid circuit | Check the actuator solenoid and circuit |
| **Code 24**<br>(2 flashes, pause, 4 flashes) | Problem in left rear wheel solenoid circuit | Check the actuator solenoid and circuit |
| **Code 31**<br>(3 flashes, pause, 1 flash) | Sensor signal problem - right front wheel | Check the speed sensor, sensor rotors, wire harness and connector of the speed sensor |
| **Code 32**<br>(3 flashes, pause, 2 flashes) | Sensor signal problem - left front wheel | Check the speed sensor, sensor rotors, wire harness and connector of the speed sensor |
| **Code 33**<br>(3 flashes, pause, 3 flashes) | Sensor signal problem - right rear wheel | Check the speed sensor, sensor rotors, wire harness and connector of the speed sensor |
| **Code 34**<br>(3 flashes, pause, 4 flashes) | Sensor signal problem - left rear wheel | Check the speed sensor, sensor rotors, wire harness and connector of the speed sensor |
| **Code 35**<br>(3 flashes, pause, 5 flashes) | Open circuit - left front or right rear speed sensor or circuit | Check the speed sensor, wire harness and electrical connector |
| **Code 36**<br>(3 flashes, pause, 6 flashes) | Open circuit - right front or left rear speed sensor or circuit | Check the speed sensor, wire harness and electrical connector |
| **Code 37**<br>(3 flashes, pause, 7 flashes) | Speed sensor rotor has incorrect number of teeth | Check for a damaged sensor rotor |
| **Code 41**<br>(4 flashes, pause, 1 flash) | Abnormally low or high battery voltage | Check the charging system (alternator, battery and voltage regulator) for any problems (see Chapter 5) |
| **Code 49**<br>(4 flashes, pause, 9 flashes) | Open circuit - brake light switch or circuit | Check the brake light switch and circuit |
| **Code 51**<br>(5 flashes, pause, 1 flash) | Pump motor locked | Check the pump motor and relay battery for shorts or abnormalities |
| **Light always ON** | ECU malfunction | ECU problem |

**9**

## Speed sensor check (Toyota models only)

16   The diagnostic connector and jumper wire can also be used to check the operation of the wheel speed sensors. This procedure involves driving the vehicle in a straight line, so find a suitable isolated area where you can perform this check without interfering with traffic.

17   Make sure the ignition switch is Off.

Using a three-way jumper wire **(see illustration 10.13)**, connect terminals E1, Ts and Tc of the data link connector **(see illustration 10.14)**.

18   Start the engine and make sure the ABS light on the instrument panel blinks rapidly. If it doesn't, the warning light circuit or the ABS ECU may have a problem; have the vehicle diagnosed by a dealer service department or other qualified repair shop.

19   Drive the vehicle in a straight line, faster

than 28 mph, for several seconds, then stop the vehicle (but leave the engine running).

20   Unplug the jumper wire from terminal Ts only, then read the number of blinks of the ABS light, as described in Step 13. If the speed sensors are operating normally, the ABS light will blink at 1/4-second intervals.

21   If any codes are obtained, compare them with the accompanying table.

## Wheel speed sensor trouble codes

| Code number | Trouble area | Action to take |
| --- | --- | --- |
| **Code 71**<br>(7 flashes, pause, 1 flash) | Right front speed sensor - low output voltage | Faulty or improperly installed sensor, faulty sensor rotor |
| **Code 72**<br>(7 flashes, pause, 2 flashes) | Left front speed sensor - low output voltage | Faulty or improperly installed sensor, faulty sensor rotor |
| **Code 73**<br>(7 flashes, pause, 3 flashes) | Right rear speed sensor - low output voltage | Faulty or improperly installed sensor, faulty sensor rotor |
| **Code 74**<br>(7 flashes, pause, 4 flashes) | Left rear speed sensor - low output voltage | Faulty or improperly installed sensor, faulty sensor rotor |
| **Code 75**<br>(7 flashes, pause, 5 flashes) | Right front speed sensor - abnormal change in output voltage | Faulty sensor rotor |
| **Code 76**<br>(7 flashes, pause, 6 flashes) | Left front speed sensor - abnormal change in output voltage | Faulty sensor rotor |
| **Code 77**<br>(7 flashes, pause, 7 flashes) | Right rear speed sensor - abnormal change in output voltage | Faulty sensor rotor |
| **Code 78**<br>(7 flashes, pause, 8 flashes) | Left rear speed sensor - abnormal change in output voltage | Faulty sensor rotor |

3.5 Before removing the caliper, be sure to depress the piston into the bottom of its bore in the caliper with a large C-clamp to make room for the new pads

3.6a Always wash the brakes with brake cleaner before disassembling anything

3.6b To remove the caliper, remove the bolts indicated by the upper and lower arrows (the middle arrow points to the brake hose banjo bolt, which shouldn't be unscrewed unless the caliper is being removed for overhaul or replacement, or for hose replacement)

## 3    Disc brake pads - replacement

*Refer to illustrations 3.5 and 3.6a through 3.6u*

**Warning:** *Disc brake pads must be replaced on both front wheels at the same time - never replace the pads on only one wheel. Also, the dust created by the brake system is harmful to your health. Never blow it out with compressed air and don't inhale any of it. An approved filtering mask should be worn when*

working on the brakes. Do not, under any circumstances, use petroleum-based solvents to clean brake parts. Use brake system cleaner only!

1    Remove the cap from the brake fluid reservoir.

2    Loosen the wheel lug nuts, raise the front of the vehicle and support it securely on jackstands. Block the wheels at the opposite end.

3    Remove the wheels. Work on one brake assembly at a time, using the assembled brake for reference if necessary.

4    Inspect the brake disc carefully as outlined in Section 5. If machining is necessary, follow the information in that Section to remove the disc, at which time the pads can be removed as well.

5    Push the piston back into its bore to provide room for the new brake pads. A C-clamp can be used to accomplish this **(see illustration)**. As the piston is depressed to

3.6c  Remove the caliper . . .

3.6d  . . . and hang it from the strut coil spring with a piece of coat hanger or wire; do not allow the caliper to hang by the brake hose

3.6e  Remove the upper anti-squeal spring (1997 and earlier models only) . . .

the bottom of the caliper bore, the fluid in the master cylinder will rise. Make sure that it doesn't overflow. If necessary, siphon off some of the fluid.

6    Follow the accompanying photos (illustrations 3.6a through 3.6u), for the actual pad replacement procedure. Be sure to stay

in order and read the caption under each illustration.

7    When reinstalling the caliper, be sure to tighten the mounting bolts to the torque listed in this Chapter's Specifications. After the job

has been completed, firmly depress the brake pedal a few times to bring the pads into contact with the disc. Check the level of the brake fluid, adding some if necessary. Check the operation of the brakes carefully before placing the vehicle into normal service.

3.6f  . . . and the lower anti-squeal spring (1997 and earlier models only)

3.6g  Remove the outer shim . . .

3.6h  . . . and the inner shim from the outer brake pad

3.6i  Remove the outer brake pad

3.6j  Remove the outer shim . . .

3.6k  . . . and the inner shim from the inner brake pad

9

3.6l   Remove the inner brake pad

3.6m   Remove the pad support plates; inspect the plates for damage and replace as necessary (they should fit snugly); 1997 and earlier models have four plates - 1998 and later models have two

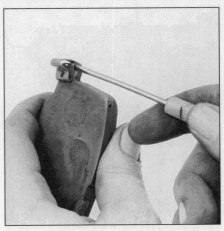

3.6n   If equipped, pry the wear indicator off the old inner brake pad and transfer it to the new inner pad (if the wear indicator is worn or bent, replace it)

3.6o   Install the pad support plates, the new inner brake pads and the shims; make sure the ears on the pad are properly engaged with the pad support plates as shown

3.6p   Install the outer pad and the shims

3.6q   Install the upper and lower anti-squeal springs; make sure both springs are properly engaged with the pads as shown (1997 and earlier models only)

3.6r   Pull out the upper and lower sliding pins and clean them off (if either boot is damaged, remove it by prying the flange of the metal bushing that retains the boot) . . .

3.6s   . . . apply a coat of high-temperature grease to the pins before installing them

3.6t   Install the caliper and tighten the caliper bolts to the torque listed in this Chapter's Specifications

**3.6u If you have difficulty installing the caliper over the new pads, use a C-clamp to bottom the piston in its bore, then try again - it should now slip over the pads**

**4.2 Using a piece of rubber hose of the appropriate size, plug the brake line; this will prevent brake fluid from leaking out and dirt and moisture from contaminating the system**

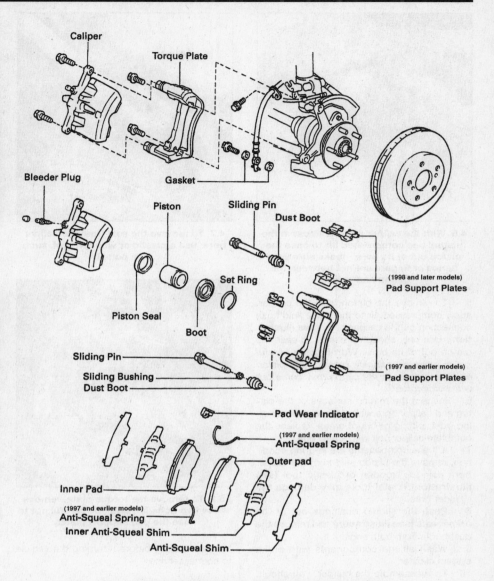

**4.4a An exploded view of a typical brake caliper assembly**

Caliper
Torque Plate
Bleeder Plug
Gasket
Piston
Sliding Pin
Dust Boot
Set Ring
Piston Seal
Boot
(1998 and later models) Pad Support Plates
Sliding Pin
Sliding Bushing
Dust Boot
(1997 and earlier models) Pad Support Plates
Pad Wear Indicator
(1997 and earlier models) Anti-Squeal Spring
Outer pad
Inner Pad
(1997 and earlier models) Anti-Squeal Spring
Inner Anti-Squeal Shim
Anti-Squeal Shim

## 4 Disc brake caliper - removal, overhaul and installation

**Warning:** *Dust created by the brake system is harmful to your health. Never blow it out with compressed air and don't inhale any of it. An approved filtering mask should be worn when working on the brakes. Do not, under any circumstances, use petroleum-based solvents to clean brake parts. Use brake system cleaner only.*
**Note:** *If an overhaul is indicated (usually because of fluid leakage), explore all options before beginning the job. New and factory rebuilt calipers are available on an exchange basis, which makes this job quite easy. If it's decided to rebuild the calipers, make sure a rebuild kit is available before proceeding. Always rebuild the calipers in pairs - never rebuild just one of them.*

### Removal
*Refer to illustration 4.2*
1    Loosen the front wheel lug nuts, raise the front of the vehicle and place it securely on jackstands. Remove the wheel.
2    Remove the bolt and disconnect the brake hose from the caliper **(see illustration 3.6b)**. Plug the brake hose to keep contaminants out of the brake system and to prevent losing any more brake fluid than is necessary **(see illustration)**.
3    Refer to Section 3 for the caliper removal procedure (it's part of the brake pad replacement procedure).

### Overhaul
*Refer to illustrations 4.4a, 4.4b, 4.5, 4.7 and 4.8*
4    To overhaul the caliper, remove the boot set ring and the boot **(see illustrations)**. Before you remove the piston, place a wood block or some rags between the piston and caliper to prevent damage as it is removed.

**4.4b Using a screwdriver, remove the cylinder boot set ring**

9

**4.5  With the caliper padded to catch the piston, use compressed air to ease the piston out of its bore - make sure your hands or fingers are not between the piston and caliper**

**4.7  To remove the seal from the caliper bore, use a plastic or wooden tool, such as a pencil**

**4.8  Push out each sliding bushing through the boot, pull it free, then remove the dust boots**

5    To remove the piston from the caliper, apply compressed air to the brake fluid hose connection on the caliper body **(see illustration)**. Use only enough pressure to ease the piston out of its bore. **Warning:** *Be careful not to place your fingers between the piston and the caliper, as the piston may come out with some force.*

6    Inspect the mating surfaces of the piston and caliper bore wall. If there is any scoring, rust, pitting or bright areas, replace the complete caliper unit with a new one.

7    If these components are in good condition, remove the piston seal from the caliper bore using a wooden or plastic tool **(see illustration)**. Metal tools may damage the cylinder bore.

8    Push the sliding bushings out of the caliper ears **(see illustration)** and remove the dust boots from both ends.

9    Wash all the components with brake system cleaner.

10    To reassemble the caliper, you should already have the correct rebuild kit for your vehicle.

11    Submerge the new piston seal and the piston in brake fluid and install them into the caliper bore. Do not force the piston into the bore, but make sure it is squarely in place, then apply firm (but not excessive) pressure to install it.

12    Install the new piston dust boot and set ring.

13    Lubricate the sliding bushings with silicone-based grease (supplied in the kit) and push them into the caliper ears. Install the dust boots.

## *Installation*

14    Install the caliper by reversing the removal procedure. Remember to replace the sealing washers (gaskets) at the brake hose-to-caliper connection (new washers normally come with the rebuild kit).

15    Bleed the brake circuit according to the procedure in Section 10. Make sure there are no leaks from the hose connections. Test the

**5.2  To remove the torque plate, remove these two bolts (arrows); be careful not to lose the pad support plates**

brakes carefully before returning the vehicle to normal service.

---

## 5    Brake disc - inspection, removal and installation

### *Inspection*

*Refer to illustrations 5.2, 5.3, 5.4a, 5.4b, 5.5a and 5.5b*

1    Loosen the wheel lug nuts, raise the vehicle and support it securely on jackstands. Remove the wheel and install the lug nuts to hold the disc in place. **Note:** *If the lug nuts don't contact the disc when screwed on all the way, install washers under them.*

2    Remove the brake caliper as outlined in Section 4. It isn't necessary to disconnect the brake hose. After removing the caliper bolts, suspend the caliper out of the way with a piece of wire **(see illustration 3.6d)**. Remove the two torque plate-to-steering knuckle bolts **(see illustration)** and detach the torque plate.

3    Visually inspect the disc surface for score marks and other damage. Light scratches and shallow grooves are normal

**5.3  The brake pads on this vehicle were obviously neglected, as they wore down to the rivets and cut deep grooves into the disc - wear this severe means the disc must be replaced**

after use and may not always be detrimental to brake operation, but deep scoring - over 0.039-inch (1.0 mm) - requires disc removal and refinishing by an automotive machine shop. Be sure to check both sides of the disc **(see illustration)**. If pulsating has been noticed during application of the brakes, suspect disc runout.

4    To check disc runout, place a dial indicator at a point about 1/2-inch from the outer edge of the disc **(see illustration)**. Set the indicator to zero and turn the disc. The indicator reading should not exceed the specified allowable runout limit. If it does, the disc should be refinished by an automotive machine shop. **Note:** *The discs should be resurfaced regardless of the dial indicator reading, as this will impart a smooth finish and ensure a perfectly flat surface, eliminating any brake pedal pulsation or other undesirable symptoms related to questionable discs. At the very least, if you elect not to have the discs resurfaced, remove the glaze from the surface with emery cloth or sandpaper, using a swirling motion* **(see illustration)**.

5.4a  To check disc runout, mount a dial indicator as shown and rotate the disc

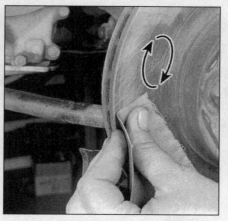

5.4b  Using a swirling motion, remove the glaze from the disc surface with sandpaper or emery cloth

5.5a  The minimum wear dimension is cast into the back side of the disc (typical)

5     It's absolutely critical that the disc not be machined to a thickness under the specified minimum thickness. The minimum wear (or discard) thickness is cast or stamped into the inside of the disc **(see illustration)**. The disc thickness can be checked with a micrometer **(see illustration)**.

## Removal

6     Remove the lug nuts which were installed to hold the disc in place and remove the disc from the hub.

## Installation

7     Place the disc in position over the threaded studs.
8     Install the torque plate and caliper, tightening the bolts to the torque listed in this Chapter's Specifications.
9     Install the wheel, then lower the vehicle to the ground. Tighten the lug nuts to the torque listed in the Chapter 1 Specifications. Depress the brake pedal a few times to bring the brake pads into contact with the disc. Bleeding won't be necessary unless the brake hose was disconnected from the caliper. Check the operation of the brakes carefully before driving the vehicle.

## 6     Drum brake shoes - replacement

*Refer to illustrations 6.4a through 6.4ff and 6.5*
**Warning:** *Drum brake shoes must be replaced on both wheels at the same time - never replace the shoes on only one wheel. Also, the dust created by the brake system is harmful to your health. Never blow it out with compressed air and don't inhale any of it. An approved filtering mask should be worn when working on the brakes. Do not, under any circumstances, use petroleum-based solvents to clean brake parts. Use brake system cleaner only!*
**Caution:** *Whenever the brake shoes are replaced, the return and hold-down springs should also be replaced. Due to the continu-*

*ous heating/cooling cycle the springs are subjected to, they lose tension over a period of time and may allow the shoes to drag on the drum and wear at a much faster rate than normal.*
1     Loosen the wheel lug nuts, raise the rear of the vehicle and support it securely on jackstands. Block the front wheels to keep the vehicle from rolling.
2     Release the parking brake.
3     Remove the wheel. **Note:** *All four rear brake shoes must be replaced at the same time, but to avoid mixing up parts, work on only one brake assembly at a time.*
4     Follow the accompanying illustrations for the brake shoe replacement procedure **(see illustrations 6.4a through 6.4ff)**. Be sure to stay in order and read the caption under each illustration. **Note:** *If the brake drum cannot be easily pulled off the axle and shoe assembly, make sure the parking brake is completely released. If the drum still cannot be pulled off, the brake shoes will have to be retracted. This is done by first removing the plug from the backing plate. With the plug removed, push the lever off the adjuster star wheel with a narrow screwdriver while turning*

6.4a  Mark the relationship of the drum to the hub, so the drum will retain its dynamic balance after reassembly

5.5b  Use a micrometer to measure disc thickness

*the adjuster wheel with another screwdriver, moving the shoes away from the drum* **(see illustration 6.4b)**. *The drum should now come off.*

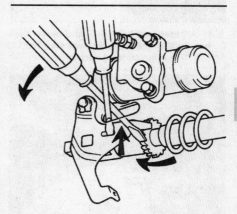

6.4b  If the brake drum is hanging up on the shoes because of excessive wear, insert two screwdrivers through the hole in the backing plate, push the adjuster lever away from the star wheel and turn the star wheel to retract the brake shoes

**9**

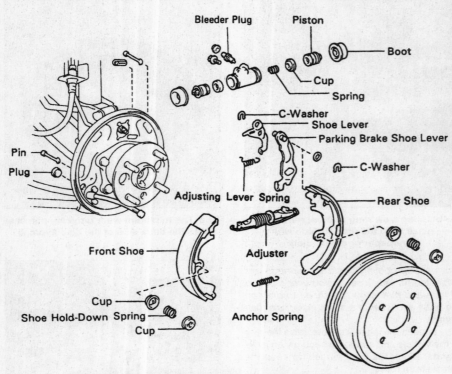

Bleeder Plug

Piston

Boot

Cup

Spring

C-Washer

Shoe Lever

Parking Brake Shoe Lever

C-Washer

Pin

Plug

Adjusting Lever Spring

Rear Shoe

Front Shoe

Adjuster

Cup

Shoe Hold-Down Spring

Cup

Anchor Spring

**6.4c  An exploded view of the drum brake assembly**

**6.4d  Before removing anything, place a drain pan under the brake assembly, clean the brake assembly with brake cleaner and allow it to dry; DO NOT USE COMPRESSED AIR TO BLOW OFF BRAKE DUST! (hub removed for clarity)**

**6.4e  Unhook the return spring from its hole in the front shoe . . .**

**6.4f  . . . then pull the other end out of the hole in the rear shoe and remove the adjuster and spring**

**6.4g  Using a hold-down spring tool, remove the hold-down spring by pushing in and rotating it 1/4-turn . . .**

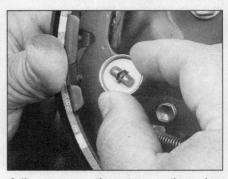

**6.4h  . . . remove the outer cup, the spring and the inner cup . . .**

**6.4i  . . . and pull the pin through the backing plate**

**6.4j  Remove the front shoe and unhook the anchor spring from the rear shoe**

**6.4k  Remove the rear shoe hold-down spring, cups and pin**

6.4l  Flip the rear shoe over, force the spring back from the parking brake lever as shown . . .

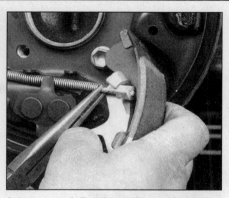

6.4m  . . . and disengage the parking brake cable from the parking brake lever

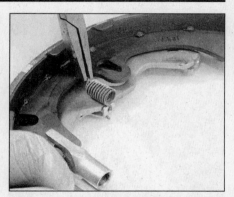

6.4n  Take the rear shoe to a clean work bench and unhook the adjusting lever spring from the shoe

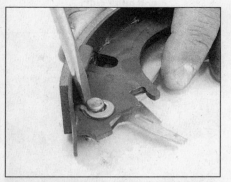

6.4o  Pry off the C-washer (don't lose the shim underneath) . . .

6.4p  . . . and remove the parking brake lever and the adjusting lever from the old rear shoe

6.4q  Attach the parking brake lever and adjusting lever to the new rear shoe and secure them with a new C-washer (don't forget the shim)

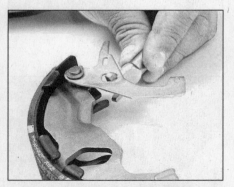

6.4r  Engage the rear part of the adjuster with the shoe lever as shown . . .

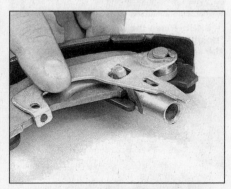

6.4s  . . . rotate the shoe lever back against the rear shoe . . .

6.4t  . . . hook the short end of the adjusting lever spring into the shoe lever . . .

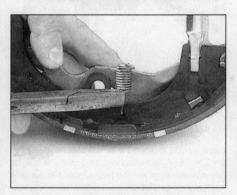

6.4u  . . . and hook the long end of the spring into the hole in the rear shoe

6.4v  Apply high-temperature grease to the friction points of the backing plate

6.4w  Insert the pin for the rear shoe hold-down spring through the hole in the backing plate

9

6.4x  Attach the parking brake cable to the parking brake lever

6.4y  Bring the rear shoe into position, insert the hold-down pin through the shoe, install the inner cup on the pin . . .

6.4z  . . . install the hold-down spring and outer cup . . .

6.4aa  . . . compress the spring with a brake spring tool, give the outer cup a 1/4-turn twist and lock it down

6.4bb  Place the adjuster assembly in position and insert it into the rear part of the adjuster (that you installed on the rear shoe on the bench)

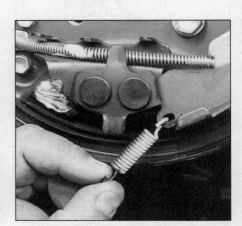

6.4cc  Attach the anchor spring to the rear shoe . . .

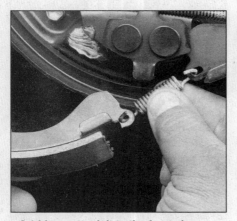

6.4dd  . . . attach it to the front shoe . . .

6.4ee  . . . place the front shoe in position and hook both ends of the return spring into their holes in the front and rear shoes

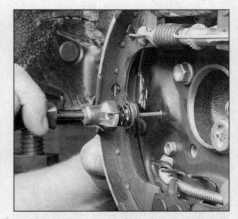

6.4ff  Install the front shoe hold-down pin, spring and inner and outer cups and lock the hold-down assembly into place with a brake spring tool

5    Before reinstalling the drum, it should be checked for cracks, score marks, deep scratches and hard spots, which will appear as small discolored areas. If the hard spots cannot be removed with fine emery cloth or if any of the other conditions listed above exist, the drum must be taken to an automotive machine shop to have it resurfaced. **Note:**

*Professionals recommend resurfacing the drums each time a brake job is done. Resurfacing will eliminate the possibility of out-of-round drums. If the drums are worn so much that they can't be resurfaced without exceeding the maximum allowable diameter (stamped or cast into the drum), then new*

*ones will be required* (**see illustration**). *At the very least, if you elect not to have the drums resurfaced, remove the glaze from the surface with emery cloth using a swirling motion.*

6    Install the brake drum on the axle flange.

7    Mount the wheel and install the lug nuts.

**6.5  The maximum drum diameter is cast into the drum (typical)**

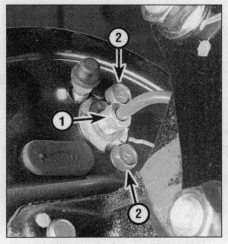
**7.4  Disconnect the brake line fitting (1), then remove the two wheel cylinder bolts (2)**

**8.2  Unplug the electrical connector for the fluid level warning switch (arrow points to the left master cylinder mounting nut)**

Using a screwdriver inserted through the adjusting hole in the backing plate **(see illustration 6.4b)**, turn the adjuster star wheel until the brake shoes drag on the drum as the drum is rotated, then back off the star wheel until the shoes don't drag. Lower the vehicle and tighten the lug nuts to the torque listed in the Chapter 1 Specifications.

8    Make a number of forward and reverse stops and operate the parking brake to adjust the brakes until satisfactory pedal action is obtained.

9    Check the operation of the brakes carefully before driving the vehicle.

## 7    Wheel cylinder - removal, overhaul and installation

**Note:** *If an overhaul is indicated (usually because of fluid leaks or sticky operation), explore all options before beginning the job. New wheel cylinders are available, which makes this job quite easy. If it's decided to rebuild the wheel cylinder, make sure a rebuild kit is available before proceeding. Never overhaul only one wheel cylinder - always rebuild both of them at the same time.*

### Removal

*Refer to illustration 7.4*

1    Raise the rear of the vehicle and support it securely on jackstands. Block the front wheels to keep the vehicle from rolling.

2    Remove the brake shoe assembly (see Section 6).

3    Remove all dirt and foreign material from around the wheel cylinder.

4    Disconnect the brake line **(see illustration)** with a flare-nut wrench, if available. Don't pull the brake line away from the wheel cylinder.

5    Remove the wheel cylinder mounting bolts.

6    Detach the wheel cylinder from the brake backing plate and place it on a clean workbench. Immediately plug the brake line to prevent fluid loss and contamination.

### Overhaul

7    Remove the bleeder screw, cups, pistons, boots and spring assembly from the wheel cylinder body **(see illustration 6.4c)**.

8    Clean the wheel cylinder with brake system cleaner. **Warning:** *Do not, under any circumstances, use petroleum-based solvents to clean brake parts!*

9    Use filtered, unlubricated compressed air to dry the wheel cylinder and blow out the passages.

10    Check the bore for corrosion and score marks. Crocus cloth can be used to remove light corrosion and stains, but the cylinder must be replaced with a new one if the defects cannot be removed easily, or if the bore is scored.

11    Lubricate the new cups with brake fluid.

12    Assemble the brake cylinder components. Make sure the cup lips face in.

### Installation

13    Place the wheel cylinder in position and install the bolts finger tight. Connect the brake line to the cylinder, being careful not to cross-thread the fitting. Tighten the wheel cylinder bolts to the torque listed in this Chapter's Specifications.

14    Tighten the brake line securely and install the brake shoe assembly (see Section 6).

15    Bleed the brakes (see Section 10).

16    Check the operation of the brakes carefully before driving the vehicle.

## 8    Master cylinder - removal, overhaul and installation

**Note:** *Before deciding to overhaul the master cylinder, check on the availability and cost of a new or factory rebuilt unit and also the availability of a rebuild kit. If you decide to rebuild the cylinder, inspect the bore as described in Step 12 before purchasing parts.*

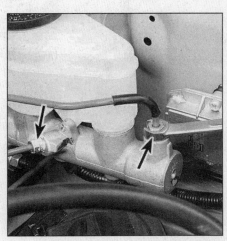
**8.4  Loosen the brake line fittings (arrows) with a flare-nut wrench**

### Removal

*Refer to illustrations 8.2, 8.4 and 8.6*

1    Remove the air cleaner assembly (see Chapter 4).

2    Unplug the electrical connector for the fluid level warning switch **(see illustration)**.

3    Remove as much fluid as possible from the reservoir with a syringe.

4    Place rags under the fittings and prepare caps or plastic bags to cover the ends of the lines once they're disconnected. **Caution:** *Brake fluid will damage paint. Cover all body parts and be careful not to spill fluid during this procedure. Loosen the fittings at the ends of the brake lines where they enter the master cylinder* **(see illustration)**. *To prevent rounding off the flats, use a flare-nut wrench, which wraps around the fitting hex.*

5    Pull the brake lines away from the master cylinder and plug the ends to prevent contamination.

6    Remove the three nuts attaching the

**9**

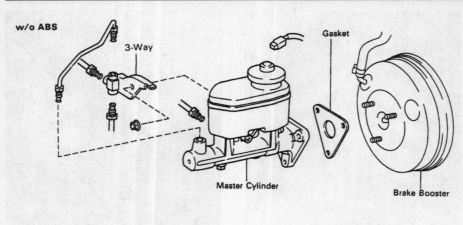

w/o ABS

3-Way

Gasket

Master Cylinder

Brake Booster

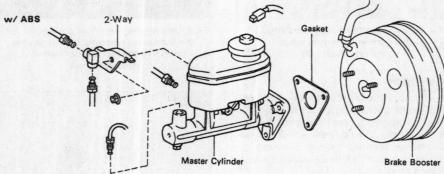

w/ ABS

2-Way

Gasket

Master Cylinder

Brake Booster

**8.6  Master cylinder mounting details**

**8.8a  The brake fluid reservoir is retained by a screw**

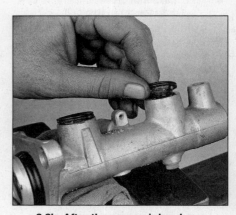

**8.8b  After the reservoir has been removed, pull the grommets from the master cylinder body; if they're hard, cracked or damaged, or have been leaking, replace them**

master cylinder to the power booster (see illustration). Pull the master cylinder off the studs to remove it. Again, be careful not to spill the fluid as this is done. Remove and discard the old gasket between the master cylinder and the power brake booster.

## Overhaul

*Refer to illustrations 8.8a, 8.8b, 8.9, 8.10, 8.11a, 8.11b, 8.11c and 8.18*

7    Before attempting the overhaul of the master cylinder, obtain the proper rebuild kit, which will contain the necessary replacement parts and also any instructions which may be specific to your model.

8    Remove the reservoir retaining screw, pull off the reservoir and remove the grommets (see illustrations).

9    Place the cylinder in a vise and use a punch or Phillips screwdriver to depress the pistons until they bottom against the other end of the master cylinder. Hold the pistons

in this position and remove the stopper bolt from the master cylinder (see illustration).

10   Carefully remove the snap-ring at the end of the master cylinder (see illustration).

11   The internal components can now be removed from the bore (see illustrations). Make a note of the proper order of the components so they can be returned to their original locations. **Note:** *The two springs are different, so pay particular attention to their installed order.*

12   Carefully inspect the bore of the master cylinder. Any deep score marks or other damage will mean a new master cylinder is required. DO NOT attempt to hone the bore.

13   Replace all parts included in the rebuild kit, following any instructions in the kit. Clean all re-used parts with brake system cleaner. **Warning:** *Do not use any petroleum-based solvents. During reassembly, lubricate all parts liberally with clean brake fluid.*

14   Push the assembled components into the bore, bottoming them against the end of the master cylinder, then install the stopper bolt.

15   Install the new snap-ring, making sure it's seated properly in the groove.

16   Install the reservoir grommets, reservoir and screw.

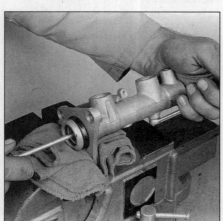

**8.9  Using a Phillips screwdriver, depress the pistons, then remove the stopper bolt; be sure to replace the sealing washer for the stopper bolt**

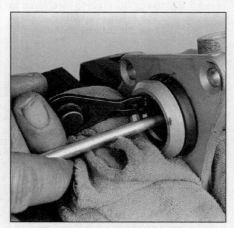

**8.10  Depress the pistons again and remove the snap-ring with a pair of snap-ring pliers**

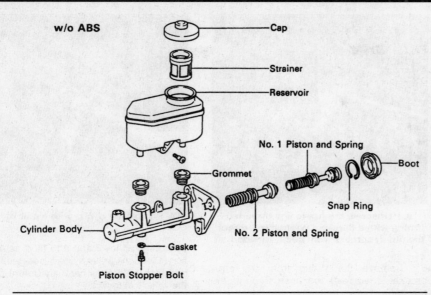

w/o ABS

Cap
Strainer
Reservoir
No. 1 Piston and Spring
Boot
Grommet
Snap Ring
Cylinder Body
No. 2 Piston and Spring
Gasket
Piston Stopper Bolt

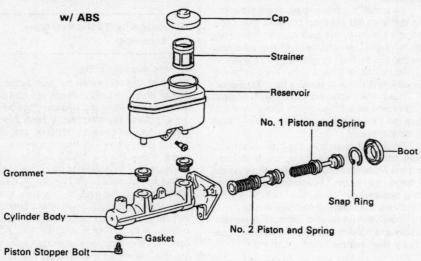

w/ ABS

Cap
Strainer
Reservoir
No. 1 Piston and Spring
Boot
Grommet
Snap Ring
Cylinder Body
No. 2 Piston and Spring
Gasket
Piston Stopper Bolt

**8.11a An exploded view of the master cylinder assembly**

**8.11b After the snap-ring has been removed, the primary (No. 1) piston assembly can be removed**

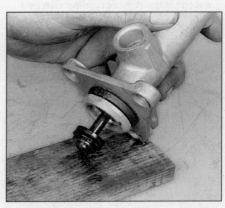

**8.11c Remove the cylinder from the vise and tap it against a block of wood until the secondary (No. 2) piston is exposed. Pull the piston assembly STRAIGHT OUT - if it becomes even slightly cocked, the bore may be damaged**

**8.18 The best way to bleed air from the master cylinder before installing it on the vehicle is with a pair of bleeder tubes that direct brake fluid into the reservoir during bleeding**

17    Before installing the master cylinder, it should be bench bled. Since you'll have to apply pressure to the master cylinder piston and, at the same time, control flow from the brake line outlets, the master cylinder should be mounted in a vise, with the jaws of the vise clamping on the mounting flange.

18    Attach a pair of master cylinder bleeder tubes to the outlet ports of the master cylinder **(see illustration)**.

19    Fill the reservoir with brake fluid of the recommended type (see Chapter 1).

20    Slowly push the pistons into the master cylinder (a large Phillips screwdriver can be used for this) - air will be expelled from the pressure chambers and into the reservoir. Because the tubes are submerged in fluid, air can't be drawn back into the master cylinder when you release the pistons.

21    Repeat the procedure until no more air bubbles are present.

22    Remove the bleed tubes, one at a time, and install plugs in the open ports to prevent fluid leakage and air from entering. Install the reservoir cap.

## Installation

*Refer to illustration 8.26*

**Note:** *Before installing a new or rebuilt master cylinder, check and, if necessary, adjust the length of the power brake booster pushrod (see Section 11).*

23    Install the master cylinder over the studs on the power brake booster and tighten the nuts only finger-tight at this time. Don't forget to use a new gasket.

24    Thread the brake line fittings into the master cylinder. Since the master cylinder is still a bit loose, it can be moved slightly so the fittings thread in easily. Don't strip the threads as the fittings are tightened.

25    Tighten the mounting nuts to the torque listed in this Chapter's Specifications. Tighten the brake line fittings securely.

26    Fill the master cylinder reservoir with fluid, then bleed the master cylinder and the brake system (see Section 10). To bleed the master cylinder on the vehicle, have an assistant depress the brake pedal and hold it down. Loosen the fitting to allow air and fluid

**9**

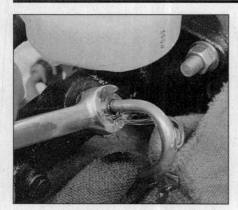

**8.26 Have an assistant depress the brake pedal and hold it down, then loosen the fitting nut, allowing the air and fluid to escape; repeat this procedure on both fittings until the fluid is clear of air bubbles**

**9.3 Unscrew the brake line threaded fitting with a flare-nut wrench to protect the fitting corners from being rounded off**

**9.4 Pull off the U-clip with a pair of pliers**

to escape **(see illustration)**. Tighten the fitting, then allow your assistant to return the pedal to its rest position. Repeat this procedure on both fittings until the fluid is free of air bubbles. Check the operation of the brake system carefully before driving the vehicle.

## 9   Brake hoses and lines - inspection and replacement

### *Inspection*

1   About every six months, with the vehicle raised and supported securely on jackstands, the rubber hoses which connect the steel brake lines with the front and rear brake assemblies should be inspected for cracks, chafing of the outer cover, leaks, blisters and other damage. These are important and vulnerable parts of the brake system and inspection should be complete. A light and mirror will be helpful for a thorough check. If a hose exhibits any of the above conditions, replace it with a new one.

### *Replacement*

#### Front brake hose

*Refer to illustrations 9.3 and 9.4*

2   Loosen the wheel lug nuts, raise the vehicle and support it securely on jackstands. Remove the wheel.
3   At the frame bracket, unscrew the brake line fitting from the hose **(see illustration)**. Use a flare-nut wrench to prevent rounding off the corners. If the bracket begins to bend, hold the hose fitting with an open-end wrench.
4   Remove the U-clip from the female fitting at the bracket with a pair of pliers **(see illustration)**, then pass the hose through the bracket.
5   At the caliper end of the hose, remove the banjo fitting bolt, then separate the hose from the caliper. Note that there are two copper sealing washers on either side of the fitting - they should be replaced with new ones during installation.

6   Remove the U-clip from the strut bracket, then feed the hose through the bracket.
7   To install the hose, pass the caliper fitting end through the strut bracket, then connect the fitting to the caliper with the banjo bolt and copper washers. Make sure the locating lug on the fitting is engaged with the hole in the caliper, then tighten the bolt to the torque listed in this Chapter's Specifications.
8   Push the metal support into the strut bracket and install the U-clip. Make sure the hose isn't twisted between the caliper and the strut bracket.
9   Route the hose into the frame bracket, again making sure it isn't twisted, then connect the brake line fitting, starting the threads by hand. Install the U-clip, then tighten the fitting securely.
10   Bleed the caliper (see Section 10).
11   Install the wheel and lug nuts, lower the vehicle and tighten the lug nuts to the torque listed in the Chapter 1 Specifications.

#### Rear brake hose

12   The rear brake hose serves as the flexible connection between two rigid metal lines, one on the body and the other on the axle. Both ends of the hose are attached to these metal lines with threaded fittings and U-clips. Refer to Steps 2, 3 and 4. Be sure to bleed the wheel cylinder when you're done (see Section 10).

#### Metal brake lines

13   When replacing brake lines, be sure to use the correct parts. Don't use copper tubing for any brake system components. Purchase steel brake lines from a dealer or auto parts store.
14   Prefabricated brake line, with the tube ends already flared and fittings installed, is available at auto parts stores and dealer parts departments. These lines can bent to the proper shape with a tubing bender.
15   When installing the new line, make sure it's securely supported in the brackets and has plenty of clearance between moving or hot components.
16   After installation, check the master

cylinder fluid level and add fluid as necessary. Bleed the brake system (see Section 10) and test the brakes carefully before driving the vehicle in traffic.

## 10   Brake hydraulic system - bleeding

*Refer to illustration 10.8*

**Warning 1:** *If you are working on a Geo Prizm model, this procedure might not completely purge the air from the hydraulic system. The manufacturer specifies that a Tech 1 or Tech II scan tool be used to "rehone" the pistons (which brings them to their uppermost positions) within the modulator before bleeding the system. The normal state of the pistons is in the uppermost position, but if there is a malfunction in the system they may not have returned to this position, in which case it may not be possible to successfully bleed the brakes. If the BRAKE, ABS or ABS ACTIVE lights on the dash are illuminated after bleeding the brakes and will not go out, or if the brake pedal feels spongy or unusually hard, or if you have any doubts as to the effectiveness of the brake system, DO NOT drive the vehicle. Instead, have it towed to a dealer service department or other repair shop equipped with a Tech 1 or Tech II scan tool.*

**Warning 2:** *Wear eye protection when bleeding the brake system. If the fluid comes in contact with your eyes, immediately rinse them with water and seek medical attention.*

**Note:** *Bleeding the hydraulic system is necessary to remove any air that manages to find its way into the system when it's been opened during removal and installation of a hose, line, caliper or master cylinder.*

1   You'll have to bleed the system at all four brakes if air has entered it due to low fluid level, or if the brake lines have been disconnected the master cylinder or ABS hydraulic actuator.
2   If a brake line was disconnected only at a wheel, then only that caliper or wheel cylinder must be bled.
3   If a brake line is disconnected at a fitting located between the master cylinder and any of the brakes, the entire system must be bled.

**10.8  When bleeding the brakes, a hose is connected to the bleed screw at the caliper or wheel cylinder and then submerged in brake fluid - air will be seen as bubbles in the tube and container (all air must be expelled before moving to the next wheel)**

And, if the master cylinder has run dry, has been overhauled or replaced, bleed the master cylinder as described in Section 8, Step 26.

4    Remove any residual vacuum from the brake power booster by applying the brake several times with the engine off.

5    Remove the master cylinder reservoir cap and fill the reservoir with brake fluid. Reinstall the cover. **Note:** *Check the fluid level often during the bleeding operation and add fluid as necessary to prevent the fluid level from falling low enough to allow air bubbles into the master cylinder.*

6    Have an assistant on hand, as well as a supply of new brake fluid, a clear plastic container partially filled with clean brake fluid, a length of clear tubing to fit over the bleeder valve and a wrench to open and close the bleeder valve.

7    Beginning at the right rear wheel, loosen the bleeder valve slightly, then tighten it to a point where it's snug but can still be loosened quickly and easily.

8    Place one end of the tubing over the bleeder valve and submerge the other end in brake fluid in the container **(see illustration)**.

9    Have the assistant depress the brake pedal slowly and hold the pedal down firmly.

10   While the pedal is held down, open the bleeder valve just enough to allow a flow of fluid to leave the valve. Watch for air bubbles to exit the submerged end of the tube. When the fluid flow slows after a couple of seconds, close the valve and have your assistant release the pedal.

11   Repeat Steps 9 and 10 until no more air is seen leaving the tube, then tighten the bleeder valve and proceed to the left front wheel, the left rear wheel and the right front wheel, in that order, and perform the same procedure. Be sure to check the fluid in the master cylinder reservoir frequently.

12   Never use old brake fluid. It contains moisture which can boil, rendering the brakes inoperative.

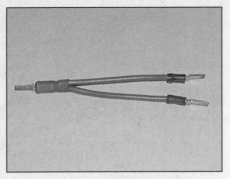

**10.13  To bleed the ABS actuator on Corolla models, you'll have to fabricate jumper wire like this**

## Corolla models with ABS

*Refer to illustrations 10.13 and 10.14*

13   If the vehicle is equipped with ABS, now bleed the ABS actuator. To do this, you'll first have to fabricate a three-way jumper wire, using small spade connectors that will fit into the terminals of the data link connector **(see illustration)**.

14   Insert the ends of the jumper wires into terminals Tc, Ts and E1 of the data link connector **(see illustration)**. **Note:** *The data link connector is mounted to the left strut tower in the engine compartment* **(see illustration 2.12c)**.

15   Turn the ignition switch to the On position.

16   Check the ABS warning light on the instrument panel - if the ABS ECU has entered the air bleed mode, the light will blink steadily every 1.5 seconds.

17   Depress the brake pedal one time and hold it down. After one second, you should hear the pressure reduction solenoid operate and the pedal should sink to the floor.

18   Release the brake pedal and listen for the pump motor - it should run for 1.5 seconds. **Note:** *Don't push on the brake pedal at this time.*

19   Check the ABS warning light once again. This time, it should be blinking at 1/2-second intervals. If it blinks at one-second intervals, turn the ignition Off and repeat Steps 14 through 19 again.

20   If the light is flashing at the proper rate, turn the ignition Off, then bleed the brakes at the each individual wheel, repeating Steps 14 through 19 before moving to the next wheel in the bleeding sequence. Be sure to check the fluid level in the reservoir frequently.

## All models

21   Refill the master cylinder with fluid at the end of the operation.

22   Check the operation of the brakes. The pedal should feel solid when depressed, with no sponginess. If necessary, repeat the entire process. **Warning:** *Do not operate the vehicle if the ABS light, BRAKE light or ABS ACTIVE light fails to go out, if the brakes feel low or spongy, or if you have any doubts as to the effectiveness of the brake system.*

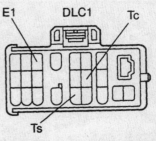

**10.14  Insert the ends of the jumper wire into terminals E1, Tc and Ts**

## 11   Power brake booster - check, removal and installation

### *Operating check*

1    Depress the brake pedal several times with the engine off and make sure there's no change in the pedal reserve distance.

2    Depress the pedal and start the engine. If the pedal goes down slightly, operation is normal.

### *Airtightness check*

3    Start the engine and turn it off after one or two minutes. Depress the brake pedal slowly several times. If the pedal depresses less each time, the booster is airtight.

4    Depress the brake pedal while the engine is running, then stop the engine with the pedal depressed. If there's no change in the pedal reserve travel after holding the pedal for 30 seconds, the booster is airtight.

### *Removal*

*Refer to illustrations 11.6, 11.10a and 11.10b*

5    Power brake booster units shouldn't be disassembled. They require special tools not normally found in most automotive repair stations or shops. Because of its critical relationship to brake performance, the booster should be replaced with a new or rebuilt one.

6    Disconnect the hose leading from the engine to the booster **(see illustration)**. Be careful not to damage the hose when remov-

**11.6  Detach this hose (arrow) from the power brake booster; make sure you don't puncture or tear the hose during removal**

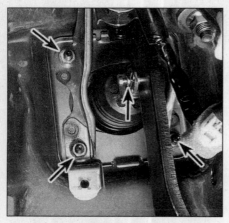

**11.10a To disconnect the power brake booster pushrod from the brake pedal, remove the retaining clip and clevis pin (center arrow); to detach the booster from the firewall, remove the four mounting nuts (arrows, upper right nut not visible in this photo)**

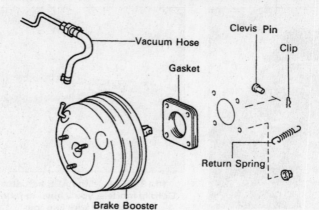

**11.10b Power brake booster installation details**

ing it from the booster fitting.

7    Remove the brake master cylinder (see Section 8). On 1998 and later models with ABS, also remove the ABS actuator.

8    Remove the steering column lower finish panel (see Chapter 11).

9    Remove the pedal return spring.

10   Locate the pushrod clevis connecting the booster to the brake pedal **(see illustrations)**. Remove the clevis pin retaining clip with pliers and pull out the pin.

11   Remove the four nuts and washers holding the brake booster to the firewall **(see illustration 11.10a)**; you may need a light to see them.

12   Slide the booster straight out from the firewall until the studs clear the holes.

## Installation

*Refer to illustrations 11.14a, 11.14b, 11.14c, 11.14d and 11.14e*

13   Installation procedures are basically the reverse of removal. Tighten the clevis locknut securely and the booster mounting nuts to the torque listed in this Chapter's Specifications.

14   If a new power brake booster unit is

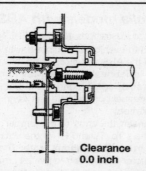

**11.14a There should be no clearance between the booster pushrod and the master cylinder pushrod, but no interference either; if there is interference between the two, the brakes may drag; if there is clearance, there will be excessive brake pedal travel**

**11.14b Measure the distance that the pushrod protrudes from the brake booster at the master cylinder mounting surface (including the gasket)**

being installed, check the pushrod clearance **(see illustration)** as follows:

a)  *Measure the distance that the pushrod protrudes from the master cylinder mounting surface on the front of the power brake booster, including the gasket. Write down this measurement* **(see illustration)**. *This is "dimension A."*

b)  *Measure the distance from the mounting flange to the end of the master cylinder* **(see illustration)**. *Write down this measurement. This is "dimension B."*

c)  *Measure the distance from the end of the master cylinder to the bottom of the pocket in the piston* **(see illustration)**. *Write down this measurement. This is "dimension C."*

d)  *Subtract measurement B from measurement C, then subtract measurement A from the difference between B and C. This the pushrod clearance.*

e)  *Compare your calculated pushrod clearance to the pushrod clearance listed in this Chapter's Specifications. If neces-*

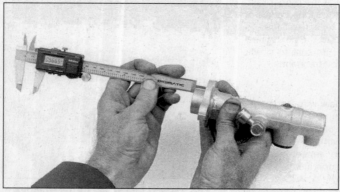

**11.14c Measure the distance from the mounting flange to the end of the master cylinder**

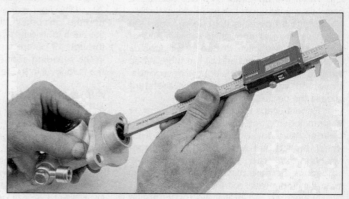

**11.14d Measure the distance from the piston pocket to the end of the master cylinder**

**11.14e  To adjust the length of the booster pushrod, hold the serrated portion of the rod with a pair of pliers and turn the adjusting screw in or out, as necessary, to achieve the desired setting**

**12.3  Loosen the locknut, then turn the adjusting nut until the desired handle travel is obtained**

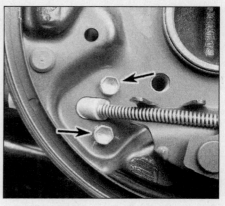

**13.4  Unscrew the bolts (arrows) from the backing plate and pass the cable through**

*sary, adjust the pushrod length to achieve the correct clearance (see illustration).*

15    After the final installation of the master cylinder and brake hoses and lines, the brake pedal height and freeplay must be adjusted and the system must be bled. See the appropriate Sections of this Chapter for the procedures.

## 12  Parking brake - adjustment

*Refer to illustration 12.3*

1    The parking brake lever, when properly adjusted, should travel four to seven clicks, when a moderate pulling force is applied. If it travels less than the specified minimum number of clicks, there's a chance the parking brake might not be releasing completely and might be dragging on the drum. If the lever can be pulled up more than the specified maximum number of clicks, the parking brake may not hold adequately on an incline, allowing the car to roll.

2    To gain access to the parking brake cable adjuster, remove the center console (see Chapter 11).

3    Loosen the locknut (the upper nut) while holding the adjusting nut (lower nut) with a wrench **(see illustration)**. Turn the adjusting

nut until the desired travel is attained. Tighten the locknut.

4    Install the center console.

## 13  Parking brake cables - replacement

### Equalizer-to-parking brake cable

*Refer to illustrations 13.4, 13.5, 13.6 and 13.8*

1    Loosen the rear wheel lug nuts, raise the rear of the vehicle and support it securely on jackstands. Block the front wheels. Remove the wheel.

2    Make sure the parking brake is completely released, then remove the brake drum.

3    Remove the brake shoes and disconnect the cable from the parking brake lever (see Section 6).

4    Remove the two cable retaining bolts from the backing plate **(see illustration)** and pull the cable through the backing plate.

5    Unbolt the cable clamp near the forward end of the strut rod **(see illustration)**.

6    Unbolt the cable clamp from the floor pan, just to the left of the center tunnel **(see illustration)**.

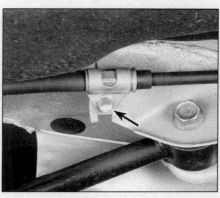

**13.5  Remove this nut (arrow) and detach this cable bracket from the frame near the forward end of the strut rod**

7    Remove the exhaust pipe and catalytic converter heat shields (see Chapter 6).

8    Unclamp the cable from the rear retaining bracket, pull the nylon bushing out of the front bracket and disconnect the cable from the equalizer **(see illustration)**.

9    Installation is the reverse of removal. Apply a light coat of grease to the portion of the cable end that engages with the equalizer.

10    Adjust the parking brake when you're done (see Section 12).

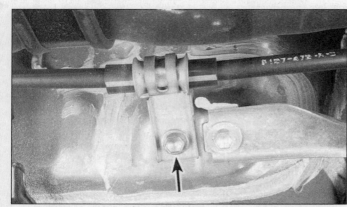

**13.6  Remove this bolt (arrow) and detach the cable bracket from the floor pan**

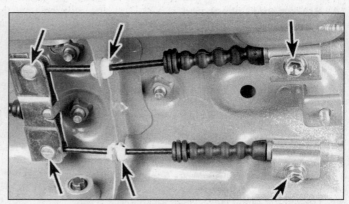

**13.8  Remove the clamp retaining bolt (right arrows), pry the nylon bushing out of the bracket (center arrows) and disengage the forward end of the cable from the equalizer (left arrows)**

**9**

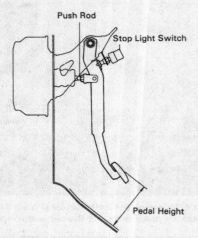

**14.1  Brake pedal height is the distance between the pedal and the firewall when the pedal is released**

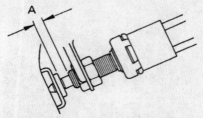

**14.10  Clearance "A" is the distance between the brake light switch and the pedal arm**

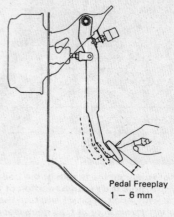

**14.16  Brake pedal freeplay is the distance between the pedal when it's released and the point at which some resistance is first felt when the pedal is depressed**

### Equalizer-to-brake lever cable

11    Remove the center console (see Chapter 11).
12    With the lever in the down (off) position, remove the locknut and the adjusting nut (see Section 12) and detach the cable from the lever.
13    Raise the rear of the vehicle and place it securely on jackstands.
14    Remove the exhaust pipe and catalytic converter heat shields (see Chapter 4).
15    Turn the cable end 90-degrees and disconnect it from the equalizer **(see illustration 13.8)**.
16    Pry out the rubber grommet from the floorpan and pull the cable through the hole in the pan.
17    Installation is the reverse of removal. Apply a light coat of grease to the portion of the cable end that engages with the equalizer, and coat the sealing edge of the rubber grommet with silicone to ensure that it remains watertight.
18    Adjust the parking brake lever when you're done (see Section 12).

## 14   Brake pedal - check and adjustment

### Pedal height

*Refer to illustrations 14.1 and 14.10*

1    Measure the pedal height **(see illustration)** and compare your measurement to the pedal height listed in this Chapter's Specifications. If the pedal height is incorrect, adjust it as follows:
2    Remove the steering column lower finish panel and air duct.
3    Unplug the electrical connector from the brake light switch.
4    Loosen the brake light switch locknut and remove the brake light switch.
5    Loosen the pushrod locknut.
6    Adjust the pedal height by turning the pedal pushrod.
7    Tighten the pushrod locknut.
8    Install the brake light switch and turn it until it lightly contacts the pedal stopper.
9    Back off the brake light switch one turn.
10    Measure the distance (clearance "A") between the threaded portion of the brake light switch and the pedal **(see illustration)** and compare your measurement to the clearance listed in this Chapter's Specifications. If the clearance is not as specified, repeat the previous two steps and try again.
11    Tighten the brake light switch locknut.
12    Plug in the brake light switch electrical connector.
13    Verify that brake lights come on when the brake pedal is depressed, and go off when the brake pedal is released.
14    Check the pedal freeplay (see below).

### Pedal freeplay

*Refer to illustration 14.16*

15    Stop the engine, if it's running, and depress the brake pedal several times until there's no more vacuum left in the booster.
16    Push in the pedal until you feel some resistance, then measure the distance between the release pedal and this point at which you can feel resistance **(see illustration)**. Compare your measurement with the pedal freeplay listed in this Chapter's Specifications. If the pedal freeplay is incorrect, adjust it as follows:
17    Check the brake light switch clearance. If the brake light switch clearance is okay, loosen the pedal pushrod locknut and turn the pushrod until the freeplay is correct. Tighten the locknut and recheck the brake pedal height.

### Pedal reserve

18    Start the engine, depress the brake pedal a few times, then press down hard and hold it.
19    Pedal reserve travel is measured from the floor to the top of the pedal while it's being depressed. Compare your measurement to the pedal reserve listed in this Chapter's Specifications.
20    If the pedal reserve is less than specified, check the adjustment of the rear brake shoes and/or the power brake booster pushrod-to-master cylinder piston clearance. If the brake pedal feels spongy, bleed the brake system (see Section 10).

## 15   Brake light switch - check and replacement

### Check

1    The brake light switch is located on a bracket at the top of the brake pedal **(see illustration 14.1)**. The switch activates the brake lights at the rear of the vehicle when the pedal is depressed.
2    To check the brake light switch, simply note whether the brake lights come on when the pedal is depressed and go off when the pedal is released. If they don't, check the fuse first (see Chapter 12). If the fuse is good, adjust the switch as described in Section 14 (adjusting the switch is part of brake pedal adjustment).
3    If the lights still don't come on, either the switch is not getting voltage, the switch itself is defective, or the circuit between the switch and the lights is defective. There is always the remote possibility that all of the brake light bulbs are burned out, but this is not very likely.
4    Use a voltmeter or test light to verify that there's voltage present at one side of the switch connector. If no voltage is present, troubleshoot the circuit from the switch to the fuse box. If there is voltage present, check for voltage on the other terminal when the brake pedal is depressed. If no voltage is present, replace the switch. If there is voltage present, troubleshoot the circuit from the switch to the brake lights (see the *Wiring diagrams* at the end of Chapter 12).

### Replacement

5    Disconnect the negative battery cable from the battery.
6    Unplug the electrical connector for the brake light switch.
7    Loosen the locknut **(see illustration 14.10)** and unscrew the switch from the pedal bracket.
8    Installation is the reverse of removal.
9    Adjust the brake pedal and brake light switch (see Section 14).

# Chapter 10
# Suspension and steering systems

## Contents

## Specifications

### Torque specifications

Ft-lbs (unless otherwise indicated)

#### Front suspension

Balljoints
| | |
|---|---|
| Balljoint-to-control arm bolt/nuts | 105 |
| Balljoint-to-steering knuckle nut | |
| 1997 and earlier models | 87 |
| 1998 and later models | 91 |

Control arm
| | |
|---|---|
| Front pivot bolt | |
| 1997 and earlier models | 161 |
| 1998 and later models | 158 |
| Rear pivot stud nut (1993 through 1995) | 101 |
| Rear bushing bolt (1996 and later) | 129 |
| Rear bushing bracket bolts | 108 |

Stabilizer bar
| | |
|---|---|
| Bracket | |
| Front bolt | |
| 1997 and earlier | 108 |
| 1998 and later | 167 |
| Rear bolt | |
| 1997 and earlier | 37 |
| 1998 and later | 109 |
| Nut | 168 in-lbs |
| Link nuts | 33 |

Struts
| | |
|---|---|
| Strut-to-steering knuckle bolts/nuts | 203 |
| Strut upper mounting nuts | 29 |
| Damper shaft nut | 34 |

Suspension crossmember bolts
| | |
|---|---|
| 1995 and earlier | 152 |
| 1996 and later | 167 |

#### Rear suspension

Suspension arm nuts/bolts (either end)
| | |
|---|---|
| 1995 and earlier | 87 |
| 1996 and 1997 | 92 |
| 1998 and later | 89 |

**10**

## Torque specifications

Ft-lbs (unless otherwise indicated)

### Rear suspension

| | |
|---|---|
| Hub and bearing assembly-to-rear axle carrier........................................ | 59 |
| Hub and bearing assembly lock nut (1995 and earlier models only)....... | 90 |
| Rear stabilizer bar | |
| Link nuts......................................................................................... | 33 |
| Bushing retainer bolts................................................................... | 168 in-lbs |
| Rear struts | |
| Strut-to-axle carrier nuts/bolts..................................................... | 105 |
| Strut upper mounting nuts............................................................. | 29 |
| Suspension support-to-piston rod nut........................................... | 36 |
| Strut rod nuts/bolts............................................................................. | 67 |
| Suspension crossmember-to-body bolts.............................................. | 55 |

### Steering

| | |
|---|---|
| Airbag module Torx screws.................................................................. | 78 in-lbs |
| Steering gear bracket bolts/nuts | |
| 1997 and earlier.............................................................................. | 43 |
| 1998 and later | |
| Bracket bolts/nuts..................................................................... | 52 |
| Bolt "A"....................................................................................... | 48 in-lbs |
| Steering wheel nut............................................................................... | 25 |
| Tie-rod end-to-steering knuckle nut.................................................... | 36 |
| U-joint-to-pinion shaft pinch bolt ...................................................... | 26 |
| Power steering pressure line-to-pump union bolt................................. | 40 |
| Power steering pulley nut ................................................................... | 32 |

**1.1 Front suspension and steering components**

| | | | | | |
|---|---|---|---|---|---|
| 1 | Front stabilizer bar | 4 | Balljoint | 7 | Support brace |
| 2 | Stabilizer bar bushing clamp | 5 | Strut/coil spring assembly | 8 | Suspension crossmember |
| 3 | Control arm | 6 | Rack-and-pinion steering gear | | |

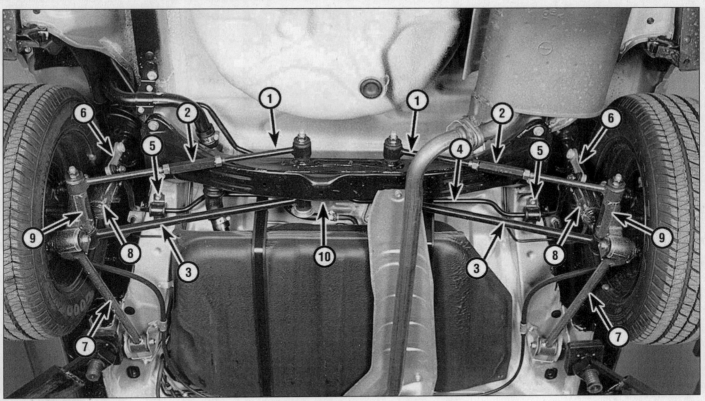

**1.2  Typical rear suspension components**

| | | | | | |
|---|---|---|---|---|---|
| 1 | Rear lateral arm | 5 | Stabilizer bar bushing clamp | 8 | Strut/coil spring assembly |
| 2 | Rear arm toe adjuster | 6 | Stabilizer link | 9 | Rear axle carrier |
| 3 | Front lateral arm | 7 | Strut rod | 10 | Rear suspension crossmember |
| 4 | Rear stabilizer bar | | | | |

## 1  General information

*Refer to illustrations 1.1 and 1.2*

The front suspension **(see illustration)** is a MacPherson strut design. The upper end of each strut/coil spring assembly is attached to the vehicle's body strut support. The lower end of the strut assembly is connected to the upper end of the steering knuckle. The steering knuckle is attached to a balljoint mounted on the outer end of the suspension control arm. A stabilizer bar reduces body roll.

The rear suspension **(see illustration)** also utilizes strut/coil spring assemblies. The upper end of each strut is attached to the vehicle body. The lower end of each strut is attached to an axle carrier. The carrier is located by a pair of suspension arms on each side, and a longitudinally mounted strut rod between the body and each carrier.

The rack-and-pinion steering gear is located behind the engine/transaxle assembly on the firewall and actuates the tie-rods, which are attached to the steering knuckles. The inner ends of the tie-rods are protected by rubber boots which should be inspected periodically for secure attachment, tears and leaking lubricant.

The power assist system consists of a belt-driven pump and associated lines and hoses. The fluid level in the power steering pump reservoir should be checked periodically (see Chapter 1).

The steering wheel operates the steering shaft, which actuates the steering gear through universal joints. Looseness in the steering can be caused by wear in the steering shaft universal joints, the steering gear, the tie-rod ends and loose retaining bolts.

### Precautions

Frequently, when working on the suspension or steering system components, you may come across fasteners which seem impossible to loosen. These fasteners on the underside of the vehicle are continually subjected to water, road grime, mud, etc., and can become rusted or "frozen," making them extremely difficult to remove. In order to unscrew these stubborn fasteners without damaging them (or other components), be sure to use lots of penetrating oil and allow it to soak in for a while. Using a wire brush to clean exposed threads will also ease removal of the nut or bolt and prevent damage to the threads. Sometimes a sharp blow with a hammer and punch will break the bond between a nut and bolt threads, but care must be taken to prevent the punch from slipping off the fastener and ruining the threads. Heating the stuck fastener and surrounding area with a torch sometimes helps too, but isn't recommended because of the obvious dangers associated with fire. Long breaker bars and extension, or "cheater," pipes will increase leverage, but never use an extension pipe on a ratchet - the ratcheting mechanism could be damaged. Sometimes tightening the nut or bolt first will help to break it loose. Fasteners that require drastic measures to remove should always be replaced with new ones.

Since most of the procedures dealt with in this Chapter involve jacking up the vehicle and working underneath it, a good pair of jackstands will be needed. A hydraulic floor jack is the preferred type of jack to lift the vehicle, and it can also be used to support certain components during various operations. **Warning:** *Never, under any circumstances, rely on a jack to support the vehicle while working on it. Whenever any of the suspension or steering fasteners are loosened or removed they must be inspected and, if necessary, replaced with new ones of the same part number or of original equipment quality and design. Torque specifications must be followed for proper reassembly and component retention. Never attempt to heat or straighten any suspension or steering components. Instead, replace any bent or damaged part with a new one.*

**10**

**2.3a  Mark the relationship of the strut to the steering knuckle (to preserve the camber setting when reassembling)**

**2.3b  To detach the strut assembly from the steering knuckle, remove the two nuts, then knock out the bolts with a hammer and punch**

**2.5  To detach the upper end of the strut assembly from the body, remove the upper mounting nuts (arrows)**

## 2  Strut assembly (front) - removal, inspection and installation

### Removal

*Refer to illustrations 2.3a, 2.3b and 2.5*

1    Loosen the wheel lug nuts, raise the vehicle and support it securely on jackstands. Remove the wheel.

2    Unbolt the brake hose bracket from the strut. If the vehicle is equipped with ABS, detach the speed sensor wiring harness from the strut by removing the clamp bracket bolt.

3    Mark the relationship of the strut to the steering knuckle **(see illustration). Note:** *Also mark the positions of the bolts, as special camber adjusting bolts may have been fitted at some point.* Remove the strut-to-knuckle nuts **(see illustration)** and knock the bolts out with a hammer and punch.

4    Separate the strut from the steering knuckle. Be careful not to overextend the inner CV joint. Also, don't let the steering knuckle fall outward and strain the brake hose.

5    Support the strut and spring assembly with one hand and remove the three strut-to-body nuts **(see illustration)**. Remove the assembly out from the fenderwell.

### Inspection

6    Check the strut body for leaking fluid, dents, cracks and other obvious damage which would warrant repair or replacement.

7    Check the coil spring for chips or cracks in the spring coating (this will cause premature spring failure due to corrosion). Inspect the spring seat for cuts, hardness and general deterioration.

8    If any undesirable conditions exist, proceed to the strut disassembly procedure (see Section 3).

### Installation

9    Guide the strut assembly up into the fenderwell and insert the upper mounting studs through the holes in the body. Once the

studs protrude, install the nuts so the strut won't fall back through. This is most easily accomplished with the help of an assistant, as the strut is quite heavy and awkward.

10    Slide the steering knuckle into the strut flange and insert the two bolts. Install the nuts, align the previously made matchmarks and tighten them to the torque listed in this Chapter's Specifications.

11    Connect the brake hose bracket to the strut and tighten the bolt securely. If the vehicle is equipped with ABS, install the speed sensor wiring harness bracket.

12    Install the wheel and lug nuts, then lower the vehicle and tighten the lug nuts to the torque listed in the Chapter 1 Specifications.

13    Tighten the upper mounting nuts to the torque listed in this Chapter's Specifications.

14    Drive the vehicle to an alignment shop to have the front end alignment checked, and if necessary, adjusted.

## 3  Strut/spring assembly - replacement

1    If the struts or coil springs exhibit the telltale signs of wear (leaking fluid, loss of damping capability, chipped, sagging or cracked coil springs) explore all options before beginning any work. The strut/shock absorber assemblies are not serviceable and must be replaced if a problem develops. However, strut assemblies complete with springs may be available on an exchange basis, which eliminates much time and work. Whichever route you choose to take, check on the cost and availability of parts before disassembling your vehicle. **Warning:** *Disassembling a strut is potentially dangerous and utmost attention must be directed to the job, or serious injury may result. Use only a high-quality spring compressor and carefully follow the manufacturer's instructions furnished with the tool. After removing the coil spring from the strut assembly, set it aside in a safe, isolated area.*

**3.3  Install the spring compressor according to the tool manufacturer's instructions and compress the spring until all pressure is relieved from the upper spring seat**

### Disassembly

*Refer to illustrations 3.3, 3.4, 3.5, 3.6 and 3.7*

2    Remove the strut assembly following the procedure described in Section 2 (front) or Section 10 (rear). Mount the strut assembly in a vise. Line the vise jaws with wood or rags to prevent damage to the unit and don't tighten the vise excessively.

3    Following the tool manufacturer's instructions, install the spring compressor (which can be obtained at most auto parts stores or equipment yards on a daily rental basis) on the spring and compress it sufficiently to relieve all pressure from the upper spring seat **(see illustration)**. This can be verified by wiggling the spring.

4    Loosen the damper shaft nut with a socket wrench **(see illustration)**.

5    Remove the nut and suspension support **(see illustration)**. Inspect the bearing in the suspension support for smooth operation. If it doesn't turn smoothly, replace the suspension support. Check the rubber por-

**3.4 Remove the damper shaft nut - if the upper spring seat turns while loosening the nut, immobilize it with a chain wrench or strap wrench**

**3.5 Lift the suspension support off the damper shaft**

**3.6 Remove the spring seat from the damper shaft**

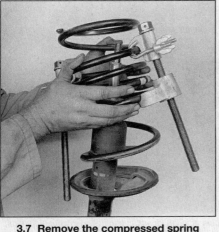

**3.7 Remove the compressed spring assembly - keep the ends of the spring pointed away from your body**

tion of the suspension support for cracking and general deterioration. If there is any separation of the rubber, replace it.

6    Lift the spring seat and upper insulator from the damper shaft **(see illustration)**. Check the rubber spring seat for cracking and hardness, replacing it if necessary.

7    Carefully lift the compressed spring from the assembly **(see illustration)** and set it in a safe place. **Warning:** *Never place your head near the end of the spring!*

8    Slide the rubber bumper off the damper shaft.

9    Check the lower insulator (if equipped) for wear, cracking and hardness and replace it if necessary.

## Reassembly

*Refer to illustrations 3.10, 3.11 and 3.12*

10   If the lower insulator is being replaced, set it into position with the dropped portion seated in the lowest part of the seat. Extend the damper rod to its full length and install the rubber bumper **(see illustration)**.

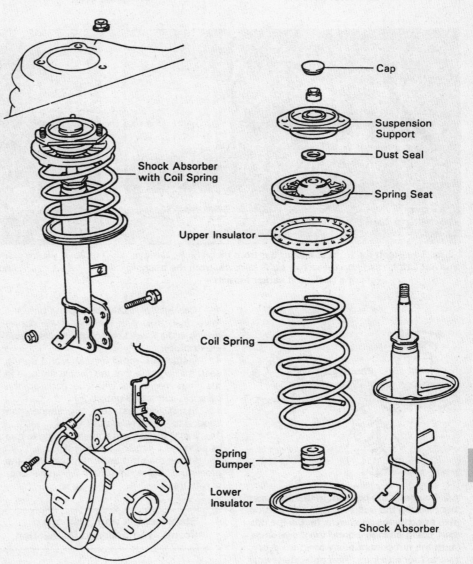

Shock Absorber with Coil Spring

Cap

Suspension Support

Dust Seal

Spring Seat

Upper Insulator

Coil Spring

Spring Bumper

Lower Insulator

Shock Absorber

**3.10 Typical strut and coil spring assembly details (1997 and earlier models shown, later models similar)**

10

**3.11  When installing the spring, make sure the end fits into the recessed portion of the lower seat (arrow)**

**3.12  The flats on the damper shaft (arrow) must match up with the flats in the spring seat**

**4.2  If you're removing the stabilizer bar, detach the bar from the link by removing the upper nut; if you're removing the control arm, remove the lower nut and detach the link from the arm**

**4.3a  To detach the front stabilizer bar from the vehicle, remove this nut (left arrow) and these two bolts (arrows) from the bushing clamps (1997 and earlier models)**

**4.3b  If you're working on a 1998 or later model, remove this nut and bolts and detach the bracket**

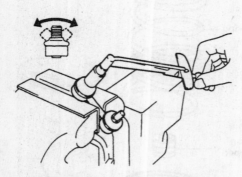

**4.5  To check the balljoint in the stabilizer bar link, flip the balljoint stud side to side five or six times as shown, install the nut and, using an inch-pound torque wrench, turn the nut continuously one turn every two to four seconds, then note the torque reading on the fifth turn. It should be about 0.4 to 8.7 in-lbs; if it isn't, replace the link assembly**

11   Carefully place the coil spring onto the lower insulator, with the end of the spring resting in the lowest part of the insulator **(see illustration)**.

12   Install the upper insulator and spring seat, making sure that the flats in the hole in the seat match up with the flats on the damper shaft **(see illustration)**.

13   Install the dust seal and suspension support to the damper shaft.

14   Install the nut and tighten it to the torque listed in this Chapter's Specifications.

15   Install the strut assembly following the procedure outlined in Section 2 (front) or Section 10 (rear).

## 4   Stabilizer bar and bushings (front) - removal and installation

### Removal

*Refer to illustrations 4.2, 4.3a, 4.3b and 4.5*

1   Loosen the front wheel lug nuts. Raise the front of the vehicle and support it securely on jackstands. Apply the parking brake and block the rear wheels to keep the vehicle from rolling off the stands. Remove the front wheels.

2   Detach the stabilizer bar link from the bar **(see illustration)**. If the ballstud turns with the nut, use an Allen wrench to hold the stud.

3   Unbolt the stabilizer bar bushing clamps **(see illustrations)**.

4   If you're working on a 1997 or earlier model, unbolt the front exhaust pipe from the flange in front of the catalytic converter (see Chapter 4) and pull it down far enough to remove the stabilizer bar.

5   While the stabilizer bar is off the vehicle, slide off the retainer bushings and inspect them. If they're cracked, worn or deteriorated, replace them. It's also a good idea to inspect the stabilizer bar link. To check it, flip the balljoint stud side-to-side five or six times as shown **(see illustration)**, then install the nut. Using an inch-pound torque wrench, turn the nut continuously one turn every two to four seconds and note the torque reading on

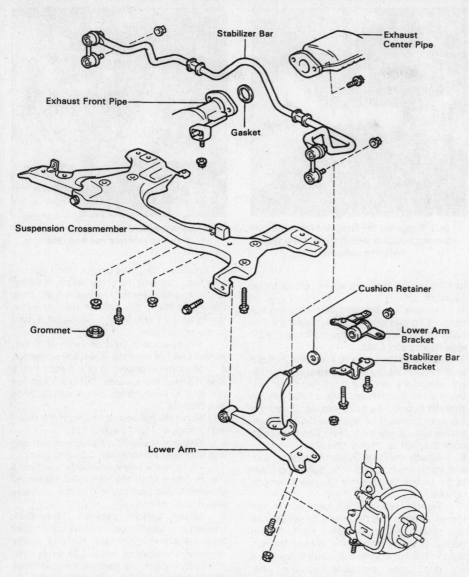

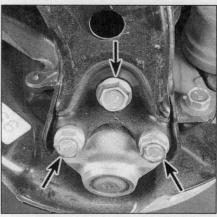

5.2a  To detach the control arm from the steering knuckle balljoint, remove this bolt and these two nuts (arrows) . . .

5.2b  . . . and pry the control arm and balljoint apart with a large prybar or screwdriver

4.8  An exploded view of the front stabilizer bar and control arm assemblies (1993 through 1995 models are as shown; 1996 and later models use a vertical bolt instead of a nut on the end of a pivot stud for attaching the rear of the control arm to the crossmember)

the fifth turn. It should be about 0.4 to 8.7 in-lbs. If it isn't, replace the link.

6    Clean the bushing area of the stabilizer bar with a stiff wire brush to remove any rust or dirt.

## Installation

*Refer to illustration 4.8*

7    Lubricate the inside and outside of the new bushing with vegetable oil (used in cooking) to simplify reassembly. **Caution:** *Don't use petroleum or mineral-based lubricants or brake fluid - they will lead to deterioration of the bushings.*

8    Installation is the reverse of removal **(see illustration).**

## 5    Control arm - removal, inspection and installation

### Removal

*Refer to illustrations 5.2a, 5.2b, 5.3 and 5.4*

1    Loosen the wheel lug nuts on the side to be dismantled, raise the front of the vehicle, support it securely on jackstands and remove the wheel.

2    Remove the bolt and two nuts holding the control arm to the steering knuckle. Use a prybar to disconnect the control arm from the steering knuckle **(see illustrations).**

3    Remove the control arm front pivot bolt **(see illustration). Note:** *If you're working on*

5.3  To detach the front of the control arm from the frame, remove this pivot bolt (arrow)

*a 1995 or earlier model with an automatic transaxle, and you're removing the left side control arm, the suspension crossmember must be supported, unbolted and then lowered.*

**10**

**5.4 To detach the rear end of the control arm from the frame on a 1996 or later model, remove this bolt (arrow); to detach the rear end of the arm on earlier models, unbolt the lower arm bracket (see illustration 4.8)**

**6.3 Separate the balljoint from the steering knuckle with a picklefork-type balljoint separator**

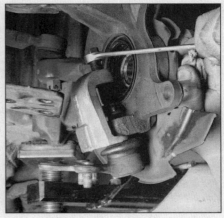

**7.8 Use a balljoint removal tool or small puller to remove the balljoint**

4    Remove the rear pivot bushing bracket (1995 and earlier models) or the rear bushing bolt **(see illustration)**.
5    Remove the control arm.

## Inspection

6    Check the control arm for distortion and the bushings for wear, replacing parts as necessary. Do not attempt to straighten a bent control arm.

## Installation

7    Installation is the reverse of removal **(see illustration 4.8)**. Tighten all of the fasteners to the torque values listed in this Chapter's Specifications. **Note:** *Before tightening the pivot bolt (and the pivot nut on the rear of the control arm, if you're working on a 1995 or earlier model), raise the outer end of the control arm with a floor jack to simulate normal ride height.*
8    Install the wheel and lug nuts, lower the vehicle and tighten the lug nuts to the torque listed in the Chapter 1 Specifications.
9    It's a good idea to have the front wheel alignment checked, and if necessary, adjusted after this job has been performed.

## 6   Balljoint - replacement

*Refer to illustration 6.3*
1    Loosen the wheel lug nuts, raise the vehicle and support it securely on jackstands. Remove the wheel.
2    Remove the cotter pin from the balljoint stud and loosen the nut (but don't remove it yet).
3    Separate the balljoint from the steering knuckle with a picklefork-type balljoint separator **(see illustration)**. Remove the balljoint stud nut. The clearance between the balljoint stud and the CV joint is very tight. To remove the stud nut, you'll have to alternately back off the nut a turn or two, pull down the stud,

turn the nut another turn or two, etc. until the nut is off.
4    Remove the bolt and nuts securing the balljoint to the control arm. Separate the balljoint from the control arm with a prybar **(see illustration 5.2b)**.
5    To install the balljoint, insert the balljoint stud through the hole in the steering knuckle and install the nut, but don't tighten it yet. Don't push the balljoint stud all the way up into and through the hole; instead, thread the nut onto the stud as soon as the stud protrudes through the hole, then turn the nut to draw the stud up through the hole.
6    Attach the balljoint to the control arm and install the bolt and nuts, tightening them to the torque listed in this Chapter's Specifications.
7    Tighten the balljoint stud nut to the torque listed in this Chapter's Specifications and install a new cotter pin. If the cotter pin hole doesn't line up with the slots on the nut, tighten the nut additionally until it does line up - don't loosen the nut to insert the cotter pin.
8    Install the wheel and lug nuts. Lower the vehicle and tighten the lug nuts to the torque listed in the Chapter 1 Specifications.

## 7   Steering knuckle and hub - removal and installation

**Warning:** *Dust created by the brake system is harmful to your health. Never blow it out with compressed air and don't inhale any of it. Do not, under any circumstances, use petroleum-based solvents to clean brake parts. Use brake system cleaner only.*

## Removal

*Refer to illustration 7.8*
1    Remove the wheel cover and loosen, but don't remove, the driveaxle/hub nut. Loosen the wheel lug nuts, raise the vehicle and support it securely on jackstands, then remove the wheel.

2    Remove the brake caliper (don't disconnect the hose) and the brake disc (see Chapter 9), and disconnect the brake hose from the strut. Hang the caliper from the coil spring with a piece of wire - don't let it hang by the brake hose.
3    If the vehicle is equipped with ABS, disconnect and remove the wheel speed sensor.
4    Mark the relationship of the strut to the steering knuckle. Loosen, but don't remove the strut-to-steering knuckle nuts and bolts (see Section 2).
5    Separate the tie-rod end from the steering knuckle arm (see Section 17).
6    Remove the balljoint-to-lower arm bolt and nuts **(see illustrations 5.2a and 5.2b)**.
7    Remove the driveaxle/hub nut and push the driveaxle from the hub as described in Chapter 8. Support the end of the driveaxle with a piece of wire.
8    Using a balljoint removal tool **(see illustration)** or a small puller, remove the balljoint from the steering knuckle. **Note:** *If you're removing the steering knuckle to replace the hub bearings, and the balljoint is in good condition, the balljoint can remain attached.*
9    The strut-to-knuckle bolts can now be removed.
10    Carefully separate the steering knuckle from the strut.

## Installation

11    Guide the knuckle and hub assembly into position, inserting the driveaxle into the hub.
12    Push the knuckle into the strut flange and install the bolts and nuts, but don't tighten them yet.
13    If you removed the balljoint from the old knuckle, and are planning to use it with the new knuckle, connect the balljoint to the knuckle and tighten the balljoint stud nut to the torque listed in this Chapter's Specifications. Install a new cotter pin.
14    Attach the balljoint to the control arm, but don't tighten the bolt and nuts yet.
15    Attach the tie-rod to the steering knuckle arm (see Section 17). Tighten the strut bolt nuts, the balljoint-to-control arm

**9.2 To detach a rear stabilizer bar link from the stabilizer bar, remove this nut (arrow) and pivot the link out of the way - remove the upper nut if you're removing the strut**

**9.3 To detach the rear stabilizer bar from the vehicle body, remove the bolts (arrows) from the bushing clamps**

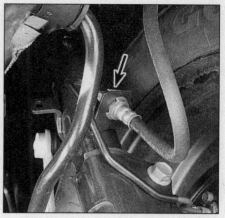

**10.3 Remove the clip (arrow) and pass the hose fitting through the slot in the bracket**

**10.6 To detach the rear strut from the axle carrier, remove these two nuts (arrows) and drive out the bolts with a hammer and punch**

**10.7 To detach the upper end of the rear strut from the vehicle, remove these three nuts (arrows)**

bolt and nuts and the tie-rod nut to the torque listed in this Chapter's Specifications.

16   Place the brake disc on the hub and install the caliper as outlined in Chapter 9.

17   Install the driveaxle/hub nut and tighten it securely, but not completely yet.

18   Install the wheel and lug nuts. Lower the vehicle and tighten the lug nuts to the torque listed in the Chapter 1 Specifications.

19   Tighten the driveaxle/hub nut to the torque listed in the Chapter 8 Specifications. Install the wheel cover.

20   Have the front-end alignment checked and, if necessary, adjusted.

## 8   Hub and bearing assembly (front) - removal and installation

Due to the special tools and expertise required to press the hub and bearing from the steering knuckle, this job should be left to a professional mechanic. However, the steering knuckle and hub may be removed and the assembly taken to an automotive machine shop or other qualified repair facility equipped with the necessary tools. See Section 7 for the steering knuckle and hub removal procedure.

## 9   Stabilizer bar and bushings (rear) - removal and installation

*Refer to illustrations 9.2 and 9.3*

1   Loosen the rear wheel lug nuts. Raise the rear of the vehicle and place it securely on jackstands. Remove the rear wheels.

2   Remove the stabilizer bar-to-link nut **(see illustration)**. If the ballstud turns with the nut, use a hex wrench to hold the stud.

3   Unbolt the stabilizer bar bushing clamps from the body **(see illustration)**.

4   The stabilizer bar can now be removed from the vehicle. Pull the retainers off the sta-

bilizer bar (if they haven't fallen off already) using a rocking motion.

5   Check the bushings for wear, hardness, distortion, cracking and other signs of deterioration, replacing them if necessary. Also check the link bushings for these signs.

6   Using a wire brush, clean the areas of the bar where the bushings ride. Installation is the reverse of the removal procedure. If necessary, use a light coat of vegetable oil to ease bushing and U-bracket installation (don't use petroleum-based products or brake fluid, as these will damage the rubber).

7   To inspect the links, refer to Step 5 in Section 4.

8   Installation is the reverse of removal.

## 10   Strut assembly (rear) - removal, inspection and installation

### Removal

*Refer to illustrations 10.3, 10.6 and 10.7*

1   On models with a one-piece rear seat, remove the seat cushions and the rear seat

back (see Chapter 11); on models with separate rear seat backs, lean the seat backs forward; on station wagon models, remove the rear strut cover (the plastic trim piece/cup holder on top of the side trim panel, attached by a single screw) (see Chapter 11).

2   Loosen the rear wheel lug nuts, raise the rear of the vehicle and support it securely on jackstands. Remove the wheel.

3   Remove the clip and detach the brake hose from the bracket on the strut **(see illustration)**. If the vehicle is equipped with ABS, detach the ABS sensor wire from the strut.

4   Disconnect the stabilizer bar link from the strut **(see illustration 9.2)**.

5   Support the axle carrier with a floor jack.

6   Loosen the strut-to-axle carrier bolt nuts **(see illustration)**.

7   Remove the three upper strut-to-body mounting nuts **(see illustration)**.

8   Lower the axle carrier with the jack and remove the two strut-to-axle carrier bolts.

9   Remove the strut assembly.

### Inspection

*Refer to illustration 10.11*

10   Follow the inspection procedures

**10**

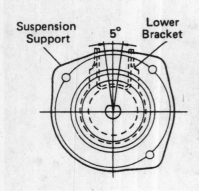

**10.11 Install the suspension support with the flat side facing forward and the strut-to-axle carrier bracket within 5-degrees of the middle of the distance between the two outer holes**

**11.2 To disconnect the strut rod from the axle carrier, remove this nut and bolt (arrows)**

described in Section 3. If you determine that the strut assembly must be disassembled for replacement of the strut or the coil spring, refer to Section 3.

11   When reassembling the strut, make sure the suspension support is aligned as shown **(see illustration)**.

## Installation

12   Maneuver the assembly up into the fenderwell and insert the mounting studs through the holes in the body. Install the nuts, but don't tighten them yet.

13   Push the axle carrier into the strut lower bracket and install the bolts and nuts, tightening them to the torque listed in this Chapter's Specifications.

14   Connect the stabilizer bar link to the strut bracket.

15   Attach the brake hose to the strut bracket and install the clip. If the vehicle is equipped with ABS, attach the ABS wire to the strut.

16   Install the wheel and lug nuts, lower the vehicle and tighten the lug nuts to the torque listed in the Chapter 1 Specifications.

17   Tighten the three strut upper mounting nuts to the torque listed in this Chapter's

Specifications.

18   Repeat Steps 2 through 17 for the other strut.

19   Install the seat, rear side seat backs or rear strut covers (see Chapter 11).

## 11   Strut rod - removal and installation

*Refer to illustrations 11.2 and 11.3*

1   Loosen the wheel lug nuts, raise the vehicle and support it securely on jackstands. Remove the wheel.

2   Remove the strut rod-to-axle carrier bolt **(see illustration)**.

3   Remove the strut rod-to-body bracket bolt **(see illustration)** and detach the rod from the vehicle.

4   Installation is the reverse of the removal procedure, but don't tighten the bolts until the suspension is raised by a jack (placed under the axle carrier) to simulate normal ride height. Be sure to tighten the bolts to the torque listed in this Chapter's Specifications. Tighten the wheel lug nuts to the torque listed in the Chapter 1 Specifications.

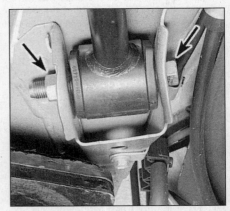

**11.3 To disconnect the strut rod from the body, remove this nut and bolt (arrows)**

## 12   Suspension arms - removal and installation

## Removal

1   Loosen the rear wheel lug nuts, raise the rear of the vehicle and support it securely on jackstands.

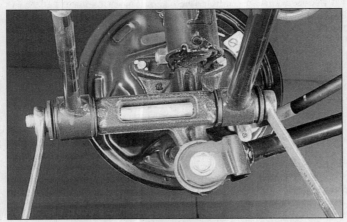

**12.3 To detach the suspension arms from the axle carrier, remove this nut and bolt (if you're only removing the no. 2 arm, it isn't necessary to pull the bolt out)**

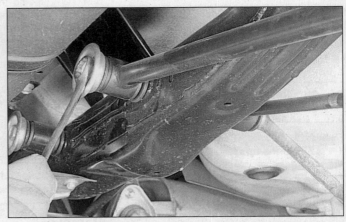

**12.4 To detach the No. 2 (rear) suspension arms from the suspension crossmember, remove the nut and washer; to remove the No. 1 (front) arms, remove the center exhaust pipe and lower the suspension crossmember so the bolt can be pulled out**

**Suspension Member**

**No. 2 Lower Suspension Arm**

**No. 1 Lower Suspension Arm**

**Strut Rod**

**12.7  An exploded view of the rear suspension arms and the suspension crossmember**

2    Block the front wheels and remove the rear wheel.

### No. 2 (rear) suspension arms
*Refer to illustrations 12.3 and 12.4*

3    Remove the nut and washer from the outer end of the suspension arm at the axle carrier **(see illustration)**.
4    Remove the nut and washer from the inner end of the suspension arm at the suspension crossmember **(see illustration)**.
5    Remove the No. 2 suspension arm. **Caution:** *Do NOT loosen the locknuts and turn the adjusting tube; moving this tube will affect the rear wheel toe adjustment.*

### No. 1 (front) suspension arms
*Refer to illustration 12.7*

6    Loosen the nut and bolt securing the inner end of the arms **(see illustration 12.4)**. Also loosen the nut and bolt attaching the outer ends of the arms to the axle carrier **(see illustration 12.3)**. Remove the exhaust system center pipe and the exhaust pipe insulator (see Chapter 4).
7    Support the suspension crossmember with a floor jack. Remove the four bolts retaining the crossmember **(see illustration)**.

Lower the crossmember far enough to allow the long through bolt at the inner end of the suspension arm to be pulled forward, out of the crossmember.
8    Remove the front suspension arm-to-rear axle carrier nut and bolt.
9    Remove the front suspension arm-to-suspension crossmember nut and bolt.
10   Remove the No. 1 suspension arm.

### Installation
11   Installation is the reverse of removal. If you removed a No. 1 (front) suspension arm, be sure to tighten the crossmember-to-body bolts to the torque listed in this Chapter's Specifications. Don't tighten the bolts/nuts at either end of the arm until the suspension is raised by a jack (placed under the axle carrier) to simulate normal ride height. Be sure to tighten the bolts/nuts to the torque listed in this Chapter's Specifications.
12   Install the wheel and lug nuts, then lower the vehicle to the ground. Tighten the wheel lug nuts to the torque listed in the Chapter 1 Specifications.
13   Have the rear wheel alignment checked and, if necessary, adjusted.

### 13  Hub and bearing assembly (rear) - removal and installation

**Warning:** *Dust created by the brake system is harmful to your health. Never blow it out with compressed air and don't inhale any of it. Do not, under any circumstances, use petroleum-based solvents to clean brake parts. Use brake system cleaner only.*
**Note:** *Due to the special tools required to replace the bearing, the hub and bearing assembly should not be disassembled by the home mechanic. The assembly can be removed, however, and taken to a dealer service department or other repair shop to have the bearing replaced.*

### Removal
*Refer to illustrations 13.3 and 13.5*

1    Loosen the wheel lug nuts, raise the vehicle and support it securely on jackstands. Remove the wheel.
2    Remove the brake drum (see Chapter 9).
3    Remove the four hub-to-axle carrier bolts, accessible by turning the hub flange so that the large circular cutout exposes each bolt **(see illustration)**.
4    Remove the hub and bearing assembly from its seat, maneuvering it out through the brake assembly.
5    Remove the old O-ring from the hub

**13.3 To remove the four bolts that attach the hub and bearing assembly to the rear axle carrier, rotate the hub flange and align one of the holes in the flange with each of the bolts**

**10**

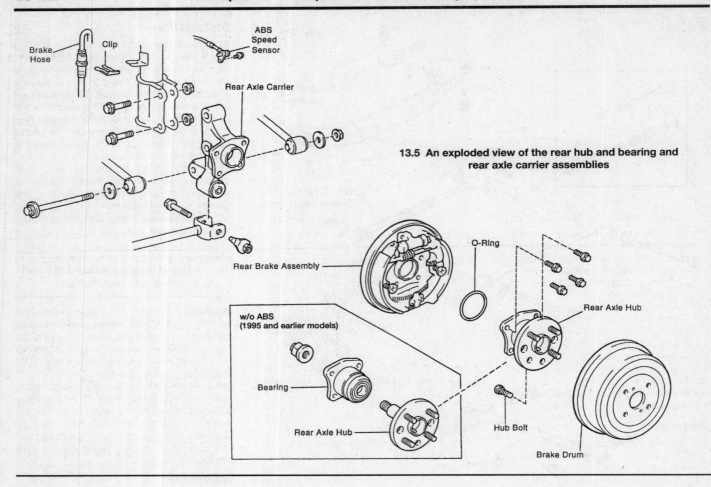

**13.5 An exploded view of the rear hub and bearing and rear axle carrier assemblies**

seat **(see illustration)**.

6    On 1995 and earlier models without anti-lock brakes the rear axle hub can be removed from the bearing to allow replace-

**13.7 Apply a light coat of oil to a new O-ring and install it on the raised hub of the backing plate; don't push the O-ring down against the face of the backing plate - when the hub/bearing assembly is installed, the O-ring will seat in its groove around the inside of the hub nut bore as the hub/bearing assembly is tightened**

ment of the bearing (on models with anti-lock brakes, and all 1996 and later models, the entire assembly must be replaced as a unit). Due to the special tools required to do this, you'll have to take the hub and bearing assembly to an automotive machine shop and have the old bearing pulled off the hub and a new bearing pressed on (you can re-use the hub itself, as long as it's in good condition). Make sure the hub retaining nut is tightened to the torque listed in this Chapter's Specifications.

## Installation

*Refer to illustration 13.7*

7    Apply a light coat of oil to the new O-ring and install it on the outer circumference of the raised hub in the backing plate **(see illustration)**.

8    Position the hub and bearing assembly on the axle carrier and align the holes in the backing plate. Install the bolts. A magnet is useful in guiding the bolts through the hub flange and into position. After all four bolts have been installed, tighten them to the torque listed in this Chapter's Specifications.

9    Install the brake drum and the wheel. Lower the vehicle and tighten the lug nuts to the torque listed in the Chapter 1 Specifications.

## 14    Rear axle carrier - removal and installation

**Warning:** *Dust created by the brake system is harmful to your health. Never blow it out with compressed air and don't inhale any of it. Do not, under any circumstances, use petroleum-based solvents to clean brake parts. Use brake system cleaner only.*

## Removal

1    Loosen the wheel lug nuts, raise the vehicle and support it on jackstands. Block the front wheels and remove the rear wheel.

2    Remove the rear brake drum and detach the brake line from the bracket on the strut **(see illustration 10.3)**.

3    Remove the rear hub and bearing assembly (see Section 13).

4    Detach the backing plate and rear brake assembly from the axle carrier. It isn't necessary to disassemble the brake shoe assembly or disconnect the parking brake cable from the backing plate. Suspend the backing plate and brake assembly from the coil spring with a piece of wire. Be careful not to kink the brake line.

5    On models with ABS, remove the wheel speed sensor from the axle carrier **(see illustration 13.5)**.

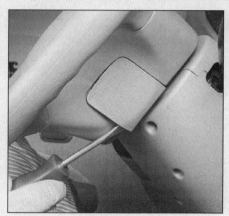

16.2a To access the airbag module retaining screws, remove the cover on each side of the steering wheel - 1997 and earlier models have large covers . . .

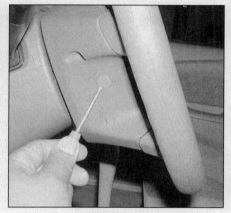

16.2b . . . while 1998 and later models have small covers

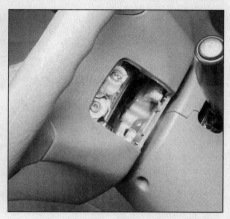

16.2c On 1997 and earlier models there is one airbag retaining screw on the right side of the steering wheel (shown), and two screws on the left side

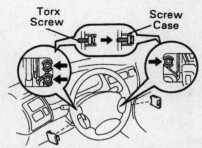

16.2d Back out the Torx screws until the airbag module is free, but don't try to remove them from the screw case (1997 and earlier models shown, later models similar)

6    Loosen, but don't remove the strut-to-axle carrier bolts **(see illustration 10.6)**.
7    Remove the suspension arm-to-axle carrier bolt, nut and washers **(see illustration 12.3)**. Also remove the rear strut rod-to-axle carrier bolt **(see illustration 11.2)**.
8    Remove the loosened strut-to-axle carrier bolts while supporting the carrier so it doesn't fall and detach the axle carrier from the strut bracket.

## Installation

9    Inspect the carrier bushing for cracks, deformation and signs of wear. If it is worn out, take the carrier to a dealer service department or other repair shop to have the old one pressed out and a new one pressed in.
10   Push the axle carrier into the strut bracket, aligning the two bolt holes. Insert the two strut-to-carrier bolts and tighten them to the torque listed in this Chapter's Specifications.
11   Install the suspension arm-to-axle carrier bolt (from the front), washers and nut. Tighten the nut by hand.
12   Connect the strut rod to the axle carrier and tighten the nut and bolt finger tight.
13   Place a jack under the carrier and raise it to simulate normal ride height.
14   Tighten the suspension arm bolt/nut and the strut rod bolt/nut to the torque values listed in this Chapter's Specifications. Remove the floor jack from under the rear axle carrier.
15   On models with ABS, reattach the wheel speed sensor to the axle carrier.
16   Attach the brake backing plate to the axle carrier, install the hub and tighten the four bolts to the torque listed in this Chapter's Specifications.
17   Connect the brake line to the strut bracket and install the clip. Make sure the hose isn't twisted.
18   Install the rear brake drum and adjust the brake shoes (see Chapter 9).
19   Install the wheel and lug nuts. Lower the vehicle and tighten the lug nuts to the torque listed in the Chapter 1 Specifications.

## 15   Steering system - general information

All models are equipped with rack-and-pinion steering. The steering gear operates the steering knuckles via tie-rods. The inner ends of the tie-rods are protected by rubber boots which should be inspected periodically for secure attachment, tears and leaking lubricant.

On models with power steering, the power assist system consists of a belt-driven pump and associated lines and hoses. The fluid level in the power steering pump reservoir should be checked periodically (see Chapter 1).

The steering wheel operates the steering shaft, which actuates the steering gear through universal joints. Looseness in the steering can be caused by wear in the steering shaft universal joints, the steering gear, the tie-rod ends, or loose retaining bolts.

## 16   Steering wheel - removal and installation

**Warning:** *These models are equipped with airbags. Always disable the airbag system before working in the vicinity of any airbag system component to avoid the possibility of accidental deployment of the airbag(s), which could cause personal injury (see Chapter 12).*

## Removal

*Refer to illustrations 16.2a, 16.2b, 16.2c, 16.3a, 16.3b, 16.3c, 16.3d, 16.5 and 16.6*
1    Turn the ignition key to Off, then disconnect the cable from the negative terminal of the battery. If the vehicle is equipped with an airbag system, wait at least two minutes before proceeding. **Caution:** *If the stereo in your vehicle is equipped with an anti-theft system, make sure you have the correct activation code before disconnecting the battery.*
2    Turn the steering wheel so the wheels are pointing straight ahead, then pry off the covers on either side of the steering wheel and loosen the Torx screws that attach the

16.3a Lift the airbag module straight out from the steering wheel . . .

airbag module to the steering wheel **(see illustrations)**. Loosen each screw until the groove in the circumference of the screw catches on the screw case **(see illustration)**. Note that on 1997 and earlier models there are two screws on the left and one screw on the right; this is normal - there is no missing screw on the right.
3    Pull the airbag module off the steering wheel **(see illustrations)** and disconnect the module electrical connector **(see illustra-**

**10**

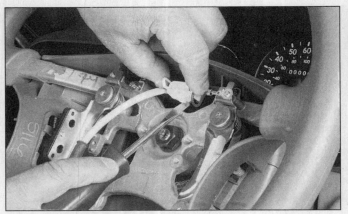

**16.3b** . . . flip up the locking tab on the electrical connector for the module and unplug the connector - this is a 1996 or earlier model . . .

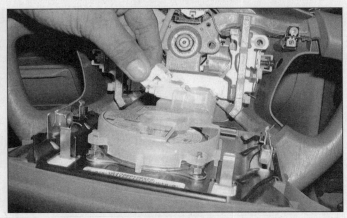

**16.3c** . . . 1998 and later models have a spring-loaded connector lock; flip up the lock . . .

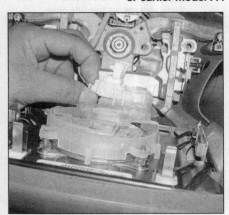

**16.3d** . . . depress the tab and unplug the connector

**16.5** After removing the steering wheel nut, mark the relationship of the steering wheel to the shaft before removing the wheel

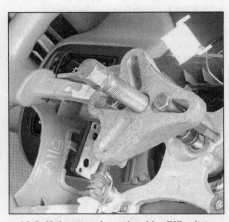

**16.6** If the steering wheel is difficult to remove from the shaft, use a steering wheel puller to remove it

tions). Set the airbag module in a safe, isolated area. **Warning:** *Carry the airbag module with the trim side facing away from you, and set the airbag module down with the trim side facing up. Don't place anything on top of the airbag module.*

4     Unplug the electrical connector for the cruise control (if equipped).

5     Remove the steering wheel retaining nut, then mark the relationship of the steering shaft to the hub (if marks don't already exist or don't line up) to simplify installation and ensure steering wheel alignment **(see illustration)**.

6     Use a puller to disconnect the steering wheel from the shaft **(see illustration)**. **Warning:** *Do not hammer on the shaft or the puller in an attempt to loosen the wheel from the shaft. Also, don't allow the steering shaft to turn with the steering wheel removed. If the shaft turns, the airbag spiral cable will become uncentered, which may cause the wire inside to break when the vehicle is returned to service.*

7     If it is necessary to remove the airbag spiral cable, follow the spiral cable wiring harness and unplug the electrical connector, then remove the four screws and detach it from the combination switch.

## Installation

*Refer to illustrations 16.8a, 16.8b and 16.12*

8     Make sure that the front wheels are pointing straight ahead. Depress the lock tab or hub of the spiral cable and turn the spiral

**16.8a** After properly centering the spiral cable as described in the text, make sure the pointers are aligned - 1997 and earlier models look like this . . .

cable counterclockwise by hand until it stops (don't apply too much force). Rotate the cable clockwise about three turns (1997 and earlier models) or 2-1/2 turns (1998 and later models) and align the two pointers **(see illustrations)**.

9     To install the wheel, align the mark on the steering wheel hub with the mark on the

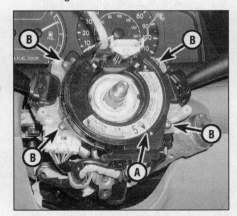

**16.8b** . . . and 1998 and later models look like this

A    *Pointers*
B    *Mounting screws*

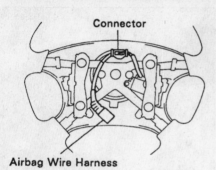

Connector

Airbag Wire Harness

**16.12  Before installing the airbag module, make sure that no wiring interferes with, or is pinched between, other parts; if you route the airbag harness as shown, it should be fine**

**17.2  Remove the cotter pin from the castle nut and loosen - but don't remove - the nut**

**17.3a  Loosen the jam nut . . .**

**17.3b  . . . then mark the position of the tie-rod end in relation to the threads**

**17.4  Disconnect the tie-rod from the steering knuckle arm with a puller**

## Installation

6    Thread the tie-rod end on to the marked position and insert the tie-rod stud into the steering knuckle arm. Tighten the jam nut securely.

7    Install the castle nut on the stud and tighten it to the torque listed in this Chapter's Specifications. Install a new cotter pin. If the hole for the cotter pin doesn't line up with one of the slots in the nut, turn the nut an additional amount until it does.

8    Install the wheel and lug nuts. Lower the vehicle and tighten the lug nuts to the torque listed in the Chapter 1 Specifications.

9    Have the alignment checked and, if necessary, adjusted.

## 18  Steering gear boots - replacement

*Refer to illustrations 18.3a and 18.3b*

1    Loosen the lug nuts, raise the vehicle and support it securely on jackstands. Remove the wheel.

2    Remove the tie-rod end and jam nut (see Section 17).

3    Remove the outer steering gear boot clamp with a pair of pliers (see illustration).

shaft and slip the wheel onto the shaft. Install the nut and tighten it to the torque listed in this Chapter's Specifications.

10    Plug in the cruise control connector.

11    Plug in the electrical connector for the airbag module and flip down the locking tab.

12    Make sure the airbag module electrical connector is positioned correctly and that the wires don't interfere with anything **(see illustration)**, then install the airbag module and tighten the retaining screws to the torque listed in this Chapter's Specifications.

13    Connect the negative battery cable.

## 17  Tie-rod ends - removal and installation

### Removal

*Refer to illustrations 17.2, 17.3a, 17.3b and 17.4*

1    Loosen the wheel lug nuts. Raise the front of the vehicle, support it securely on jackstands, block the rear wheels and set the parking brake. Remove the front wheel.

2    Remove the cotter pin **(see illustration)** and loosen the nut on the tie-rod end stud.

3    Hold the tie-rod with a pair of locking pliers or wrench and loosen the jam nut

enough to mark the position of the tie-rod end in relation to the threads **(see illustrations)**.

4    Disconnect the tie-rod from the steering knuckle arm with a puller **(see illustration)**. Remove the nut and detach the tie-rod.

5    Unscrew the tie-rod end from the tie-rod.

**18.3a  The outer ends of the steering gear boots are secured by spring-type clamps; they're easily released with a pair of pliers**

**18.3b  The inner ends of the steering gear boots are retained by boot clamps which must be cut off and discarded**

**10**

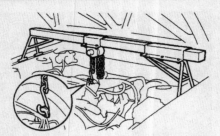

19.3 If you're working on a 1998 or later model, support the engine from above with an engine support fixture

Cut off the inner boot clamp with a pair of diagonal cutters **(see illustration)**. Slide off the boot.

4    Before installing the new boot, wrap the threads and serrations on the end of the steering rod with a layer of tape so the small end of the new boot isn't damaged.

5    Slide the new boot into position on the steering gear until it seats in the groove in the steering rod and install new clamps.

6    Remove the tape and install the tie-rod end (see Section 17).

7    Install the wheel and lug nuts. Lower the vehicle and tighten the lug nuts to the torque listed in the Chapter 1 Specifications.

## 19   Steering gear - removal and installation

**Warning:** *These models are equipped with airbags. Always disable the airbag system before working in the vicinity of any airbag system component to avoid the possibility of accidental deployment of the airbag(s), which could cause personal injury* (see Chapter 12).

### *Removal*

*Refer to illustrations 19.3, 19.5, 19.6a, 19.6b, 19.11a and 19.11b*

1    Disconnect the cable from the negative

19.6b  Mark the relationship of the universal joint to the steering gear input shaft and loosen the U-joint pinch bolt (arrow)

19.5  Disconnect the power steering line fittings (arrows)

terminal of the battery.

2    Remove the steering wheel (see Section 16). **Note:** *This will prevent the spiral cable from being damaged in the event the steering gear is not centered when it is installed.*

3    If you're working on a 1998 or later model, support the engine from above with an engine support fixture **(see illustration)**. This tool can be obtained at most equipment rental yards.

4    Loosen the front wheel lug nuts, raise the front of the vehicle and support it securely on jackstands. Apply the parking brake and remove the wheels. Remove the engine under covers.

5    If equipped with power steering, place a drain pan under the steering gear. Detach the power steering pressure and return lines **(see illustration)** and cap the ends to prevent excessive fluid loss and contamination.

6    Remove the universal joint cover **(see illustration)**. Mark the relationship of the lower universal joint to the steering gear input shaft and remove the lower intermediate shaft pinch bolt **(see illustration)**.

7    Separate the tie-rod ends from the steering knuckle arms (see Section 17).

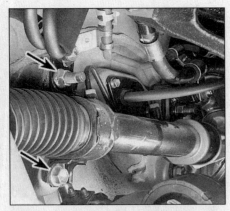

19.11a  Remove the lower bolt (arrow) and the upper nut (arrow) from the right steering gear bracket, then remove the nut and bolt from the left bracket (not visible in this photo)

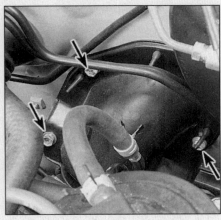

19.6a  Remove the universal joint cover bolts (arrows) and remove the cover (only three cover bolts are visible in this photo - there are actually five)

### Models with power steering only

8    Disconnect the balljoints from the control arms **(see illustrations 5.2a and 5.2b)**.

9    Unbolt the suspension crossmember and lower it along with the control arms. Also remove the front exhaust pipe support.

### All models

10    Remove the rear engine mount and bracket (see Chapter 2A).

11    Support the steering gear and remove the steering gear bracket bolts/nuts **(see illustrations)**. Separate the intermediate shaft from the steering gear input shaft, then move the steering gear to the right side of the vehicle, lower the left tie-rod, then maneuver the steering gear out to the left side of the vehicle.

12    Check the steering gear mounting grommets for excessive wear or deterioration, replacing them if necessary.

### *Installation*

13    Raise the steering gear into position and connect the U-joint, aligning the marks.

14    Install the mounting brackets and bolts and tighten them to the torque listed in this Chapter's Specifications.

15    Connect the tie-rod ends to the steering

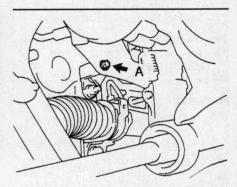

19.11b  On 1998 and later models, this bolt (A) must also be removed from the right-side bracket

**20.3 Loosen the fluid return hose clamp (left arrow) and detach the return hose from the power steering pump; remove the pressure line union bolt (right arrow) and disconnect the pressure line from the pump (1997 and earlier models)**

knuckle arms (see Section 17).

16   Install the U-joint pinch bolt and tighten it to the torque listed in this Chapter's Specifications. Install the universal joint cover.

17   If equipped with power steering, connect the power steering pressure and return lines to the steering gear.

18   Install the rear engine mount bracket and mount (see Chapter 2A).

19   On power steering equipped models, install the suspension crossmember, tightening the fasteners to the torque listed in this Chapter's Specifications. Connect the balljoints to the control arms, tightening the fasteners to the torque listed in this Chapter's Specifications.

20   Install the wheels and lug nuts. Lower the vehicle and tighten the lug nuts to the torque listed in the Chapter 1 Specifications.

21   Set the front wheels in the straight-ahead position, then center the spiral cable and install the steering wheel and airbag module (if equipped) (see Section 16).

22   On models equipped with power steer-

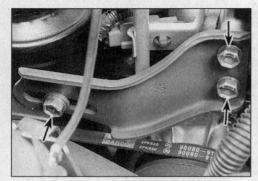

**20.7a Besides the pivot bolt, you'll need to remove the adjuster bolt (left arrow) to remove the pump; if you want to remove the pump bracket, remove the two bolts (arrows) at the right as well (1997 and earlier models)**

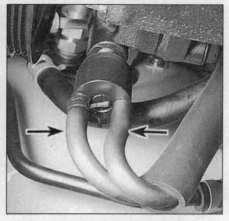

**20.6 Working from underneath, detach these two vacuum lines (arrows) from the power steering pump (1997 and earlier models)**

ing, fill the power steering fluid reservoir with the recommended fluid (see Chapter 1). Bleed the steering system (see Section 21).

## 20   Power steering pump - removal and installation

1   Disconnect the cable from the negative battery terminal. **Caution:** *If the stereo in your*

vehicle is equipped with an anti-theft system, make sure you have the correct activation code before disconnecting the battery.

2   Using a large syringe or suction gun, suck as much fluid out of the power steering fluid reservoir as possible. Place a drain pan under the vehicle to catch any fluid that spills out when the hoses are disconnected.

### 1997 and earlier models

#### Removal

*Refer to illustrations 20.3, 20.6, 20.7a and 20.7b*

3   Loosen the clamp and disconnect the fluid return hose from the pump **(see illustration)**.

4   Remove the pressure line-to-pump union bolt **(see illustration 20.3)**, then detach the line from the pump. Remove and discard the copper sealing washers. They must be replaced when installing the pump.

5   Raise the front of the vehicle and place it securely on jackstands. Remove the right-side engine under cover.

6   Working from underneath the vehicle, detach the two vacuum lines from the pump **(see illustration)**.

7   Loosen the pivot bolt (underneath) and adjuster bolt (above) and remove the drivebelt (see Chapter 1). Remove the pivot, adjuster and mounting bolts **(see illustrations)**, then remove the pump from the vehicle.

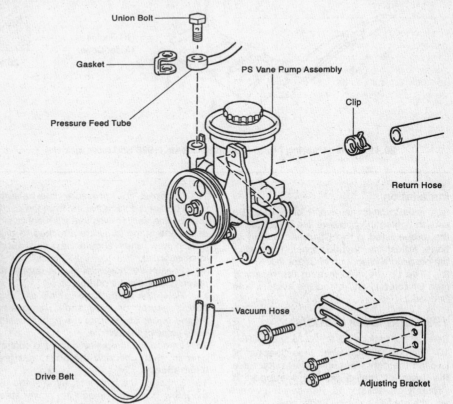

**20.7b Power steering pump mounting details (1997 and earlier models)**

**10**

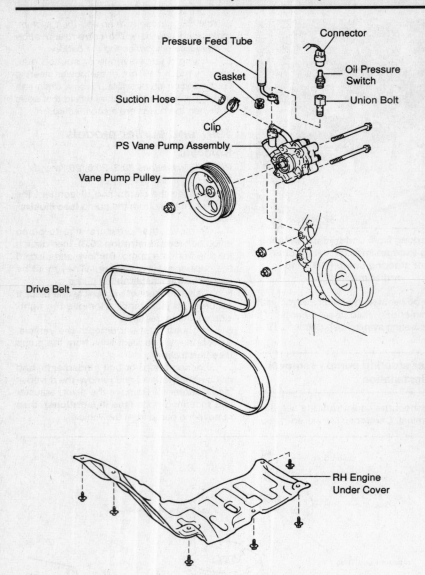

Pressure Feed Tube
Connector
Oil Pressure Switch
Gasket
Union Bolt
Suction Hose
Clip
PS Vane Pump Assembly
Vane Pump Pulley
Drive Belt
RH Engine Under Cover

**20.11  Power steering pump mounting details (1998 and later models)**

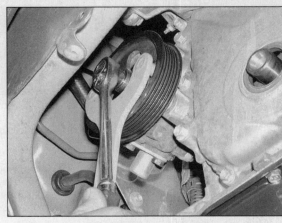

**20.15  Prevent the pulley from turning with a pin spanner wrench, then remove the pulley nut (1998 and later models)**

**20.16  Unscrew these two nuts, push the bolts out, then remove the power steering pump**

## Installation

8    Installation is the reverse of removal. Be sure to tighten the pressure line union bolt to the torque listed in this Chapter's Specifications. Adjust the drivebelt tension following the procedure described in Chapter 1.

9    Top up the fluid level in the reservoir (see Chapter 1) and bleed the system (see Section 21).

### 1998 and later models

*Refer to illustrations 20.11, 20.15 and 20.16*

10    Remove the drivebelt (see Chapter 1). Loosen the right front wheel lug nuts, raise the front of the vehicle and support it securely on jackstands. Remove the wheel and the right side engine under cover.

11    Unplug the electrical connector from the power steering fluid pressure switch **(see illustration)**.

12    Unscrew the pressure line-to-pump union bolt and detach the line from the pump. Discard the sealing washers from either side of the line fitting (they are attached to each other); new ones should be used during installation.

13    Loosen the hose clamp and detach the return hose from the pump.

14    Raise the front of the vehicle and support it securely on jackstands. Remove the engine under cover from the right (passenger's) side of the vehicle.

15    Immobilize the pulley with a pin spanner wrench, then unscrew the pulley nut **(see illustration)**.

16    Unscrew the two nuts, remove the two mounting bolts and detach the power steering pump from the engine **(see illustration)**.

## Installation

17    Installation is the reverse of removal. Be

sure to tighten the mounting bolts/nuts, the pulley nut and the pressure line union bolt to the torque listed in this Chapter's Specifications.

18    Install the engine under cover, wheel and lug nuts. Lower the vehicle and tighten the lug nuts to the torque listed in the Chapter 1 Specifications.

19    Install the drivebelt (see Chapter 1).

20    Top up the fluid level in the reservoir (see Chapter 1) and bleed the system (see Section 21).

## 21   Power steering system - bleeding

1    Following any operation in which the power steering fluid lines have been disconnected, the power steering system must be bled to remove all air and obtain proper steering performance.

2    With the front wheels in the straight ahead position, check the power steering fluid level and, if low, add fluid until it reaches the Cold mark on the dipstick.

3    Start the engine and allow it to run at fast idle. Recheck the fluid level and add

**22.3  Use a press tool to push the stud out of the flange**

| 1 | Hub flange | 2 | Lug nut on stud | 3 | Press tool |

**22.4  Install a spacer and a lug nut on the stud, then tighten the nut to draw the stud into place**

| 1 | Hub flange | 2 | Spacer |

more if necessary to reach the Cold mark on the dipstick.

4    Bleed the system by turning the wheels from side to side, without hitting the stops. This will work the air out of the system. Keep the reservoir full of fluid as this is done.

5    When the air is worked out of the system, return the wheels to the straight ahead position and leave the vehicle running for several more minutes before shutting it off.

6    Road test the vehicle to be sure the steering system is functioning normally and noise free.

7    Recheck the fluid level to be sure it is up to the Hot mark on the dipstick while the engine is at normal operating temperature. Add fluid if necessary (see Chapter 1).

## 22  Wheel studs - replacement

*Refer to illustrations 22.3 and 22.4*
**Note:** *This procedure applies to both the* front and rear wheel studs.

1    Loosen the wheel lug nuts, raise the vehicle and support it securely on jackstands. Remove the wheel.

2    Remove the brake disc or drum (see Chapter 9).

3    Install a lug nut part way onto the stud being replaced. Push the stud out of the hub flange with a press tool **(see illustration)**.

4    Insert the new stud into the hub flange from the back side and install some flat washers and a lug nut on the stud **(see illustration)**.

5    Tighten the lug nut until the stud is seated in the flange.

6    Reinstall the brake drum or disc. Install the wheel and lug nuts. Lower the vehicle and tighten the lug nuts to the torque listed in the Chapter 1 Specifications.

## METRIC TIRE SIZES

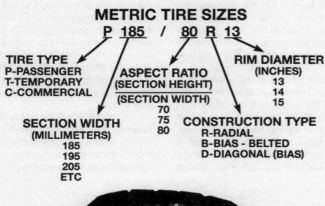

P 185 / 80 R 13

**TIRE TYPE**
P-PASSENGER
T-TEMPORARY
C-COMMERCIAL

**ASPECT RATIO**
(SECTION HEIGHT)
(SECTION WIDTH)

**RIM DIAMETER**
(INCHES)
13
14
15

**SECTION WIDTH**
(MILLIMETERS)
185
195
205
ETC

70
75
80

**CONSTRUCTION TYPE**
R-RADIAL
B-BIAS - BELTED
D-DIAGONAL (BIAS)

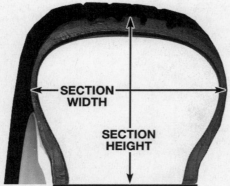

SECTION WIDTH

SECTION HEIGHT

**23.1  Metric tire size code**

## 23  Wheels and tires - general information

*Refer to illustration 23.1*

1    All vehicles covered by this manual are equipped with metric-sized fiberglass or steel belted radial tires **(see illustration)**. Use of other size or type of tires may affect the ride and handling of the vehicle. Don't mix different types of tires, such as radials and bias belted, on the same vehicle as handling may be seriously affected. It's recommended that tires be replaced in pairs on the same axle, but if only one tire is being replaced, be sure it's the same size, structure and tread design as the other.

2    Because tire pressure has a substantial effect on handling and wear, the pressure on all tires should be checked at least once a month or before any extended trips (see Chapter 1).

3    Wheels must be replaced if they are bent, dented, leak air, have elongated bolt holes, are heavily rusted, out of vertical symmetry or if the lug nuts won't stay tight.

**10**

Wheel repairs that use welding or peening are not recommended.

4    Tire and wheel balance is important in the overall handling, braking and performance of the vehicle. Unbalanced wheels can adversely affect handling and ride characteristics as well as tire life. Whenever a tire is installed on a wheel, the tire and wheel should be balanced by a shop with the proper equipment.

## 24   Wheel alignment - general information

*Refer to illustration 24.1*

A wheel alignment refers to the adjustments made to the wheels so they are in proper angular relationship to the suspension and the ground. Wheels that are out of proper alignment not only affect vehicle control, but also increase tire wear. The front end angles normally measured are camber, caster and toe-in **(see illustration)**. Toe-in and camber are adjustable on 1996 and later models. Toe-in is the only adjustable angle on 1995 and earlier models. The only adjustment possible on the rear is toe-in. The other angles should be measured to check for bent or worn suspension parts.

Getting the proper wheel alignment is a very exacting process, one in which complicated and expensive machines are necessary to perform the job properly. Because of this, you should have a technician with the proper equipment perform these tasks. We will, however, use this space to give you a basic idea of what is involved with a wheel alignment so you can better understand the process and deal intelligently with the shop that does the work.

Toe-in is the turning in of the wheels. The purpose of a toe specification is to ensure parallel rolling of the wheels. In a vehicle with zero toe-in, the distance between the front edges of the wheels will be the same as the distance between the rear edges of the wheels. The actual amount of toe-in is normally only a fraction of an inch. On the front end, toe-in is controlled by the tie-rod end

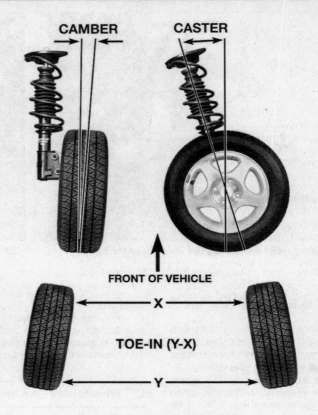

24.1  Camber, caster and toe-in angles

position on the tie-rod. On the rear end, it's controlled by a cam on the inner end of the rear (number two) suspension arm. Incorrect toe-in will cause the tires to wear improperly by making them scrub against the road surface.

Camber is the tilting of the wheels from vertical when viewed from one end of the vehicle. When the wheels tilt out at the top, the camber is said to be positive (+). When the wheels tilt in at the top the camber is negative (-). The amount of tilt is measured in degrees from vertical and this measurement is called the camber angle. This angle affects

the amount of tire tread which contacts the road and compensates for changes in the suspension geometry when the vehicle is cornering or traveling over an undulating surface. On the front end it is adjusted using special camber adjusting bolts, which alter the relationship between the strut and the steering knuckle (it isn't adjustable on the rear end).

Caster is the tilting of the top of the front steering axis from vertical. A tilt toward the rear is positive caster and a tilt toward the front is negative caster. Caster isn't adjustable on these vehicles.

# Chapter 11   Body

## Contents

## 1   General information

The models covered by this manual feature a "unibody" construction, using a floor pan with front and rear frame side rails which support the body components, front and rear suspension systems and other mechanical components. Certain components are particularly vulnerable to accident damage and can be unbolted and repaired or replaced. Among these parts are the body moldings, bumpers, hood and trunk lids and all glass.

Only general body maintenance practices and body panel repair procedures within the scope of the do-it-yourselfer are included in this Chapter.

## 2   Body - maintenance

1   The condition of your vehicle's body is very important, because the resale value depends a great deal on it. It's much more difficult to repair a neglected or damaged body than it is to repair mechanical components. The hidden areas of the body, such as the wheel wells, the frame and the engine compartment, are equally important, although they don't require as frequent attention as the rest of the body.
2   Once a year, or every 12,000 miles, it's a good idea to have the underside of the body steam cleaned. All traces of dirt and oil will be removed and the area can then be inspected carefully for rust, damaged brake lines, frayed electrical wires, damaged cables and other problems. The front suspension components should be greased after completion of this job.
3   At the same time, clean the engine and the engine compartment with a steam cleaner or water soluble degreaser.
4   The wheel wells should be given close attention, since undercoating can peel away and stones and dirt thrown up by the tires can cause the paint to chip and flake, allowing rust to set in. If rust is found, clean down to the bare metal and apply an anti-rust paint.
5   The body should be washed about once a week. Wet the vehicle thoroughly to soften the dirt, then wash it down with a soft sponge and plenty of clean soapy water. If the surplus dirt is not washed off very carefully, it can wear down the paint.
6   Spots of tar or asphalt thrown up from the road should be removed with a cloth soaked in solvent.
7   Once every six months, wax the body and chrome trim. If a chrome cleaner is used to remove rust from any of the vehicle's plated parts, remember that the cleaner also removes part of the chrome, so use it sparingly.

## 3   Vinyl trim - maintenance

Don't clean vinyl trim with detergents, caustic soap or petroleum-based cleaners. Plain soap and water works just fine, with a soft brush to clean dirt that may be ingrained. Wash the vinyl as frequently as the rest of the vehicle.

After cleaning, application of a high quality rubber and vinyl protectant will help prevent oxidation and cracks. The protectant can also be applied to weatherstripping, vacuum lines and rubber hoses, which often fail as a result of chemical degradation, and to the tires.

## 4   Upholstery and carpets - maintenance

1   Every three months remove the carpets or mats and clean the interior of the vehicle (more frequently if necessary). Vacuum the upholstery and carpets to remove loose dirt and dust.
2   Leather upholstery requires special care. Stains should be removed with warm water and a very mild soap solution. Use a clean, damp cloth to remove the soap, then wipe again with a dry cloth. Never use alcohol, gasoline, nail polish remover or thinner to clean leather upholstery.
3   After cleaning, regularly treat leather upholstery with a leather wax. Never use car wax on leather upholstery.
4   In areas where the interior of the vehicle is subject to bright sunlight, cover leather seats with a sheet if the vehicle is to be left out for any length of time.

## 5   Body repair - minor damage

*See photo sequence*

### Repair of minor scratches

1   If the scratch is superficial and does not penetrate to the metal of the body, repair is very simple. Lightly rub the scratched area with a fine rubbing compound to remove loose paint and built-up wax. Rinse the area with clean water.
2   Apply touch-up paint to the scratch, using a small brush. Continue to apply thin layers of paint until the surface of the paint in the scratch is level with the surrounding paint. Allow the new paint at least two weeks to harden, then blend it into the surrounding paint by rubbing with a very fine rubbing compound. Finally, apply a coat of wax to the scratch area.
3   If the scratch has penetrated the paint and exposed the metal of the body, causing the metal to rust, a different repair technique is required. Remove all loose rust from the bottom of the scratch with a pocket knife,

**11**

These photos illustrate a method of repairing simple dents. They are intended to supplement *Body repair - minor damage* in this Chapter and should not be used as the sole instructions for body repair on these vehicles.

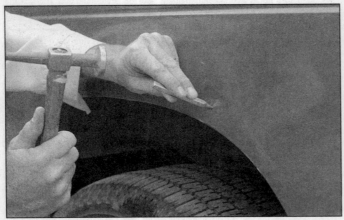

1   If you can't access the backside of the body panel to hammer out the dent, pull it out with a slide-hammer-type dent puller. In the deepest portion of the dent or along the crease line, drill or punch hole(s) at least one inch apart . . .

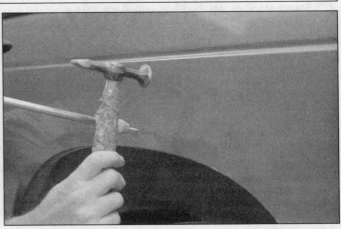

2   . . . then screw the slide-hammer into the hole and operate it. Tap with a hammer near the edge of the dent to help 'pop' the metal back to its original shape. When you're finished, the dent area should be close to its original contour and about 1/8-inch below the surface of the surrounding metal

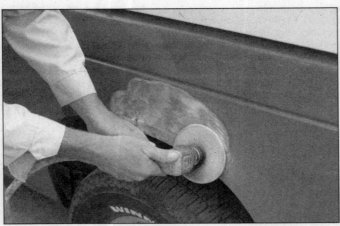

3   Using coarse-grit sandpaper, remove the paint down to the bare metal. Hand sanding works fine, but the disc sander shown here makes the job faster. Use finer (about 320-grit) sandpaper to feather-edge the paint at least one inch around the dent area

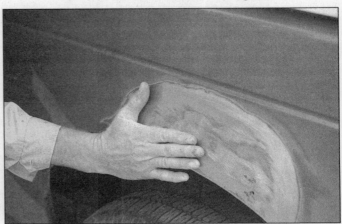

4   When the paint is removed, touch will probably be more helpful than sight for telling if the metal is straight. Hammer down the high spots or raise the low spots as necessary. Clean the repair area with wax/silicone remover

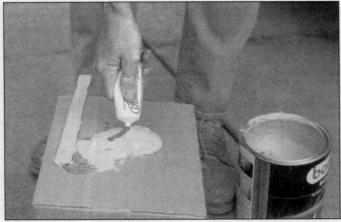

5   Following label instructions, mix up a batch of plastic filler and hardener. The ratio of filler to hardener is critical, and, if you mix it incorrectly, it will either not cure properly or cure too quickly (you won't have time to file and sand it into shape)

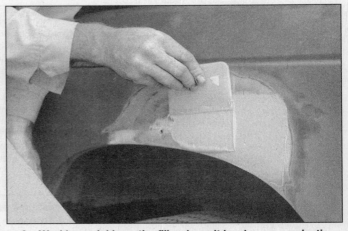

6   Working quickly so the filler doesn't harden, use a plastic applicator to press the body filler firmly into the metal, assuring it bonds completely. Work the filler until it matches the original contour and is slightly above the surrounding metal

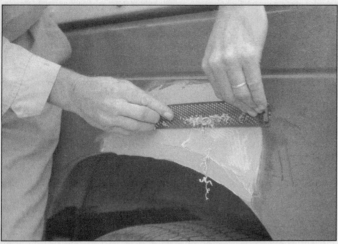

7   Let the filler harden until you can just dent it with your fingernail. Use a body file or Surform tool (shown here) to rough-shape the filler

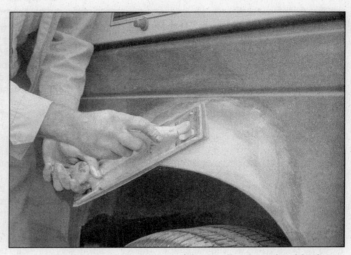

8   Use coarse-grit sandpaper and a sanding board or block to work the filler down until it's smooth and even. Work down to finer grits of sandpaper - always using a board or block - ending up with 360 or 400 grit

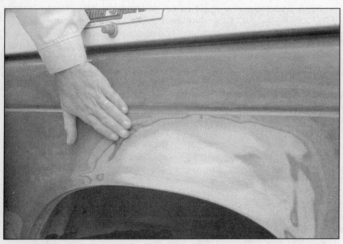

9   You shouldn't be able to feel any ridge at the transition from the filler to the bare metal or from the bare metal to the old paint. As soon as the repair is flat and uniform, remove the dust and mask off the adjacent panels or trim pieces

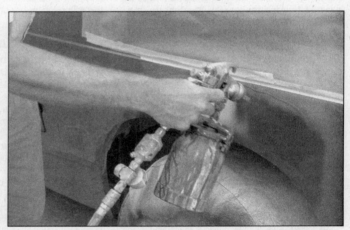

10   Apply several layers of primer to the area. Don't spray the primer on too heavy, so it sags or runs, and make sure each coat is dry before you spray on the next one. A professional-type spray gun is being used here, but aerosol spray primer is available inexpensively from auto parts stores

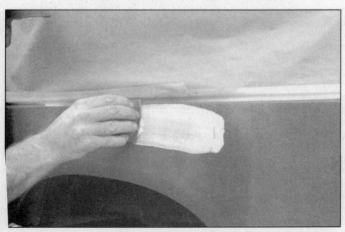

11   The primer will help reveal imperfections or scratches. Fill these with glazing compound. Follow the label instructions and sand it with 360 or 400-grit sandpaper until it's smooth. Repeat the glazing, sanding and respraying until the primer reveals a perfectly smooth surface

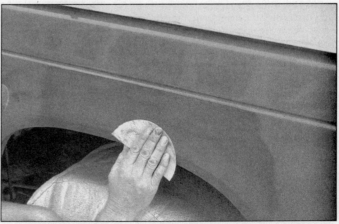

12   Finish sand the primer with very fine sandpaper (400 or 600-grit) to remove the primer overspray. Clean the area with water and allow it to dry. Use a tack rag to remove any dust, then apply the finish coat. Don't attempt to rub out or wax the repair area until the paint has dried completely (at least two weeks)

then apply rust inhibiting paint to prevent the formation of rust in the future. Using a rubber or nylon applicator, coat the scratched area with glaze-type filler. If required, the filler can be mixed with thinner to provide a very thin paste, which is ideal for filling narrow scratches. Before the glaze filler in the scratch hardens, wrap a piece of smooth cotton cloth around the tip of a finger. Dip the cloth in thinner and then quickly wipe it along the surface of the scratch. This will ensure that the surface of the filler is slightly hollow. The scratch can now be painted over as described earlier in this section.

## Repair of dents

4    When repairing dents, the first job is to pull the dent out until the affected area is as close as possible to its original shape. There is no point in trying to restore the original shape completely as the metal in the damaged area will have stretched on impact and cannot be restored to its original contours. It is better to bring the level of the dent up to a point which is about 1/8-inch below the level of the surrounding metal. In cases where the dent is very shallow, it is not worth trying to pull it out at all.

5    If the back side of the dent is accessible, it can be hammered out gently from behind using a soft-face hammer. While doing this, hold a block of wood firmly against the opposite side of the metal to absorb the hammer blows and prevent the metal from being stretched.

6    If the dent is in a section of the body which has double layers, or some other factor makes it inaccessible from behind, a different technique is required. Drill several small holes through the metal inside the damaged area, particularly in the deeper sections. Screw long, self-tapping screws into the holes just enough for them to get a good grip in the metal. Now the dent can be pulled out by pulling on the protruding heads of the screws with locking pliers.

7    The next stage of repair is the removal of paint from the damaged area and from an inch or so of the surrounding metal. This is done with a wire brush or sanding disk in a drill motor, although it can be done just as effectively by hand with sandpaper. To complete the preparation for filling, score the surface of the bare metal with a screwdriver or the tang of a file, or drill small holes in the affected area. This will provide a good grip for the filler material. To complete the repair, see the subsection on filling and painting later in this Section.

## Repair of rust holes or gashes

8    Remove all paint from the affected area and from an inch or so of the surrounding metal using a sanding disk or wire brush mounted in a drill motor. If these are not available, a few sheets of sandpaper will do the job just as effectively.

9    With the paint removed, you will be able to determine the severity of the corrosion and decide whether to replace the whole panel, if possible, or repair the affected area. New body panels are not as expensive as most people think and it is often quicker to install a new panel than to repair large areas of rust.

10   Remove all trim pieces from the affected area except those which will act as a guide to the original shape of the damaged body, such as headlight shells, etc. Using metal snips or a hacksaw blade, remove all loose metal and any other metal that is badly affected by rust. Hammer the edges of the hole in to create a slight depression for the filler material.

11   Wire brush the affected area to remove the powdery rust from the surface of the metal. If the back of the rusted area is accessible, treat it with rust inhibiting paint.

12   Before filling is done, block the hole in some way. This can be done with sheet metal riveted or screwed into place, or by stuffing the hole with wire mesh.

13   Once the hole is blocked off, the affected area can be filled and painted. See the following subsection on filling and painting.

## Filling and painting

14   Many types of body fillers are available, but generally speaking, body repair kits which contain filler paste and a tube of resin hardener are best for this type of repair work. A wide, flexible plastic or nylon applicator will be necessary for imparting a smooth and contoured finish to the surface of the filler material. Mix up a small amount of filler on a clean piece of wood or cardboard (use the hardener sparingly). Follow the manufacturer's instructions on the package, otherwise the filler will set incorrectly.

15   Using the applicator, apply the filler paste to the prepared area. Draw the applicator across the surface of the filler to achieve the desired contour and to level the filler surface. As soon as a contour that approximates the original one is achieved, stop working the paste. If you continue, the paste will begin to stick to the applicator. Continue to add thin layers of paste at 20-minute intervals until the level of the filler is just above the surrounding metal.

16   Once the filler has hardened, the excess can be removed with a body file. From then on, progressively finer grades of sandpaper should be used, starting with a 180-grit paper and finishing with 600-grit wet-or-dry paper. Always wrap the sandpaper around a flat rubber or wooden block, otherwise the surface of the filler will not be completely flat. During the sanding of the filler surface, the wet-or-dry paper should be periodically rinsed in water. This will ensure that a very smooth finish is produced in the final stage.

17   At this point, the repair area should be surrounded by a ring of bare metal, which in turn should be encircled by the finely feathered edge of good paint. Rinse the repair area with clean water until all of the dust produced by the sanding operation is gone.

18   Spray the entire area with a light coat of primer. This will reveal any imperfections in the surface of the filler. Repair the imperfections with fresh filler paste or glaze filler and once more smooth the surface with sandpaper. Repeat this spray-and-repair procedure until you are satisfied that the surface of the filler and the feathered edge of the paint are perfect. Rinse the area with clean water and allow it to dry completely.

19   The repair area is now ready for painting. Spray painting must be carried out in a warm, dry, windless and dust free atmosphere. These conditions can be created if you have access to a large indoor work area, but if you are forced to work in the open, you will have to pick the day very carefully. If you are working indoors, dousing the floor in the work area with water will help settle the dust which would otherwise be in the air. If the repair area is confined to one body panel, mask off the surrounding panels. This will help minimize the effects of a slight mismatch in paint color. Trim pieces such as chrome strips, door handles, etc., will also need to be masked off or removed. Use masking tape and several thickness of newspaper for the masking operations.

20   Before spraying, shake the paint can thoroughly, then spray a test area until the spray painting technique is mastered. Cover the repair area with a thick coat of primer. The thickness should be built up using several thin layers of primer rather than one thick one. Using 600-grit wet-or-dry sandpaper, rub down the surface of the primer until it is very smooth. While doing this, the work area should be thoroughly rinsed with water and the wet-or-dry sandpaper periodically rinsed as well. Allow the primer to dry before spraying additional coats.

21   Spray on the top coat, again building up the thickness by using several thin layers of paint. Begin spraying in the center of the repair area and then, using a circular motion, work out until the whole repair area and about two inches of the surrounding original paint is covered. Remove all masking material 10 to 15 minutes after spraying on the final coat of paint. Allow the new paint at least two weeks to harden, then use a very fine rubbing compound to blend the edges of the new paint into the existing paint. Finally, apply a coat of wax.

## 6    Body repair - major damage

1    Major damage must be repaired by an auto body shop specifically equipped to perform unibody repairs. These shops have the specialized equipment required to do the job properly.

2    If the damage is extensive, the body must be checked for proper alignment or the vehicle's handling characteristics may be adversely affected and other components may wear at an accelerated rate.

3    Due to the fact that all of the major body components (hood, fenders, etc.) are separate and replaceable units, any seriously damaged components should be replaced

**9.1  Before removing the hood, make marks around the hinge plate**

**9.10  Loosen the hood latch bolts, move the latch and retighten bolts, then close the hood to check the fit - repeat the procedure until the hood is flush with the fenders**

**9.11  Adjust the hood height by screwing the hood bumpers in-or-out**

rather than repaired. Sometimes the components can be found in a wrecking yard that specializes in used vehicle components, often at considerable savings over the cost of new parts.

## 7  Hinges and locks - maintenance

Once every 3000 miles, or every three months, the hinges and latch assemblies on the doors, hood and trunk should be given a few drops of light oil or lock lubricant. The door latch strikers should also be lubricated with a thin coat of grease to reduce wear and ensure free movement. Lubricate the door and trunk locks with spray-on graphite lubricant.

## 8  Windshield and fixed glass - replacement

Replacement of the windshield and fixed glass requires the use of special fast-setting adhesive/caulk materials and some specialized tools. It is recommended that these operations be left to a dealer or a shop specializing in glass work.

## 9  Hood - removal, installation and adjustment

*Refer to illustrations 9.1, 9.10 and 9.11*
**Note:** *The hood is heavy and somewhat awkward to remove and install - at least two people should perform this procedure.*

### Removal and installation

1    Make marks around the hinge plate to ensure proper alignment during installation **(see illustration)**.
2    Use blankets or pads to cover the cowl area of the body and fenders. This will protect the body and paint as the hood is lifted off.
3    Disconnect any cables or wires that will

interfere with removal. On some models, it may be necessary to remove the lower two hood insulation retaining pins and peel the insulation back enough to disconnect the windshield wiper hoses.
4    Have an assistant support the hood. Remove the hinge-to-hood bolts.
5    Lift off the hood.
6    Installation is the reverse of removal.

### Adjustment

7    Fore-and-aft and side-to-side adjustment of the hood is done by moving the hinge plate slot after loosening the bolts.
8    Scribe a line around the entire hinge plate so you can judge the amount of movement **(see illustration 9.1)**
9    Loosen the bolts or nuts and move the hood into correct alignment. Move it only a little at a time. Tighten the hinge bolts and carefully lower the hood to check the position.
10    If necessary after installation, the entire hood latch assembly can be adjusted up-and-down as well as from side-to-side on the radiator support so the hood closes securely, flush with the fenders. To make the adjustment,

scribe a line around the hood latch mounting bolts to provide a reference point, then loosen them and reposition the latch assembly, as necessary **(see illustration)**. Following adjustment, retighten the mounting bolts.
11    Finally, adjust the hood bumpers on the radiator support so the hood, when closed, is flush with the fenders **(see illustration)**.
12    The hood latch assembly, as well as the hinges, should be periodically lubricated with white, lithium-base grease to prevent binding and wear.

## 10  Trunk lid - removal, installation and adjustment

*Refer to illustrations 10.2, 10.3 and 10.8*
1    Open the trunk lid and cover the edges of the trunk compartment with pads or cloths to protect the painted surfaces when the lid is removed.
2    Open the trunk lid and remove the trunk lid trim cover if equipped **(see illustration)**.
3    Make alignment marks around the hinge mounting bolts **(see illustration)**.
4    While an assistant supports the lid, remove the lid-to-hinge bolts on both sides and lift it off.

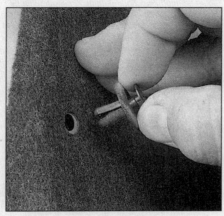

**10.2  Unscrew the plastic retainer, then pull it out**

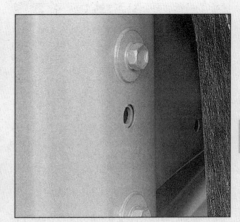

**10.3  Scribe a mark around the bolt heads to help with lid realignment on installation**

**11**

**10.8 Loosen the bolts, then adjust the latch and striker position**

**11.3 Draw a line around the hinge plate on the liftgate before removing the bolts**

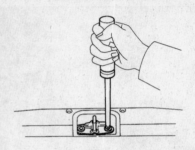

**11.9 Adjust the lock striker by loosening the mounting screws slightly and tapping the striker with a soft-faced hammer**

5    Installation is the reverse of removal. **Note:** *When reinstalling the trunk lid, align the lid-to-hinge bolts with the marks made during removal.*

6    After installation, close the lid and make sure it's in proper alignment with the surrounding panels.

7    Forward-and-backward and side-to-side adjustments are made by loosening the hinge-to-lid bolts and gently moving the lid into correct alignment.

8    To adjust the lid so it is flush with the body when closed, loosen the mounting bolts and move the lock and striker **(see illustration)**.

## 11   Liftgate (station wagon) - removal, installation and adjustment

*Refer to illustrations 11.3 and 11.9*
**Note:** *The liftgate is heavy and somewhat awkward to remove and install - at least two people should perform this procedure.*

1    Open the liftgate and cover the edges of the compartment with pads or cloths to protect the painted surfaces when the lid is removed.

2    Disconnect any cables or wire harness connectors attached to the liftgate that would interfere with removal.

3    Use a marking pen to make alignment marks around the hinge mounting flanges **(see illustration)**.

4    Have an assistant support the liftgate and detach the support struts (see Section 12).

5    While an assistant supports the liftgate, remove the lid-to-hinge bolts on both sides and lift it off.

6    Installation is the reverse of removal. **Note:** *When reinstalling the liftgate, align the hinges with the marks made during removal.*

7    After installation, close the liftgate and make sure it's in proper alignment with the surrounding panels.

8    Adjustments to the liftgate position are made by loosening the hinge-to-liftgate bolts or nuts and gently moving the liftgate into correct alignment.

9    The liftgate latch position can be adjusted by loosening the adjusting bolts and moving the latch. The latch striker can be adjusted by loosening the mounting screws and gently tapping it into position with a plastic hammer **(see illustration)**.

## 12   Liftgate support strut - replacement

*Refer to illustration 12.1*
**Warning:** *The support strut is filled with pressurized gas - do not disassemble this component. If it is faulty replace it with a new one.*
**Note:** *The liftgate is heavy and somewhat awkward to hold securely while replacing the struts - at least two people should perform this procedure.*

1    Open the liftgate and support it in the open position. Remove the bolts at the ends and detach the strut from the liftgate **(see illustration)**.

2    Installation is the reverse of the removal procedure.

## 13   Door trim panel - removal and installation

*Refer to illustrations 13.2a, 13.2b, 13.2c, 13.4a, 13.4b, 13.6 and 13.7*

### Removal

1    Disconnect the negative cable from the battery. **Caution:** *If the stereo in your vehicle is equipped with an anti-theft system, make sure you have the correct activation code*

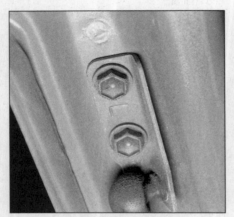

**12.1 After supporting the liftgate, remove the bolts at each end and detach the support strut**

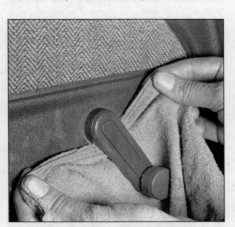

**13.2a Work a cloth up behind the regulator handle and move it back-and-forth . . .**

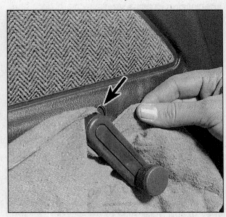

**13.2b . . . until the retainer (arrow) is pushed up so you can remove it**

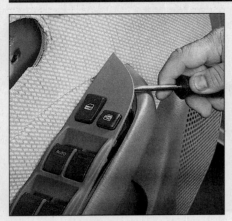

13.2c  Pry up on the power window switch and remove it

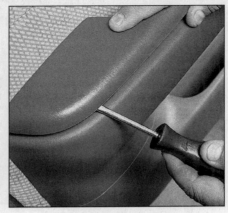

13.4a  Pry the arm rest up with a screwdriver

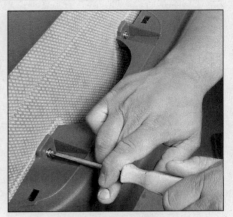

13.4b  Remove the armrest screws

13.6  Use a trim panel removal tool to detach the trim panel retaining clips, then pull the door trim up and out to remove it

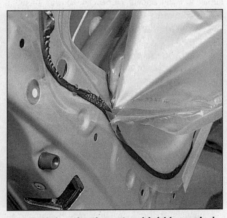

13.7  If the plastic watershield is peeled off carefully it can be reused

14.2  Use two jackstands padded with rags (to protect the paint) to support the door during the removal and installation procedures

*before disconnecting the battery.*

2    On manual window regulator equipped models, remove the window crank by working a cloth back-and-forth behind the handle to dislodge the retainer **(see illustrations)**. A special tool is available for this purpose but it's not essential. With the retainer removed, pull off the handle. On power window models, pry out the switch assembly, unplug the electrical connector and remove it **(see illustration)**.

3    Remove the outside mirror trim cover (see Section 18).

4    Remove the inside door handle (see Section 15). Remove the door trim panel retaining screws and door pull/armrest assemblies **(see illustrations)**.

5    Insert a wide putty knife, a thin screwdriver or a special trim panel removal tool between the trim panel and door to disengage the retaining clips. Work around the outer edge until the panel is free.

6    Once all of the clips are disengaged, detach the trim panel, unplug any electrical connectors and remove the trim panel from the vehicle by gently pulling it up and out **(see illustration)**.

7    For access to the inner door remove the plastic watershield. Peel back the plastic

cover, taking care not to tear it **(see illustration)**. Remove the plastic grommets, if necessary.

### Installation

8    To install the trim panel, first press the watershield back into place. If necessary, add more sealant to hold it in place.

9    Prior to installation of the door panel, be sure to reinstall any clips in the panel which may have come out during the removal procedure and stayed in the door.

10   Plug in any electrical connectors and place the panel in position. Press it into place until the clips are seated and install any retaining screws and armrest/door pulls. Install the manual regulator window crank or power switch assembly.

## 14  Door - removal, installation and adjustment

### Removal and installation

*Refer to illustrations 14.2, 14.3 and 14.4*

1    Remove the door trim panel (see Section 13). Disconnect any electrical connectors and push them through the door opening so they won't interfere with removal.

2    Position a jack or jackstands under the door or have an assistant on hand to support the door when the hinge bolts are removed **(see illustration)**. **Note:** *If a jack or stand is used, place a rag between it and the door to protect the door's paint.*

3    Remove the door stop strut bolt **(see illustration)**.

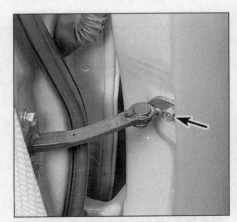

14.3  Remove the bolt (arrow) and detach the stop strut

**11**

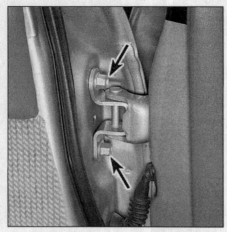

**14.4  Before loosening or removing them, mark the door bolt locations (arrows)**

4    Scribe around the door bolts **(see illustration)**.
5    Remove the hinge-to-door bolts and carefully detach the door. Installation is the reverse of removal.

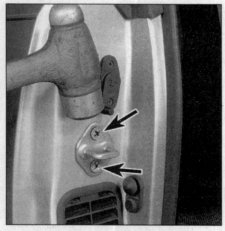

**14.6c  Adjust the door lock striker by loosening the mounting screws (arrows) and gently tapping the striker in the desired direction**

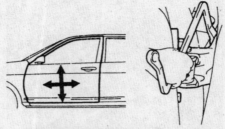

**14.6a  When adjusting the door up-and-down or forward-and-backward a special wrench such as this one will make the job easier**

**14.6b  Adjust the door up-and-down or in-and-out after loosening the hinge-to-door bolts**

## Adjustment

*Refer to illustrations 14.6a, 14.6b and 14.6c*

6    Following installation, make sure the door is aligned properly. Adjust it if necessary as follows:

a)  *Up-and-down and forward-and-backward adjustments are made by loosening the hinge-to-body bolts and moving the door, as necessary. A special offset tool may be required to reach some of the bolts* **(see illustration)**.

b)  *In-and-out and up-and-down adjustments are made by loosening the door side hinge bolts and moving the door, as necessary. A special offset tool may be required to reach some of the bolts* **(see illustration)**.

c)  *The door lock striker can also be adjusted both up-and-down and sideways to provide a positive engagement with the locking mechanism. This is done by loosening the screws and moving the striker, as necessary* **(see illustration)**.

---

**15    Door latch, lock cylinder and handles - removal and installation**

---

1    Remove the door trim panel and the plastic water shield (Section 13).

## Door latch

*Refer to illustration 15.3*

2    Reach inside the door and disconnect the control links from the latch.
3    Remove the latch retaining screws from the end of the door **(see illustration)**.
4    Detach the door latch and (if equipped) the door lock solenoid.
5    Installation is the reverse of removal.

## Lock cylinder and outside handle

*Refer to illustration 15.7*

6    Disconnect the control link from the lock cylinder and outside handle.
7    Remove the outside handle retention bolts and pull the handle and lock cylinder from the door **(see illustration)**.
8    Use a screwdriver to pry the retaining clip off or remove the lock cylinder retaining bolt and remove the lock cylinder from the handle.
9    Installation is the reverse of removal.

## Inside handle

*Refer to illustrations 15.10, 15.11a and 15.11b*

10    Remove the retaining screw **(see illustration)**.
11    Pull the handle free, disconnect the link from the inside handle control and remove

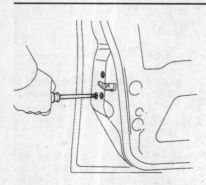

**15.3  Remove the latch screws from the end of the door**

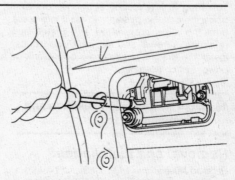

**15.7  The outside handle retention bolts can be reached through access holes in the door frame**

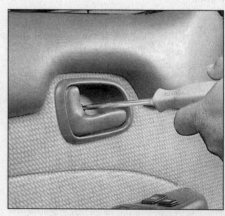

**15.10  Remove the inside handle screw**

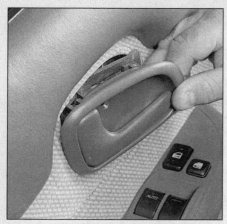

**15.11a  Rotate the handle out for access to the link**

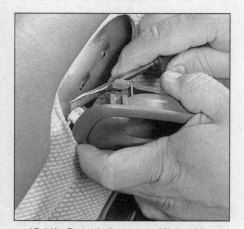

**15.11b  Detach the control link with a small screwdriver**

the handle from the door (see illustrations).

12    Installation is the reverse of removal.

## 16   Door window glass - removal and installation

1    Remove the door trim panel and the plastic watershield (Section 13).

2    Lower the window glass. Remove the door access plate from the door frame, if equipped.

3    Carefully pry the inner weatherstrip out of the door window opening.

4    Place a rag inside the door panel to help prevent scratching the glass, then remove the two glass mounting bolts.

5    Remove the glass by pulling it up.

6    Installation is the reverse of the removal procedure.

## 17   Bumpers - removal and installation

*Refer to illustrations 17.4a, 17.4b and 17.4c*
**Warning:** *These models are equipped with*

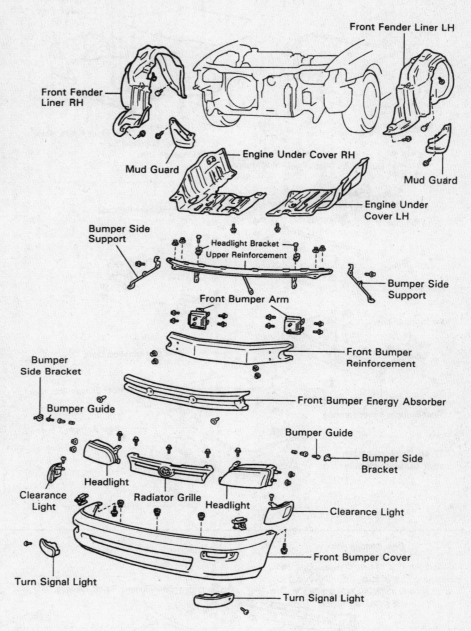

**17.4a  Typical front bumper details**

*airbags. The airbag is armed and can deploy (inflate) anytime the battery is connected. To prevent accidental deployment (and possible injury), turn the ignition key to LOCK and disconnect the negative battery cable whenever working near airbag components. After the battery is disconnected, wait at least two minutes before beginning work (the system has a back-up capacitor that must fully discharge). For more information see Chapter 12.*

1    Apply the parking brake, raise the vehicle and support it securely on jackstands.

2    Disconnect the cable from the negative battery terminal and disconnect any wiring that would interfere with bumper removal.

3    Remove the bumper cover if equipped, taking care to avoid damaging the cover and the fender.

4    Working under the vehicle, remove the bumper retention bolts (see illustrations).

5    Pull the bumper assembly from the vehicle.

6    Installation is the reverse of the removal procedure.

**11**

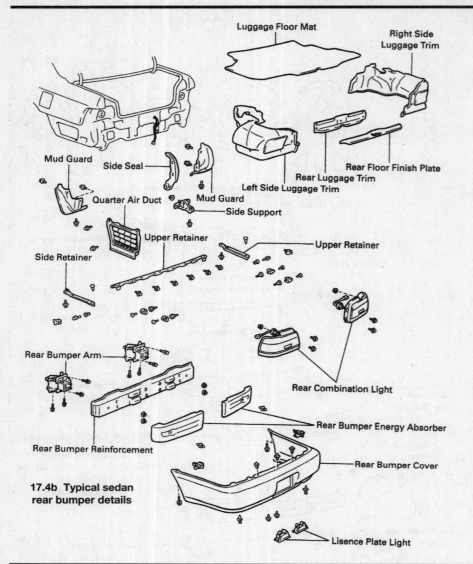

Luggage Floor Mat

Right Side Luggage Trim

Mud Guard

Side Seal

Quarter Air Duct

Mud Guard

Side Support

Side Retainer

Upper Retainer

Upper Retainer

Rear Floor Finish Plate

Rear Luggage Trim

Left Side Luggage Trim

Rear Combination Light

Rear Bumper Arm

Rear Bumper Energy Absorber

Rear Bumper Cover

Rear Bumper Reinforcement

**17.4b  Typical sedan rear bumper details**

Lisence Plate Light

## 18  Outside mirror - removal and installation

*Refer to illustrations 18.1a, 18.1b, 18.2a and 18.2b*

1    On manually operated mirrors, remove the control handle **(see illustration)**. On all models, detach the mirror cover by using a small screwdriver to pry the retainers free from the door **(see illustration)**.

2    Remove the three retaining nuts and detach the mirror **(see illustration)**. On power mirrors, unplug the electrical connector **(see illustration)**.

3    Installation is the reverse of removal.

## 19  Seats - removal and installation

### *Front seats*

*Refer to illustration 19.1*

1    Pry off the seat bolt covers if equipped and remove the retaining bolts, unplug any electrical connectors and lift the seats from the vehicle **(see illustration)**.

2    Installation is the reverse of removal.

### *Rear seats*

*Refer to illustration 19.3*

3    On sedan models, lift the front of the cushion up, then pull it out toward the front of the vehicle. Remove the seat back retaining bolts, then lift up on the back to release the seat back from the body **(see illustration)**.

4    On station wagon models, remove the retaining bolts at the base of the seat cushion, pull the back of the cushion up and remove it from the vehicle. Remove the seat back pivot bolts and remove the seat back from the vehicle.

5    Installation is the reverse of removal.

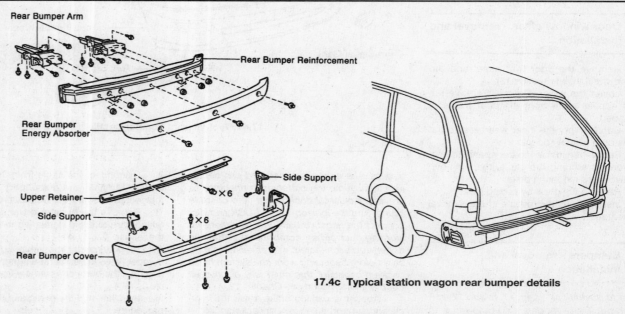

Rear Bumper Arm

Rear Bumper Reinforcement

Rear Bumper Energy Absorber

Upper Retainer

Side Support ×6

×6

Side Support

Rear Bumper Cover

**17.4c  Typical station wagon rear bumper details**

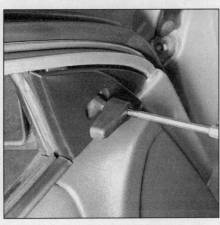

18.1a On manually operated side view mirrors, remove the screw and detach the mirror control handle

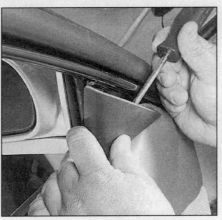

18.1b Detach the cover and lift it out (power mirror shown)

18.2a Remove the three bolts securing the outside mirror to the door

18.2b On vehicles equipped with power side view mirrors, unplug the electrical connector (arrow)

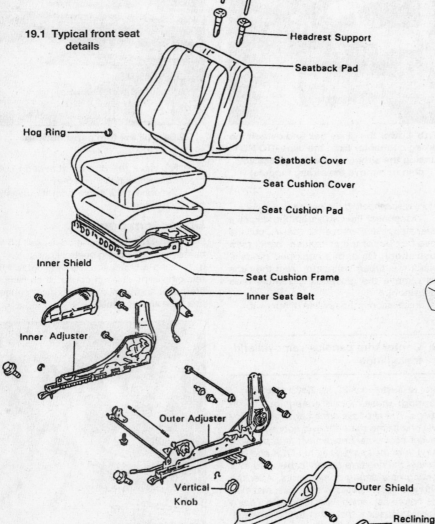

19.1 Typical front seat details

- Headrest
- Headrest Support
- Seatback Pad
- Hog Ring
- Seatback Cover
- Seat Cushion Cover
- Seat Cushion Pad
- Inner Shield
- Seat Cushion Frame
- Inner Seat Belt
- Inner Adjuster
- Outer Adjuster
- Vertical Knob
- Outer Shield
- Reclining Knob

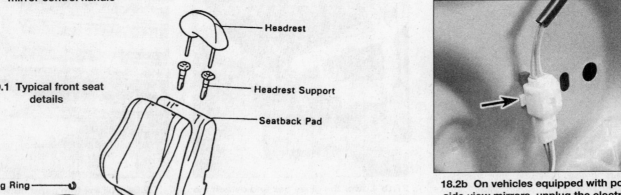

19.3 Typical rear seat details

## 20 Instrument cluster bezel - removal and installation

*Refer to illustrations 20.2a, 20.2b and 20.3*
**Warning:** *These models are equipped with airbags. The airbag is armed and can deploy (inflate) anytime the battery is connected. To prevent accidental deployment (and possible injury), turn the ignition key to LOCK and disconnect the negative battery cable whenever working near airbag components. After the battery is disconnected, wait at least two minutes before beginning work (the system has a*

**11**

**20.2a Remove the two screws at the top of the bezel**

**20.2b Use a screwdriver to detach the clips at each end of the bezel**

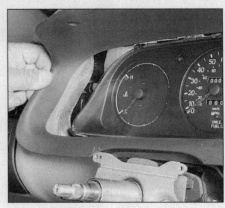

**20.3 Detach the bezel and lift it off**

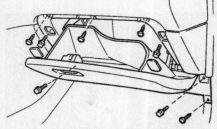

**21.1a Glove box installation details**

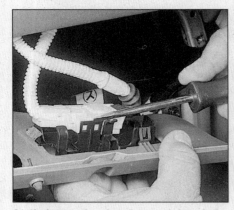

**21.1b Lower the glove box and detach the airbag connector from the clips (DO NOT unplug the airbag connector unless you plan to remove the airbag module)**

**22.3a Pull the trim panel outward . . .**

*back-up capacitor that must fully discharge). For more information see Chapter 12.*
**Caution:** *If the stereo in your vehicle is equipped with an anti-theft system, make sure you have the correct activation code before disconnecting the battery.*
1   Disconnect the cable from the negative battery terminal.
2   Tilt the steering column to its lowest position. Remove the two screws along the top of the bezel, then detach the two clips at the lower edges by prying with a screwdriver **(see illustrations)**.
3   Grasp the bezel securely and remove it **(see illustration)**.
4   Installation is the reverse of the removal procedure.

## 21   Glove box - removal and installation

*Refer to illustrations 21.1a and 21.1b*
**Warning:** *These models are equipped with airbags. The airbag is armed and can deploy (inflate) anytime the battery is connected. To prevent accidental deployment (and possible injury), turn the ignition key to LOCK and disconnect the negative battery cable whenever working near airbag components. After the battery is disconnected, wait at least two minutes before beginning work (the system has a back-up capacitor that must fully discharge). For more information see Chapter 12.*
**Caution:** *If the stereo in your vehicle is equipped with an anti-theft system, make sure you have the correct activation code*

*before disconnecting the battery.*
1   Disconnect the cable from the negative battery terminal. Remove the screws, pull the glove box out of the instrument panel **(see illustration)**. On airbag-equipped models, detach the airbag connector from the clips and remove the glovebox assembly **(see illustration)**.
2   Installation is the reverse of removal.

## 22   Center trim panels - removal and installation

*Refer to illustrations 22.3a, 22.3b and 22.7*
**Warning:** *These models are equipped with airbags. The airbag is armed and can deploy (inflate) anytime the battery is connected. To prevent accidental deployment (and possible injury), turn the ignition key to LOCK and disconnect the negative battery cable whenever working near airbag components. After the battery is disconnected, wait at least two minutes before beginning work (the system has a back-up capacitor that must fully discharge). For more information see Chapter 12.*
**Caution:** *If the stereo in your vehicle is equipped with an anti-theft system, make*

*sure you have the correct activation code before disconnecting the battery.*
1   Disconnect the cable from the negative battery terminal.

### *Upper trim panel*

2   On 1997 and earlier models, pull off the heater/air conditioning control knobs.
3   Using a small screwdriver, detach the retaining clips at each corner, then remove the panel and unplug the electrical connectors **(see illustrations)**.

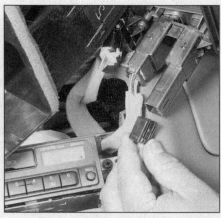

**22.3b . . . and unplug the electrical connectors from the backside**

**22.7 After the lower center trim panel is detached from the instrument panel, disconnect the electrical connectors from the ashtray and the lighter**

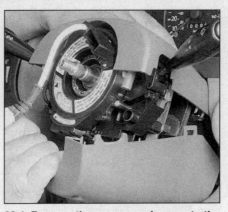

**23.1 Remove the screws and separate the steering column cover halves (steering wheel removed for clarity)**

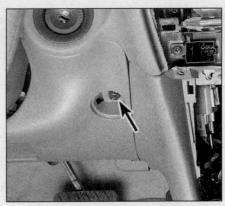

**24.1 Remove the screws (some of which are under covers like this one [arrow]), then detach the finish panel**

## Lower trim panel

4    Remove the ashtray from the lower trim panel.

5    On 1997 and earlier models, remove the two screws along the upper edge, then tilt the top of the trim panel rearward to detach the two clips at the bottom.

6    On 1998 and later models, use a small screwdriver to pry out the bezel retaining clips. On Corolla models, there are two clips at the bottom of the bezel. On Prizm models, there are six clips around the perimeter of the bezel.

7    In either case, pull the lower trim panel out and unplug the electrical connectors from the cigarette lighter and the ashtray light **(see illustration)**.

8    Installation is the reverse of the removal procedure.

## 23   Steering column covers - removal and installation

*Refer to illustration 23.1*

**Warning:** *These models are equipped with airbags. The airbag is armed and can deploy (inflate) anytime the battery is connected. To prevent accidental deployment (and possible injury), turn the ignition key to LOCK and disconnect the negative battery cable whenever working near airbag components. After the battery is disconnected, wait at least two minutes before beginning work (the system has a back-up capacitor that must fully discharge). For more information see Chapter 12.*

**Caution:** *If the stereo in your vehicle is equipped with an anti-theft system, make sure you have the correct activation code before disconnecting the battery.*

1    Remove the steering column cover screws. **Note:** *1997 and earlier models have four screws mounted vertically in the lower cover. 1998 and later models have three screws in the lower cover (two mounted horizontally and one mounted vertically). They also have one screw securing the upper cover that*

must be accessed from below, up through the steering column. On these models it may be necessary to rotate the steering wheel in order to access the forward facing (horizontally mounted) screws on the lower cover.

2    Separate the cover halves and detach them from the steering column **(see illustration)**.

3    Disconnect any electrical connections and the covers.

4    Installation is the reverse of the removal procedure.

## 24   Steering column lower finish panel - removal and installation

*Refer to illustration 24.1*

**Warning:** *These models are equipped with airbags. The airbag is armed and can deploy (inflate) anytime the battery is connected. To prevent accidental deployment (and possible injury), turn the ignition key to LOCK and disconnect the negative battery cable whenever working near airbag components. After the battery is disconnected, wait at least two minutes before beginning work (the system has a back-up capacitor that must fully discharge). For more information see Chapter 12.*

**Caution:** *If the stereo in your vehicle is equipped with an anti-theft system, make sure you have the correct activation code before disconnecting the battery.*

1    On 1997 and earlier models, remove the two screws and detach the hood release handle. Then pry out the plastic trim covering the upper mounting bolts. Remove the retaining bolts, disconnect any electrical connections and pull the panel off **(see illustration)**.

2    On 1998 and later models, remove the two lower mounting bolts and pull the trim panel outward enough to detach the hood release cable from the rear of the hood release handle. On some models it may be necessary to remove the left kick panel before removing the steering column lower finish panel.

3    Installation is the reverse of the removal procedure.

## 25   Console - removal and installation

*Refer to illustrations 25.2a, 25.2b, 25.3 and 25.7*

**Warning:** *These models are equipped with airbags. The airbag is armed and can deploy (inflate) anytime the battery is connected. To prevent accidental deployment (and possible injury), turn the ignition key to LOCK and disconnect the negative battery cable whenever working near airbag components. After the battery is disconnected, wait at least two minutes before beginning work (the system has a back-up capacitor that must fully discharge). For more information see Chapter 12.*

**Caution:** *If the stereo in your vehicle is equipped with an anti-theft system, make sure you have the correct activation code before disconnecting the battery.*

1    Disconnect the cable from the negative battery terminal.

### Rear console

2    Open the rear console and remove the retaining screws from the storage compart-

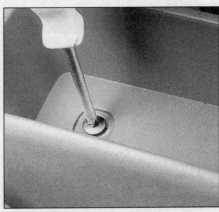

**25.2a Remove the carpet from the rear console storage compartment (if equipped), then remove the retaining screws**

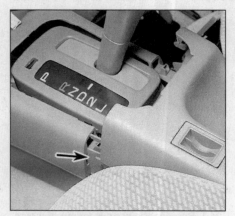

**25.2b  Pull the rear console back to remove it from the vehicle – note that some models have retaining clips (arrow) that secure the rear console to the front console, while others have retaining screws in this area**

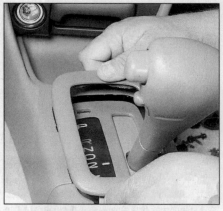

**25.3  On automatic transaxle models, detach the shifter bezel with a screwdriver, then lift it off**

**25.7  Remove the screws on each side and detach the front console from the instrument panel**

ment, then remove the screws or clips securing the rear console to the front console. Lift the console up, detaching any clips as necessary and remove the console from the vehicle **(see illustrations)**.

## Front console

3    On manual shift models, unscrew the shift knob. On automatic models detach the shift bezel **(see illustration)**.

4    Remove the lower center trim panel (Section 22). Depending on the model year of the vehicle, it may be necessary to remove the heater control panel (see Chapter 3), the radio (see Chapter 12) and/or the storage compartment from the front console. Refer to the appropriate Chapters and remove these components, only if they're located in the front console! Do not remove these items if they're part of the upper instrument panel.

5    Remove the glove box (see Section 21).

6    Remove the steering column lower finish panel (see Section 24).

7    Remove the screws retaining the front console to the instrument panel, then lift to detach the console **(see illustration)**.

8    Installation is the reverse of the removal procedure.

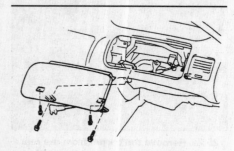

**26.3  Typical passenger side airbag module installation details**

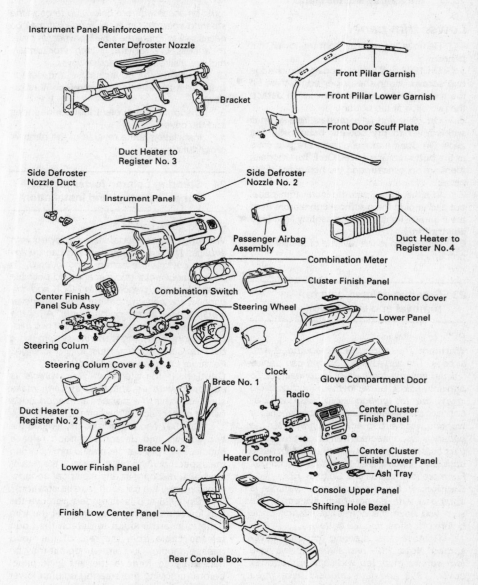

Instrument Panel Reinforcement
Center Defroster Nozzle
Bracket
Duct Heater to Register No. 3
Side Defroster Nozzle Duct
Instrument Panel
Center Finish Panel Sub Assy
Steering Colum
Steering Colum Cover
Duct Heater to Register No. 2
Lower Finish Panel
Brace No. 2
Finish Low Center Panel
Rear Console Box

Front Pillar Garnish
Front Pillar Lower Garnish
Front Door Scuff Plate
Side Defroster Nozzle No. 2
Passenger Airbag Assembly
Duct Heater to Register No.4
Combination Meter
Cluster Finish Panel
Connector Cover
Lower Panel
Combination Switch
Steering Wheel
Glove Compartment Door
Brace No. 1
Clock
Radio
Center Cluster Finish Panel
Heater Control
Center Cluster Finish Lower Panel
Ash Tray
Console Upper Panel
Shifting Hole Bezel

**26.8  Typical instrument panel and related components - exploded view**

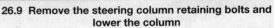

**26.9  Remove the steering column retaining bolts and lower the column**

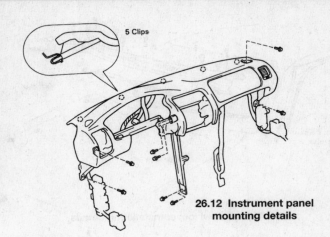

5 Clips

**26.12  Instrument panel mounting details**

## 26  Instrument panel - removal and installation

**Warning:** *These models are equipped with airbags. The airbag is armed and can deploy (inflate) anytime the battery is connected. To prevent accidental deployment (and possible injury), turn the ignition key to LOCK and disconnect the negative battery cable whenever working near airbag components. After the battery is disconnected, wait at least two minutes before beginning work (the system has a back-up capacitor that must fully discharge). For more information see Chapter 12.*

**Caution:** *If the stereo in your vehicle is equipped with an anti-theft system, make sure you have the correct activation code before disconnecting the battery.*

### Removal

*Refer to illustrations 26.3, 26.8, 26.9 and 26.12*

1    Disconnect the cable from the negative battery terminal.
2    Remove driver's airbag module and the steering wheel (see Chapter 10).
3    Remove the glove box assembly (see Section 21), disconnect the passenger airbag module electrical connector and remove the airbag module **(see illustration)**. The electrical connectors used in the airbag system are a twin-lock design; use the proper method for disconnecting these connectors or damage to the connector may occur (see Chapter 12).
**Warning:** *Store the airbag modules in a safe place with the airbag face (the trim side) pointing up.*
4    Remove the instrument cluster bezel (see Section 20) and remove the instrument cluster (see Chapter 12).
5    Remove the center trim panels (see Section 22), remove the radio (see Chapter 12) and the heater and air conditioning control panel (see Chapter 3).
6    Remove the steering column covers (see Section 23), the steering column lower finish panel and metal reinforcement panel behind the finish panel (if equipped) (see Section 24).

**27.1  Remove the three Phillips-head screws along the top of the grille**

7    Remove the front and rear floor consoles (see Section 25).
8    Remove the front door scuff plates and the front-pillar kick panels from both sides **(see illustration)**.
9    Remove the bolts retaining the steering column to the instrument panel and reinforcement brace, then lower the steering column **(see illustration)**. On 1997 and earlier models, remove the key lock cylinder from the instrument panel (see Chapter 12).
10   Pry out the left-hand switch bezel and disconnect the electrical connectors. On 1998 and later Prizm models, remove the hazard and rear defogger switch bezel at the center of the instrument panel. Remove the heating and ventilation ducts from below the instrument panel.
11   Disconnect the main junction block at the left side of the instrument panel. Detach the fuse box from the body and remove it with the panel and harness. Disconnect the electrical connectors from the heater case and remove the ground wire terminals from the center brace. The main instrument panel wiring harness will remain with the panel when you remove it.
12   Disconnect any remaining connectors and remove the remaining instrument panel retaining bolts, grasp the instrument panel firmly and pull it sharply to the rear to release

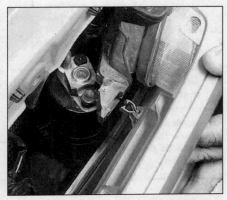

**27.2  Reach behind the grill and detach the clips by squeezing the sides together as you pull the grille out**

the five clips along the base of the windshield **(see illustration)**. Remove the instrument panel from the vehicle.

### Installation

13   Carefully remove the wiring harness and the ventilation and defroster ducts and transfer them to the new instrument panel.
14   Guide the instrument panel into position, press the tabs into the five retaining clips and install the retaining bolts.
15   The remainder of installation is the reverse of removal.

## 27  Radiator grille - removal and installation

*Refer to illustrations 27.1 and 27.2*
**Note:** *On 1998 and later models, the front grille is an integral part of the front bumper fascia and cannot be removed separately. Refer to Section 17 for bumper cover removal procedures.*

1    Remove three screws along the top of the grille **(see illustration)**.
2    Pull the top of the grille out for access and disengage the two retaining clips with a screwdriver **(see illustration)**.

**11**

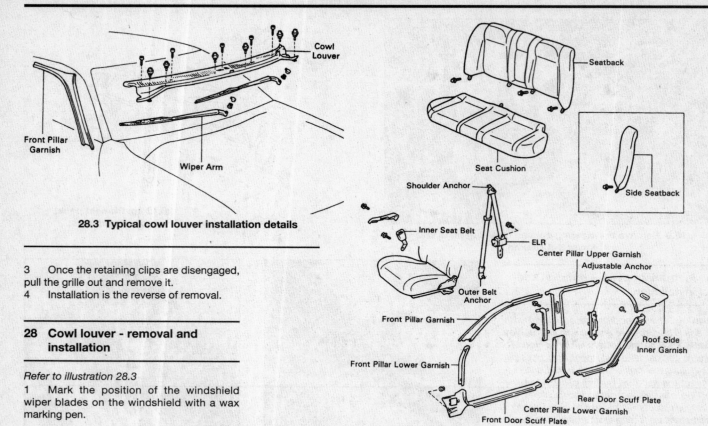

**28.3  Typical cowl louver installation details**

3    Once the retaining clips are disengaged, pull the grille out and remove it.
4    Installation is the reverse of removal.

## 28  Cowl louver - removal and installation

*Refer to illustration 28.3*
1    Mark the position of the windshield wiper blades on the windshield with a wax marking pen.

**29.4a  Typical sedan seat belt details**

2    Remove the wiper arms.
3    Remove the cowl louver retaining screws, disconnect the windshield washer hoses and detach the cowl from the vehicle **(see illustration)**.
4    Installation is the reverse of removal. Make sure to align the wiper blades with the marks made during removal.

## 29  Seat belts - check

*Refer to illustrations 29.4a and 29.4b*
1    Check the seat belts, buckles, latch plates and guide loops for any obvious damage or signs of wear.
2    Make sure the seat belt reminder light comes on when the key is turned on.
3    The seat belts are designed to lock up during a sudden stop or impact, yet allow free movement during normal driving. The retractors should hold the belt against your chest while driving and rewind the belt when the buckle is unlatched.
4    If any of the above checks reveal problems with the seat-belt system, replace parts as necessary **(see illustrations)**.

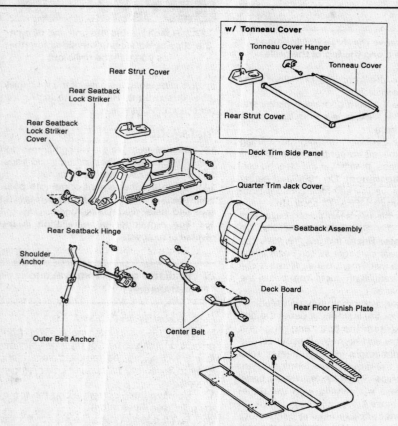

**29.4b  Typical station wagon rear seat belt details**

# Chapter 12
# Chassis electrical system

## Contents

## 1 General information

The electrical system is a 12-volt, negative ground type. Power for the lights and all electrical accessories is supplied by a lead/acid-type battery which is charged by the alternator.

This Chapter covers repair and service procedures for the various electrical components not associated with the engine. Information on the battery, charging system, ignition system and starting system can be found in Chapter 5.

It should be noted that when portions of the electrical system are serviced, the cable should be disconnected from the negative battery terminal to prevent electrical shorts and/or fires. **Caution:** *If the stereo in your vehicle is equipped with an anti-theft system, make sure you have the correct activation code before disconnecting the battery.*

## 2 Electrical troubleshooting - general information

*Refer to illustrations 2.5a, 2.5b, 2.6, 2.9 and 2.15*

A typical electrical circuit consists of an electrical component, any switches, relays, motors, fuses, fusible links or circuit breakers related to that component and the wiring and connectors that link the component to both the battery and the chassis. To help you pinpoint an electrical circuit problem, wiring diagrams are included at the end of this Chapter.

Before tackling any troublesome electrical circuit, first study the appropriate wiring diagrams to get a complete understanding of what makes up that individual circuit. Trouble spots, for instance, can often be narrowed down by noting if other components related to the circuit are operating properly. If several components or circuits fail at one time, chances are the problem is in a fuse or ground connection, because several circuits are often routed through the same fuse and ground connections.

Electrical problems usually stem from simple causes, such as loose or corroded connections, a blown fuse, a melted fusible link or a failed relay. Visually inspect the condition of all fuses, wires and connections in a problem circuit before troubleshooting the circuit.

If test equipment and instruments are going to be utilized, use the diagrams to plan ahead of time where you will make the necessary connections in order to accurately pinpoint the trouble spot.

The basic tools needed for electrical troubleshooting include a voltmeter, a test light (a 12-volt bulb with a set of test leads can also be used) or a continuity tester (which includes a bulb, battery and set of test

**12**

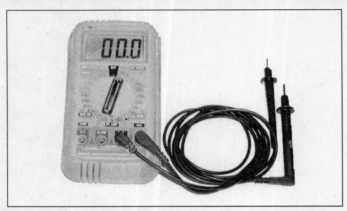

2.5a  The most useful tool for electrical troubleshooting is a digital multimeter that can check volts, amps, and test continuity

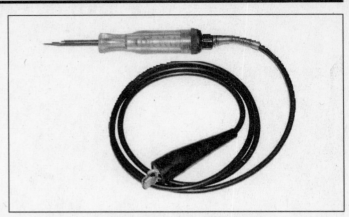

2.5b  A test light is a very handy, tool for testing voltage

leads) **(see illustrations)**. Also useful is a pair of jumper wires, preferably with a circuit breaker incorporated into one, which can be used to apply power and ground to electrical components. Before attempting to locate a problem with test instruments, use the wiring diagram(s) to decide where to make the connections.

## Voltage checks

Voltage checks should be performed if a circuit is not functioning properly. Connect one lead of a circuit tester to either the negative battery terminal or a known good ground. Connect the other lead to a connector in the circuit being tested, preferably nearest to the battery or fuse **(see illustration)**. If the bulb of the tester lights, voltage is present, which means that the part of the circuit between the connector and the battery is problem free. Continue checking the rest of the circuit in the same fashion. When you reach a point at which no voltage is present, the problem lies between that point and the last test point with voltage. Most of the time the problem can be traced to a loose connection. **Note:**

*Keep in mind that some circuits receive voltage only when the ignition key is in the Accessory or Run position.*

## Finding a short

A short-to-ground in a live circuit causes the fuse protecting the circuit to blow. When the fuse is replaced it will immediately blow again.

One method of finding the location of a short is to remove the fuse protecting the problem circuit and connect a test light in place of the fuse. Turn the ignition key on, but make sure the components in the circuit are turned off (or disconnected if they aren't controlled by a switch), there should be voltage present in the circuit and the test light should not light. Move the suspected wiring harness from side-to-side while watching the test light. If the bulb goes on, there is a short to ground somewhere in that area, probably where the insulation has rubbed through allowing the bare wire to contact the body. The same test can be performed on each component in the circuit, even a switch.

## Ground check

Perform a ground test to check whether a component is properly grounded. Disconnect the battery and connect one lead of a continuity tester or multimeter (set to the ohms scale), to a known good ground. Connect the other lead to the wire or ground connection being tested. If the resistance is low (less than 5 ohms), the ground is good. If using a self-powered continuity tester, the bulb will light if the ground is good.

## Continuity check

A continuity check is done to determine if there are any breaks in a circuit - if it is passing electricity properly. With the circuit off (no power in the circuit), a self-powered continuity tester or multimeter can be used to check the circuit. Connect the test leads to both ends of the circuit (or to the "power" end and a good ground), and if the test light comes on the circuit is passing current properly **(see illustration)**. If the resistance is low (less than 5 ohms), there is continuity; if the reading is 10,000 ohms or higher, there is a

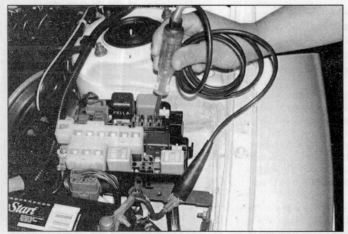

2.6  To use a test light, clip the lead to a good ground point on the body or engine block, then probe the connector, wire or electrical socket with the pointed probe - if the bulb lights, battery voltage is present at the test point

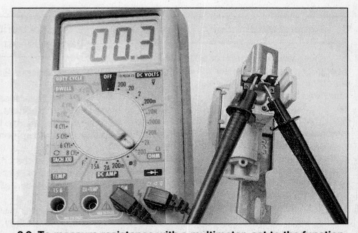

2.9  To measure resistance with a multimeter, set to the function dial to ohms scale and connect the leads to two terminals of the connector or component - when checking for continuity, a low reading indicates continuity, a very high reading (infinite) indicates lack of continuity or open circuit

**2.15 To backprobe a connector, insert a small, sharp probe (such as a straight-pin) into the back of the connector alongside the desired wire until it contacts the metal terminal inside; connect your meter leads to the probes - this allows you to test a functioning circuit**

**3.1a On 1997 and earlier models, the main fuse panel is located in the driver's side kick panel (on 1998 and later models, it's located at the left end of the instrument panel under a cover)**

**3.1b A fuse and relay panel is located in the engine compartment next to the air filter housing**

break somewhere in the circuit. The same procedure can be used to test a switch, by connecting the continuity tester to the switch terminals. With the switch turned On, the test light should come on (or low resistance should be indicated on a meter).

### Finding an open circuit

When diagnosing for possible open circuits, it is often difficult to locate them by sight because the connectors hide oxidation or terminal misalignment. Merely wiggling a connector on a sensor or in the wiring harness may correct the open circuit condition. Remember this when an open circuit is indicated when troubleshooting a circuit. Intermittent problems may also be caused by oxidized or loose connections.

Electrical troubleshooting is simple if you keep in mind that all electrical circuits are basically electricity running from the battery, through the wires, switches, relays, fuses and fusible links to each electrical component (light bulb, motor, etc.) and to ground, from which it is passed back to the battery. Any electrical problem is an interruption in the flow of electricity to and from the battery.

### Connectors

Most electrical connections on these vehicles are made with multi-wire plastic connectors. The mating halves of many connectors are secured with locking clips molded into the plastic connector shells. The mating halves of large connectors, such as some of those under the instrument panel, are held together by a bolt through the center of the connector.

To separate a connector with locking clips, use a small screwdriver to pry the clips apart carefully, then separate the connector halves. Pull only on the shell, never pull on the wiring harness as you may damage the individual wires and terminals inside the con-

**3.1c An auxiliary fuse and relay panel is located next to the battery**

nectors. Look at the connector closely before trying to separate the halves. Often the locking clips are engaged in a way that is not immediately clear. Additionally, many connectors have more than one set of clips.

Each pair of connector terminals has a male half and a female half. When you look at the end view of a connector in a diagram, be sure to understand whether the view shows the harness side or the component side of the connector. Connector halves are mirror images of each other, and a terminal shown on the right side end-view of one half will be on the left side end view of the other half.

### Backprobing a connector

It is often necessary to take circuit voltage measurements with a connector connected. Whenever possible, carefully insert a small straight pin (not your meter probe) into the rear of the connector shell to contact the terminal inside, then clip your meter lead to the pin. This kind of connection is called "backprobing" **(see illustration)**. When inserting a test probe into a male terminal, be careful not to distort the terminal opening.

**3.1d On some 1997 and earlier models and all 1998 and later models, an auxiliary fuse/relay panel is located behind the right headlight housing**

Doing so can lead to a poor connection and corrosion at that terminal later. Using the small straight pin instead of a meter probe results in less chance of deforming the terminal connector.

### 3 Fuses - general information

*Refer to illustrations 3.1a, 3.1b, 3.1c, 3.1d and 3.3*

The electrical circuits of the vehicle are protected by a combination of fuses, circuit breakers and fusible links. The fuse panels are located is various locations, depending on model. The interior fuse panel is located behind the left kick panel (1997 and earlier models) or at the left end of the instrument panel (1998 and later models). Underhood fuse/relay panels are located in the engine compartment next to the air filter housing, next to the battery and behind the right headlight housing **(see illustrations)**.

Each of the fuses is designed to protect a specific circuit, and the various circuits are

**12**

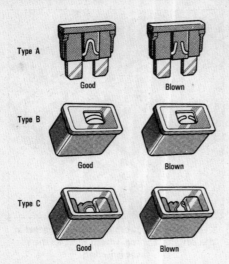

**3.3 Three types of fuses are used on these models - all can be visually checked**

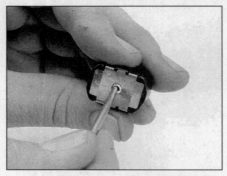

**5.2a Insert a pin or paper clip into the circuit breaker reset hole and push it in to reset it**

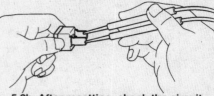

**5.2b After resetting, check the circuit breaker for continuity**

identified on the fuse panel itself.

Three types of miniaturized fuses are employed in the fuse panel. These compact fuses, with blade terminal design, allow fingertip removal and replacement. If an electrical component fails, always check the fuse first. A blown fuse is easily identified through the clear plastic body. Visually inspect the element for evidence of damage (see illustration). If a continuity check is called for on the type A fuse, the blade terminal tips are exposed in the fuse body.

Be sure to replace blown fuses with the correct type. Fuses of different ratings are physically interchangeable, but only fuses of the proper rating should be used. Replacing a fuse with one of a higher or lower value than specified is not recommended. Each electrical circuit needs a specific amount of protec-

tion. The amperage value of each fuse is molded into the fuse body.

If the replacement fuse immediately fails, don't replace it again until the cause of the problem is isolated and corrected. In most cases, this will be a short circuit in the wiring caused by a broken or deteriorated wire.

## 4  Fusible links - general information

Some circuits are protected by fusible links. The links are used in circuits which are not ordinarily fused, such as the ignition circuit.

The fusible links on these models are similar to fuses in that they can be visually checked to determine if they are melted.

To replace a fusible link, first disconnect the negative cable from the battery. **Caution:** *If the stereo in your vehicle is equipped with an anti-theft system, make sure you have the correct activation code before disconnecting*

*the battery.* Disconnect the burned-out link and replace it with a new one (available from your dealer or auto parts store). Always determine the cause for the overload which melted the fusible link before installing a new one.

## 5  Circuit breakers - general information

*Refer to illustrations 5.2a and 5.2b*

Because on some models the circuit breaker resets itself automatically, an electrical overload in a circuit breaker protected system will cause the circuit to fail momentarily, then come back on. If the circuit does not come back on, check it immediately. Note, however, that some circuit breakers must be reset manually. Once the condition is corrected, the circuit breaker will resume its normal function.

To reset a manual circuit breaker, first disconnect the cable from the negative battery terminal. **Caution:** *If the stereo in your vehicle is equipped with an anti-theft system, make sure you have the correct activation code before disconnecting the battery.* Remove the circuit breaker, insert a pin into the reset hole and push in until you hear a click (see illustration). Once the circuit breaker is reset, it's a good idea to use an

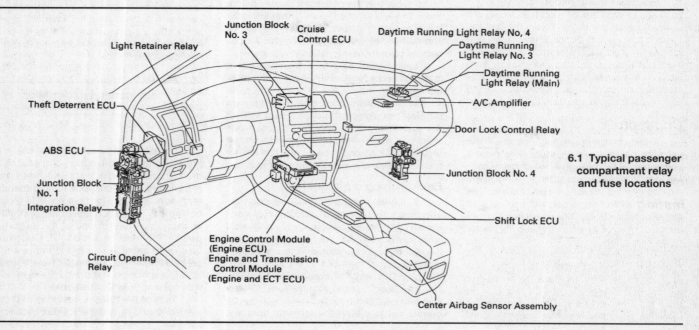

**6.1 Typical passenger compartment relay and fuse locations**

Junction Block No. 3
Cruise Control ECU
Light Retainer Relay
Daytime Running Light Relay No. 4
Daytime Running Light Relay No. 3
Daytime Running Light Relay (Main)
Theft Deterrent ECU
A/C Amplifier
Door Lock Control Relay
ABS ECU
Junction Block No. 4
Junction Block No. 1
Integration Relay
Shift Lock ECU
Circuit Opening Relay
Engine Control Module (Engine ECU)
Engine and Transmission Control Module (Engine and ECT ECU)
Center Airbag Sensor Assembly

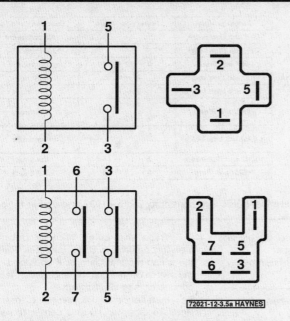

6.3a  These two relays are typical normally open types. The upper relay completes a single circuit (terminal 5 to terminal 3) when energized - the lower relay completes two circuits (6 and 7, 3 and 5)

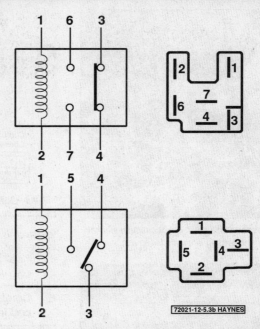

6.3b  These relays are normally closed types, where current flows though one circuit until the relay is energized, which interrupts that circuit and completes the second circuit

ohmmeter to make sure there is continuity across the terminals before reinstalling it (see illustration).

## 6   Relays - general information and testing

### General information

*Refer to illustration 6.1*

1   Several electrical accessories in the vehicle, such as the fuel injection system, horns, starter, and fog lamps use relays to transmit the electrical signal to the component. Relays use a low-current circuit (the control circuit) to open and close a high-current circuit (the power circuit). If the relay is defective, that component will not operate properly. Most relays are mounted in the engine compartment and interior fuse/relay boxes, with some specialized relays located in other locations around the vehicle (see illustration). If a faulty relay is suspected, it can be removed and tested using the procedure below or by a dealer service department or a repair shop. Defective relays must be replaced as a unit.

### Testing

*Refer to illustrations 6.3a, 6.3b and 6.6*

2   Refer to the wiring diagrams for the circuit to determine the proper connections for the relay you're testing. If you can't determine the correct connection from the wiring diagrams, however, you may be able to determine the test connections from the information that follows.

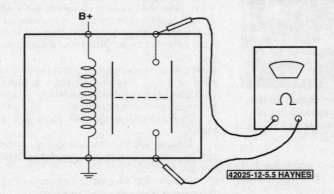

6.6  To test a typical four-terminal normally open relay, connect an ohmmeter to the two terminals on the power side; continuity will be indicated when the relay is energized

3   There are several basic types of relays used on most models (see illustrations). Some are normally open type and some normally closed, while others include a circuit of each type.

4   On most relays, two of the terminals are the relay control circuit (they connect to the relay coil which, when energized, closes the large contacts to complete the circuit). The other terminals are the power circuit (they are connected together within the relay when the control-circuit coil is energized).

5   Some relays may be marked as an aid to help you determine which terminals are the control circuit and which are the power circuit. If the relay is not marked, refer to the wiring diagrams at the end of this Chapter to determine the proper hook-ups for the relay you're testing.

6   To test a relay, connect an ohmmeter across the two terminals of the power circuit; continuity should not be indicated (see illustration). Now connect a fused jumper wire

between one of the two control circuit terminals and the positive battery terminal. Connect another jumper wire between the other control circuit terminal and ground. When the connections are made, the relay should click and continuity should be indicated on the meter. On some relays, polarity may be critical, so, if the relay doesn't click, try swapping the jumper wires on the control circuit terminals.

7   If the relay fails the above test, replace it.

## 7   Turn signal/hazard flashers - check and replacement

**Warning:** *These models are equipped with airbags. Always disable the airbag system before working near airbag system components. After the battery is disconnected, wait at least two minutes before beginning work*

**12**

**8.4  Remove the four combination switch mounting screws (arrows)**

**8.5  Slip the combination switch off the steering column and disconnect the connector (arrow)**

*(the system has a back-up capacitor that must fully discharge). For more information see Section 27.*

**Caution:** *If the stereo in your vehicle is equipped with an anti-theft system, make sure you have the correct activation code before disconnecting the battery.*

1    The turn signal/hazard flasher, a relay located in the main fuse panel, flashes the turn signals **(see illustration 3.1a)**.

2    When the flasher is functioning properly, an audible click can be heard during its operation. If the turn signals fail on one side or the other and the flasher does not make its characteristic clicking sound, a faulty turn signal bulb is indicated.

3    If both turn signals fail to blink, the problem may be due to a blown fuse, a faulty flasher, a broken switch or a loose or open connection. If a quick check of the fuse panel indicates that the turn signal fuse has blown, check the wiring for a short before installing a new fuse.

4    To replace the flasher, simply pull it out of the fuse panel.

5    Make sure that the replacement unit is

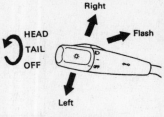

**Combination Switch Connector "A"**

| Switch position | Tester connection | Specified condition |
|---|---|---|
| OFF and Low beam | – | No continuity |
| OFF and High beam | – | No continuity |
| OFF and Flash | 9 – 12 – 14 | Continuity |
| TAIL and Low beam | 2 – 11 | Continuity |
| TAIL and High beam | 2 – 11 | Continuity |
| TAIL and Flash | 2 – 11 <br> 9 – 12 – 14 | Continuity |
| HEAD and Low beam | 2 – 11 – 13 <br> 3 – 9 | Continuity |
| HEAD and High beam | 2 – 11 – 14 | Continuity |
| HEAD and Flash | 2 – 11 – 14 | Continuity |

**9.4a  Light control switch terminal guide and continuity table - 1997 and earlier models**

identical to the original. Compare the old one to the new one before installing it.

6    Installation is the reverse of removal.

## 8    Combination switch - removal and installation

*Refer to illustrations 8.4 and 8.5*

**Warning:** *These models are equipped with airbags. Always disable the airbag system before working near airbag system components. After the battery is disconnected, wait at least two minutes before beginning work (the system has a back-up capacitor that must fully discharge). For more information see Section 27.*

**Caution:** *If the stereo in your vehicle is equipped with an anti-theft system, make sure you have the correct activation code before disconnecting the battery.*

1    Disconnect the negative cable at the battery.

2    Disable the drivers airbag (see Section 27) and remove the steering wheel (see Chapter 10).

3    Remove the steering column cover and the lower finish panel (see Chapter 11).

4    Remove the combination switch retaining screws **(see illustration)**.

5    Trace the wiring harness down the steering column to the connector. Release the wiring retainer clamps, if equipped, slide the switch off the column and disconnect the connector **(see illustration)**.

6    Installation is the reverse of removal. Refer to Chapter 10 and center the spiral cable before installing the steering wheel.

## 9    Steering column switches - check and replacement

**Warning:** *These models are equipped with airbags. Always disable the airbag system before working near airbag system components. After the battery is disconnected, wait at least two minutes before beginning work (the system has a back-up capacitor that must fully discharge). For more information see Section 27.*

**Caution:** *If the stereo in your vehicle is equipped with an anti-theft system, make sure you have the correct activation code before disconnecting the battery.*

1    Disconnect the negative cable at the battery.

2    Remove the steering column covers and lower finish panel (see Chapter 11).

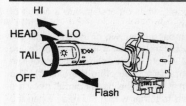

| Switch position | Tester connection | Specified condition |
|---|---|---|
| OFF | – | No continuity |
| TAIL | 14 – 16 | Continuity |
| HEAD | 13 – 14 – 16 | Continuity |

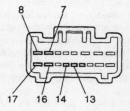

| Switch position | Tester connection | Specified condition |
|---|---|---|
| Low beam | 16 – 17 | Continuity |
| High beam | 7 – 16 | Continuity |
| Flash | 7 – 8 – 16 | Continuity |

**9.4b  Light control switch terminal guide and continuity table - 1998 and later models**

### Check

*Refer to illustrations 9.4a, 9.4b, 9.4c, 9.4d, 9.4e and 9.4f*

3    On 1997 and earlier models, trace the wiring harness from the switch to be checked to the harness connector and disconnect the connector. On 1998 and later models, remove the switch and check it off the vehicle (see *Replacement* procedure).

4    Using an ohmmeter, check for continuity between the indicated terminals with the various switches in each of the indicated positions **(see illustrations)**.

5    If the continuity is not as specified, replace the defective switch.

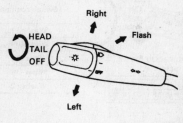

| Switch position | Tester connection | Specified condition |
|---|---|---|
| Left turn | 1 — 5 | Continuity |
| Neutral | — | No continuity |
| Right turn | 1 — 8 | Continuity |

Turn Signal Switch
(Combination Switch Connector "A")

**9.4c Turn signal switch terminal guide and continuity table - 1997 and earlier models**

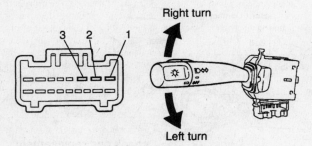

| Switch position | Tester connection | Specified condition |
|---|---|---|
| Left turn | 1 — 2 | Continuity |
| Neutral | — | No continuity |
| Right turn | 2 — 3 | Continuity |

**9.4d  Turn signal switch terminal guide and continuity table - 1998 and later models**

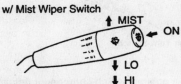

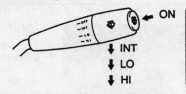

Combination Switch
Connector "B"

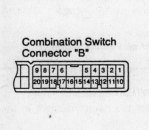

| Switch position | Tester connection | Specified condition |
|---|---|---|
| MIST | 7 — 18 | Continuity |
| OFF | 4 — 7 | Continuity |
| LO | 7 — 18 | Continuity |
| HI | 13 — 18 | Continuity |

### w/ Intermittent Wiper:

| Switch position | Tester connection | Specified condition |
|---|---|---|
| OFF | 4 — 7 | Continuity |
| INT | 4 — 7 | Continuity |
| LO | 7 — 18 | Continuity |
| HI | 13 — 18 | Continuity |

**9.4e  Windshield wiper and washer switch terminal guide and continuity table - 1997 and earlier models**

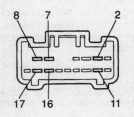

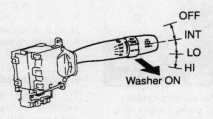

| Switch position | Tester connection | Specified condition |
|---|---|---|
| OFF | 7 — 16 | Continuity |
| MIST (w/ Mist wiper) | 3 — 11 | Continuity |
| INT (w/ Intermittent wiper) | 7 — 16 | Continuity |
| LO | 7 — 17 | Continuity |
| HI | 8 — 17 | Continuity |
| Washer ON | 2 — 11 | Continuity |

**9.4f  Windshield wiper and washer switch terminal guide and continuity table - 1998 and later models**

**12**

## Replacement
*Refer to illustrations 9.7, 9.9 and 9.13*

### 1997 and earlier models

6    Remove the combination switch (see Section 8). Remove the four screws retaining the airbag spiral cable and remove the spiral cable from the combination switch.

7    Remove the retaining screws from the switch being replaced and remove the defective switch from the switch body **(see illustration)**.

8    Separate the wiring harness of the switch being replaced from the main wiring harness and remove the terminals from the connector.

9    Release the tabs from the terminal cover at the rear of the connector. From the front of the connector, insert a small screwdriver or pick, pry down on the locking lug and remove the terminal from the rear **(see illustration)**.

10    Insert the terminals from the new switch into the connector, pushing in until they are securely locked in place.

11    The remainder of installation is the reverse of removal. Refer to Chapter 10 and center the spiral cable before installing the steering wheel.

### 1998 and later models

12    Disconnect the electrical connector from the switch.

13    Remove the screws retaining the switch to the switch body and remove the switch **(see illustration)**.

14    Installation is the reverse of removal.

## 10   Ignition switch and key lock cylinder - check and replacement

**Warning:** *These models are equipped with airbags. Always disable the airbag system before working near airbag system components. After the battery is disconnected, wait at least two minutes before beginning work (the system has a back-up capacitor that must fully discharge). For more information see Section 27.*

**Caution:** *If the stereo in your vehicle is equipped with an anti-theft system, make sure you have the correct activation code before disconnecting the battery.*

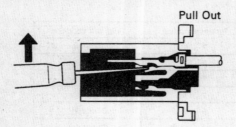

**9.9  To remove a terminal from the connector, pry down on the locking lug and pull the terminal out from the rear**

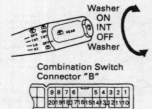

Combination Switch Connector "B"

| Switch position | Tester connection to terminal number | Specified condition |
|---|---|---|
| Washer 1 | 2 — 16 | Continuity |
| Wiper OFF | — | No continuity |
| Wiper INT | 10 — 16 | Continuity |
| Wiper ON | 1 — 16 | Continuity |
| Washer 2 | 1 — 2 — 16 | Continuity |

**9.4g  Rear window wiper and washer switch terminal guide and continuity table - Wagon models**

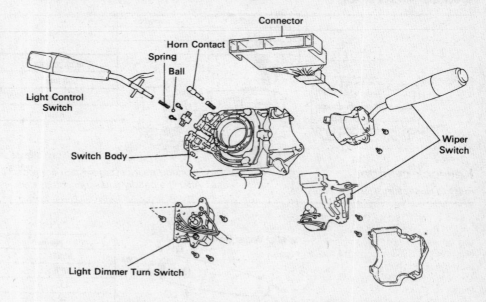

**9.7  Steering column switch component details - 1997 and earlier models**

**9.13  To remove the lighting system switch or windshield wiper switch on a 1998 and later model, remove the screws (arrows) (wiper switch shown, light switch similar)**

## Ignition switch

1    Disconnect the negative cable at the battery.

2    Remove the steering wheel (see Chapter 10).

3    Remove the steering column cover and lower finish panel (see Chapter 11).

### Check
*Refer to illustrations 10.5a and 10.5b*

4    Trace the wire from the switch to the connector and disconnect the connector.

5    Use an ohmmeter to check for continuity at the indicated terminals with the switch in each indicated position **(see illustrations)**.

6    Replace the switch if continuity is not as specified.

### Replacement

7    Remove the screw(s) retaining the switch to the rear of the lock cylinder housing and remove the switch.

8    Installation is the reverse of removal. Refer to Chapter 10 and center the spiral cable before installing the steering wheel.

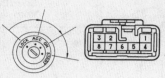

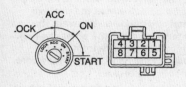

| Terminal Switch position | 1 | 2 | 3 | 4 | 5 | 7 | 8 |
|---|---|---|---|---|---|---|---|
| LOCK | | | | | | | |
| ACC | | | | | o—o | | |
| ON | | o—o—o | | | o—o—o | | |
| START | o—o—o | | | | o | o—o | |

If continuity is not as specified, replace the switch.

**10.5a  Ignition switch terminal guide and continuity table - 1997 and earlier models**

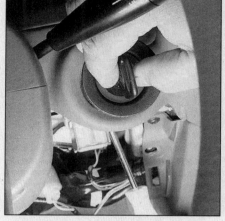

**10.9  With the key in the ACC position, push in on the release button and pull the lock cylinder straight out**

| Switch position | Tester connection | Specified condition |
|---|---|---|
| LOCK | – | No continuity |
| ACC | 2 – 3 | Continuity |
| ON | 2 – 3 – 4<br>6 – 7 | Continuity |
| START | 1 – 2 – 4<br>6 – 7 – 8 | Continuity |

**10.5b  Ignition switch terminal guide and continuity table - 1998 and later models**

## Key lock cylinder

*Refer to illustration 10.9*

9   With the key in the Accessory position, insert a small screwdriver or punch in the hole in the casting and press the release button while pulling the lock cylinder straight out, then remove it from the steering column **(see illustration)**.

10   Installation is the reverse of removal. Refer to Chapter 10 and center the spiral cable before installing the steering wheel.

## 11   Rear window defogger switch - check and replacement

*Refer to illustrations 11.3a, 11.3b and 11.4*

**Warning:** *These models are equipped with airbags. Always disable the airbag system*

before working near airbag system components. After the battery is disconnected, wait at least two minutes before beginning work (the system has a back-up capacitor that must fully discharge). For more information see Section 27.

**Caution:** *If the stereo in your vehicle is equipped with an anti-theft system, make sure you have the correct activation code before disconnecting the battery.*

1   Detach the cable from the negative battery terminal.

2   Remove the center trim panel and turn it over for access to the defogger switch.

3   On 2000 and earlier models, use an ohmmeter to check for continuity at the indicated terminals with the switch in the indicated position **(see illustrations)**. Replace the switch if the continuity is not as specified.

4   On 2001 models, use jumper wires to

apply battery voltage and ground to the indicated terminals **(see illustration)**. Replace the switch if the indicator light fails to illuminate.

## 12   Rear window defogger - check and repair

*Refer to illustrations 12.4, 12.5 and 12.7*

1   The rear window defogger consists of a number of horizontal elements baked onto the glass surface.

2   Small breaks in the element can be repaired without removing the rear window.

## Check

3   Turn the ignition switch and defogger system switches to On. Using a voltmeter, place the positive probe against the defogger grid positive terminal and the negative probe against the ground terminal. If battery voltage is not indicated, check the fuse, defogger switch, defogger relay and related wiring. If voltage is indicated, but all or part of the defogger doesn't heat, proceed with the test.

4   When measuring voltage during the remainder of the test, wrap a piece of aluminum foil around the tip of the voltmeter

| Condition | Tester connection to terminal number | Specified condition |
|---|---|---|
| Switch OFF | – | No continuity |
| Switch ON | 4 – 6 | Continuity |
| Illumination circuit | 1 – 3 | Continuity |

**11.3a  Defogger switch terminal guide and continuity table - 1997 and earlier models**

| Switch position | Tester connection | Specified condition |
|---|---|---|
| OFF | – | No continuity |
| ON | 3 – 6 | Continuity |
| Illumination circuit | 1 – 4 | Continuity |

**11.3b  Defogger switch terminal guide and continuity table - 1998 through 2000 models**

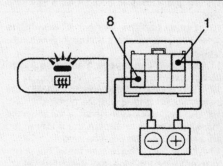

**11.4  Defogger switch test - 2001 models**

**12**

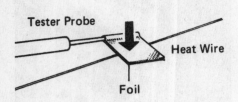

**12.4  When measuring the voltage at the rear window defogger grid, wrap a piece of aluminum foil around the negative probe of the voltmeter and press the foil against the wire with your finger**

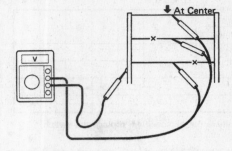

**12.5  To determine if a wire has broken, place the voltmeter positive lead against the defogger positive terminal and check the voltage with the negative lead at the *center* of each wire - if the voltage is approximately 5-volts, the wire is unbroken; if the voltage is approximately 10-volts, the wire is broken between center of the wire and the positive end; if the voltage is 0-volts, the wire is broken between the center of the wire and ground**

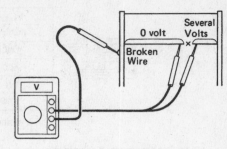

**12.7  To find the break, place the voltmeter positive lead against the defogger positive terminal, place the voltmeter negative lead with the foil strip against the heat wire at the positive terminal end and slide it toward the negative terminal end - the point at which the voltmeter deflects from zero to several volts is the point at which the wire is broken**

negative probe and press the foil against the wire with your finger **(see illustration)**.

5    Check the voltage at the center of each heat wire **(see illustration)**. If the voltage is approximately 5-volts, the wire is okay (there is no break). If the voltage is approximately 10-volts, the wire is broken between the center of the element and the positive end. If the voltage is 0-volts the wire is broken between the center of the element and ground.

6    Connect the negative lead to a good body ground. The reading should stay the same.

7    To find the break, place the voltmeter positive lead against the defogger positive terminal. Place the voltmeter negative lead with the foil strip against the heat wire at the positive terminal end and slide it toward the negative terminal end. The point at which the voltmeter deflects from zero to several volts is the point at which the heat element is broken **(see illustration)**. **Note:** *If the heat element is not broken, the voltmeter will indicate no voltage at the positive end of the heat element but gradually increase to about 12-volts.*

## *Repair*

8    Repair the break in the element using a repair kit specifically recommended for this purpose, available at most auto parts stores. Included in this kit is plastic conductive epoxy.

9    Prior to repairing a break, turn off the system and allow it to cool off for a few minutes.

10    Lightly buff the element area with fine steel wool, then clean it thoroughly with rubbing alcohol.

11    Use masking tape to mask off the area being repaired.

12    Thoroughly mix the epoxy, following the instructions provided with the repair kit.

13    Apply the epoxy material to the slit in the masking tape, overlapping the undamaged area about 3/4-inch on either end.

14    Allow the repair to cure for 24 hours before removing the tape and using the system.

## 13    Radio and speakers - removal and installation

**Warning:** *These models are equipped with airbags. Always disable the airbag system before working near airbag system components. After the battery is disconnected, wait at least two minutes before beginning work (the system has a back-up capacitor that must fully discharge). For more information see Section 27.*
**Caution:** *If the stereo in your vehicle is equipped with an anti-theft system, make sure you have the correct activation code before disconnecting the battery.*

## *Radio*

*Refer to illustrations 13.3a and 13.3b*

1    Disconnect the negative cable at the battery.

2    Remove the center trim panel (see Chapter 11).

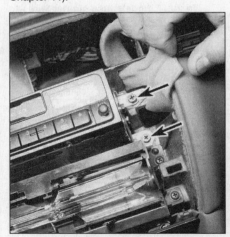

**13.3a  Remove the radio mounting screws (two on each side)**

3    Remove the mounting screws and pull the radio out of the dash as far as possible **(see illustrations)**.

4    Disconnect the electrical connector and the antenna lead and remove the radio.

5    Installation is the reverse of removal.

## *Speakers*

*Refer to illustrations 13.7, 13.13 and 13.15*

### Front speakers

6    Remove the front door trim panel (see Chapter 11).

7    Remove the speaker retaining screws/nuts. Disconnect the electrical connector and remove the speaker **(see illustration)**.

8    Installation is the reverse of removal.

### Rear speakers

9    Using a trim panel removal tool, carefully pry the retainers loose and remove the

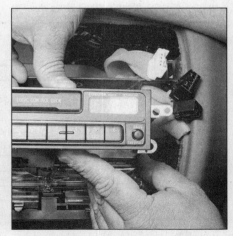

**13.3b  Pull the radio unit out, disconnect the electrical connectors and antenna cable**

13.7  Typical front door speaker screws and electrical connector (arrows)

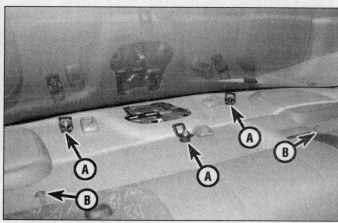

13.13  To remove the package tray on 1998 and later models, remove the plastic covers, remove the tie-downs (A), and pry out the fasteners (B)

13.15  Rear speaker mounting screw (arrow) - 1998 and later models

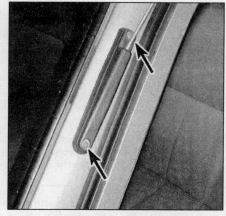

14.4  Roof-mounted antenna screws (arrows)

### 1998 and later Prizm models

8    Using a small wrench, remove the antenna mast from the antenna base.
9    Open the trunk lid and remove the right-side luggage compartment trim panel.
10   Disconnect the antenna cable from the antenna housing and detach the drain hose.
11   Remove the nut from the antenna housing bracket.
12   Apply masking tape to the quarter panel around the antenna base bezel to protect the paint.
13   Using a pair of adjustable pliers or other appropriate tool, unscrew the bezel nut from the antenna housing and withdraw the antenna housing from the vehicle.
14   Installation is the reverse of removal.

### 15   Headlight bulb - replacement

*Refer to illustrations 15.3a, 14.3b, 15.4, 15.5 and 15.6*

**Warning:** *These models are equipped with halogen gas-filled bulbs which are under pressure and may shatter if the surface is scratched or the bulb is dropped. Wear eye protection and handle the bulbs carefully, grasping only the base whenever possible. Do not touch the surface of the bulb with your fingers because the oil from your skin could cause it to overheat and fail prematurely. If you do touch the bulb surface, clean it with rubbing alcohol.*
**Caution:** *If the stereo in your vehicle is equipped with an anti-theft system, make sure you have the correct activation code before disconnecting the battery.*

1    Open the hood.
2    Disconnect the cable from the negative battery terminal.
3    On 1997 and earlier models, the bulbs are accessible from the back of the housing; remove the air intake duct for access to the driver's-side headlight bulb (see illustra-

---

pillar trim panels from each end of the rear package tray.
10   On 1997 and earlier models, pry the retaining clips loose and remove the cover from the high-mounted brake light. Remove the mounting bolts, disconnect the electrical connector and remove the brake light assembly.
11   On 1998 and later models, carefully pry the high-mounted brake light housing from the rear window trim panel, disconnect the electrical connector and remove the brake light assembly (see Section 18).
12   Remove the rear seat headrests (if equipped).
13   Pry the retainers loose, remove the bolts (if equipped) and remove the package tray **(see illustration)**.
14   On 1997 and earlier models, remove the push-nuts and separate the speaker grilles from the package tray. Disconnect the electrical connectors, remove the mounting bolts and remove the speakers from the package tray.
15   On 1998 and later models, disconnect the electrical connectors, remove the mounting screws and remove the speakers from the

vehicle **(see illustration)**.
16   Installation is the reverse of removal.

---

### 14   Antenna - removal and installation

#### Corolla models and 1997 and earlier Prizm models

*Refer to illustration 14.4*
1    Remove the driver's-side kick panel.
2    Disconnect the antenna cable from the cable extension leading to the radio.
3    Attach a length of wire to the antenna cable end.
4    Remove the antenna mounting screws from the roof **(see illustration)**. Remove the antenna and pull the cable and wire through the pillar.
5    Remove the wire from the cable end, leaving the wire in the pillar. Attach the new cable to the wire.
6    Pull the wire and cable through the pillar and install the antenna base to the roof.
7    Connect the cable to the cable extension and install the kick panel.

**12**

**15.3a  Use a small screwdriver to pry up on the air intake retainer**

**15.3b  Lift the air intake duct out**

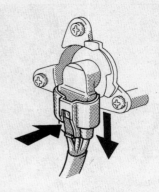

**15.4 Depress the lever and disconnect connector**

tions). On 1998 and later models, remove the headlight housing mounting screws (see Section 17) and pull the housing forward to access the bulbs.

4    Press the lock release and disconnect the electrical connector from the bulb assembly **(see illustration)**.

5    On 1997 and earlier models, rotate bulb assembly counterclockwise and withdraw it from the headlight housing **(see illustration)**.

6    On 1998 and later models, remove the rubber insulator and release the locking clip and withdraw the headlight bulb from the housing **(see illustration)**.

7    Without touching the glass with your bare fingers, insert the new bulb assembly into the headlight housing and lock it in place.

8    The remainder of installation is the reverse of removal. Test the headlight operation. It is not necessary to re-adjust the headlights unless the original position has been disturbed.

## 16   Headlights - adjustment

*Refer to illustrations 16.1a, 16.1b and 16.5*
**Note:** *It is important that the headlights are*

**15.5  Rotate the bulb holder counterclockwise (when viewed from behind) and withdraw it from the housing**

*aimed correctly. If adjusted incorrectly they could blind the driver of an oncoming vehicle and cause a serious accident or seriously reduce your ability to see the road. The headlights should be checked for proper aim every 12 months and any time a new headlight housing is installed or front end body work is performed. It should be emphasized that the following procedure is only an interim step which will provide temporary adjustment until the headlights can be adjusted by a properly equipped shop.*

1    These models have two adjustment

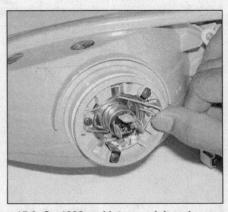

**15.6  On 1998 and later models, release the locking clip to remove the headlight bulb**

screws, one controlling left-and-right (horizontal) movement and one controlling up-and-down (vertical) movement that are accessible with the hood open **(see illustrations)**.

2    There are several methods of adjusting the headlights. The most accurate method requires special headlight adjusting equipment. A simple method, requiring an open area with a blank wall and a level floor, may be used in lieu of the special equipment to bring the headlights close to optimum adjustment. For best results, the area should be

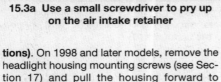

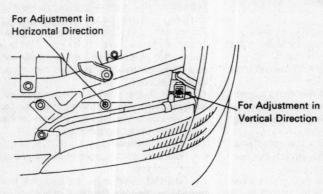

**16.1a  Typical headlight adjustment screws - 1997 and earlier models**

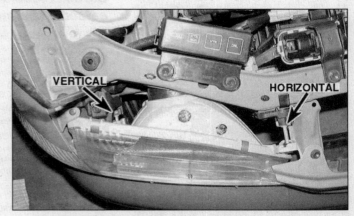

**16.1b  Headlight adjustment screws - 1998 and earlier models (use a small open-end wrench to turn the nuts)**

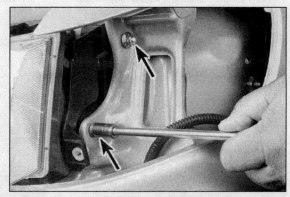

**17.4a  On 1997 and earlier models, remove the headlight-to-radiator crossmember bolt**

**17.4b  After removing the parking light assembly, the two side retaining nuts (arrows) are accessible**

**16.5  Headlight aiming details**

---

dimly light and the wall should be light colored with a non-glossy finish.

3    Position masking tape vertically on the wall in reference to the vehicle centerline and the centerlines of both headlights. Position a horizontal tape line in reference to the centerline of the headlights. **Note:** It *may be easier to position the tape on the wall with the vehicle parked a few feet from the wall.*

4    Adjustment should be made with the vehicle parked 25 feet from the wall, sitting level, the gas tank half-full and no unusually heavy load in the vehicle.

5    Starting with the low beam adjustment, position the center of the high intensity zone so it is two inches below the horizontal line and two inches to the side of the headlight vertical line away from oncoming traffic **(see illustration)**. Turn the adjustment screws until the desired level has been achieved.

6    With the high beams on, the high intensity zone should be vertically centered with the exact center just below the horizontal line. **Note:** *It may not be possible to position the headlight aim exactly for both high and low beams. If a compromise must be made, keep in mind that the low beams are the most used and have the greatest effect on driver safety.*

7    Have the headlights adjustment checked by a dealer service department or other properly equipped facility at the earliest opportunity.

**17.4c  Pull the headlight straight out to detach the retainer at the lower corner (arrow)**

---

## 17   Composite headlight housing - removal and installation

1    Disconnect the cable from the negative battery terminal. **Caution:** *If the stereo in your vehicle is equipped with an anti-theft system, make sure you have the correct activation code before disconnecting the battery.*

### 1997 and earlier

*Refer to illustrations 17.4a, 17.4b and 17.4c*

2    Remove the radiator grille (Chapter 11).

**17.7  Headlight housing mounting screws (arrows) - 1998 and later models**

3    Remove the parking light housing (see Section 18).

4    Remove the headlight housing retaining bolts and nuts. Detach the headlight housing from the lower retainer by pulling the bottom straight out **(see illustrations)**.

5    Remove the headlight bulb (Section 15).

6    Installation is the reverse of removal.

### 1998 and later

*Refer to illustrations 17.7 and 17.8*

7    Remove the headlight housing retaining screws **(see illustration)**.

**12**

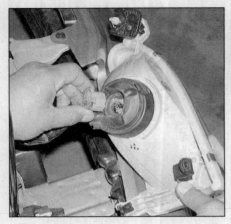

**17.8 Pull the headlight housing forward and disconnect the electrical connector from the bulb**

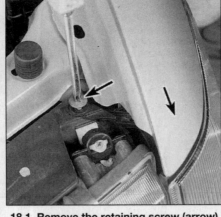

**18.1 Remove the retaining screw (arrow) and push the lens housing forward to detach it**

**18.2 Disconnect the connector and rotate the bulb holders to remove them from the housing**

**18.3 Remove the turn signal light screw and rotate the housing out of the fender**

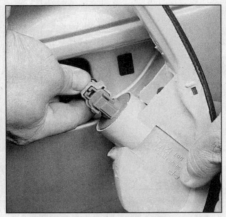

**18.4a Disconnect the electrical connector**

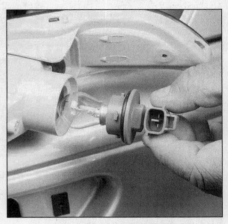

**18.4b Rotate the bulb holder counterclockwise and pull it out**

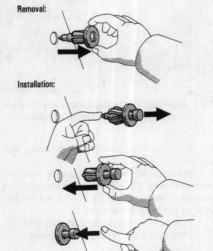

**18.5a Trunk trim panel clip retainer details**

8    Pull the housing forward and disconnect the electrical connector from the headlight bulb **(see illustration)**. Remove the rubber insulator and the headlight bulb.

9    Installation is the reverse of removal.

## 18   Bulb replacement

### *Front parking and side marker*

*Refer to illustrations 18.1 and 18.2*

1    Remove the retaining screw and detach the housing by sliding it forward **(see illustration)**.

2    Disconnect the electrical connector and rotate the bulb holders counterclockwise to replace the bulbs **(see illustration)**.

### *Turn signal*

*Refer to illustrations 18.3, 18.4a and 18.4b*

3    Remove the retaining screw and detach the housing by rotating it forward out of the fender **(see illustration)**.

4    Disconnect the electrical connector and rotate the bulb holder counterclockwise to remove it **(see illustrations)**.

### *Rear side marker, turn signal, brake, tail and back-up lights*

*Refer to illustrations 18.5a, 18.5b, 18.6, 18.7 and 18.8*

#### Sedan

5    Remove the two screws retaining the housing, open the trunk lid and detach the

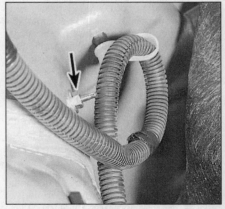

**18.5b Remove the tail light housing retaining nut (arrow)**

**18.6  Rotate the bulb holder counterclockwise to remove it**

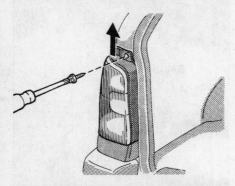

**18.7  Remove the station wagon tail light screws and lift the housing straight up**

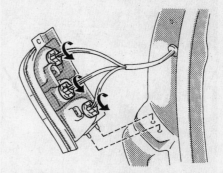

**18.8  Station wagon tail light bulb details**

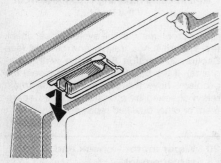

**18.9  Press the tab in and lower the sedan license plate light housing**

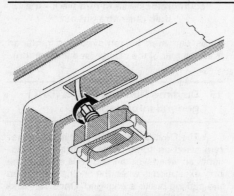

**18.10  Sedan license plate bulb details**

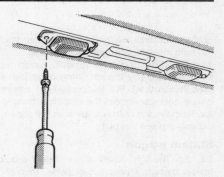

**18.11  Station wagon license plate housing screws**

trim cover for access. On 1997 and earlier models, remove the nut and pull the housing out **(see illustrations)**.

6    Rotate the bulb holders counterclockwise and replace the bulbs **(see illustration)**.

### Station wagon

7    Remove the retaining screw and lift the housing up to remove **(see illustration)**.

8    Rotate the bulb holders counterclockwise and replace the bulbs **(see illustration)**.

## License plate light

*Refer to illustrations 18.9, 18.10, 18.11 and 18.12*

### Sedan

9    Depress the tab and rotate the light housing down for access to the bulb holder

**(see illustration)**.

10    Rotate the holder counterclockwise to remove it, the pull the bulb straight out **(see illustration)**.

### Station wagon

11    Remove the screws and lower the housing **(see illustration)**.

12    Remove the bulb holder by turning it counterclockwise, then pull the bulb out of the holder **(see illustration)**.

## High mounted brake light

*Refer to illustrations 18.13a, 18.13b, 18.13c, 18.13d, 18.14 and 18.15*

### Sedan

13    On 1997 and earlier models, use a ball-point pen or similar tool to push the center of retainers in. Pry the retainers out with a small screwdriver and remove the light cover **(see illustrations)**. Remove the housing bolts, lift the housing up for access, then rotate the

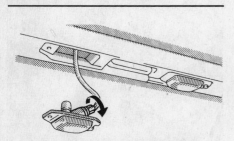

**18.12  Station wagon license plate bulb details**

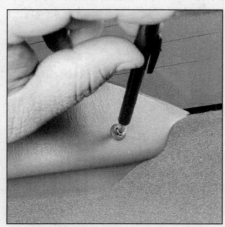

**18.13a  On 1997 and earlier models, use a ball-point pen to push in on the center of the retainers**

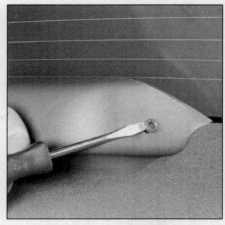

**18.13b  Use a screwdriver to pry out the retainers**

**12**

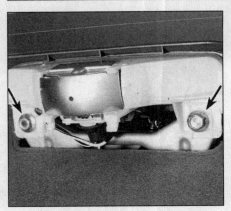

**18.13c Remove the nuts and lift the housing up for access to the bulb**

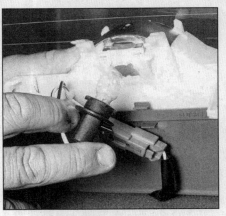

**18.13d Rotate the holder counterclockwise to remove it - the bulb pulls straight out**

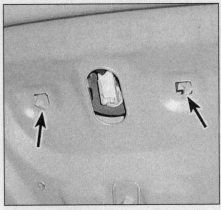

**18.14 On 1998 and later models, open the trunk lid and pry the high mounted brake light housing retaining clips loose from the body**

holder counterclockwise to remove it and pull the bulb straight out **(see illustrations)**.

14   On 1998 and later models, open the trunk lid and pry the retaining clips loose. **(see illustration)**. Pull the brake light assembly up and disconnect the electrical connector. Remove the bulb socket from the assembly and replace the bulb.

### Station wagon

15   Pry the light cover off, remove the bulb holder by turn it counterclockwise, then pull the bulb out of the holder **(see illustration)**.

## Interior lights

*Refer to illustration 18.17*

16   Use a small screwdriver to pry off the lens.

17   Detach the bulb from the terminal. It may be necessary to pry the bulb out, if this is the case, pry only at the end of the bulb (otherwise the glass may shatter **(see illustration)**.

## Instrument cluster illumination

*Refer to illustration 18.18*

18   To gain access to the instrument cluster illumination lights, the instrument cluster will have to be removed (see Section 21). The

bulbs can then be removed and replaced from the rear of the cluster **(see illustration)**.

### 19   Daytime Running Lights (DRL) - general information

The Daytime Running Lights (DRL) system, used on some models, turns the headlights on whenever the engine is started. The only exception is when the engine is on when the parking brake is engaged. Once the parking brake is released, the lights will remain on as long as the ignition switch is on, even if the parking brake is later applied. The DRL system supplies reduced power to the headlights so they won't be too bright for daytime use while prolonging headlight life.

The main components controlling the Daytime Running lights are the DRL control unit and relays. The DRL control unit (main relay) is located behind the right side of the instrument panel above the glove box, and the DRL relay is located in an auxiliary relay box under the hood behind the right headlight. On 1997 and earlier models, two additional relays are located behind the instrument panel above the main relay **(see illustration 6.2)**.

If problems with the system arise, check the system fuses, relays and circuits (see Section 3, Section 6 and the wiring diagrams). If the basic checks fail to reveal the problem, have the DRL control unit (main relay) checked by a dealer service department or other qualified repair shop.

### 20   Wiper motor - check and replacement

**Warning:** *These models are equipped with airbags. Always disable the airbag system before working near airbag system components. After the battery is disconnected, wait at least two minutes before beginning work (the system has a back-up capacitor that must fully discharge). For more information see Section 27.*
**Caution:** *If the stereo in your vehicle is equipped with an anti-theft system, make sure you have the correct activation code before disconnecting the battery.*

1   The windshield wiper motor is located on the right (passenger) side of the underhood compartment and the rear wiper motor is mounted in the liftgate.

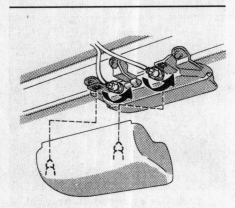

**18.15 Station wagon high mounted brake light details**

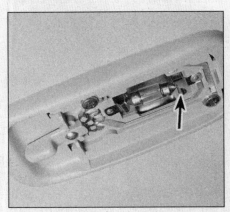

**18.17 Pry carefully at the bulb clip (arrow) and detach the bulb**

**18.18 Rotate the instrument cluster bulb counterclockwise and lift it out**

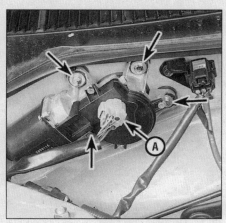

20.5a  Disconnect the wiper motor connector (A), remove the bolts (arrows) and detach the motor

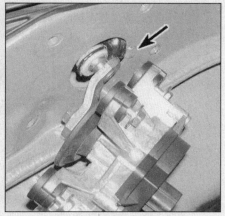

20.5b  Secure the claw on the wiper linkage to the firewall (arrow) and pry the wiper linkage arm from the wiper motor bellcrank ball

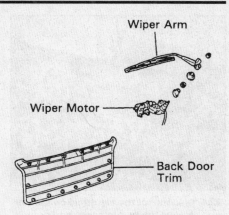

20.7  Rear wiper arm details

## Check

### Windshield washer/wiper switch

2    Refer to Section 9 for the wiper and washer switch check procedure.

### Wiper motor

3    If the wipers fail to operate when activated, check the fuse. If the fuse is OK, connect a jumper wire between the wiper motor and ground, then retest. If the motor works now, repair the ground connection. If the motor still doesn't work, turn the wiper switch to the HI position and check for voltage at the motor electrical connector (see illustration 20.5a). If there's voltage at the connector, remove the motor and check it off the vehicle with fused jumper wires from the battery. If the motor still doesn't work, replace it. If there's no voltage to the motor, check the switch and related circuits for continuity.

## Replacement

*Refer to illustrations 20.5a, 20.5b and 20.7*

### Windshield wiper motor

4    Disconnect the cable from the negative battery terminal.
5    Disconnect the electrical connector from the wiper motor and remove the mount-

ing bolts (see illustration). Secure the linkage claw on the firewall and pry the wiper linkage arm from the wiper motor bellcrank ball (see illustration). Caution: *Do not remove the bellcrank from the wiper motor. The bellcrank is positioned on the spindle to properly park the wiper arms.* Remove the wiper motor from the vehicle.
6    Installation is the reverse of removal.

### Rear wiper motor

7    Remove the wiper arm, then remove the shaft spindle nuts and washers (see illustration).
8    Disconnect the electrical connector, detach the wiper linkage, remove the retaining bolts and lower the motor through the liftgate access hole as an assembly.
9    Installation is the reverse of removal.

---

## 21  Instrument cluster - removal and installation

*Refer to illustrations 21.3 and 21.4*
**Warning:** *These models are equipped with airbags. Always disable the airbag system before working near airbag system compo-*

nents. After the battery is disconnected, wait at least two minutes before beginning work (the system has a back-up capacitor that must fully discharge). For more information see Section 27.
**Caution:** *If the stereo in your vehicle is equipped with an anti-theft system, make sure you have the correct activation code before disconnecting the battery.*
1    Disconnect the cable from the negative battery terminal.
2    Remove the instrument cluster bezel (see Chapter 11).
3    Remove the retaining screws and pull the cluster forward (see illustration).
4    Disconnect the electrical connectors and remove the cluster (see illustration).
5    Installation is the reverse of the removal procedure.

---

## 22  Horn - check and replacement

*Refer to illustration 22.4*

## Check

1    Disconnect the electrical connector from the horn.
2    Connect a voltmeter or test light to the terminal in the wire harness connector. Have an assistant depress the horn button. If bat-

21.3  Remove the instrument cluster screws

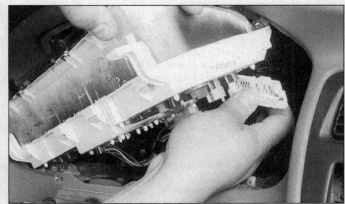

21.4  Disconnect the electrical connectors and remove the cluster

**12**

**22.4  Disconnect the electrical connector and remove the bolt (arrows) and detach the horn**

tery voltage is not present, check the fuse, relay, horn switch and related wiring.
3    If battery voltage is present at the connector and the horn doesn't sound when connected, replace the horn.

## *Replacement*

4    Disconnect the electrical connector and remove the bracket bolt **(see illustration)**.
5    Installation is the reverse of removal.

## 23  Cruise control system - description and check

*Refer to illustration 23.1*

1    The cruise control system maintains a desired vehicle speed under normal driving conditions. The main components of the cruise control system are the cruise control module and actuator. The cruise control module is located in the passenger compartment. On 1997 and earlier models, the cruise control module is located in the center of the instrument panel ahead of the center console. On 1998 and later models, the cruise control module is located at the right end of the instrument panel. The actuator is located in the right-side of the engine compartment and connected to the throttle linkage by a cable. Besides the control module and actuator, the system consists of the brake switch, clutch switch, control switches, a relay and associated wiring **(see illustration)**. The cruise control system features a self-diagnostic system. The system has the ability to store a trouble code in the control module memory if a fault is detected. Retrieving the trouble code requires special testers and diagnostic procedures, so have the system diagnosed by a dealer service department or other qualified repair shop if problems develop. Listed below are some general procedures that may be used to locate common problems.
2    Locate and check the fuse (see Section 3).
3    Have an assistant operate the brake lights while you check their operation (voltage from the brake light and if equipped, clutch

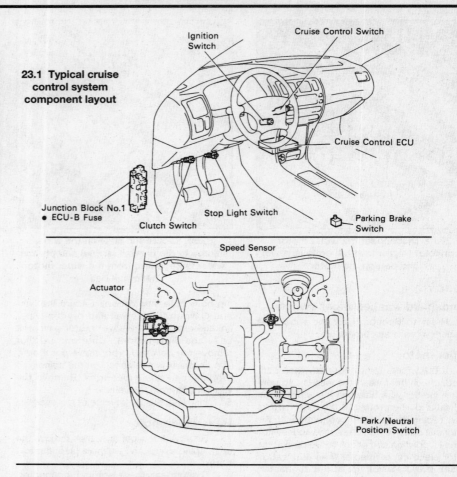

**23.1  Typical cruise control system component layout**

switch, deactivates the cruise control).
4    If the brake lights don't come on or don't shut off, correct the problem and retest the cruise control. Check the clutch switch

(Chapter 8).
5    Visually inspect the system vacuum hoses and electrical connectors. Check the control cable between the cruise control

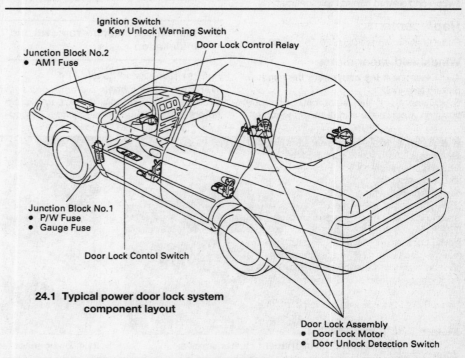

**24.1  Typical power door lock system component layout**

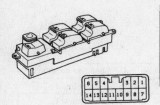

## 1996 and earlier

| Terminal<br>Switch position | 1 and 2 | 3 | 4 |
|---|---|---|---|
| LOCK | ○— | | —○ |
| OFF | | | |
| UNLOCK | ○— | —○ | |

## 1997 and later

| | 2 | 3 | 7 |
|---|---|---|---|
| | ○— | —○ | |
| | | ○— | —○ |

**24.6a Power door lock master switch terminal guide and continuity table**

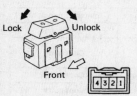

| Terminal<br>Switch position | 2 | 3 | 4 |
|---|---|---|---|
| LOCK | | ○— | —○ |
| OFF | | | |
| UNLOCK | ○— | | —○ |

**24.6b Power door lock switch terminal guide and continuity table**

repair shop.

5  If one or more (but not all) the actuators are inoperable, remove the door trim panel from the inoperable door (see Chapter 11) and disconnect the electrical connector from the actuator. Using a voltmeter or test light, check for power at the connector while operating the door lock switch. One of the terminals should have power in the Lock position; the other should have power in the unlock position. If power is present and the door lock actuator does not operate when connected, replace the door lock actuator.

6  If power is not present at the door lock actuator, test the switch for continuity. Replace the switch if there's no continuity in either switch position **(see illustrations)**.

7  If the switch tests good but the actuator does not receive power, check the wiring for continuity between the control unit and the switch and actuator. Repair the wiring if there's no continuity.

8  If the switches, actuators and wiring are good, have the control unit diagnosed by a dealer service department or other qualified repair shop.

actuator and the throttle linkage and replace as necessary.

6  The cruise control system uses the vehicle speed sensor as the speed sensing device. The vehicle speed sensor is located in the transaxle. Check the sensor for proper operation (see Chapter 6).

7  Test drive the vehicle to determine if the cruise control is now working. If it isn't, take it to a dealer service department or an automotive electrical specialist for further diagnosis and repair.

## 24  Power door lock system - description and check

*Refer to illustrations 24.1, 24.6a and 24.6b*

1  The power door lock system operates the door lock actuators mounted in each door. The system consists of the door lock control unit (relay), door switches, actuators and associated wiring **(see illustration)**. Diagnosis can usually be limited to simple checks of the wiring connections, switches and actuators for minor faults which can be easily repaired.

2  The doors are locked by bi-directional solenoids (actuators) located in the doors. The lock switches have two operating positions: Lock and Unlock. These switches signal the door lock control unit (relay) which in turn connects voltage to the door lock actuators. Depending on which way the actuator is activated, it reverses polarity, allowing the two sides of the circuit to be used alternately as the feed (positive) and ground side, locking or unlocking the doors.

3  Always check the circuit protection first. Some vehicles use a combination of circuit

breakers and fuses.

4  Operate the door lock switches in both directions (Lock and Unlock) with the engine off. Listen for the sound of the actuators operating. If the system is completely inoperable (none of the actuators operate), the door lock control unit is defective or a serious wiring problem exists in the system wiring harness. Have the control unit diagnosed by a dealer service department or other qualified

## 25  Power window system - description and check

*Refer to illustrations 25.1, 25.9a, 25.9b, 25.10a and 25.10b*

1  The power window system operates the electric motors mounted in the doors which lower and raise the windows. The system consists of the control switches, the motors (regulators), glass mechanisms and associated wiring **(see illustration)**.

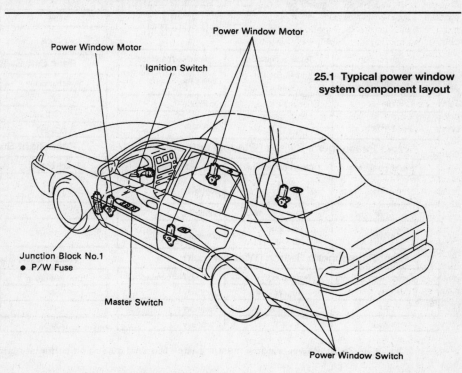

**25.1 Typical power window system component layout**

**12**

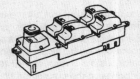
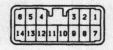

### Front Driver's Switch (Window unlock and lock)

| Switch position | Tester connection | Specified condition |
|---|---|---|
| UP | 6 – 7 – 8 | Continuity |
| | 1 – 2 – 13 | |
| OFF | 1 – 2 – 6 – 13 | Continuity |
| DOWN | 1 – 2 – 6 | Continuity |
| | 7 – 8 – 13 | |

### Front Passenger's Switch

| Switch position | Tester connection | Specified condition |
|---|---|---|
| UP | 7 – 8 – 12 | Continuity |
| UP and window lock | 1 – 2 – 5 | Continuity |
| | 7 – 8 – 12 | |
| OFF | 5 – 12 | Continuity |
| OFF and window lock | 1 – 2 – 5 – 12 | Continuity |
| DOWN | 5 – 7 – 8 | Continuity |
| DOWN and window lock | 5 – 7 – 8 | Continuity |
| | 1 – 2 – 12 | |

### Rear Left Switch

| Switch position | Tester connection | Specified condition |
|---|---|---|
| UP | 7 – 8 – 10 | Continuity |
| UP and window lock | 7 – 8 – 10 | Continuity |
| OFF | 9 – 10 | Continuity |
| OFF and window lock | 1 – 2 – 9 | Continuity |
| DOWN | 7 – 8 – 9 | Continuity |
| DOWN and window lock | 1 – 2 – 9 – 10 | Continuity |

### Rear Right Switch

| Switch position | Tester connection | Specified condition |
|---|---|---|
| UP | 7 – 8 – 11 | Continuity |
| UP and window lock | 7 – 8 – 11 | Continuity |
| | 1 – 2 – 14 | |
| OFF | 11 – 14 | Continuity |
| OFF and window lock | 1 – 2 – 11 – 14 | Continuity |
| DOWN | 7 – 8 – 14 | Continuity |
| DOWN and window lock | 1 – 2 – 11 | Continuity |
| | 7 – 8 – 14 | |

**25.9a Power window master switch terminal guide and continuity table - 1996 and earlier models**

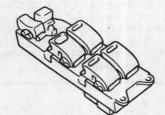
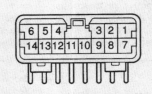

| Switch position | Tester connection | Specified condition |
|---|---|---|
| UP | 1 – 10 – 11 | Continuity |
| | 3 – 4 – 5 | |
| OFF | 1 – 3 – 4 – 5 | Continuity |
| DOWN | 1 – 3 – 4 | Continuity |
| | 5 – 10 – 11 | |

### Front Passenger's Switch (Window unlock):

| Switch position | Tester connection | Specified condition |
|---|---|---|
| UP | 3 – 4 – 6 | Continuity |
| | 10 – 11 – 14 | |
| OFF | 3 – 4 – 6 – 14 | Continuity |
| DOWN | 3 – 4 – 14 | Continuity |
| | 6 – 10 – 11 | |

### Front Passenger's Switch (Window lock):

| Switch position | Tester connection | Specified condition |
|---|---|---|
| UP | 10 – 11 – 14 | Continuity |
| OFF | 6 – 14 | Continuity |
| DOWN | 6 – 10 – 11 | Continuity |

### Rear Left Switch (Window unlock):

| Switch position | Tester connection | Specified condition |
|---|---|---|
| UP | 3 – 4 – 12 | Continuity |
| | 9 – 10 – 11 | |
| OFF | 3 – 4 – 9 – 12 | Continuity |
| DOWN | 3 – 4 – 9 | Continuity |
| | 10 – 11 – 12 | |

### Rear Left Switch (Window lock):

| Switch position | Tester connection | Specified condition |
|---|---|---|
| UP | 9 – 10 – 11 | Continuity |
| OFF | 9 – 12 | Continuity |
| DOWN | 10 – 11 – 12 | Continuity |

### Rear Right Switch (Window unlock):

| Switch position | Tester connection | Specified condition |
|---|---|---|
| UP | 3 – 4 – 13 | Continuity |
| | 8 – 10 – 11 | |
| OFF | 3 – 4 – 8 – 13 | Continuity |
| DOWN | 3 – 4 – 8 | Continuity |
| | 10 – 11 – 13 | |

### Rear Right Switch (Window lock):

| Switch position | Tester connection | Specified condition |
|---|---|---|
| UP | 8 – 10 – 11 | Continuity |
| OFF | 8 – 13 | Continuity |
| DOWN | 10 – 11 – 13 | Continuity |

**25.9b Power window master switch terminal guide and continuity table - 1997 and later models**

| Switch position | Tester connection to terminal number | Specified condition |
|---|---|---|
| UP | 1 — 5<br>3 — 4 | Continuity |
| OFF | 1 — 2<br>3 — 4 | Continuity |
| DOWN | 1 — 2<br>4 — 5 | Continuity |

**25.10a Door window switch terminal guide and continuity table - 1996 and earlier models**

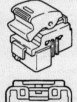

| Switch position | Tester connection | Specified condition |
|---|---|---|
| UP | 1 — 2<br>3 — 4 | Continuity |
| OFF | 1 — 2<br>3 — 5 | Continuity |
| DOWN | 1 — 4<br>3 — 5 | Continuity |

**25.10b Door window switch terminal guide and continuity table - 1997 and later models**

2   The power window system is designed so the windows can be lowered and raised from the master control switch by the driver or by remote switches located at the individual windows. Each window has a separate motor which is reversible. The position of the control switch determines the polarity and therefore the direction of operation.

3   The system is equipped with a main relay that provides power to the system in addition to the fuse protecting the entire circuit.

4   The system is equipped with a separate circuit breaker incorporated into each motor.

This prevents one stuck window from disabling the entire system.

5   The power window system will only operate when the ignition switch is ON. In addition, when activated the window lockout switch at the master control switch disables the switches at the passenger's window. Always check these items before troubleshooting a window problem.

6   These procedures are general in nature, so if you can't find the problem using them, take the vehicle to a dealer service department.

7   If the power windows don't work at all, check the fuse or fusible link. Make sure power is present at the master switch and each door switch with the ignition key On.

8   If voltage isn't reaching the master switch and/or door switch, check the wiring in the circuit for continuity between the fuse panel and switches. You'll need to consult the wiring diagram for the vehicle. The power window circuit is equipped with a main relay. Make sure the relay is grounded properly and receiving voltage from the fuse panel. Check the relay operation (see Section 6). Replace the relay if defective.

9   If one or more of the switches in the master switch are inoperative, check each switch in the master switch for continuity in each position (see illustrations). Replace the master switch if continuity isn't as specified.

10   If the window works from the master switch, but not the door switch, check the door switch for continuity (see illustrations). **Note:** *This doesn't apply to the driver's door window.*

11   If the switch tests OK, check for a short or open in the wiring between the affected switch and the window motor.

12   If one window is inoperative from both switches and the switches test good, remove the trim panel from the affected door and check for voltage at the motor while the switch is operated.

13   If voltage is reaching the motor, disconnect the glass from the regulator (see Chapter 11). Move the window up and down by hand while checking for binding and damage. Also check for binding and damage to the regulator. If the regulator is not damaged and the window moves up and down smoothly, replace the motor. If there's binding or damage, lubricate, repair or replace parts, as necessary.

14   Test the windows after you are done to confirm proper repairs.

## 26   Electric rear view mirrors - description and check

*Refer to illustrations 26.1, 26.6, 26.7a and 26.7b*

1   The electric rear view mirrors use two motors to move the glass; one for up and down adjustments and one for left-right adjustments (see illustration).

2   The control switch has a selector portion which sends voltage to the left or right side mirror. With the ignition key in the ACC position, roll down the windows and operate the mirror control switch through all functions (left-right and up-down) for both the left and right side mirrors.

3   Listen carefully for the sound of the electric motors running in the mirrors.

4   If the motors can be heard but the mirror glass doesn't move, there's probably a problem with the drive mechanism inside the mirror. Replace the mirror.

5   If the mirrors don't operate and no

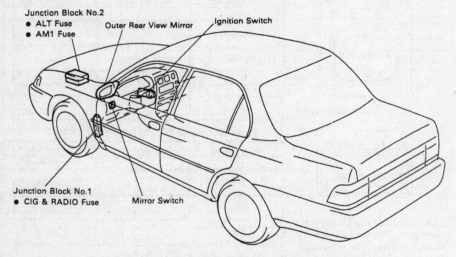

**26.1 Typical electric mirror system component layout**

**12**

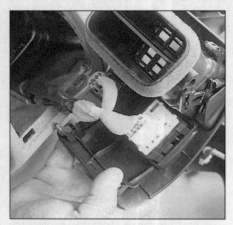

**26.6  Use a screwdriver to detach the air register/mirror switch from the instrument panel; press in on the clips and push the mirror switch out**

sound comes from the mirrors, check the CIG/Radio fuse in the passenger compartment fuse panel (see Section 3).

6   If the fuse is good, remove the mirror control switch from the instrument panel for access the back of the switch without disconnecting the wires attached to it **(see illustration)**. Turn the ignition On and check for voltage at the switch. There should be voltage at one terminal. If there's no voltage at the switch, check for an open or short in the wiring between the fuse panel and the switch.

7   If there's voltage at the switch, disconnect it. Check the switch for continuity in all its operating positions **(see illustrations)**. If the switch does not have continuity, replace it.

8   Re-connect the switch. Locate the wire going from the switch to ground. Leaving the switch connected, connect a jumper wire between this wire and ground. If the mirror works normally with this wire in place, repair the faulty ground connection.

9   If the mirror still doesn't work, remove the mirror and check the wires at the mirror for voltage. Check with ignition ON and the mirror selector switch on the appropriate side. Operate the mirror switch in all its positions. There should be voltage at one of the switch-to-mirror wires in each switch position (except the neutral "off" position).

10   If there's no voltage in each switch position, check the wiring between the mirror and control switch for opens and shorts.

11   If there's voltage, remove the mirror and test it off the vehicle with jumper wires. Replace the mirror if it fails this test.

## 27  Airbag system - general information

*Refer to illustration 27.1*

The models covered by this manual are equipped with a Supplemental Restraint System (SRS), more commonly known as

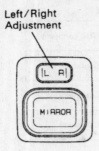

Left/Right
Adjustment

**26.7a  Power mirror switch terminal guide and continuity table - 1997 and earlier models**

| Switch position | Tester connection to terminal number | Specified condition |
|---|---|---|
| OFF | – | No continuity |
| UP | 3 – 7 | Continuity |
|    | 4 – 8 |            |
| DOWN | 3 – 4 | Continuity |
|      | 7 – 8 |            |
| LEFT | 1 – 8 | Continuity |
|      | 3 – 7 |            |
| RIGHT | 1 – 3 | Continuity |
|       | 7 – 8 |            |

### RIGHT SIDE

| Switch position | Tester connection to terminal number | Specified condition |
|---|---|---|
| OFF | – | No continuity |
| UP | 3 – 7 | Continuity |
|    | 6 – 8 |            |
| DOWN | 3 – 6 | Continuity |
|      | 7 – 8 |            |
| LEFT | 3 – 7 | Continuity |
|      | 5 – 8 |            |
| RIGHT | 3 – 5 | Continuity |
|       | 7 – 8 |            |

"airbags." This system is designed to protect the driver and front seat passenger from serious injury in the event of a head-on or frontal collision. It consists of airbag modules in the center of the steering wheel and the right side of the instrument panel, two impact sensors mounted at the front of the vehicle (except 1996 and 1997 models) and a center airbag sensor assembly located inside the passenger compartment **(see illustration)**. On 1998 and later models, additional side airbags are incorporated into the system. The side airbag modules are built into the outer side of the two front seats and are designed to protect the front seat passengers from side impact collisions. Side impact sensors are located in the center pillars on each side of the vehicle. On 1998 and later models, the center airbag sensor assembly has been relocated from the rear of the center console to the center of the instrument panel, ahead of the center console.

### *Airbag modules*

The airbag modules contain a housing incorporating the cushion (airbag) and inflator

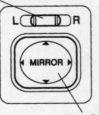

Left/Right Adjustment Switch

Mirror Switch

**26.7b  Power mirror switch terminal guide and continuity table - 1998 and later models**

#### Left Side for Left/Right Adjustment Switch

| Switch position | Tester connection | Specified condition |
|---|---|---|
| OFF | – | No continuity |
| UP | 1 – 9 | Continuity |
|    | 6 – 10 |            |
| DOWN | 1 – 10 | Continuity |
|      | 6 – 9 |            |
| LEFT | 5 – 9 | Continuity |
|      | 6 – 10 |            |
| RIGHT | 5 – 10 | Continuity |
|       | 6 – 9 |            |

#### Right Side for Left/Right Adjustment Switch

| Switch position | Tester connection | Specified condition |
|---|---|---|
| OFF | – | No continuity |
| UP | 6 – 10 | Continuity |
|    | 7 – 9 |            |
| DOWN | 6 – 9 | Continuity |
|      | 7 – 10 |            |
| LEFT | 6 – 10 | Continuity |
|      | 8 – 9 |            |
| RIGHT | 6 – 9 | Continuity |
|       | 8 – 10 |            |

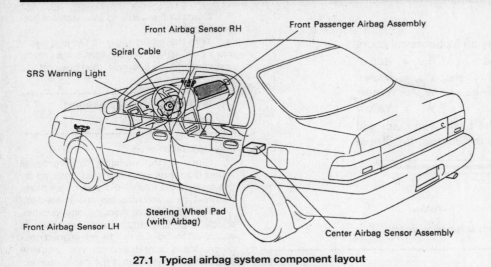

Front Airbag Sensor RH
Front Passenger Airbag Assembly
Spiral Cable
SRS Warning Light
Front Airbag Sensor LH
Steering Wheel Pad (with Airbag)
Center Airbag Sensor Assembly

**27.1 Typical airbag system component layout**

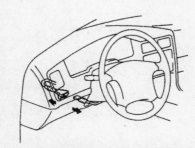

Secondary Lock
Ribs
Primary Lock

- Primary Lock Incomplete (Secondary Lock Prevented)

Lock

- Primary Lock Complete (Secondary Lock Permitted)

Lock

- Twin-Lock Completed

**27.6 Each connector used in the airbag system uses a two stage locking system for security purposes - use care when disconnecting the airbag module connectors**

unit. The inflator assembly is mounted on the back of the housing over a hole through which gas is expelled, inflating the bag almost instantaneously when an electrical signal is sent from the system. The specially wound wire that carries this signal to the driver's module is called a spiral cable. The spiral cable is a flat, ribbon-like electrically conductive tape which is wound many times so that it can transmit an electrical signal regardless of steering wheel position.

### Sensors

On 1995 earlier models, the system has three sensors: two impact sensors at the front of the vehicle behind the bumper and above the wheel arches and a safing sensor in the center airbag sensor assembly located in the center console. On 1996 and 1997 models, all sensors are located in the center airbag sensor assembly. On 1998 and later models, the system uses two front impact sensors the front of the vehicle, two side impact sensors in the center pillars and a safing sensor in the center airbag sensor assembly.

The impact sensors are basically pressure sensitive switches that complete an

electrical circuit during an impact of sufficient G force. The electrical signal from the impact sensors is sent to the safing sensor in the center airbag sensor assembly, which then completes the circuit and inflates the airbag.

### Center airbag sensor assembly

The center airbag sensor contains the safing sensor and an on-board microproces-

sor which monitors the operation of the system. It checks this system every time the vehicle is started, causing the "AIRBAG" warning light to go on, then off, if the system is operating properly. If there is a fault in the system, the light will go on and stay on and the center airbag sensor assembly will store fault codes indicating the nature of the fault. If the AIRBAG light goes on and stays on, the vehicle should be taken to your dealer immediately for service.

### Servicing components near the SRS system

*Refer to illustration 27.6*

Nevertheless, there are times when you need to remove the steering wheel, radio or service other components on or near the instrument panel. At these times, you'll be working around components and wiring harnesses for the SRS system. SRS system wiring is easy to identify; they're all covered by a bright yellow conduit. Do not disconnect the connectors for the SRS system wiring, except to disable the system **(see illustration)**. And do not use electrical test equipment on the SRS system wiring. **ALWAYS DISABLE THE SRS SYSTEM BEFORE WORKING NEAR THE SRS SYSTEM COMPONENTS OR RELATED WIRING**.

### Disabling the SRS system

*Refer to illustrations 27.9 and 27.11*

Turn the steering wheel to the straight ahead position, place the ignition switch in Lock and remove the key. Disconnect the cable from the negative battery terminal. Wait two minutes for the back-up capacitor to discharge.

Disconnect the yellow connectors at the base of the steering column, under the right side of the instrument panel and on 1998 and later models, under the front seats, as described in the following steps.

#### Driver's side airbag

Remove the steering column lower finish panel below the instrument panel (see Chapter 11) and disconnect the yellow steering column harness connector **(see illustration)**.

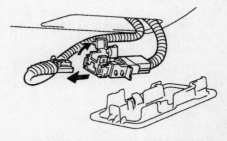

**27.9 Remove the steering column lower finish panel for access to the driver's airbag module connector**

**27.11 On 1997 and earlier models, remove the access cover inside the glove box for access to the passenger's airbag module connector**

Wire colors are indicated by an alphabetical code.

| B | = | Black | L | = | Blue | R | = | Red |
| BR | = | Brown | LG | = | Light Green | V | = | Violet |
| G | = | Green | O | = | Orange | W | = | White |
| GR | = | Gray | P | = | Pink | Y | = | Yellow |

The first letter indicates the basic wire color and the second letter indicates the color of the stripe.

Example: L—Y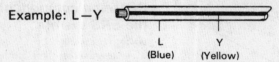

L
(Blue)

Y
(Yellow)

**28.4  Wiring diagram color code chart**

### Passenger's side airbag

Open the glove box. On 1997 and earlier models, remove the access cover. On 1998 and later models, remove the glove box (see Chapter 11).

Disconnect the yellow electrical connector from the passenger inflator module **(see illustration)**.

### Side impact airbag

A side impact airbag connector is located under each front seat. Disconnect the yellow electrical connector from the side impact airbag module.

### Enabling the SRS system

After you've disabled the airbag and performed the necessary service, plug in the steering column (driver's side), passenger side and side impact airbag connectors as necessary. Reinstall the lower finish panel and glove box access cover or glove box.

Connect the cable to the negative battery terminal.

Turn the ignition key to On and verify that the "AIRBAG" warning light comes on for approximately six seconds, then goes off.

## 28  Wiring diagrams - general information

*Refer to illustration 28.4*

Since it isn't possible to include all wiring diagrams for every year covered by this manual, the following diagrams are those that are typical and most commonly needed.

Prior to troubleshooting any circuits, check the fuse and circuit breakers (if equipped) to make sure they are in good condition. Make sure the battery is properly charged and has clean, tight cable connections (see Chapter 1).

When checking the wiring system, make sure that all electrical connectors are clean, with no broken or loose pins. When disconnecting an electrical connector, do not pull on the wires, only on the connector housings themselves.

Refer to the **accompanying illustration** for the wiring diagram color code.

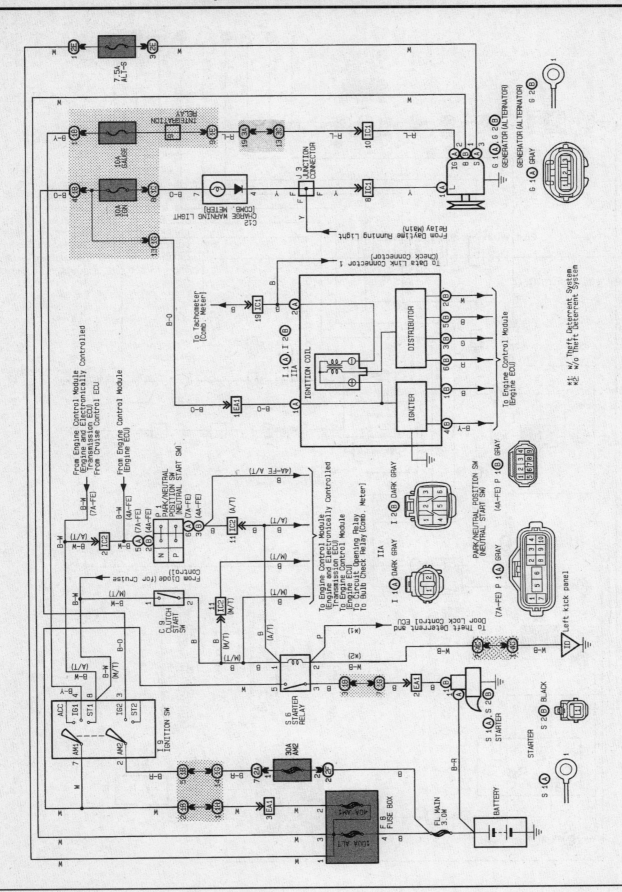

Typical starting, charging and ignition system wiring diagram - 1997 and earlier models

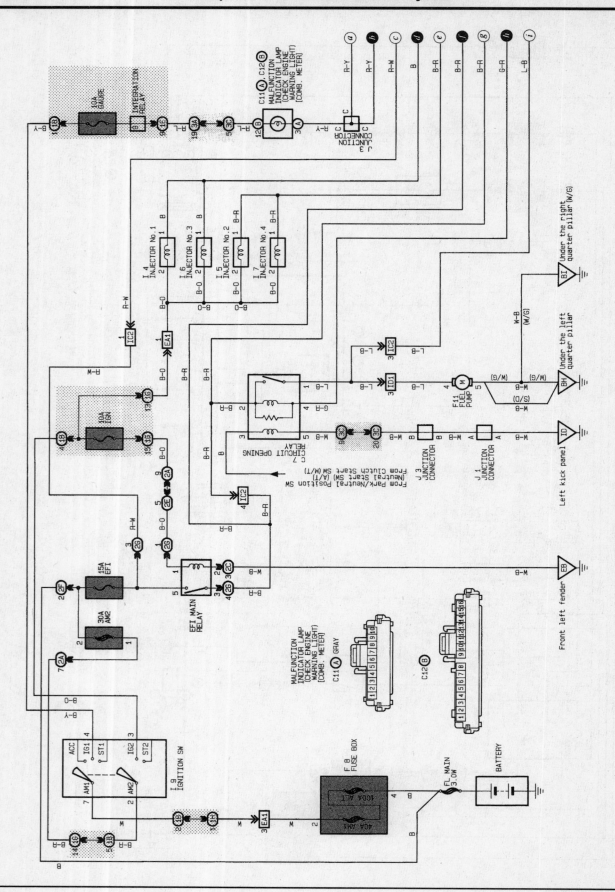

**Typical engine control system wiring diagram (1 of 2) - 1997 and earlier models**

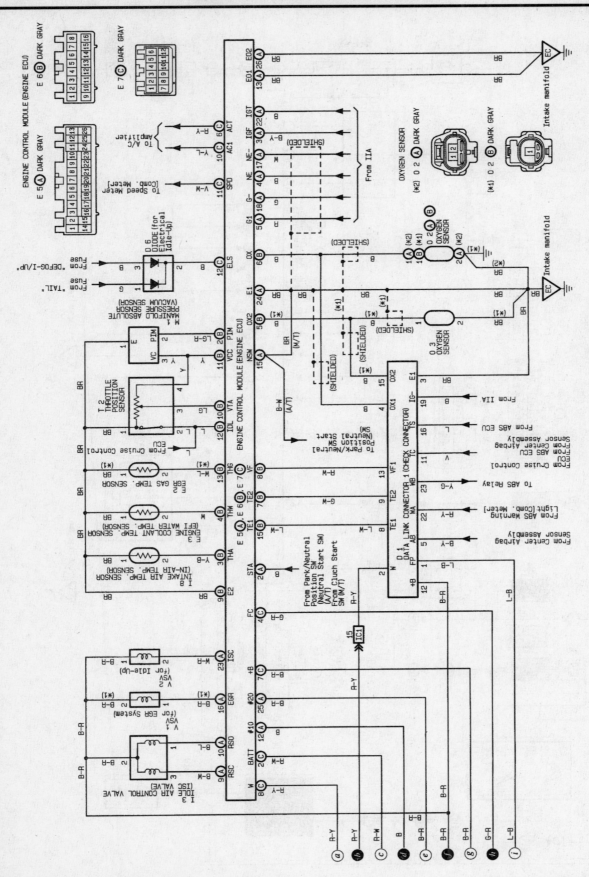

Typical engine control system wiring diagram (2 of 2) - 1997 and earlier models

12

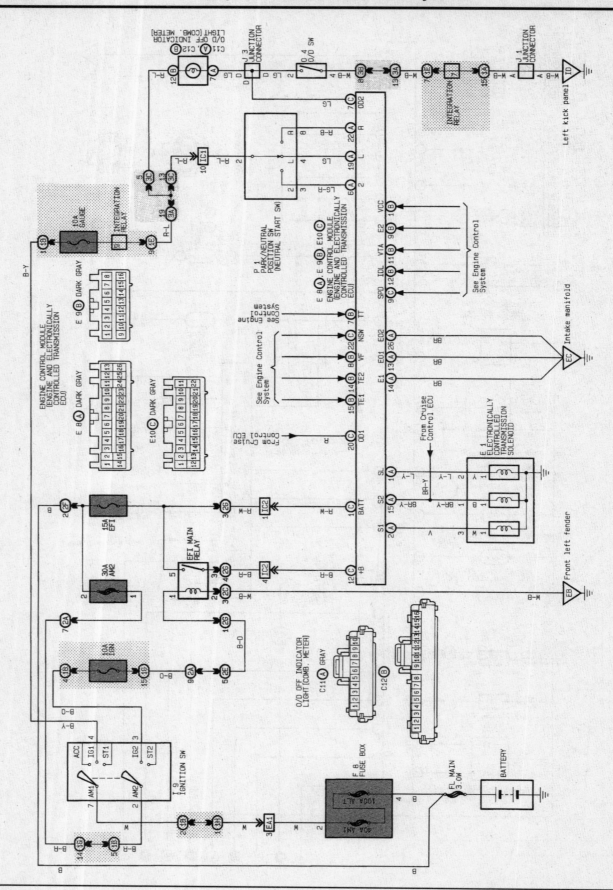

**Typical electronically-controlled transaxle system wiring diagram - 1997 and earlier models**

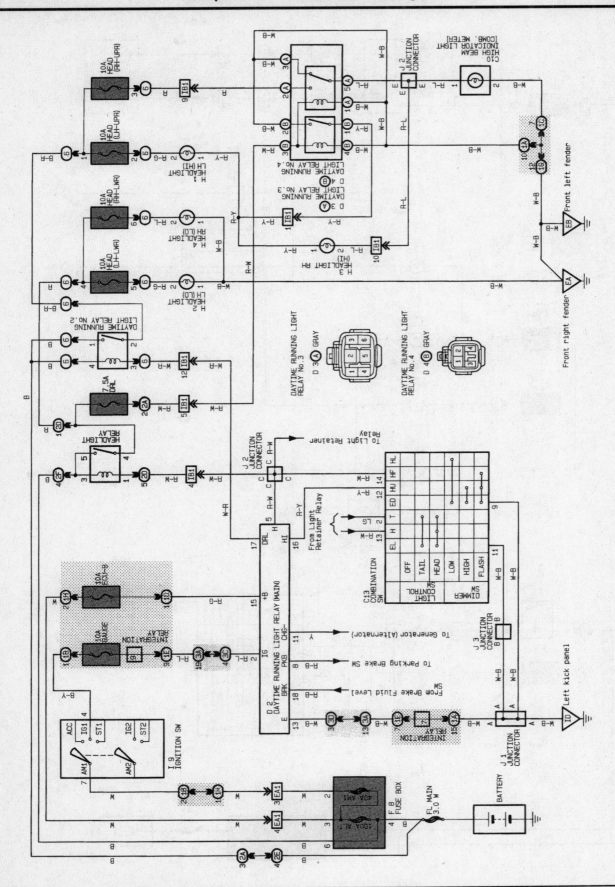

**Typical Daytime Running Light (DRL) system wiring diagram - 1997 and earlier models**

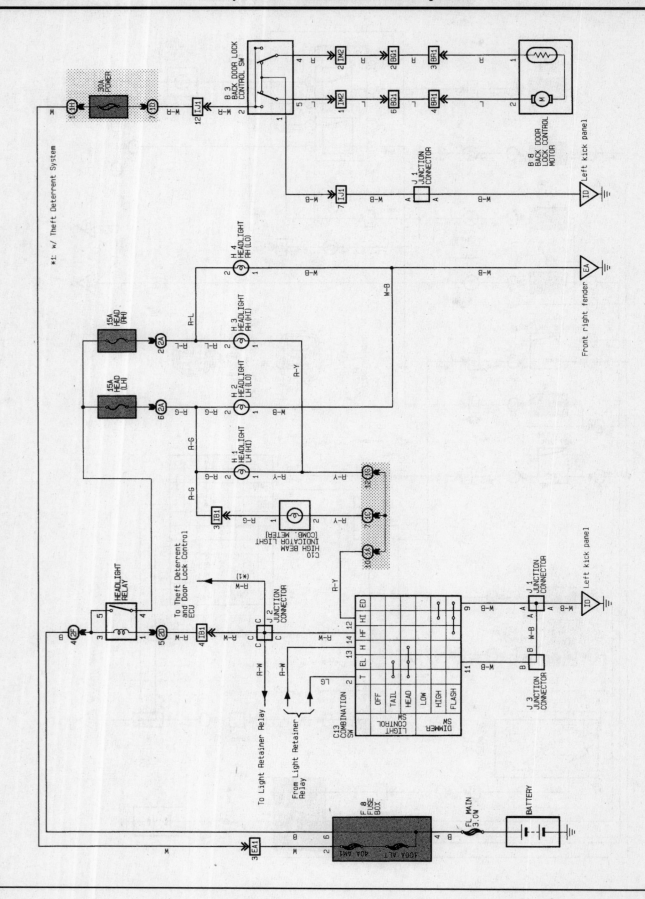

**Typical headlight system wiring diagram (without DRL) - 1997 and earlier models**

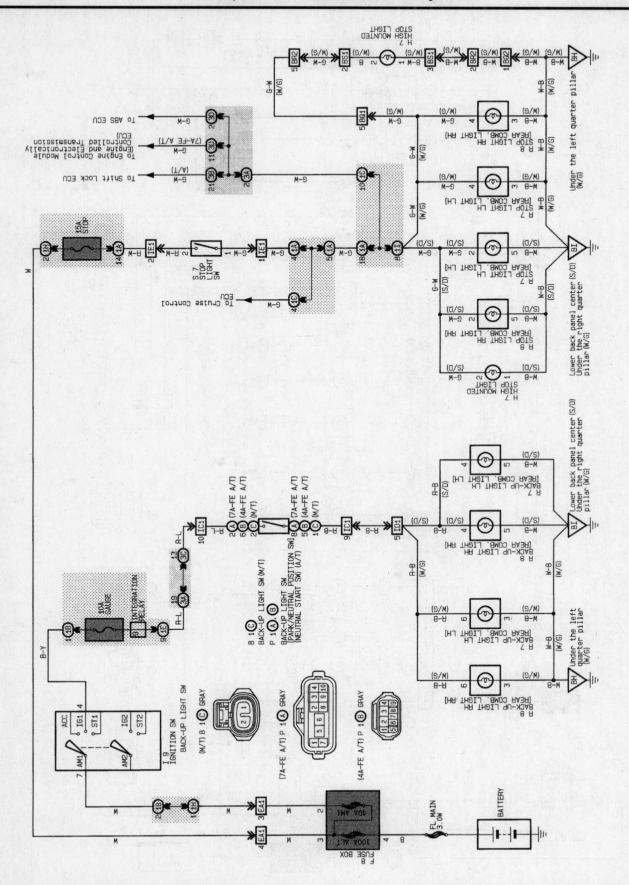

**Typical backup and stoplight system wiring diagram - 1997 and earlier models**

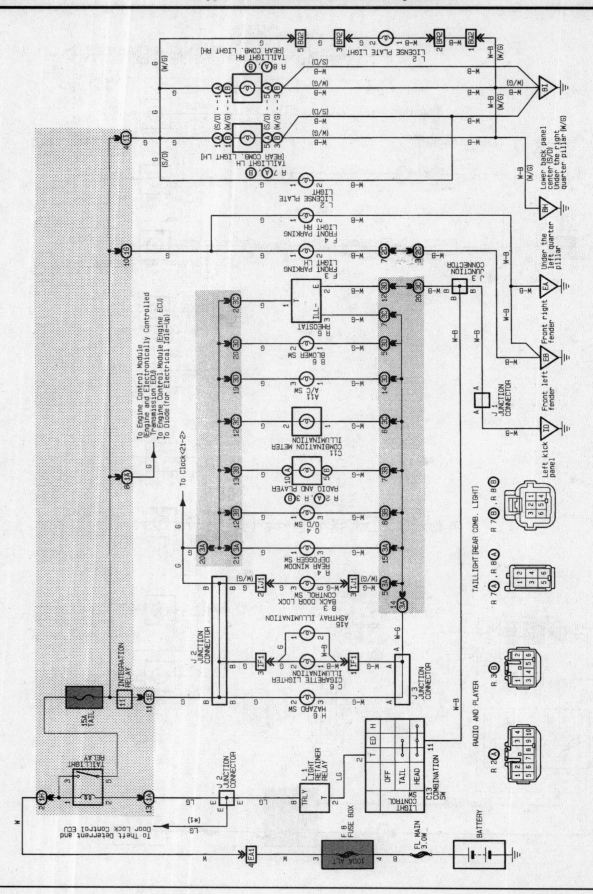

**Typical tail light system wiring diagram - 1997 and earlier models**

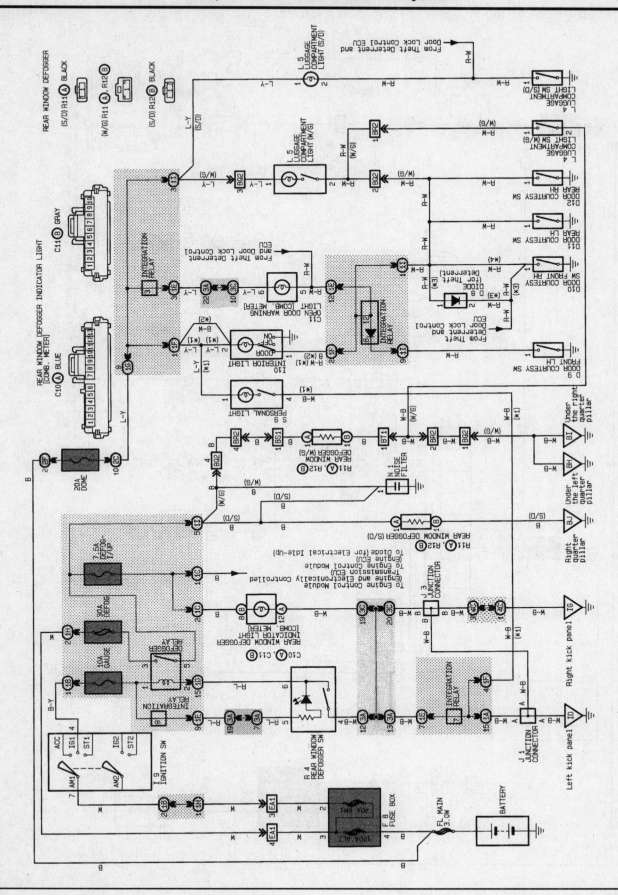

Typical rear window defogger and interior light system wiring diagram - 1997 and earlier models

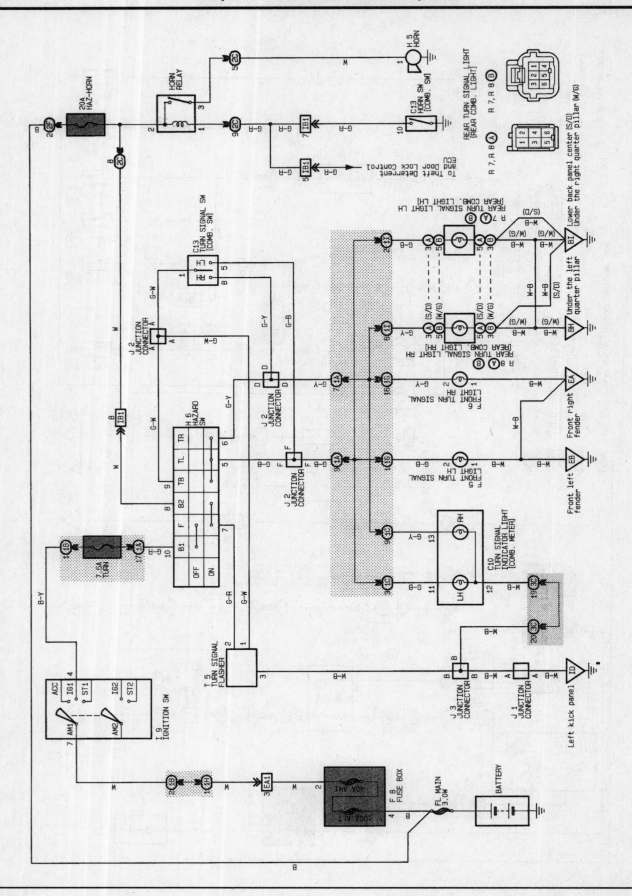

**Typical turn signal, hazard warning and horn system wiring diagram - 1997 and earlier models**

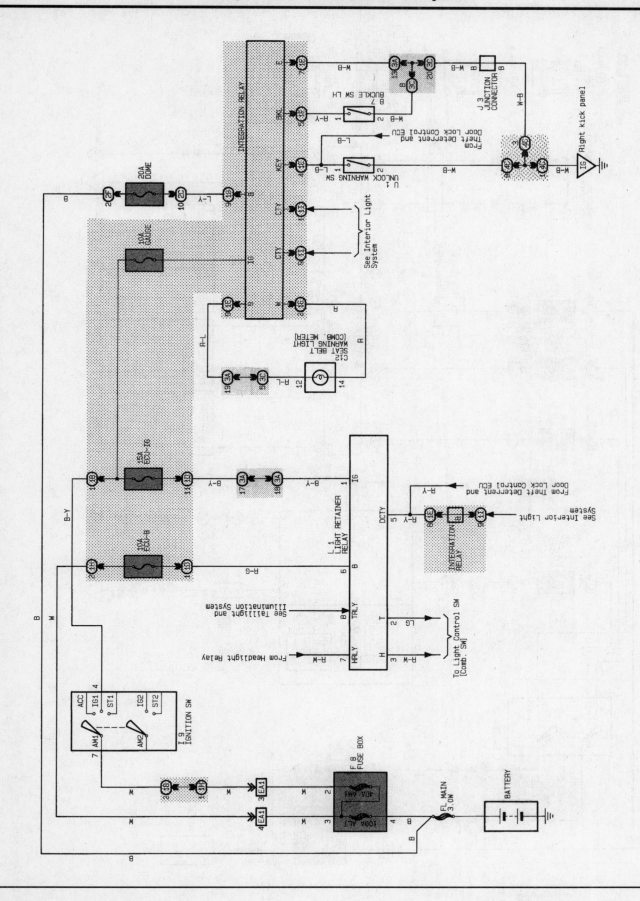

Typical automatic light and seat belt warning system wiring diagram - 1997 and earlier models

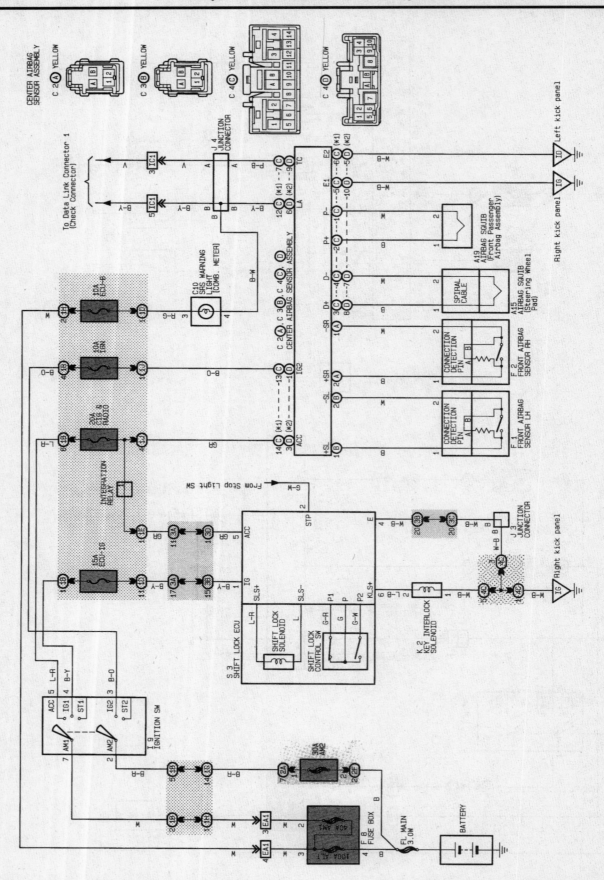

**Typical shift lock and airbag system wiring diagram – 1997 and earlier models**

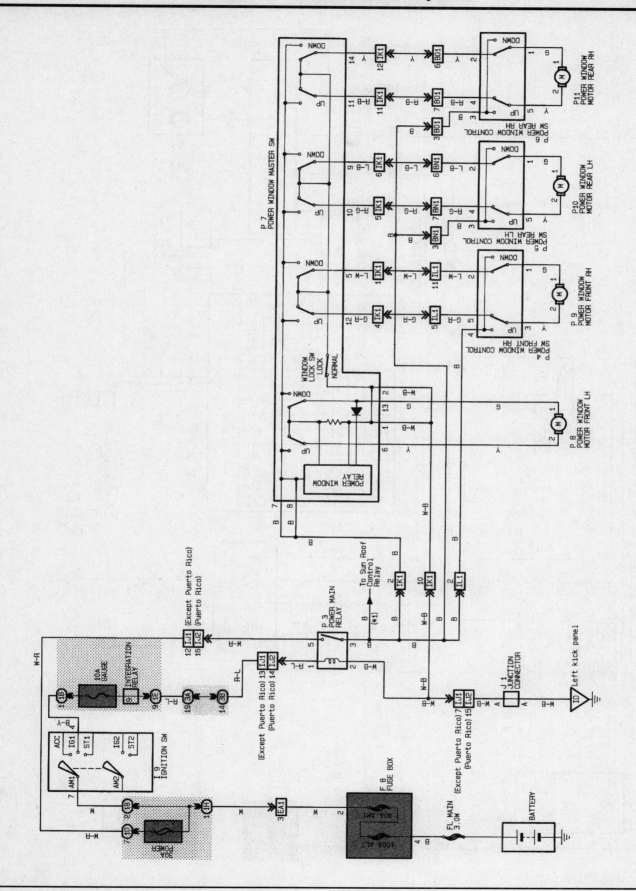

Typical power window system wiring diagram - 1997 and earlier models

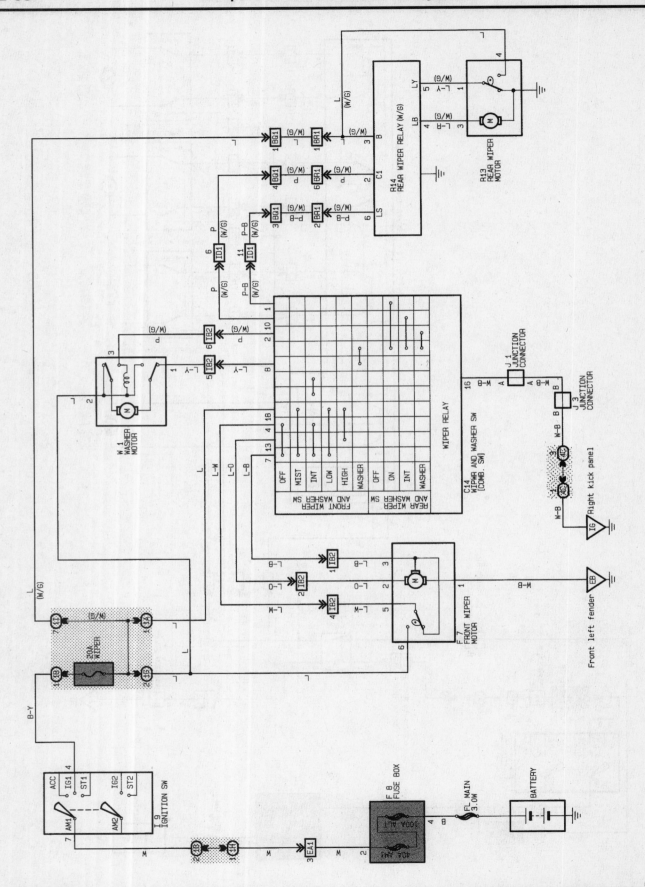

**Typical wiper and washer system wiring diagram - 1997 and earlier models**

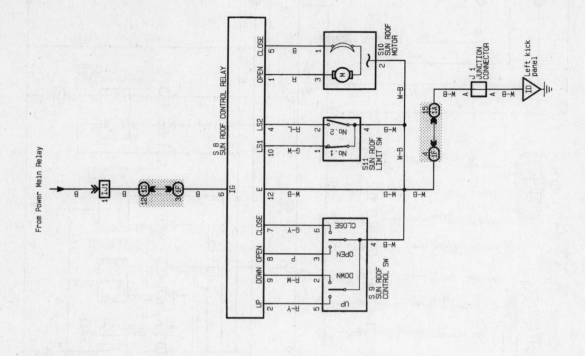

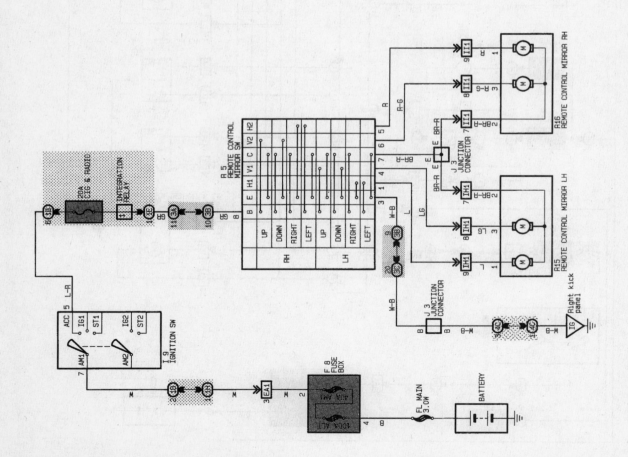

Typical electric mirror and sun roof system wiring diagram - 1997 and earlier models

12

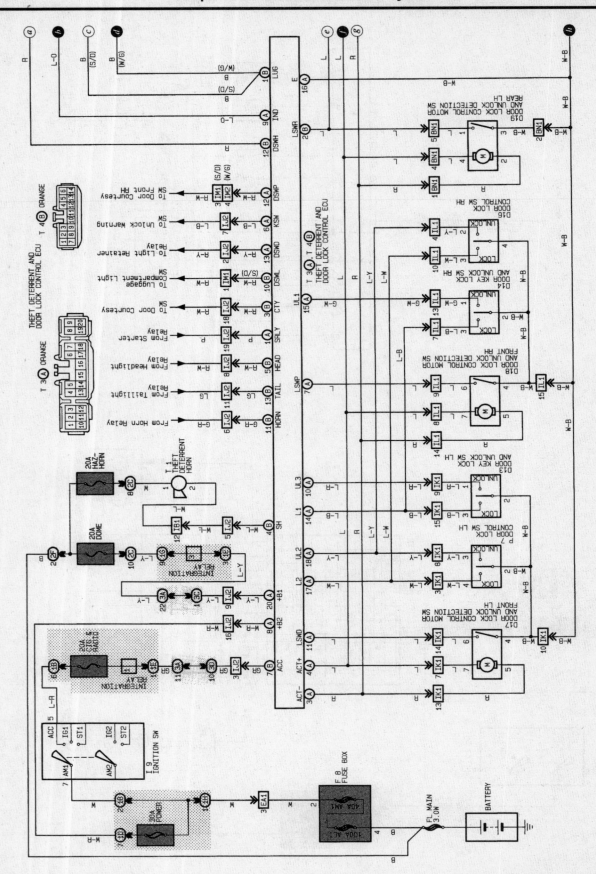

**Typical power door lock system wiring diagram (with anti-theft - 1997 and earlier models**

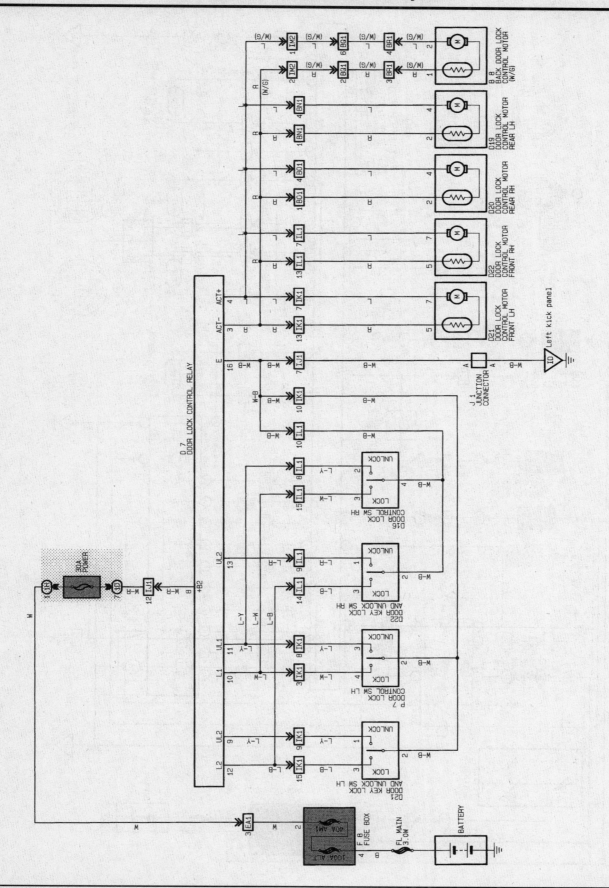

Typical power door lock system wiring diagram (without anti-theft) - 1997 and earlier models

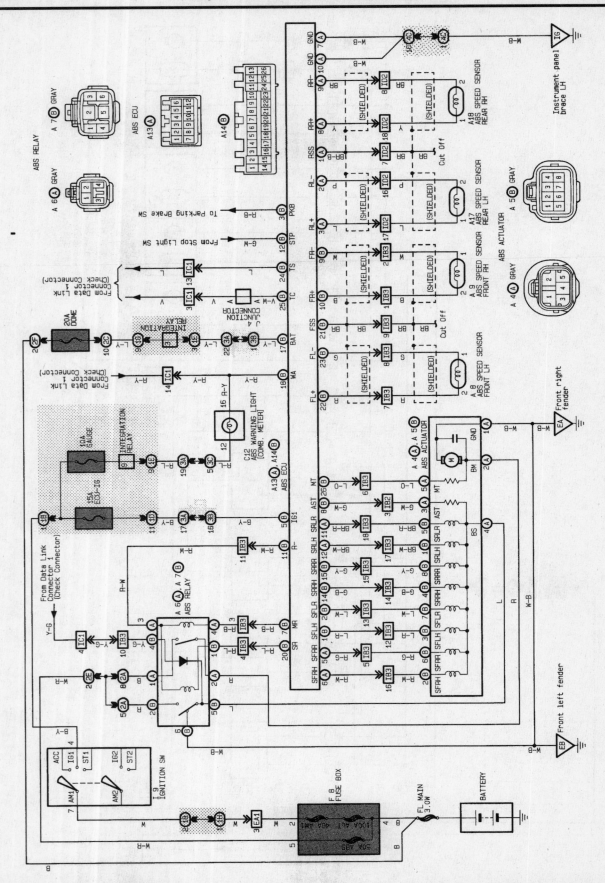

Typical Anti-lock Brake System (ABS) wiring diagram - 1997 and earlier models

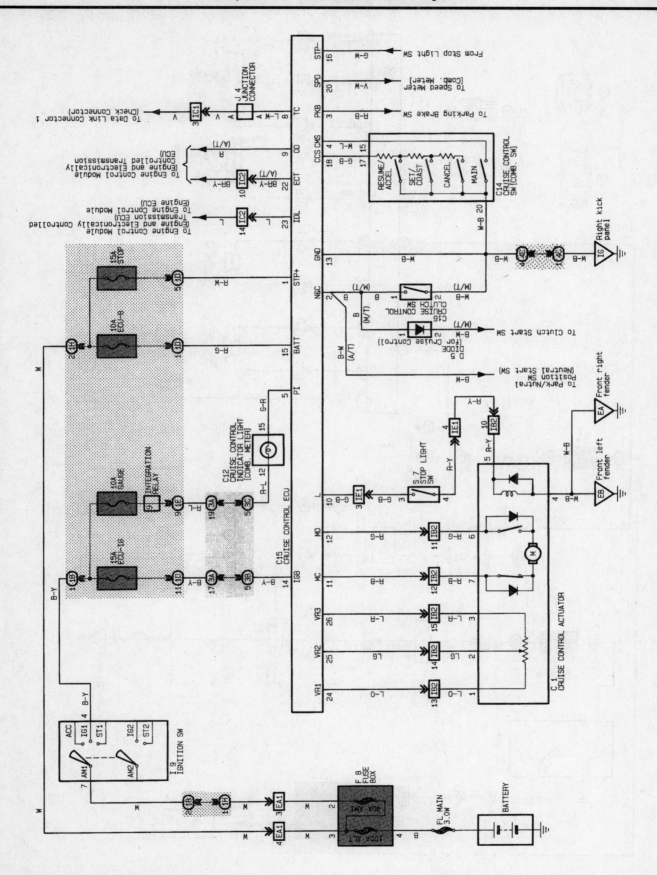

Typical cruise control system wiring diagram - 1997 and earlier models

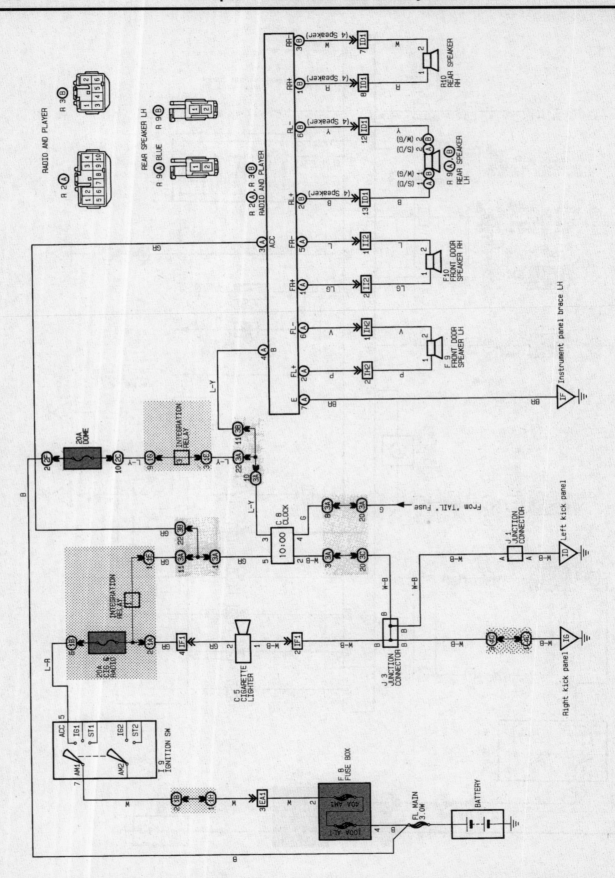

**Typical cigarette lighter, clock and audio system wiring diagram - 1997 and earlier models**

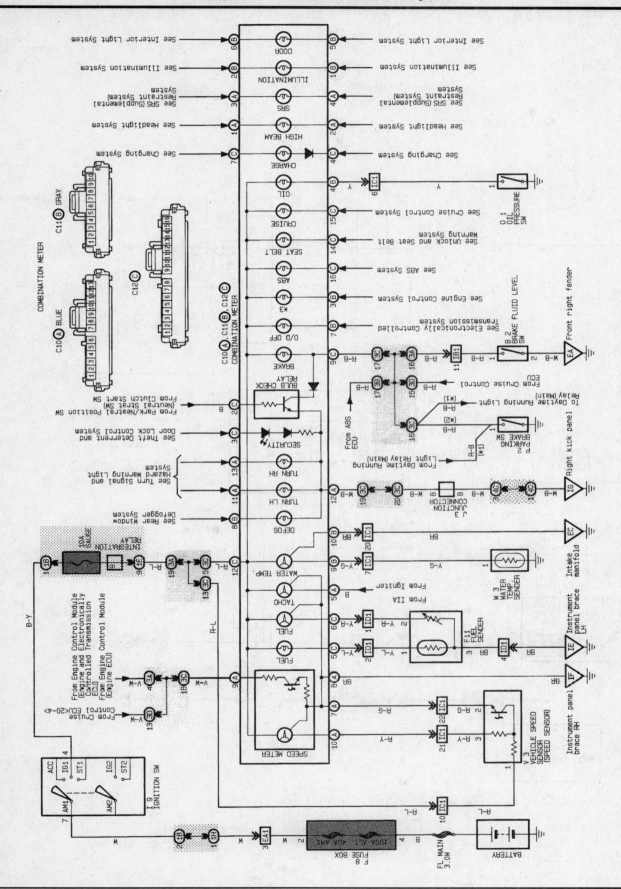

Typical instrument cluster system wiring diagram - 1997 and earlier models

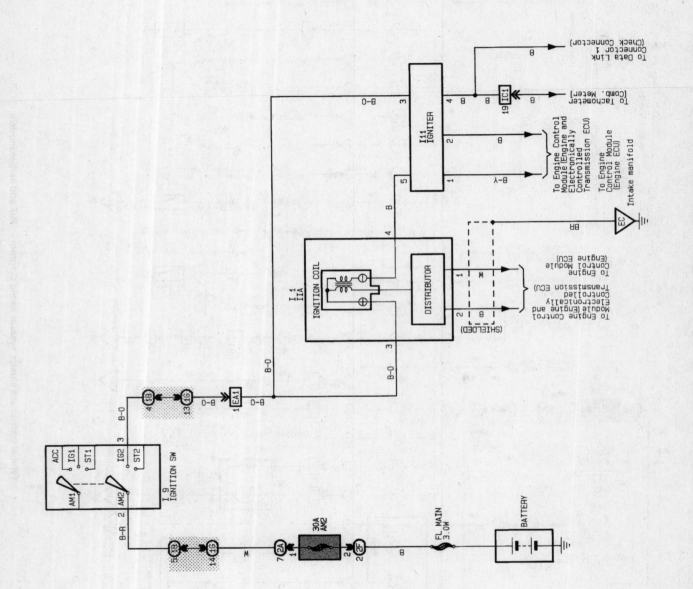

**Typical 1.8L engine ignition system wiring diagram - 1997 and earlier models**

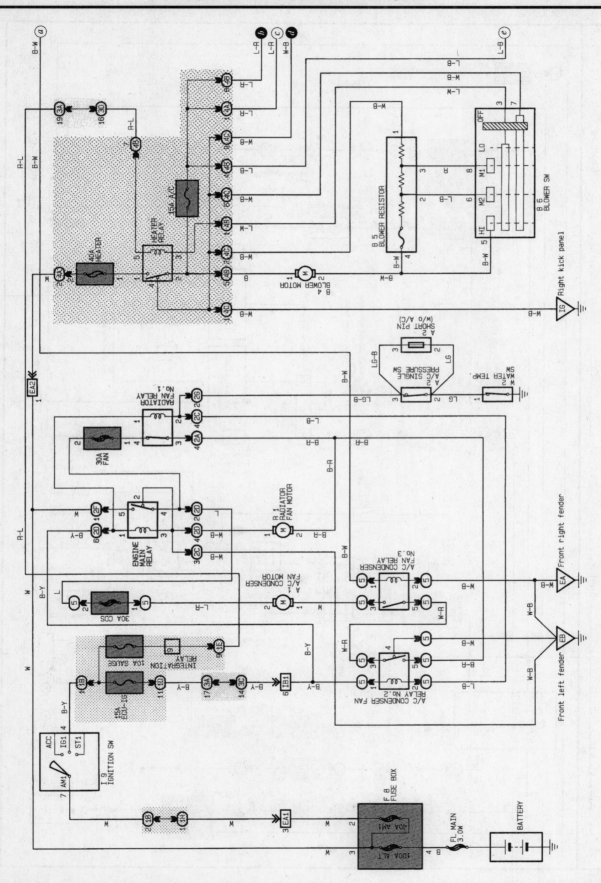

**Typical radiator cooling fan and air conditioning system wiring diagram - 1997 and earlier models**

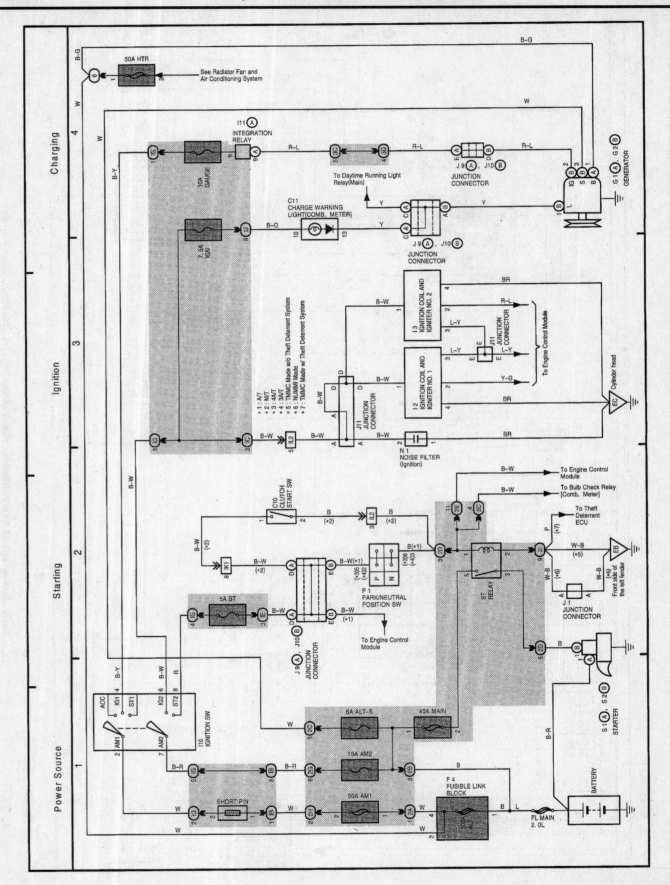

Typical overall wiring diagram - 1998 and later models (1 of 23)

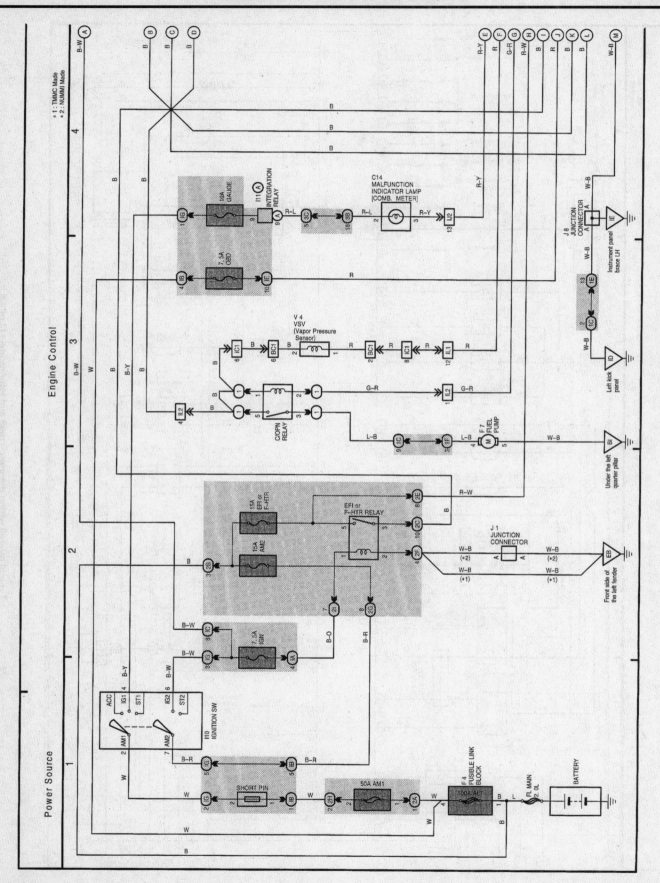

Typical overall wiring diagram - 1998 and later models (2 of 23)

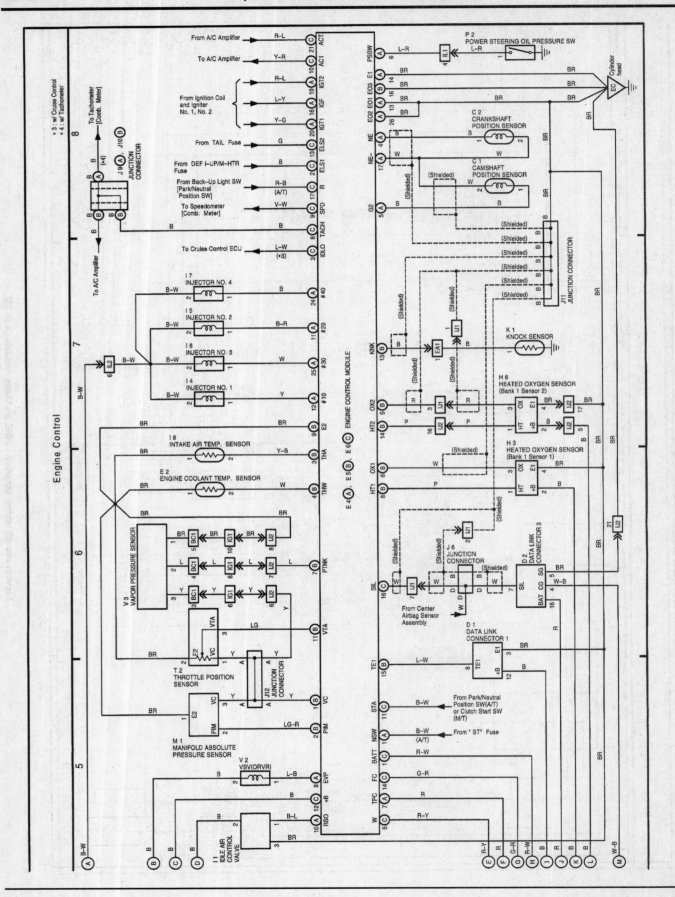

Typical overall wiring diagram - 1998 and later models (3 of 23)

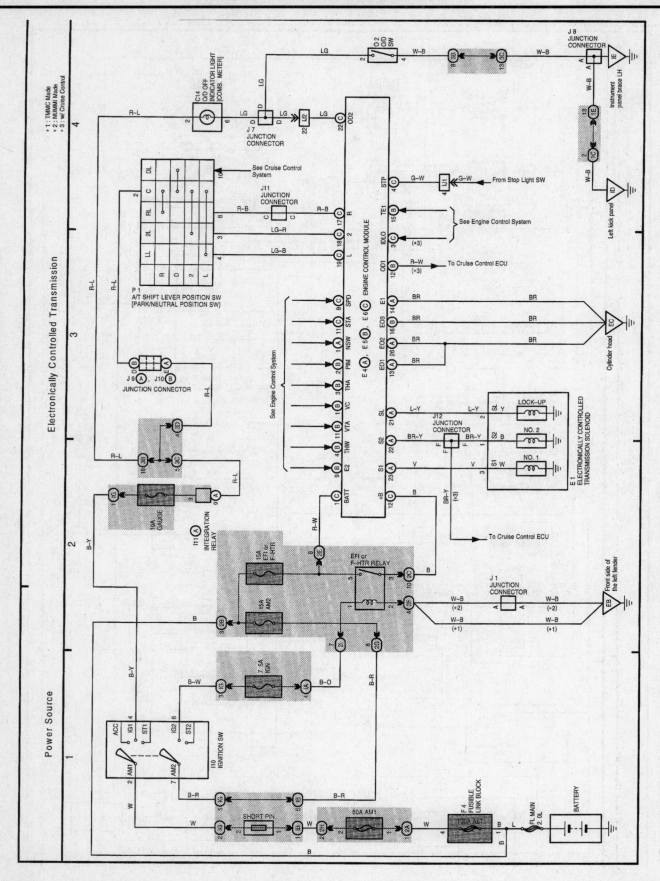

**Typical overall wiring diagram - 1998 and later models (4 of 23)**

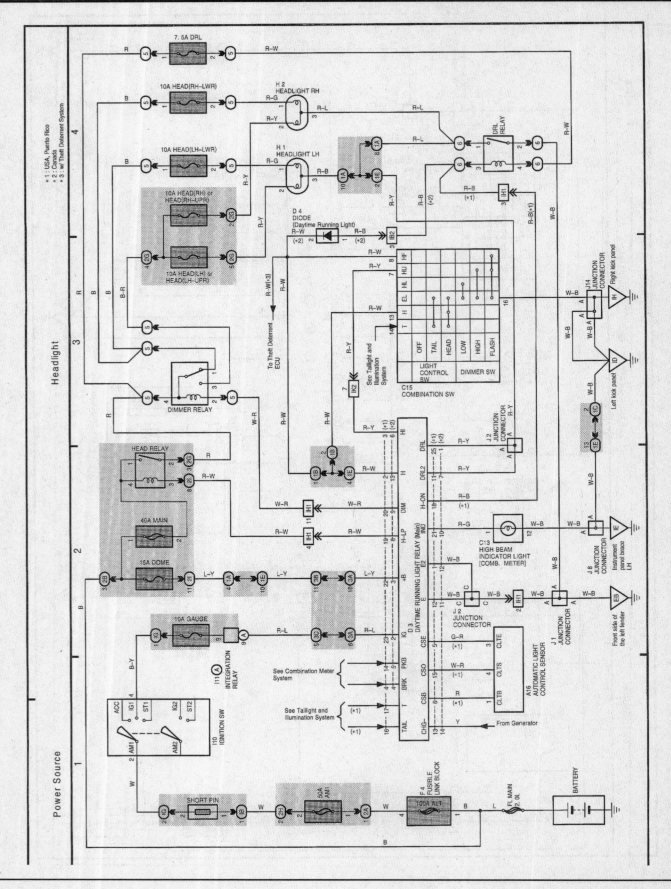

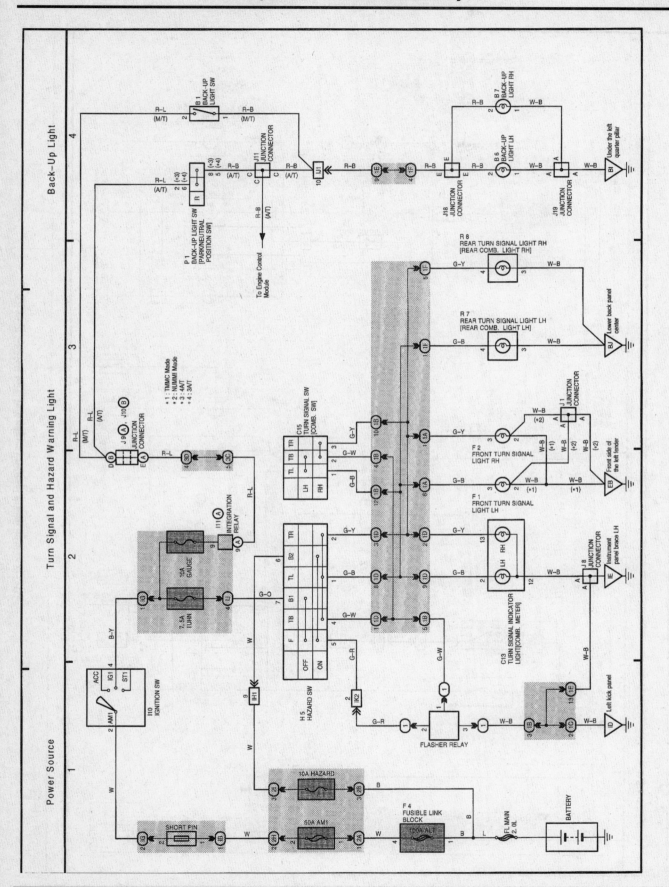

Typical overall wiring diagram - 1998 and later models (6 of 23)

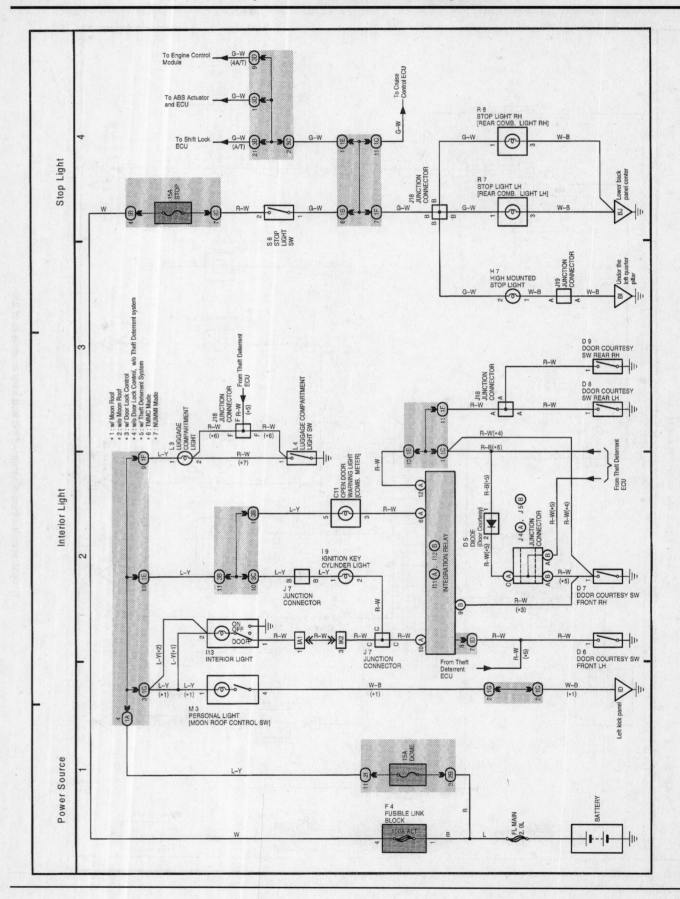

Typical overall wiring diagram - 1998 and later models (7 of 23)

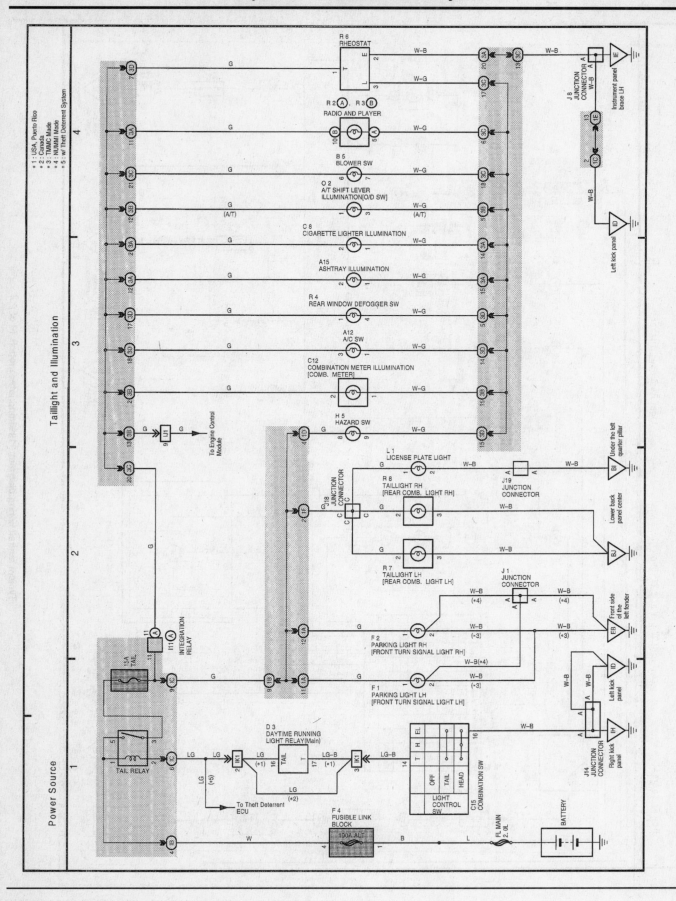

Typical overall wiring diagram - 1998 and later models (8 of 23)

**12**

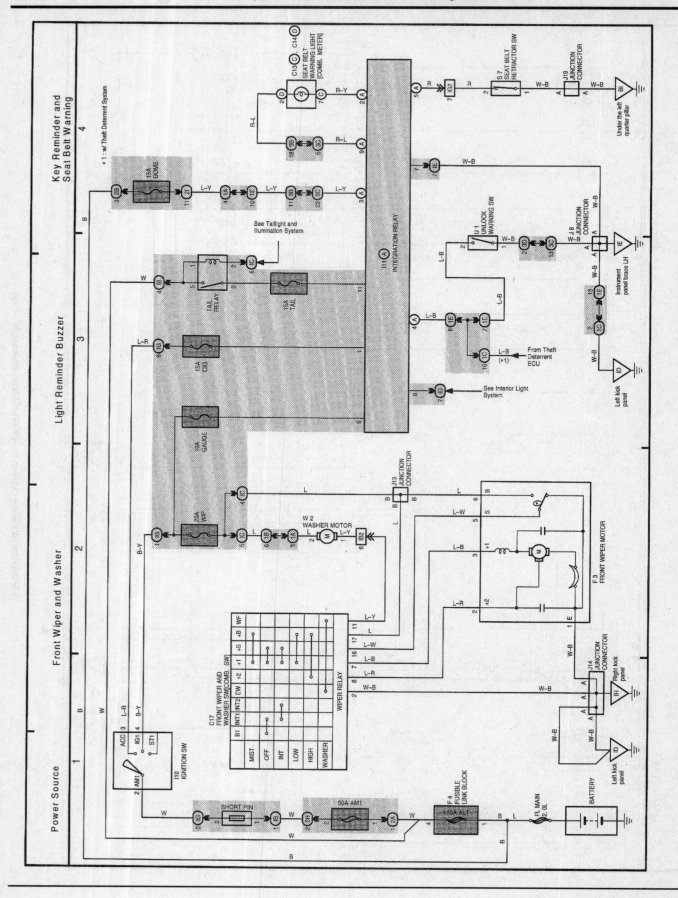

Typical overall wiring diagram - 1998 and later models (9 of 23)

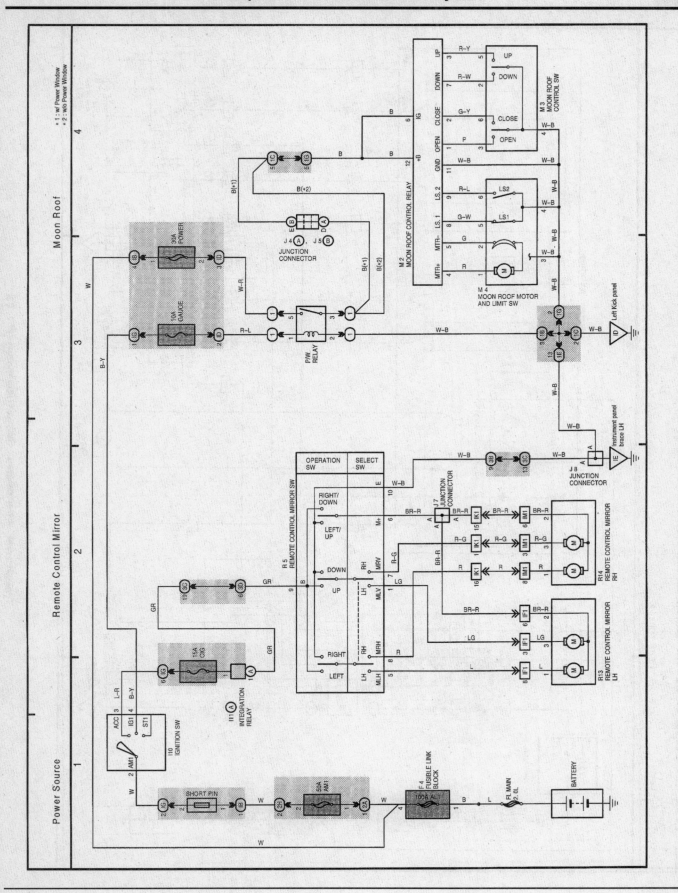

Typical overall wiring diagram - 1998 and later models (10 of 23)

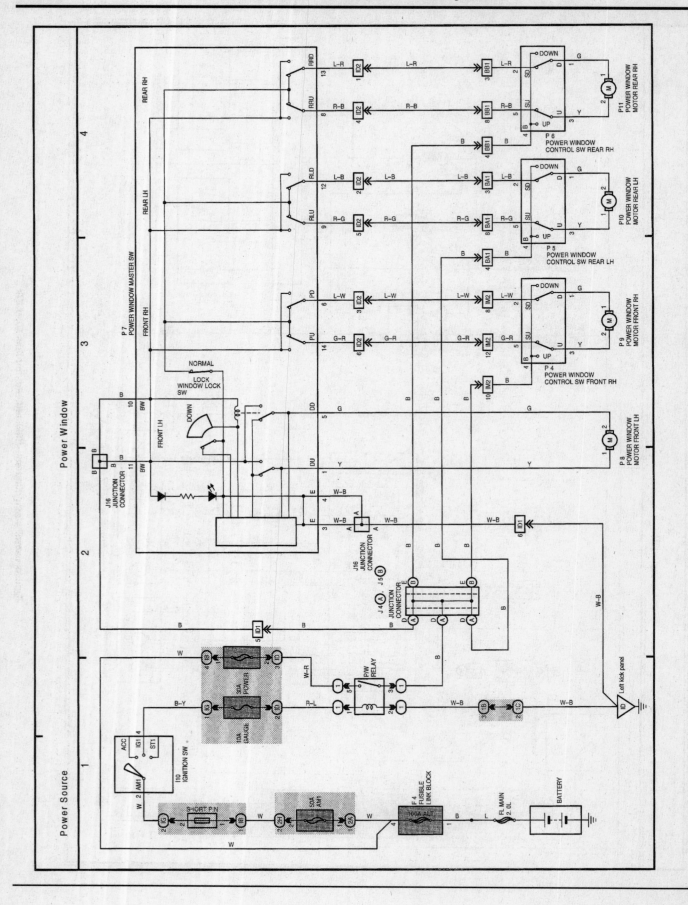

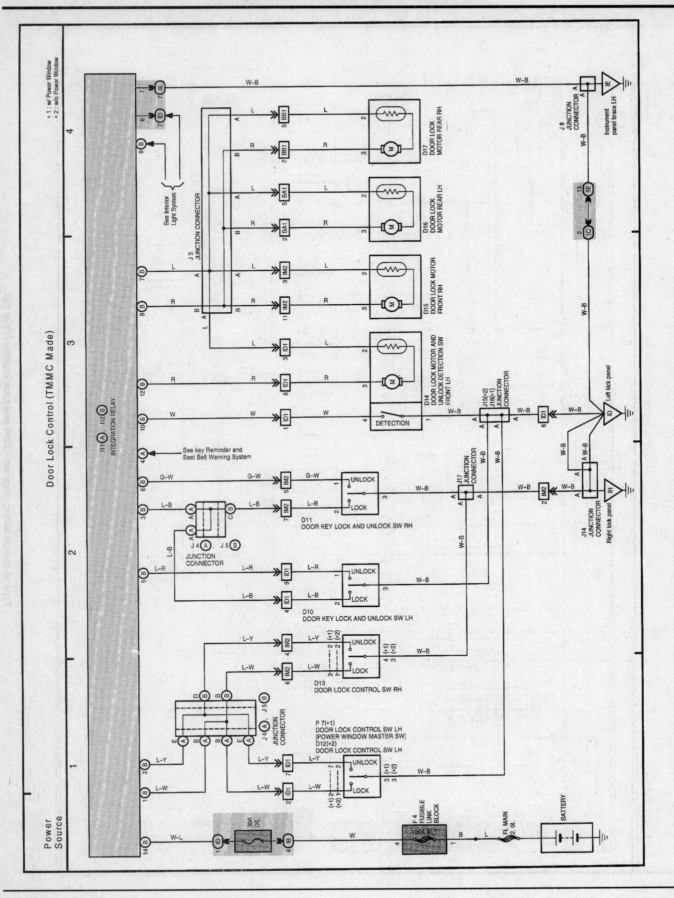

Typical overall wiring diagram - 1998 and later models (12 of 23)

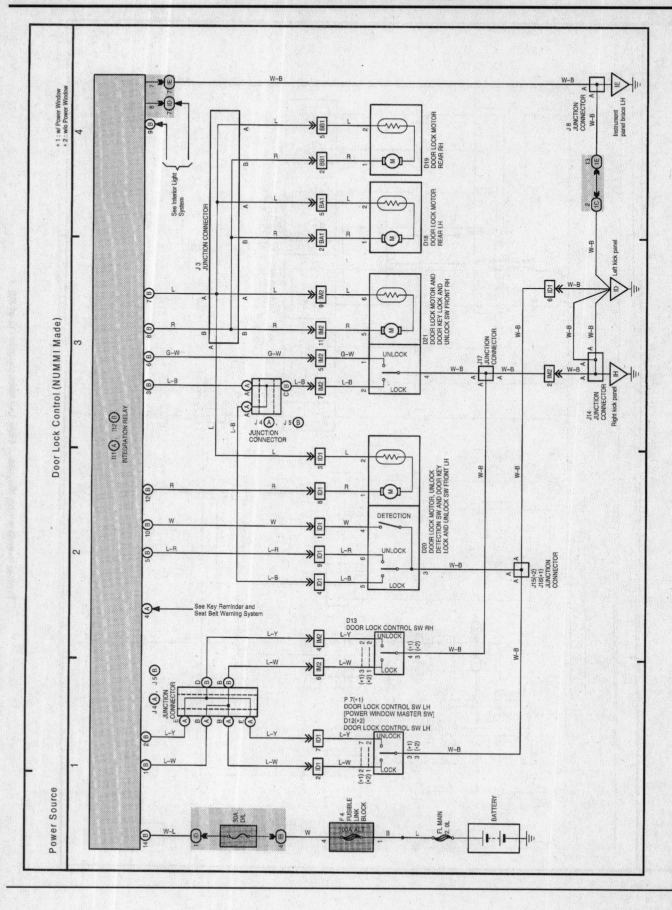

Typical overall wiring diagram - 1998 and later models (13 of 23)

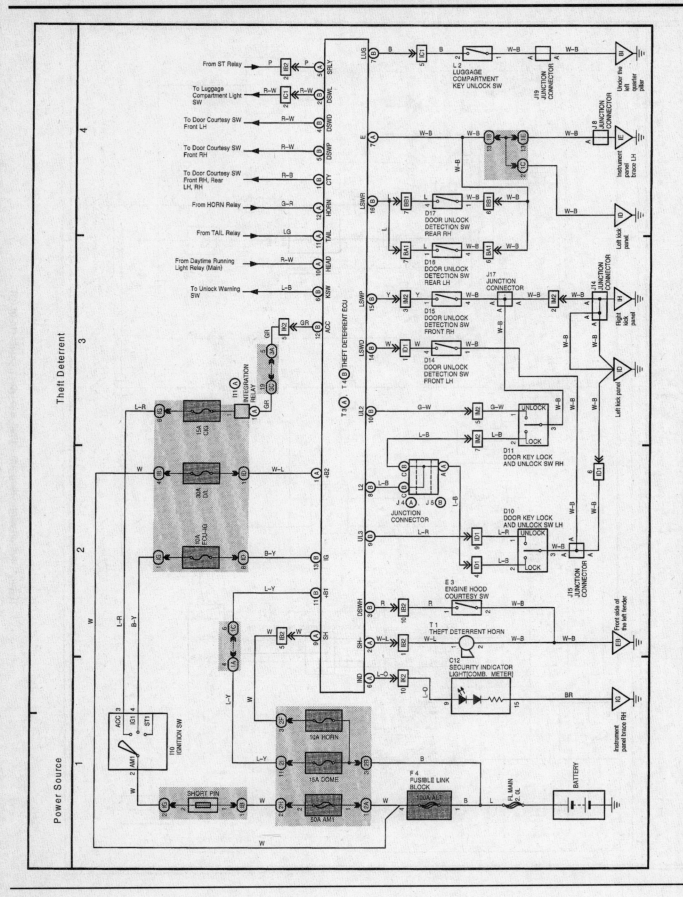

Typical overall wiring diagram - 1998 and later models (14 of 23)

12

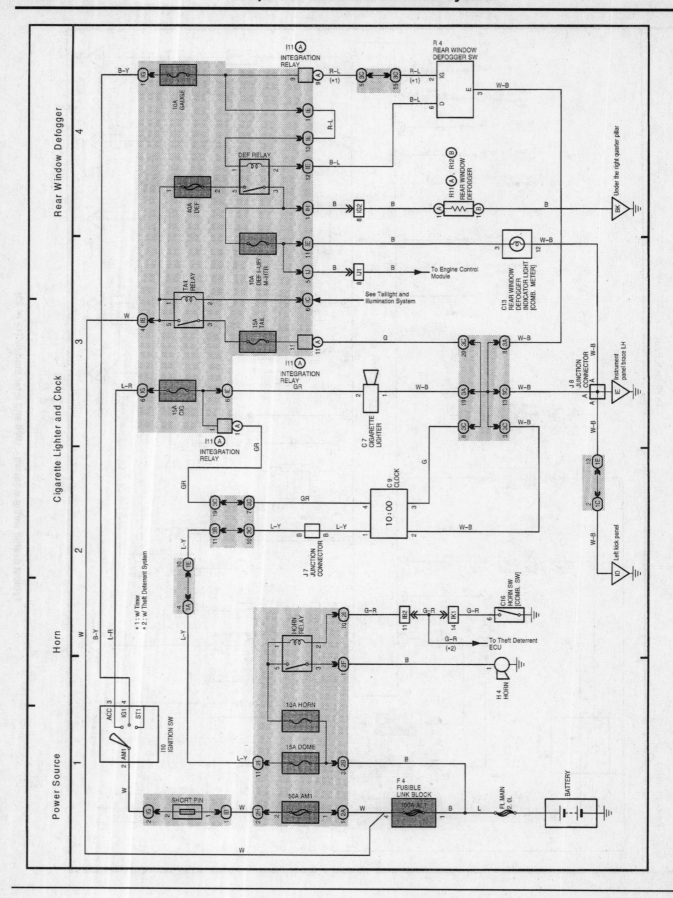

Typical overall wiring diagram - 1998 and later models (15 of 23)

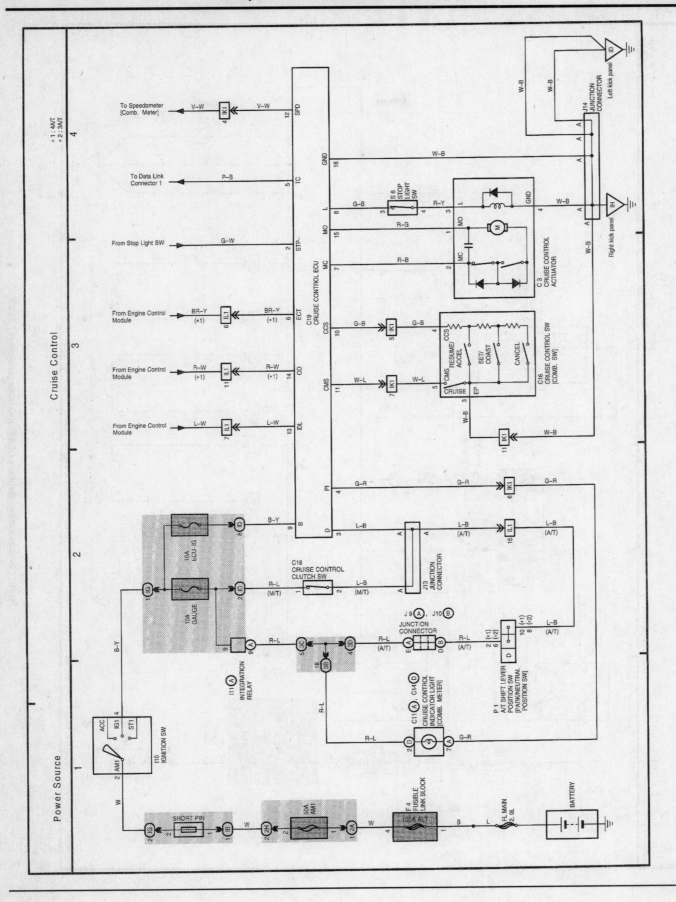

Typical overall wiring diagram - 1998 and later models (16 of 23)

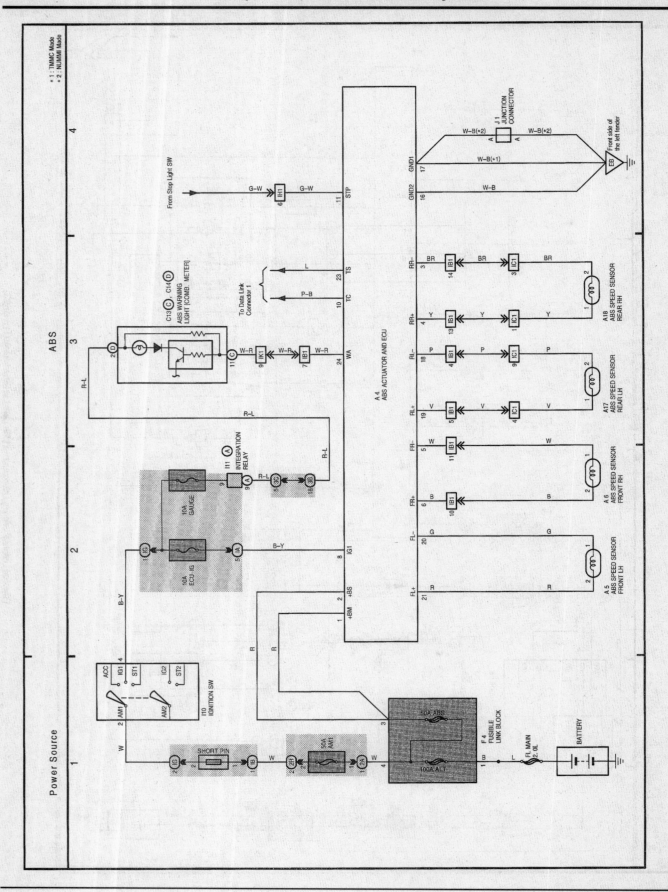

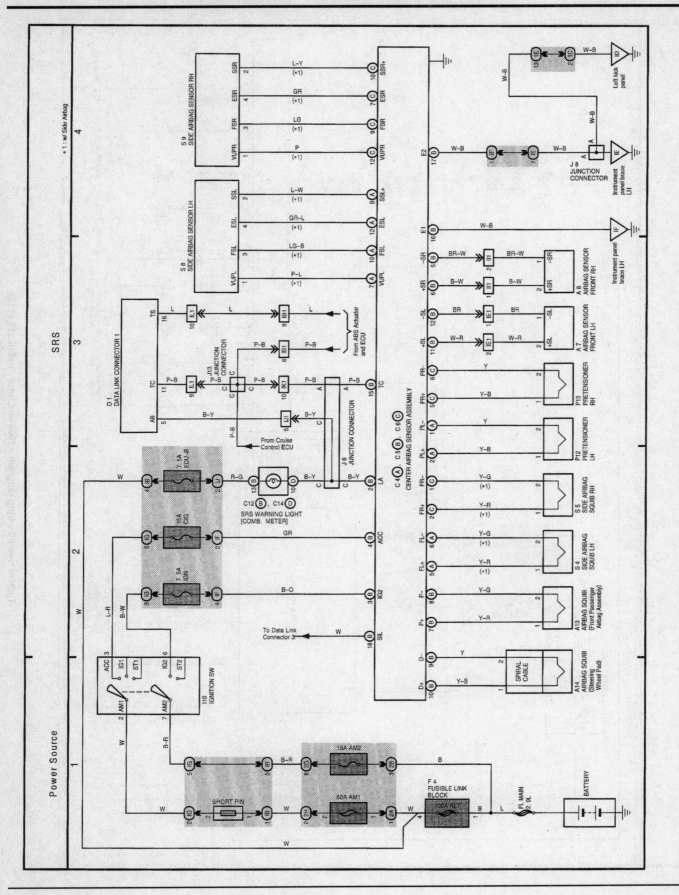

Typical overall wiring diagram - 1998 and later models (18 of 23)

**12**

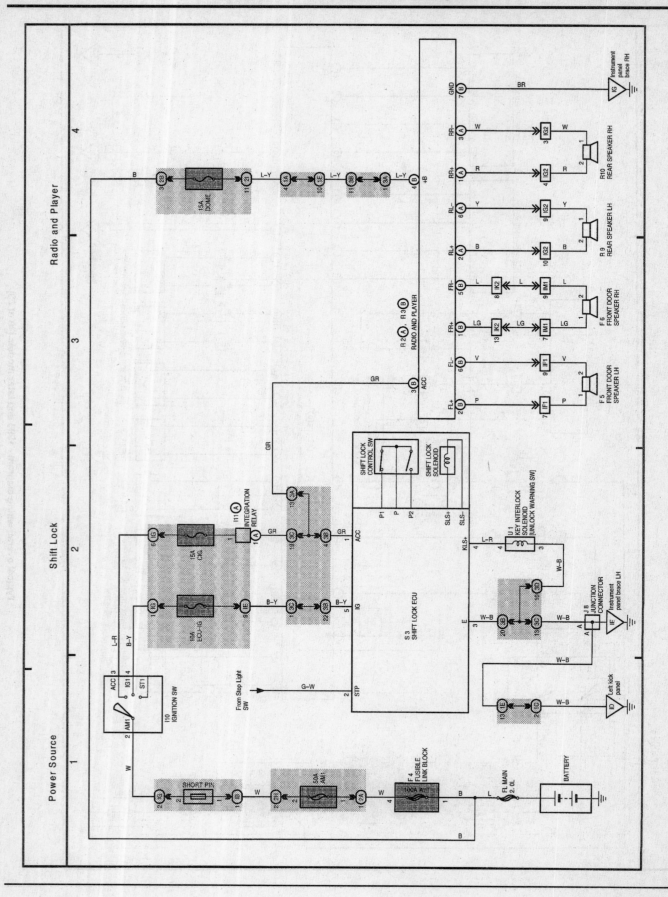

**Typical overall wiring diagram - 1998 and later models (19 of 23)**

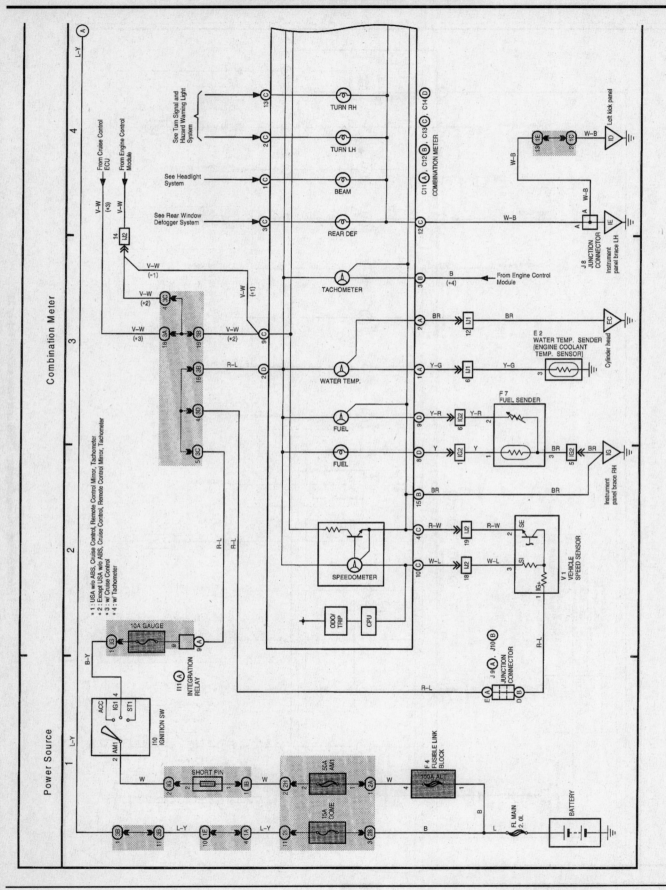

**Typical overall wiring diagram - 1998 and later models (20 of 23)**

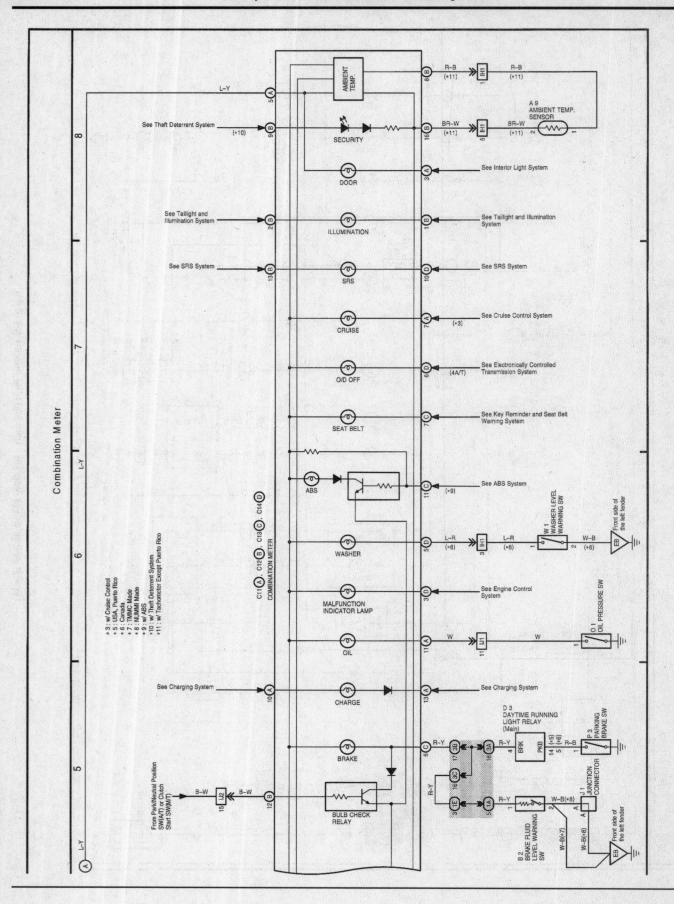

Typical overall wiring diagram - 1998 and later models (21 of 23)

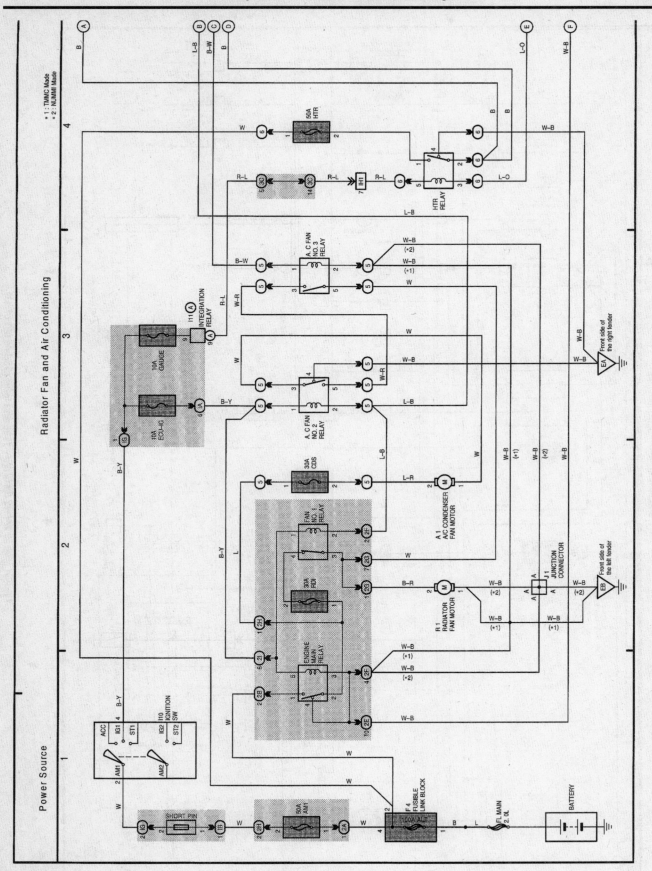

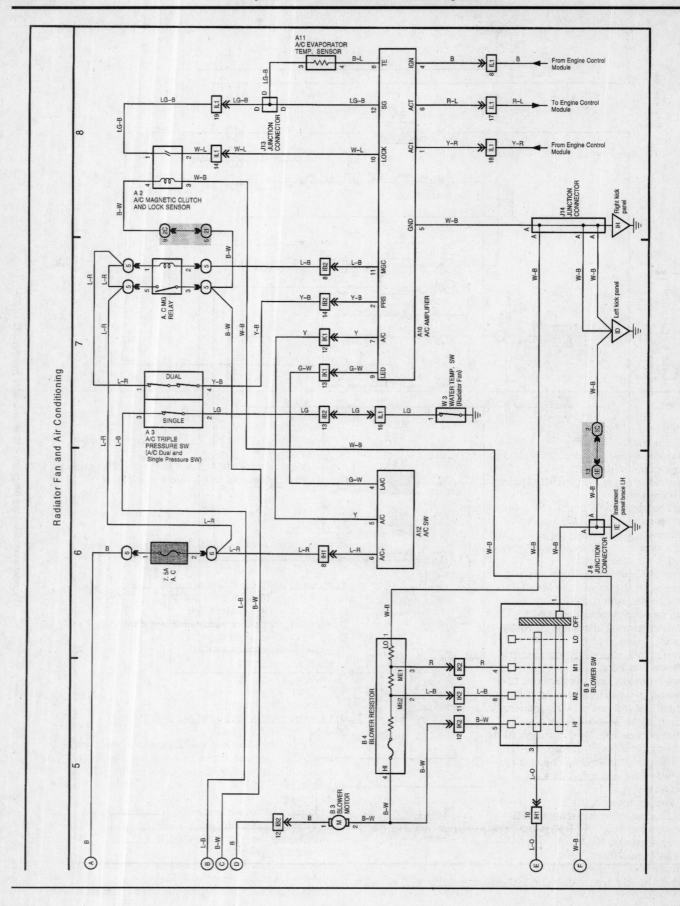

Radiator Fan and Air Conditioning

Typical overall wiring diagram – 1998 and later models (23 of 23)

# Index

# Haynes Automotive Manuals

*NOTE: New manuals are added to this list on a periodic basis. If you do not see a listing for your vehicle, consult your local Haynes dealer for the latest product information.*

## ACURA
*12020 Integra '86 thru '89 & Legend '86 thru '90

## AMC
     Jeep CJ - see JEEP (50020)
14020 Mid-size models, Concord, Hornet, Gremlin & Spirit '70 thru '83
14025 (Renault) Alliance & Encore '83 thru '87

## AUDI
15020 4000 all models '80 thru '87
15025 5000 all models '77 thru '83
15026 5000 all models '84 thru '88

## AUSTIN-HEALEY
     Sprite - see MG Midget (66015)

## BMW
*18020 3/5 Series not including diesel or all-wheel drive models '82 thru '92
*18021 3 Series except 325iX models '92 thru '97
18025 320i 4 cyl all models '75 thru '83
18035 528i & 530i all models '75 thru '80
18050 1500 thru 2002 except Turbo '59 thru '77

## BUICK
     Century (front wheel drive) - see GM (829)
*19020 Buick, Oldsmobile & Pontiac Full-size (Front wheel drive) all models '85 thru '98
     Buick Electra, LeSabre and Park Avenue; Oldsmobile Delta 88 Royale, Ninety Eight and Regency; Pontiac Bonneville
19025 Buick Oldsmobile & Pontiac Full-size (Rear wheel drive)
     Buick Estate '70 thru '90, Electra '70 thru '84, LeSabre '70 thru '85, Limited '74 thru '79
     Oldsmobile Custom Cruiser '70 thru '90, Delta 88 '70 thru '85, Ninety-eight '70 thru '84
     Pontiac Bonneville '70 thru '81, Catalina '70 thru '81, Grandville '70 thru '75, Parisienne '83 thru '86
19030 Mid-size Regal & Century all rear-drive models with V6, V8 and Turbo '74 thru '87
     Regal - see GENERAL MOTORS (38010)
     Riviera - see GENERAL MOTORS (38030)
     Roadmaster - see CHEVROLET (24046)
     Skyhawk - see GENERAL MOTORS (38015)
     Skylark '80 thru '85 - see GM (38020)
     Skylark '86 on - see GM (38025)
     Somerset - see GENERAL MOTORS (38025)

## CADILLAC
*21030 Cadillac Rear Wheel Drive all models '70 thru '93
     Cimarron - see GENERAL MOTORS (38015)
     Eldorado - see GENERAL MOTORS (38030)
     Seville '80 thru '85 - see GM (38030)

## CHEVROLET
*24010 Astro & GMC Safari Mini-vans '85 thru '93
24015 Camaro V8 all '70 thru '81
24016 Camaro all models '82 thru '92
     Cavalier - see GENERAL MOTORS (38015)
     Celebrity - see GENERAL MOTORS (38005)
24017 Camaro & Firebird '93 thru '97
24020 Chevelle, Malibu & El Camino '69 thru '87
24024 Chevette & Pontiac T1000 '76 thru '87
     Citation - see GENERAL MOTORS (38020)
*24032 Corsica/Beretta all models '87 thru '96
24040 Corvette all V8 models '68 thru '82
*24041 Corvette all models '84 thru '96
10305 Chevrolet Engine Overhaul Manual
24045 Full-size Sedans Caprice, Impala, Biscayne, Bel Air & Wagons '69 thru '90
24046 Impala SS & Caprice and Buick Roadmaster '91 thru '96
     Lumina - see GENERAL MOTORS (38010)

24048 Lumina & Monte Carlo '95 thru '98
     Lumina APV - see GM (38035)
24050 Luv Pick-up all 2WD & 4WD '72 thru '82
*24055 Monte Carlo all models '70 thru '88
     Monte Carlo '95 thru '98 - see LUMINA (24048)
24059 Nova all V8 models '69 thru '79
*24060 Nova and Geo Prizm '85 thru '92
24064 Pick-ups '67 thru '87 - Chevrolet & GMC, all V8 & in-line 6 cyl, 2WD & 4WD '67 thru '87; Suburbans, Blazers & Jimmys '67 thru '91
*24065 Pick-ups '88 thru '98 - Chevrolet & GMC, all full-size pick-ups, '88 thru '98; Blazer & Jimmy '92 thru '94; Suburban '92 thru '98; Tahoe & Yukon '98
24070 S-10 & S-15 Pick-ups '82 thru '93, Blazer & Jimmy '83 thru '94,
*24071 S-10 & S-15 Pick-ups '94 thru '96 Blazer & Jimmy '95 thru '96
*24075 Sprint & Geo Metro '85 thru '94
*24080 Vans - Chevrolet & GMC, V8 & in-line 6 cylinder models '68 thru '96

## CHRYSLER
25015 Chrysler Cirrus, Dodge Stratus, Plymouth Breeze '95 thru '98
25025 Chrysler Concorde, New Yorker & LHS, Dodge Intrepid, Eagle Vision, '93 thru '97
10310 Chrysler Engine Overhaul Manual
*25020 Full-size Front-Wheel Drive '88 thru '93
     K-Cars - see DODGE Aries (30008)
     Laser - see DODGE Daytona (30030)
*25030 Chrysler & Plymouth Mid-size front wheel drive '82 thru '95
     Rear-wheel Drive - see Dodge (30050)

## DATSUN
28005 200SX all models '80 thru '83
28007 B-210 all models '73 thru '78
28009 210 all models '79 thru '82
28012 240Z, 260Z & 280Z Coupe '70 thru '78
28014 280ZX Coupe & 2+2 '79 thru '83
     300ZX - see NISSAN (72010)
28016 310 all models '78 thru '82
28018 510 & PL521 Pick-up '68 thru '73
28020 510 all models '78 thru '81
28022 620 Series Pick-up all models '73 thru '79
     720 Series Pick-up - see NISSAN (72030)
28025 810/Maxima all gasoline models, '77 thru '84

## DODGE
     400 & 600 - see CHRYSLER (25030)
*30008 Aries & Plymouth Reliant '81 thru '89
30010 Caravan & Plymouth Voyager Mini-Vans all models '84 thru '95
*30011 Caravan & Plymouth Voyager Mini-Vans all models '96 thru '98
30012 Challenger/Plymouth Saporro '78 thru '83
30016 Colt & Plymouth Champ (front wheel drive) all models '78 thru '87
*30020 Dakota Pick-ups all models '87 thru '96
30025 Dart, Demon, Plymouth Barracuda, Duster & Valiant 6 cyl models '67 thru '76
*30030 Daytona & Chrysler Laser '84 thru '89
     Intrepid - see CHRYSLER (25025)
*30034 Neon all models '95 thru '97
*30035 Omni & Plymouth Horizon '78 thru '90
*30040 Pick-ups all full-size models '74 thru '93
*30041 Pick-ups all full-size models '94 thru '96
*30045 Ram 50/D50 Pick-ups & Raider and Plymouth Arrow Pick-ups '79 thru '93
30050 Dodge/Plymouth/Chrysler rear wheel drive '71 thru '89
*30055 Shadow & Plymouth Sundance '87 thru '94
*30060 Spirit & Plymouth Acclaim '89 thru '95
*30065 Vans - Dodge & Plymouth '71 thru '96

## EAGLE
     Talon - see Mitsubishi Eclipse (68030)
     Vision - see CHRYSLER (25025)

## FIAT
34010 124 Sport Coupe & Spider '68 thru '78
34025 X1/9 all models '74 thru '80

## FORD
10355 Ford Automatic Transmission Overhaul
*36004 Aerostar Mini-vans all models '86 thru '96
*36006 Contour & Mercury Mystique '95 thru '98
36008 Courier Pick-up all models '72 thru '82
36012 Crown Victoria & Mercury Grand Marquis '88 thru '96
10320 Ford Engine Overhaul Manual
*36016 Escort/Mercury Lynx all models '81 thru '90
*36020 Escort/Mercury Tracer '91 thru '96
*36024 Explorer & Mazda Navajo '91 thru '95
36028 Fairmont & Mercury Zephyr '78 thru '83
36030 Festiva & Aspire '88 thru '97
36032 Fiesta all models '77 thru '80
36036 Ford & Mercury Full-size, Ford LTD & Mercury Marquis ('75 thru '82); Ford Custom 500,Country Squire, Crown Victoria & Mercury Colony Park ('75 thru '87); Ford LTD Crown Victoria & Mercury Gran Marquis ('83 thru '87)
36040 Granada & Mercury Monarch '75 thru '80
36044 Ford & Mercury Mid-size, Ford Thunderbird & Mercury Cougar ('75 thru '82); Ford LTD & Mercury Marquis ('83 thru '86); Ford Torino,Gran Torino, Elite, Ranchero pick-up, LTD II, Mercury Montego, Comet, XR-7 & Lincoln Versailles ('75 thru '86)
36048 Mustang V8 all models '64-1/2 thru '73
36049 Mustang II 4 cyl, V6 & V8 models '74 thru '78
36050 Mustang & Mercury Capri all models Mustang, '79 thru '93; Capri, '79 thru '86
*36051 Mustang all models '94 thru '97
36054 Pick-ups & Bronco '73 thru '79
36058 Pick-ups & Bronco '80 thru '96
36059 Pick-ups, Expedition & Mercury Navigator '97 thru '98
36062 Pinto & Mercury Bobcat '75 thru '80
36066 Probe all models '89 thru '92
36070 Ranger/Bronco II gasoline models '83 thru '92
*36071 Ranger '93 thru '97 & Mazda Pick-ups '94 thru '97
36074 Taurus & Mercury Sable '86 thru '95
*36075 Taurus & Mercury Sable '96 thru '98
*36078 Tempo & Mercury Topaz '84 thru '94
36082 Thunderbird/Mercury Cougar '83 thru '88
36086 Thunderbird/Mercury Cougar '89 and '97
36090 Vans all V8 Econoline models '69 thru '91
*36094 Vans full size '92-'95
*36097 Windstar Mini-van '95-'98

## GENERAL MOTORS
*10360 GM Automatic Transmission Overhaul
*38005 Buick Century, Chevrolet Celebrity, Oldsmobile Cutlass Ciera & Pontiac 6000 all models '82 thru '96
*38010 Buick Regal, Chevrolet Lumina, Oldsmobile Cutlass Supreme & Pontiac Grand Prix front-wheel drive models '88 thru '95
*38015 Buick Skyhawk, Cadillac Cimarron, Chevrolet Cavalier, Oldsmobile Firenza & Pontiac J-2000 & Sunbird '82 thru '94
*38016 Chevrolet Cavalier & Pontiac Sunfire '95 thru '98
38020 Buick Skylark, Chevrolet Citation, Olds Omega, Pontiac Phoenix '80 thru '85
38025 Buick Skylark & Somerset, Oldsmobile Achieva & Calais and Pontiac Grand Am all models '85 thru '95
38030 Cadillac Eldorado '71 thru '85, Seville '80 thru '85, Oldsmobile Toronado '71 thru '85 & Buick Riviera '79 thru '85
*38035 Chevrolet Lumina APV, Olds Silhouette & Pontiac Trans Sport all models '90 thru '95
     General Motors Full-size Rear-wheel Drive - see BUICK (19025)

*(Continued on other side)*

---

*\* Listings shown with an asterisk (*) indicate model coverage as of this printing. These titles will be periodically updated to include later model years - consult your Haynes dealer for more information.*

## Haynes North America, Inc., 861 Lawrence Drive, Newbury Park, CA 91320-1514 • (805) 498-6703

# Haynes Automotive Manuals (continued)

*NOTE: New manuals are added to this list on a periodic basis. If you do not see a listing for your vehicle, consult your local Haynes dealer for the latest product information.*

## GEO
**Metro** - *see CHEVROLET Sprint (24075)*
**Prizm** - '85 thru '92 see CHEVY (24060), '93 thru '96 see TOYOTA Corolla (92036)
*40030 **Storm** all models '90 thru '93
**Tracker** - *see SUZUKI Samurai (90010)*

## GMC
**Safari** - *see CHEVROLET ASTRO (24010)*
**Vans & Pick-ups** - *see CHEVROLET*

## HONDA
42010 **Accord CVCC** all models '76 thru '83
42011 **Accord** all models '84 thru '89
42012 **Accord** all models '90 thru '93
42013 **Accord** all models '94 thru '95
42020 **Civic 1200** all models '73 thru '79
42021 **Civic 1300 & 1500 CVCC** '80 thru '83
42022 **Civic 1500 CVCC** all models '75 thru '79
42023 **Civic** all models '84 thru '91
*42024 **Civic & del Sol** '92 thru '95
*42040 **Prelude CVCC** all models '79 thru '89

## HYUNDAI
*43015 **Excel** all models '86 thru '94

## ISUZU
**Hombre** - *see CHEVROLET S-10 (24071)*
*47017 **Rodeo** '91 thru '97; **Amigo** '89 thru '94; **Honda Passport** '95 thru '97
*47020 **Trooper & Pick-up**, all gasoline models Pick-up, '81 thru '93; Trooper, '84 thru '91

## JAGUAR
*49010 **XJ6** all 6 cyl models '68 thru '86
*49011 **XJ6** all models '88 thru '94
*49015 **XJ12 & XJS** all 12 cyl models '72 thru '85

## JEEP
*50010 **Cherokee, Comanche & Wagoneer Limited** all models '84 thru '96
50020 **CJ** all models '49 thru '86
*50025 **Grand Cherokee** all models '93 thru '98
50029 **Grand Wagoneer & Pick-up** '72 thru '91 Grand Wagoneer '84 thru '91, Cherokee & Wagoneer '72 thru '83, Pick-up '72 thru '88
*50030 **Wrangler** all models '87 thru '95

## LINCOLN
**Navigator** - *see FORD Pick-up (36059)*
59010 **Rear Wheel Drive** all models '70 thru '96

## MAZDA
61010 **GLC Hatchback (rear wheel drive)** '77 thru '83
61011 **GLC (front wheel drive)** '81 thru '85
*61015 **323 & Protegé** '90 thru '97
*61016 **MX-5 Miata** '90 thru '97
*61020 **MPV** all models '89 thru '94
**Navajo** - *see Ford Explorer (36024)*
61030 **Pick-ups** '72 thru '93
**Pick-ups** '94 thru '96 - *see Ford Ranger (36071)*
61035 **RX-7** all models '79 thru '85
*61036 **RX-7** all models '86 thru '91
61040 **626 (rear wheel drive)** all models '79 thru '82
*61041 **626/MX-6 (front wheel drive)** '83 thru '91

## MERCEDES-BENZ
63012 **123 Series Diesel** '76 thru '85
*63015 **190 Series** four-cyl gas models, '84 thru '88
63020 **230/250/280** 6 cyl sohc models '68 thru '72
63025 **280 123 Series** gasoline models '77 thru '81
63030 **350 & 450** all models '71 thru '80

## MERCURY
*See FORD Listing.*

## MG
66010 **MGB** Roadster & GT Coupe '62 thru '80
66015 **MG Midget, Austin Healey Sprite** '58 thru '80

## MITSUBISHI
*68020 **Cordia, Tredia, Galant, Precis & Mirage** '83 thru '93
*68030 **Eclipse, Eagle Talon & Ply. Laser** '90 thru '94
*68040 **Pick-up** '83 thru '96 & **Montero** '83 thru '93

## NISSAN
72010 **300ZX** all models including Turbo '84 thru '89
*72015 **Altima** all models '93 thru '97
*72020 **Maxima** all models '85 thru '91
*72030 **Pick-ups** '80 thru '96 **Pathfinder** '87 thru '95
72040 **Pulsar** all models '83 thru '86
*72050 **Sentra** all models '82 thru '94
*72051 **Sentra & 200SX** all models '95 thru '98
*72060 **Stanza** all models '82 thru '90

## OLDSMOBILE
*73015 **Cutlass** V6 & V8 gas models '74 thru '88
*For other OLDSMOBILE titles, see BUICK, CHEVROLET or GENERAL MOTORS listing.*

## PLYMOUTH
*For PLYMOUTH titles, see DODGE listing.*

## PONTIAC
79008 **Fiero** all models '84 thru '88
79018 **Firebird** V8 models except Turbo '70 thru '81
79019 **Firebird** all models '82 thru '92
*For other PONTIAC titles, see BUICK, CHEVROLET or GENERAL MOTORS listing.*

## PORSCHE
*80020 **911** except Turbo & Carrera 4 '65 thru '89
80025 **914** all 4 cyl models '69 thru '76
80030 **924** all models including Turbo '76 thru '82
*80035 **944** all models including Turbo '83 thru '89

## RENAULT
**Alliance & Encore** - *see AMC (14020)*

## SAAB
*84010 **900** all models including Turbo '79 thru '88

## SATURN
87010 **Saturn** all models '91 thru '96

## SUBARU
89002 **1100, 1300, 1400 & 1600** '71 thru '79
*89003 **1600 & 1800** 2WD & 4WD '80 thru '94

## SUZUKI
*90010 **Samurai/Sidekick & Geo Tracker** '86 thru '96

## TOYOTA
92005 **Camry** all models '83 thru '91
92006 **Camry** all models '92 thru '96
92015 **Celica Rear Wheel Drive** '71 thru '85
*92020 **Celica Front Wheel Drive** '86 thru '93
92025 **Celica Supra** all models '79 thru '92
92030 **Corolla** all models '75 thru '79
92032 **Corolla** all rear wheel drive models '80 thru '87
92035 **Corolla** all front wheel drive models '84 thru '92
*92036 **Corolla & Geo Prizm** '93 thru '97
92040 **Corolla Tercel** all models '80 thru '82
92045 **Corona** all models '74 thru '82
92050 **Cressida** all models '78 thru '82
92055 **Land Cruiser** FJ40, 43, 45, 55 '68 thru '82
92056 **Land Cruiser** FJ60, 62, 80, FZJ80 '80 thru '96
*92065 **MR2** all models '85 thru '87
92070 **Pick-up** all models '69 thru '78
*92075 **Pick-up** all models '79 thru '95
*92076 **Tacoma** '95 thru '98, **4Runner** '96 thru '98, **& T100** '93 thru '98
*92080 **Previa** all models '91 thru '95
92085 **Tercel** all models '87 thru '94

## TRIUMPH
94007 **Spitfire** all models '62 thru '81
94010 **TR7** all models '75 thru '81

## VW
96008 **Beetle & Karmann Ghia** '54 thru '79
96012 **Dasher** all gasoline models '74 thru '81
*96016 **Rabbit, Jetta, Scirocco, & Pick-up** gas models '74 thru '91 & Convertible '80 thru '92
96017 **Golf & Jetta** all models '93 thru '97
96020 **Rabbit, Jetta & Pick-up** diesel '77 thru '84
96030 **Transporter 1600** all models '68 thru '79
96035 **Transporter 1700, 1800 & 2000** '72 thru '79
96040 **Type 3 1500 & 1600** all models '63 thru '73
96045 **Vanagon** all air-cooled models '80 thru '83

## VOLVO
97010 **120, 130 Series & 1800 Sports** '61 thru '73
97015 **140 Series** all models '66 thru '74
*97020 **240 Series** all models '76 thru '93
97025 **260 Series** all models '75 thru '82
*97040 **740 & 760 Series** all models '82 thru '88

## TECHBOOK MANUALS
10205 **Automotive Computer Codes**
10210 **Automotive Emissions Control Manual**
10215 **Fuel Injection Manual, 1978 thru 1985**
10220 **Fuel Injection Manual, 1986 thru 1996**
10225 **Holley Carburetor Manual**
10230 **Rochester Carburetor Manual**
10240 **Weber/Zenith/Stromberg/SU Carburetors**
10305 **Chevrolet Engine Overhaul Manual**
10310 **Chrysler Engine Overhaul Manual**
10320 **Ford Engine Overhaul Manual**
10330 **GM and Ford Diesel Engine Repair Manual**
10340 **Small Engine Repair Manual**
10345 **Suspension, Steering & Driveline Manual**
10355 **Ford Automatic Transmission Overhaul**
10360 **GM Automatic Transmission Overhaul**
10405 **Automotive Body Repair & Painting**
10410 **Automotive Brake Manual**
10415 **Automotive Detailing Manual**
10420 **Automotive Eelectrical Manual**
10425 **Automotive Heating & Air Conditioning**
10430 **Automotive Reference Manual & Dictionary**
10435 **Automotive Tools Manual**
10440 **Used Car Buying Guide**
10445 **Welding Manual**
10450 **ATV Basics**

## SPANISH MANUALS
98903 **Reparación de Carrocería & Pintura**
98905 **Códigos Automotrices de la Computadora**
98910 **Frenos Automotriz**
98915 **Inyección de Combustible 1986 al 1994**
99040 **Chevrolet & GMC Camionetas** '67 al '87 Incluye Suburban, Blazer & Jimmy '67 al '91
99041 **Chevrolet & GMC Camionetas** '88 al '95 Incluye Suburban '92 al '95, Blazer & Jimmy '92 al '94, Tahoe y Yukon '95
99042 **Chevrolet & GMC Camionetas Cerradas** '68 al '95
99055 **Dodge Caravan & Plymouth Voyager** '84 al '95
99075 **Ford Camionetas y Bronco** '80 al '94
99077 **Ford Camionetas Cerradas** '69 al '91
99083 **Ford Modelos de Tamaño Grande** '75 al '87
99088 **Ford Modelos de Tamaño Mediano** '75 al '86
99091 **Ford Taurus & Mercury Sable** '86 al '95
99095 **GM Modelos de Tamaño Grande** '70 al '90
99100 **GM Modelos de Tamaño Mediano** '70 al '88
99110 **Nissan Camionetas** '80 al '96, Pathfinder '87 al '95
99118 **Nissan Sentra** '82 al '94
99125 **Toyota Camionetas y 4Runner** '79 al '95

Over 100 Haynes motorcycle manuals also available

5-98

*Listings shown with an asterisk (*) indicate model coverage as of this printing. These titles will be periodically updated to include later model years - consult your Haynes dealer for more information.*